ANIMAL GENOCIDE AND ITS AFTERMATH

ANIMAL GENOCIDE AND ITS AFTERMATH

The Tasmanian Tiger and the Newfoundland Wolf

Nicholas Chare

berghahn
NEW YORK • OXFORD
www.berghahnbooks.com

First published in 2025 by
Berghahn Books
www.berghahnbooks.com

Library of Congress Cataloging-in-Publication Data
A C.I.P. cataloging record is available from the Library of Congress
Library of Congress Cataloging in Publication Control Number: 2025018000

British Library Cataloguing in Publication Data
A catalogue record for this book is available from the British Library

EU GPSR Authorized Representative
LOGOS EUROPE, 9 rue Nicolas Poussin, 17000, LA ROCHELLE, France
Email: Contact@logoseurope.eu

ISBN 978-1-83695-125-4 hardback
ISBN 978-1-83695-126-1 epub
ISBN 978-1-83695-127-8 web pdf

https://doi.org/10.3167/9781836951254

Contents

Figures

Acknowledgements

I acknowledge that this book was written on the unceded traditional territory of the Kanien'kehà:ka, a place that has long served as a site of meeting and exchange among many First Nations including the Kanien'kehà:ka, Huron-Wendat, Abenaki, and Anishinaabeg. I recognise and respect the Kanien'kehà:ka as the traditional custodians of the lands and waters of Tiohtiá:ke/Montréal where I now have the good fortune to live and work.

This book was many years in the making and I am very grateful to Angela Mortimer and Esther Chare for their unstinting support and great patience during its researching and writing. They have enabled me to see the project through to successful completion and it is to them that it is dedicated. I have been researching the thylacine for approximately a decade but the initial idea for this book arose in 2018 while I was Diane and Howard Wohl Fellow at the Jack, Joseph and Morton Mandel Center for Advanced Holocaust Studies at the United States Holocaust Memorial Museum, and I was able to conduct important preliminary research using the extensive library resources there. I am particularly grateful to Megan Lewis, Vincent Slatt and Elliott Wrenn of the Library and Archives Reference desk in the Mandel Center. During my free time in Washington, I visited the Smithsonian National Museum of Natural History and viewed their thylacine holdings. I want to thank Darrin Lunde and Ingrid Rochon for their assistance during that visit.

I recognise the support of the Social Sciences and Humanities Research Council of Canada for providing funding for research necessary to the project. In that context, I am indebted to Michel Ledoux for his aid with the initial framing of the proposed study. I also wish to thank the Faculté des arts et des sciences here at the Université de Montréal for granting me research leave that made it possible for me to write the initial chapters. During that time, my colleague Louise Vigneault generously covered my administrative responsibilities, for which I thank her deeply.

Projects develop over many years and a number of people have helped to shape this one through conversations and shared insights, including, most notably, Nicolas Barnett, Katharina Bonzel, Abigail Celis, Justin Clemens, Ersy Contogouris, Barbara Creed, Paul Davies, Alex Fitzgerald, Mitchell

Frank, Michelle Gewurtz, Julie Gough, Jeanette Hoorn, András Lénárt, Justin Linger, Allegrius Little, Susan Lowish, Silvestra Mariniello, David McInnis, Nick Mooney, Smokey Nougu, James Overstall, Suzanne Paquet, Maree Pardy, Bernard Perron, Manon des Petits Frères, Griselda Pollock, Nathanael Price, Éduardo Ralickas, Adrian Rifkin, Tweed Roosevelt, Andrew Smith, Monika Vrzgulova, Anthony White, Dominic Williams, Ika Willis and the late Carol Zemel. Additionally, I thank Steven Allen for sharing aspects of his research on cinematic portrayals of Lutruwita/Tasmania. Catherine Girard generously allowed me to trial trace elements of Chapter 5 as a paper at the 2016 Universities Art Association of Canada annual conference here in Tiohtiá:ke/Montréal.

Many libraries have assisted me during the research and writing of this book, with some, most notably Cambridge University Library, going to considerable lengths to continue to provide access to resources during the COVID-19 pandemic. I am grateful to Rosalind Grooms of the Cambridge University Press Archive at the University of Cambridge, Tara Cuthbert and Julie Kaja of the Allport Library and Museum of Fine Arts, Kylie Mallett of Launceston Library, Caroline Homer of the State Library and Archive Service, Tasmania, Heather Excell and Wendy Rimon of the Special and Rare Collections at the Morris Miller Library of the University of Tasmania, Elizabeth 'Liz' Jack of Libraries Tasmania, Renee Holman and Kathi Spinks of the State Library of New South Wales, the staff of the Heritage Collections Reading Room at the State Library of Victoria, Colleen Field of the Centre for Newfoundland Studies at the Queen Elizabeth II Library of Memorial University and Debbie Maynard, Susan Rubin and, especially, John Griffin of the Newfoundland and Labrador Collection of the Provincial Resource Library in St John's, Newfoundland.

At the Université de Montréal, I would like to thank all the library staff including Marie-Ève Mesnard, Margie De Champlain, Émilie Dalpé, Nancy Desforges, Jason Gerlick, Diane Moreau, Frédérick Savard, Josée Vachon and, particularly, Valérie Rioux of the Bibliothèque des lettres et sciences humaines. Chantal Lajoie of the Bibliothèque de médecine vétérinaire, Édouard-Alexandre Bureau of the Bibliothèque de la santé, Jérémy Prime of the Bibliothèque des sciences and Christine Chénier of the Bibliothèque Thérèse-Gouin-Décarie also offered valuable aid. I would also like to express my gratitude to Pascale Bellemare, Vincent Perrault, Nancy Théberge and Julie Touchette of the Service du prêt entre bibliothèques.

Relating to museum visits, I wish to thank Sandy Ingleby (Manager of the Mammalogy Collection) and Harry Parnaby and Ron Lovatt of the Australian Museum, Anthony 'Tony' Gill (Curator in Natural History) at the Chau Chak Wing Museum of the University of Sydney, Katie

Date (Collection Manager) and Karen Roberts (Collections Manager) of the Melbourne Museum, Minette Brewin (Administrations Officer) and, especially, David Maynard (Senior Curator of Natural Sciences) of the Queen Victoria Museum and Art Gallery, Belinda Bauer (Collections Manager), David Hocking (Senior Curator of Vertebrate Zoology and Palaeontology), Peter Hughes (Senior Curator of Decorative Arts), Kathryn Medlock (formerly Senior Curator of Zoology) and Nicole Zehntner (Registration Technician) of the Tasmanian Museum and Art Gallery, Patricia 'Trish' Koh (Collections Manager) of the Tiegs Museum at the University of Melbourne, Burton Lim and Mark Peck of the Royal Ontario Museum in Toronto, Karen LeDrew (Site Supervisor) of the Beothuk Interpretation Centre at Boyds Cove, and Elaine Anton (Registrar), Mark Ferguson (Manager of Collections and Exhibitions), Lori Temple (Collections Manager) and, especially, Nathalie Djan-Chékar (Collections Manager) of The Rooms Museum for their assistance researching the Newfoundland wolf. I am also very grateful to the Newfoundland naturalist John Maunder for sharing his considerable knowledge of the history of the wolf mount displayed at The Rooms. I also want to thank Eleanor Hoeger, Sara Ketelson and Marisa Survoy of the Mammalogy Department of the American Museum of Natural History, who gave generously of their time during a visit to their collections by my research assistant Valérie Bienvenue. For their help in viewing some of the thylacine petroglyphs at Murujuga, I am grateful to Megan Berry, Ben Gunn and Ken Mulvaney.

For administrative assistance at different times, I thank Geneviève Bédard, Savedie Chhith, Patricia Correa-Romero, Marisa Flores, Leïla Fourmas, Marie-Janine Georges, Susadny Gonzalez Rodriguez, Laurence Labelle, Chia Tien Jessi Lee, Sébastien Lévesque, Maeva Moreno, Sylvie Rouleau, Karine Selles, Annie Tremblay and Jeff Tshibaka.

I have benefitted immeasurably from conversations about themes related to this project with colleagues in screen studies including Olivier Asselin, Marta Boni, Serge Cardinal, Frédéric Dallaire-Tremblay, Marion Froger, Michelle Garneau, André Gaudreault, André Habib, Santiago Hidalgo, Emanuel Licha, Silvestra Mariniello and Isabelle Raynauld. I am also indebted to all my art history colleagues including, most notably, Peter Krausz, who gave generously of his time to help with my analyses of drawings by Shanawdithit of the Beothuk and Kristine Tanton, who kindly shared insights into digital technologies and their application.

In relation to language issues, I am very grateful to Daisy Allan, Language and Research Project Officer at the Tasmanian Aboriginal Centre, for answering my questions relating to palawa kani. I am also deeply thankful to Yagiya Bobongua, Sara Chare and Shumpei Tachi for their help with Japanese texts relating to wolves.

For sharing general insights into farming, hunting and trapping, I am indebted to Jack Cameron of 'Kiora North' (Temora, NSW, Australia), and for providing specific information on the pastoral industry and sheep predation, I wish to thank Douglas Mortimer, Mark Mortimer and Robert Mortimer of the merino stud Centre Plus Merinos (Albert, NSW, Australia).

This book would not have been possible without the help of my research assistants Valérie Bienvenue, Camila de Oliveira Savoi, Jade Brais-Dussault and Milan Cail and I thank them for their time and effort. I am also indebted to Valérie for our frequent enriching discussions about animal studies. Additionally, I am grateful to David Sume for his assistance transcribing pertinent materials and for offering sage advice regarding historical printing processes.

Over the years many students from the courses ART3001 *Écrire l'histoire de l'art*, HAR1210 *Introduction aux cultures visuelles* and, most notably, HAR3550 *L'art du témoignage* shared important insights regarding the representation of nonhuman animals and the *enjeux* that need to be taken into account when writing about them. I am deeply indebted to the classroom discussions we had relating to those topics. Hopefully this book signals something of the value of research-led teaching (which can also become teaching-led research) at a time when the marketising of higher education places such pedagogical practices under threat.

At Berghahn, I am very grateful to Marion Berghahn, Harry Eagles, Amanda Horn, Caroline Kuhtz, Mark Stanton and Emma Wallace for believing in the project and helping see it through to fruition. I am also indebted to the invaluable comments, insights and suggestions shared by the anonymous reviewers of an earlier draft of the manuscript. Additionally, I warmly thank Marie-Soleil Provençal for granting me permission to reproduce a photograph of her mural *Shadows of an Epoch* as my cover image.

Finally, once again, my thanks to Ange and Esther. This book is for you.

Introduction

Cruelty, Thy Name Is Wolf!

> I caught dozens of those things back in the old days.
> —Bushman Bob Wainwright speaking of thylacines, cited in Col Bailey, *Lure of the Thylacine*

Beginnings

In 1830, in Lutruwita/Tasmania, the Van Diemen's Land Company (VDLC) introduced a bounty scheme aimed at wiping out all dogs (*Canis lupus familiaris*), Tasmanian devils (*Sarcophilus harrisii*) and thylacines (*Thylacinus cynocephalus*) in the vicinity of its landholdings.[1] This was the first of several private bounty schemes established on the island that included or were exclusively directed at the eradication of thylacines or Tasmanian tigers (as they are otherwise known), a species of marsupial carnivore that formed the island's largest predator.[2] The same year the VDLC bounty was established, a bounty on captured palawa (First Nations Tasmanians) was also instigated, which offered £5 for an adult and £2 for a child.[3] In 1842, the VDLC paid £50 for the capture of a palawa family of seven who were believed to be the last free Indigenous people on the island.[4] After much debate, a government bounty for the killing of thylacines was also agreed in 1886 and financing for the scheme began in 1888.[5] The first payment was made in April of that year and the last in June 1909 (the bounty was discontinued in 1908). In addition to the bounties, money was offered by menageries and zoos for live thylacines and by museums for dead ones. Animal dealers would advertise for live or dead tigers. These financial incentives continued to be offered after the bounties were discontinued.

Numerous live thylacines were exported from Lutruwita to zoos in mainland Australia, the United States and various countries in Europe. Abroad, the thylacine was rendered an exotic spectacle. As Nigel Rothfels explains of London Zoo, it 'was a place designed by the bourgeoisie for its own education and amusement'.[6] Dead thylacines were also a valuable commodity. In 1874, for example, the London-based animal dealer William Jamrach was offering varied sums of money for Lutruwita's fauna including devils and tigers. He advertised that he was willing to pay £4 each for up to a total of twelve dead 'striped wolves', which he instructed should be pickled in 'strong brine of salt' and then forwarded to him.[7] Jamrach was also offering £6 per pair for live wombats (family *Vombatidae*). The newspaper that reported his offer, *The Mercury*, did so in an article titled 'Pickled Devils'. It used the opportunity to call upon shepherds to kill and pickle tigers and devils rather than engaging in the wholesale slaughter of kangaroos, which were 'disappearing' and formed 'valuable game'.

The last known living thylacine died in 1936 yet zoo demands for replacement animals continued after that date.[8] The naturalist David Fleay reported briefly catching a thylacine in the 1940s but the animal escaped.[9] The marsupial was almost certainly still extant in the early 1950s in the north-east of Lutruwita but searches were being conducted elsewhere in habitat often less suited to the animal. Some argue the thylacine disappeared late in the 1950s.[10] In 1980, the animal was declared extinct although there was a credible sighting by ranger Hans Naarding in 1982 near Togari in the north-west of Lutruwita. Andrew Baker and Chris Dickman note that although sightings continue to this day, 'the thylacine is considered by all authorities to be extinct'.[11] This is not entirely accurate as some recent computer modelling has indicated possible survival to the present.[12]

If thylacines do survive today, a possibility I believe is extremely unlikely, they are critically endangered. Given their iconic status, the conservation of any relict population would be a clear priority. Recognition of the rarity of the species, if not of the need for their conservation, is long-standing, with John Gould writing in 1863 that '[a] price is already put upon the head of the Native tiger'.[13] He foresaw that once Lutruwita was more densely populated 'the numbers of this singular animal will speedily diminish, extermination will have its full sway, and it will then, like the Wolf in England and Scotland, be recorded as an animal of the past'.[14] Despite such warnings, the bounty and other reward schemes remained in place until the early twentieth century.

In Newfoundland (known as Ktaqmkuk in Mi'kmaw), Canada's easternmost province, a government-funded bounty scheme intended to eradicate the island's wolves was first initiated in 1839.[15] The scheme, which was created because of fears that the wolf was killing livestock, ran intermittently for over a century. The last recorded payment was made in 1896 but the bounty remained on the statute books until 1954.[16] The Newfoundland wolf (*Canis lupus beothucus*) probably became extinct in the 1920s, but fears about their extinction were articulated at least a decade earlier.[17] Unlike the thylacine, the wolf, considered a subspecies of *Canis lupus*, was not of strong scientific interest nor sought after by zoos. The causes of both the wolf's and the thylacine's extinction are much debated with various factors used to explain their disappearance. Whether they caused their extinction or not, the bounty schemes were instigated with the intention of eliminating both carnivores. The schemes incorporated no restrictions designed to prevent 'over-harvesting', no measures to safeguard the long-term survival of either targeted animal.

For Donna Haraway, as she explains in *The Companion Species Manifesto*, humans and some nonhuman species such as the domestic dog (her chosen example) exist in relationships that are co-constitutive and continuous, emergent and changeable.[18] Reciprocity is a key element of these interactions.[19] Relationships with companion species involve communication, albeit communication across 'irreducible difference'.[20] In the case of the thylacine and wolf, however, there were not even relations of domination with humans (relations of the kind that exist, for instance, between a farmer and their flock of sheep). Relations were refused and recognition was foreclosed. To most, the thylacine and the Newfoundland wolf were not companion species in Haraway's sense of the term, they were not animal bodies that mattered, bodies to which human animals related, but were abject bodies that required removal from colonised territories, from domesticated land.[21] An easy way to remove them was to destroy them. The bounty schemes can therefore be viewed as encouraging animal genocide. Animal genocide and its aftermath, the ways in which it has been represented or its representation suppressed, form the subject of this book.

Animal genocide as it is defined here refers to the targeted destruction of a specific group of animals such as a species or subspecies. I explore the history and meaning of the term at length in Chapter 1. This includes considering the dangers that accompany unreflectively making comparisons between conceptions of animal genocide and dimensions of the Holocaust (as some thinkers, including Jacques Derrida, have done). The chapter foregrounds that, crucially, the destruction of the group does not have to be total for a genocide to exist, only the presence of the intent for such encompassing destruction. In this sense, whether genocidal intent was a major contributing factor in the extinction of the thylacine and the Newfoundland wolf is unimportant when it comes to establishing the occurrence of genocide. Recently, efforts to link the extinction of the thylacine to disease have been used to argue for the exoneration of traders who trafficked tigers (removing them from the potential breeding population) despite knowing their rarity.[22] Such people would still be responsible, culpable for genocide (as they were wilfully contributing to the destruction of the species) as I understand it.

I focus on two charismatic animals in the thylacine and the Newfoundland wolf, the first very well known and symbolically important, the second much less noted, yet there are many other examples of animal genocide through history and I discuss some of these, including the kea (*Nestor notabilis*) and the Falkland Islands wolf (*Dusicyon australis*), in this book.[23] My decision to focus on the thylacine and the Newfoundland wolf is motivated, in part, by illuminating parallels between their histories. Analysing their histories in tandem enables patterns regarding the justifications offered for the genocidal campaigns perpetrated against them to be identified. My hope is also that by discussing the Newfoundland wolf alongside the thylacine, I can accord the former and their tragic history greater visibility than they currently possess, contributing in a small way to the preservation of their memory. The archives and the scholarship devoted to the thylacine is considerably more substantial than that dedicated to the wolf and this disparity is reflected in the book. Although the Newfoundland wolf is referenced in most chapters and Chapter 6 is entirely dedicated to them, the majority of chapters, of necessity, centre on the thylacine.

The fate of the Newfoundland wolf and the thylacine, which were vilified by European settler colonists and scapegoated by them for livestock losses, is inextricably linked to imperialist expansion. They both coexisted with the Indigenous human populations of their habitats for centuries, if not millennia, before Europeans invaded and began their genocidal campaigns. These campaigns were directed at those Indigenous human populations as well as at native nonhuman animals.[24] There was crossover in terms of the perception and treatment of both and my investigation of animal genocide therefore sometimes involves also acknowledging human genocide.

Human Genocide in Newfoundland and Lutruwita

Patrick Brantlinger affirms that in the 1830s, the fates of Lutruwita's and Newfoundland's Indigenous peoples along with that of the Xhosas were perceived as clear and troubling cases of extermination in British territory.[25] In testimony to the 1837 Parliamentary Select Committee on Aboriginal Tribes, Bishop Broughton of Sydney was quoted in a discussion of First Nations Australians (with Australia referred to as New Holland) as stating that 'within a very limited period, a few years . . . those who are in most contact with Europeans will be utterly extinct – I do not say exterminated – but they will be extinct.'[26] The conclusion to the same committee's discussion of Newfoundland was that in 'the colony . . . it may . . . be stated that we have exterminated the natives'.[27] Here, the intentional destruction of the Beothuk (a First Nations people of Newfoundland) is admitted, although the committee resists doing so in an Australian context. Parallels between the treatment of the Indigenous populations of Newfoundland and Lutruwita have subsequently been examined by scholars including Terry Goldie and Fiona Polack.[28]

Later, Indigenous peoples in Australia and Turtle Island (North America) were recognised as victims of genocide. In a book manuscript he was preparing at the time of his death, the lawyer Raphael Lemkin listed the palawa along with other First Nations Australian groups and also Native American Indians as historical victims of genocide.[29] In *Genocide*, Leo Kuper also includes the 'systematic annihilation of Aborigines in Tasmania' as one of his examples.[30] The extermination of the palawa was openly discussed in newspapers of the 1820s. The anonymous author of one article debates the need for their extermination or, their favoured solution, removal, to prevent bloodshed against settler colonists. They ask of relocating the palawa to an outlying island: 'Would it not save scores upon scores of them from being shot? – which they are now, with as little remorse, as if they were so many crows!'[31] The author rejects the idea of extermination because the palawa are 'human creatures' rather than 'brute beasts'.[32] His remark about crows, however, demonstrates that he perceived parallels between the treatment of nonhuman animal pest species and Indigenous peoples who were viewed as troublesome. Often, Indigenous peoples were seen by European settler colonists as akin to nonhuman animals such as corvids and wolves (animals that, non-coincidentally, were subject in some colonies to bounty schemes aimed at their destruction).

In the context of the Beothuk, the ambiguous feelings of settler colonists towards the Indigenous population are strongly brought out in the writings of the explorer and entrepreneur William Eppes (also Epps) Cormack. He struggles to work through his distress at the violence inflicted upon them by Europeans, writing:

> a defenceless and once independent proud tribe of men have been mostly extirpated from the Earth without hardly and [*sic*] enquiring How or Why. Here still remains Man in all his primitive [illegible] with his bow

> and arrow and cloathed in skins. There on the opposite approximating point is Man improved and powerful. Barbarity and Civilization should shake hands.[33]

The rhetoric of the 'noble savage' informs Cormack's description of the Beothuk. The binary he is operating with, one in which the apex of civilisation is the European and the nadir is the Beothuk, is clearly breaking down. He goes on to state:

> Sometimes these unsullied people of the chase have been destroyed wantonly, because they have been thought more fleet and more evasive than how men ought to be. At other times, at the sight of them, the terror of the ignorant European has goaded him to murder the innocent. Civilization might weep.[34]

Here, civilisation is found wanting, seen to be uncivil, barbarous in its actions. Something of this barbarity seems linked with perceiving the Beothuk as more animal than human. Cormack identifies them as hunters (people of the chase) yet frames them as hunted, viewed as remarkably nimble and elusive prey. In another fragment, he argues 'they ought not to have been destroyed like Wild Beasts'.[35] In his earlier 'An Account of the Red Indians of Newfoundland', John Cartwright describes how settlers attacked and killed Beothuk, stating that whomever 'has shot one indian [*sic*] values himself more upon this feat than had he overcome a bear or wolf'.[36] Cartwright goes on to note:

> A Red Indian in the summer season, may with too much propriety be compared to a beast of chase such as the wolf or fox, that preys on the smaller game and in his turn is liable to fall himself prey to hunters more destructive.[37]

The Beothuk are framed as a superior kind of wolf, a lesser kind of human. Both Cartwright and Cormack signal how notions of human proximity to nonhuman animals inform the attitudes and actions of settler colonists.

Nonhuman animals were also points of comparison when the palawa were discussed in the colonial press. An 1830 newspaper article, for example, describes the palawa as watching 'like cats for days, and even weeks for their prey, and when they have struck their victim, they will murder him with more savage rage than the hyena or the ravenous wolf'.[38] Historically, human genocide has therefore often accompanied animal genocide, with the 'animality' of a human group part of the logic employed to justify violence against them. This complex relation between the two forms of genocide is examined in Chapter 1. The links between animal and human genocide also features in other chapters, including, most notably, Chapter 4, in which I analyse how Erle Wilson uses the figure of the thylacine in his novel *Coorinna* to express and seek to work through the genocide against the palawa.

Cultural representations of Indigenous people like the Beothuk and the palawa as like nonhuman animals has historically often helped to create conditions conducive to genocide against them. Representations have also significantly contributed to encouraging genocide against the thylacine and the Newfoundland wolf. I therefore explore the nature of these representations at length in the following chapters and also how those genocides they facilitated have been portrayed (or ignored) retrospectively. The role of images in the enabling of genocide was recognised by Lemkin. Using Nazi-occupied Poland as an example, he linked popular cultural images (specifically films and pornography) with the suppression of moral feelings: the 'desire for cheap individual pleasure' substituting for the 'desire for collective feelings'.[39] Lemkin focussed on how the easy pleasures of mass cultural forms distracted from the negative political situation and the suffering of others. For him, images diverted and dulled feelings of community. In this, they made genocide possible through stifling resistance to it. Here, however, I am interested in how the content of images shaped the perceptions held by perpetrators of the victims of genocide (specifically, the thylacine and the Newfoundland wolf) thereby fostering conditions conducive to their annihilation.

Ervin Staub has suggested that the cultural characteristics that contribute to group violence can be deduced from sources such as art and literature.[40] Art, literature and folklore can help to articulate and perpetuate negative images of groups.[41] Staub affirms that bystanders often fail to challenge genocidal actions because of fear but also because they accept the definitions of reality that are dominant in their culture.[42] In such circumstances, the importance of the dissemination of divergent views is clear. In the case studies discussed here, divergent views were always extant but never dominant. Belatedly, the genocide-enabling perceptions of the thylacine and the Newfoundland wolf have been largely challenged and displaced. Now, efforts are being made to process the aftermath of these animal genocides in contexts such as museums (see Chapter 3), film (see Chapter 5) and literature (see Chapter 4). In the remainder of this chapter, I want to outline the historical circumstances within which animal genocide against thylacines and wolves was enacted. This involves engaging with some of the negative perceptions and myths that circulated regarding both. Animal genocide in the context of my case studies is bound up with the creation of bounty schemes and I will therefore also give a summary of their employment in Lutruwita and Newfoundland.

Of Marsupial Wolves

The thylacine is generally accepted to have disappeared from the Australian mainland around 3,500 years ago with a relict population persisting in Lutruwita.[43] The likely population of adult female tigers living on the island at European invasion was approximately 4,500.[44] The thylacine was identified as a new species, named *Didelphis cyncocephala* in a paper by George Prideaux Robert Harris delivered on 21 April 1807 and published in the *Transactions of the Linnean Society* the following year.[45] One of the vulgar names Harris gave for the thylacine was the 'zebra wolf' (the other being the 'zebra opossum'). The Newfoundland wolf was proposed as a distinct subspecies of wolf by Glover Allen and Thomas Barbour in 1937, given the Latin trinomial *Canis lupus beothucus*.[46] The binomial *Canis lupus* was given to wolves in general by Linnaeus in 1758.

Both thylacines and wolves are large predators. The thylacine was the apex predator in Lutruwita. Apex predators are at the top of the food pyramid and are therefore relatively rare proportionate to the broader community of animals and in terms of numbers of individuals in a given area.[47] Their paucity of numbers, however, is inverse to their influence on ecosystems, which is frequently great and plays a key role in structuring faunal communities.[48] In Turtle Island, for example, wolves are crucial to regulating the health and population size of ungulates such as moose and caribou. The destruction of large predators therefore has considerable detrimental effects for associated fauna. Thylacines and wolves have been seen as 'a classic example of convergent evolution', but as Samuel Turvey and Joanne Cooper point out, the two species possess 'fundamentally different morphological adaptations for locomotion and carnivory'.[49] Marsupials and placentals last shared a common ancestor around 160 million years ago.[50]

The first unambiguous reference by a settler colonist to a thylacine in Lutruwita is from 1805. Lieutenant Governor William Paterson, who was then in charge of the settlement Yorkton (also York Town), located at Port Dalrymple, describes a hitherto unknown animal that was attacked and killed by dogs on nearby Mount Albany on 30 March.[51] In his undated report, he states of the 'brute' that it is 'very evident this species is destructive, and lives entirely on animal food'.[52] Paterson concludes that 'the form of the animal is that of a hyaena at the same time strongly reminding the observer of the appearance of a low wolf dog'. A truncated transcription of the report was published in the 21 April 1805 issue of the *Sydney Gazette and New South Wales Advertiser*, adjacent to the section 'Ship News', which mentions the ship *Sydney* landing 622 head of horned cattle in Port Dalrymple, a 'valuable flock' for the 'infant settlement'.[53] These cattle (the survivors of an original cargo of 910 animals sent from India) supplemented the two horses, eight cattle horses and one hundred sheep delivered by *HMS Buffalo* at the end of March and, possibly, earlier livestock.[54] Pastoral farming would subsequently come to negatively impact the thylacine because of false concerns that the carnivore was a routine sheep killer, which led to the bounty campaigns, and also because of the habitat destruction that accompanied making land suitable for grazing.

After unloading its animal cargo, the *Sydney* left Port Dalrymple on 9 April (arriving in Gadigal/Sydney on 18 April), transporting Paterson's report and also the thylacine's pickled reproductive organs ('the parts of generation') to the mainland.[55] Paterson's handwritten account provides extensive measurements of both male and female thylacines, suggesting that two tigers had in fact been killed.[56] The newspaper only gives the measurements of the male. Paterson's mention of the palawa (he suggests they are distinct from any other peoples) at the beginning of his report is also omitted. The newspaper does, however, include an account of a 'series of barbarities' perpetrated by 'natives' in the vicinity of Deerubbin (referred to as Hawkesbury by the paper) in New South Wales.[57] The Darug (also Dharug) had been resisting the invasion of their lands since Europeans first arrived in 1794.[58] The 1805 article refers to attacks on farms and acknowledges that some interpret the actions of the Darug as '*acts of retaliation* for injuries received'.[59] This reading is rejected by the anonymous author, who claims that First Nations Australians are well protected by the law with crimes against them severely punished. The writer also expounds upon the generosity of the settlers and suggests the First Nations Australians should 'lay claim to every grateful return, which can extend no further than to passive forbearance from rapacity'. They are condemned for stealing crops 'they are themselves too indolent to cultivate'.[60] The situation in Deerubbin reported in the *Sydney Gazette* prefigures the guerilla war that the palawa would come to fight against European oppression in Lutruwita.

Land usurpation of the kind occurring in Deerubbin was replicated in Lutruwita, with early settler colonists well aware of the detrimental effects of the practice. Edward Curr, for instance, noted that the land being seized was depriving the palawa of their hunting grounds.[61] In a letter to his wife in 1830, George Augustus Robinson condemned the European usurpation of palawa land, bemoaning the military action that had been sanctioned against the Indigenous population.[62] In his discussion of settler colonialism and genocide, Patrick Wolfe suggests that the primary motive for the elimination of Indigenous peoples is 'access to territory'.[63] Wolfe's insights are open to extension to instances of animal genocide. The need for land can be motivated by agriculture (crop cultivation, pastoralism) and also forestry, fishing, trapping and mining. In Lutruwita, farming was the primary cause of land theft.

The situation was different in Newfoundland, where settler colonists were initially interested in fishing and trapping with mining growing in importance in the nineteenth century and logging in the twentieth.[64] Commercial agriculture was (and remains) limited on the island although in the mid-nineteenth century farming developed rapidly in the environs of St. John's, with roughly four hundred farms to be found there.[65] This expansion coincided with the introduction of the bounty on wolves and the two are clearly linked. Livestock losses were a motivation for the introduction of the act that encouraged their killing. Loss of land, of suitable habitat, however, was minimal and therefore not the primary factor in the wolf's demise.[66] The impact of European expansion on the landscape is nonetheless recognised by Moses Harvey who discusses the displacement of the wolf by human industry, writing of the region near Grand Lake that: '[a] thriving town, it may be anticipated, will spring up here and the solitudes, hitherto the domains of the deer, the fox and the wolf, will be resounding with the din of human labour.'[67]

In contrast to Newfoundland, habitat deprivation was a significant factor in the extinction of the thylacine in Lutruwita. There, loss and degradation of territory impacted not just the palawa but also native fauna, such that Wolfe's observation that '[t]hrough its ceaseless expansion, agriculture . . . progressively eats into Indigenous territory, a primitive accumulation that turns native flora and fauna into a dwindling resource and curtails the production of Indigenous modes of production' is highly pertinent.[68] Habitat fragmentation has been identified as potentially the most important conservation issue for all surviving carnivorous marsupial species.[69]

As well as destroying the thylacine's habitat, the pastoral industry in Lutruwita blamed stock losses on the tiger.[70] The cause of these losses was multiple, linked primarily to poor animal husbandry, theft and predation by wild dogs.[71] The thylacine, however, provided a convenient scapegoat for failures recorded by ventures such as a massive wool-growing scheme that received a royal charter in 1825, granting the VDLC 250,000 acres of land. Much of the land the company was ultimately allocated was not well suited to sheep husbandry and some of the employees lacked suitable expertise.[72] The company's early surveyors knew little about farming and sometimes, as in the case of Henry Hellyer, based their assessment of the suitability of a milieu on the picturesque appearance of a location rather than its suitability for pastoral use.[73] A sense of Hellyer's privileging of aesthetics over utility is given by his choice of name for some of the geographical features he encountered, such as Mount Van Dyke and, most tellingly, Mount Claude (after the celebrated landscape painter Claude Lorrain).[74] As Kenneth Dallas has noted the VDLC's pastoral vision for Lutruwita 'was foredoomed in the sense that the growing of fine wool on a large scale was impossible in the regions to which it was confined'.[75] The poor climate caused heavy stock losses early on.[76] Elsewhere, duffers or sheep-stealers clearly sought to explain the disappearance of livestock by blaming tigers, but their efforts at deception were unsuccessful.[77]

To shift the balance for management failings, the VDLC grossly exaggerated predation of sheep by thylacines. Wild dogs were actually the sheep's main predator, which was why the 1830 bounty scheme included canines. Through the 1880s bounties were still being paid for killing dogs.[78] In 1833, George Augustus Robinson briefly encouraged some palawa to engage in dog catching on behalf of the government because packs of dogs were said to be killing many sheep.[79] Robinson witnessed dog skins lining the fences of a property near Campbell Town.[80] Writing in 1834, William Henry Breton calls the tiger a destroyer of lambs and remarks on its cunning yet also observes:

> Some of the dogs introduced by the Colonists have become wild (in all probability through the culpable neglect of their owners), and have destroyed a great number of sheep: once fairly established in the Colony, it would be in vain to attempt their destruction, as there are tracts of 'scrubby' country from which it would be impossible to expel them.[81]

Dogs continued to be a major threat to flocks throughout the nineteenth century. Writing on zoology in John West's *The History of Tasmania* published in 1852, Ronald Campbell Gunn acknowledged that the thylacine was not a significant sheep killer, noting that the tiger 'usually confines its attack to one at a time, and is therefore by no means so destructive to the flock as the domestic dog become wild, or as the Dingo of Australia, which both commit vast havoc in a single night'.[82] He also sought to indirectly blame the problem of feral dogs on the palawa, writing that: '[o]n the destruction of the aboriginal tribes [their dogs] escaped, hunted in large numbers, and committed great havoc, among the flocks: farmers lost five hundred sheep in a season.'[83] Of the dogs, he noted that '[b]y a single gripe these wild marauders destroyed a sheep, and a few minutes was sufficient to strew the downs with dead'.[84]

Despite petitioning vigorously for a government-funded thylacine bounty, the politician John Lyne was clearly aware that dogs caused far greater damage to livestock. While arguing for the use of poison on farms, he stated that '[t]he sheep destroying propensities of tame dogs are worse even than tigers, and poison was the only way by which they could [be reached]'.[85] In mainland Australia, dogs continue to pose a major threat to livestock. Writing in *The Land* in 2019, Stephen Burns explores the impact of wild dogs (specifically dingoes) on merino sheep flocks in New South Wales.[86] He uses the merino stud of Robert Bartlett as a case study.

Bartlett has engaged in an intensive programme of trapping, shooting and baiting to prevent 'the dogs tearing our sheep to pieces'.[87] Another merino breeder, Brian Bambrick, reports buying five hundred sheep in 2007 and within twelve months having none because of dog attacks. Dogs often hunt in packs of eight to ten animals, with multiple packs operating in the same area.

Lyne successfully argued for the introduction of a bounty in 1886, when he urged in committee at the House of Assembly that the Governor be petitioned for £500 for the destruction of tigers 'otherwise known as the Tasmanian dingoes'.[88] Earlier efforts to garner support for a bounty had been unsuccessful. In 1884 on Tuesday 28 October a petition for legislative action was made by Mr Gray on behalf of twenty-six people from Spring Bay for losses sustained by the 'Ravages of [the] Native Tiger'.[89] In 1885 another petition was made for the 'Destruction of Eagles and Native Tigers'. In 1886, Lyne used outlandish claims such as that '100 sheep were destroyed per dingo' and that '30,000 or 40,000 sheep were killed annually by dingoes'. During the debate, the Minister of Lands observed that the tigers 'had not one redeeming quality, as they were absolutely useless, and simply killed sheep for mischief'. Lyne is reported as stating that the scientific name for the 'dingo' was *lagunta*. This is, in fact, a palawa name for the animal and is therefore noteworthy as it obliquely acknowledges the palawa and their precontact relationship with the animal.[90] The thylacine is a part of Country for the palawa, part of the land, of culture and of the environment.[91] Other palawa names for the tiger include *corrinna*, *kannenner* and *laoonana*. The presence of the thylacine in palawa culture is also indirectly referenced in a newspaper article summarising a lecture given by the animal trafficker James Harrison on 15 July 1919. At a lecture given to the Methodist Guild in Wynyard, Harrison 'exhibited the jawbone of [a thylacine] which he estimated had been dead 150 years'. The jawbone 'was found among the remains of a blackfellows camping place in Sisters Hills'.[92]

During the 1886 debate, Edward Braddon suggested '[h]e would be very willing to assist [Lyne] to quell this *leghunter*', reducing the palawa name to a crude pun.[93] The deliberations also included a reference by Samuel Sutton to the need to 'exterminate' another peril, the 'yellow agony', which was interpreted by Peter Barrett to mean Chinese labour. In the 1870s there had been a significant increase in the Chinese workforce in north-east Lutruwita where they helped to mine tin.[94] The remarks show that, like the palawa before them, Lutruwita's Chinese minority were viewed as on a par with nonhuman animals, specifically the thylacine: 'one of the greatest pests the colony had' to use Lyne's description.[95] Sutton 'sticks it' to Chinese migrant labourers through linking them to a rampant predator, tainting them by association.[96] He makes negative feelings about the tigers adhere to the Chinese diaspora. In *Picture Theory*, W. J. T. Mitchell has examined the workings of animal metaphor, of the *figure* of the animal, as a mode of domination. Mitchell notes that: 'Animals stand for all forms of social otherness: race, class, and gender are frequently figured in images of subhuman brutishness, bestial appetite, and mechanical servility.'[97] Here, the *figure* of the thylacine is employed by Sutton in the service of racism. One of the examples used by Sara Ahmed to explore the cultural politics of emotion is political speeches. Through such speeches, negative (or positive) emotions may be transferred to an object. In the speeches given by settler colonist politicians in 1886, 'fear' is the emotion that is transferred to the tiger and the Chinese worker. They are identified as threatening European interests and therefore as requiring 'extermination' (although in the case of the Chinese, 'extermination' should be understood to mean 'restriction' or 'prohibition' rather than annihilation).

Lyne's continual references to the thylacine as a 'dingo' are revealing. This name was very seldom used for the tiger. The politician seems to be deliberately exploiting the negative emotions associated with the dingo (*Canis dingo*), which was viewed as a pest on the mainland. Bounties on dingoes were first introduced in Australia in 1830.[98] There are no dingoes in Lutruwita but Lyne figuratively 'introduces' them when using their name to refer to the thylacine. Amanda Stuart explores the way imagery of the dingo constructed the animal as debauched and menacing.[99] She also notes how European ideas about wolves framed how the dingo was perceived, with the latter receiving all the 'baggage' and the 'negative loadings of the European wolf'.[100] The dingo was, in fact, occasionally referred to as the New South Wales wolf.[101] In his *A General History of Quadrupeds*, Thomas Bewick uses this name for the dingo and positions his chapter about them immediately after the one devoted to the wolf. He writes of the dingo: 'its wild and savage nature seems strongly to point out its affinity to the Wolf.'[102] Lyne exploited the negative perceptions of the dingo, hoping they would 'stick' to the thylacine thereby aiding his argument for their extermination. His efforts were seemingly successful as the vote on his proposal, while initially tied (ten for, ten against), was carried by the Speaker. In 1887, £500 for the destruction of native tigers was agreed.[103] In 1888, Lyne wrote to Braddon (by then the Minister for Lands and Works), seeking advice on what he considered 'a sufficient reward to offer for each skin'.[104]

Through efforts such as Lyne's the thylacine gained a bad reputation and was viewed by many as highly aggressive and dangerous. This extended to scientists. In *Mammals of Australia*, the zoologist Gerard Krefft wrote that the Tasmanian tiger was 'a most ferocious and formidable animal, which will overpower even a Cow or Horse if driven by hunger to attack them'.[105] There is no historical evidence of thylacines attacking cattle or equines. The vision of the thylacine as highly aggressive also carries into more recent popular literature. Hesba Brinsmead's 1972 novel *Echo in the Wilderness*, for example, implies that a thylacine attacked a man, leaving his body 'torn, mangled, savaged'.[106] While there are historical instances of tigers biting humans (see my Conclusion), a fatal attack of the kind described by Brinsmead is unheard of.[107]

Both the dingo and the thylacine were frequently attributed lupine characteristics, the negative emotions directed at wolves thereby transferred to them. The thylacine was also sometimes referred to as a marsupial wolf, reinforcing the connection. Writing in 1888, Carl Lumholtz noted that Australian marsupial carnivores were usually named 'after animals of the old world'.[108] He discusses the 'marsupial wolf', which he states does 'great damage among the sheep' but is 'well-nigh exterminated'.[109] An 1882 newspaper article suggests that the 'native tiger' is a species that 'hunts animals with the pertinacity of a pack of wolves on the steppes of frozen Russia'.[110] In *Paper Tiger*, Carol Freeman discusses how wolflike images of the thylacine influenced how they were perceived, constructing them as a 'dangerous predator'.[111] She traces how the portrayal of the thylacine frequently contributed to their demonisation. They were rendered lupine, dethylacinised, their singularity as a species suppressed, the creature overwritten with the familiar trope of the wolf.

In *The Platypus and the Mermaid*, Harriet Ritvo also notes how imagery of marsupial carnivores produced by nineteenth-century European naturalists often made the animals similar in appearance to carnivores in Europe. She reproduces engravings of a Tasmanian devil and a doglike thylacine.[112] Figuring the thylacine as a wolf, in particular, did tangible damage to them as it helped foster a climate of hate. Words and images that emphasised lupine characteristics therefore contributed to fostering the social conditions that made animal genocide possible.

As well as being labelled a dingo and a wolf, the thylacine was also given other names. These multiple designations register the difficulty Europeans had making 'sense' of native fauna.[113] The term 'hyena' was common in the early years of invasion. Sometimes, Freeman notes, textual descriptions, while not directly referring to the thylacine as a hyena, imagined the marsupial's lair as dark, dirty and confined, implying 'distasteful habits' and placing them 'in the same perceptual zone as the hyena and jackal'.[114] During the 1820s and 1830s local newspapers commonly directly referred to the thylacine as a 'hyena' or, more rarely, a 'hyena tiger'.[115] In 1834, William Swainson described the thylacine using the common name the dog-faced opossum and stated that the animal 'suggests the idea of the union of the dog and the panther', granting the animal both canine and feline characteristics.[116] Earlier, in 1820, the *Hobart Town Gazette and Southern Reporter* referred to the thylacine as a 'hyena opossum'.[117] Breton found this name for the thylacine absurd but discovered nothing amiss with calling the 'dog-faced dasyuris' a tiger.[118]

It is the name 'tiger' when used for the thylacine that Mark Halsey employs as an example of 'lexical violence', of how words can negatively impact the things they refer to.[119] For Halsey, words have the capacity to 'alter the course of events'.[120] Drawing on a text from September 1886 that was published in the *Tasmanian Mail* in which the rectitude and utility of using the name 'tiger' is questioned, Halsey argues that referring to the thylacine as a tiger 'brought with it nothing less than the urge toward the decimation of [the] species'.[121] His recognition that 'names levy effects' and that the name 'tiger' had dire consequences for the thylacine has some parallels with Freeman's exploration of how the influence of lupine iconography on depictions of the thylacine shaped behaviours towards them.[122] Halsey focusses primarily on 'words', with images seemingly viewed as secondary effects to primary acts of naming, yet, to me, images are equally capable of violence towards their referent.[123]

As the thylacine is not his major object of investigation, Halsey does not consider how competing names for the animal historically complexify any straightforward reading of their designation as a tiger being to blame for their destruction. Clearly Lyne's decision to give the 'native tiger' the alias the 'Tasmanian dingo' was motivated by a desire to augment lexical violence against the species, with 'tiger' perceived as not emotive enough as a designation.[124] Tigers possess a certain mystique and nobility in the European imagination that dingoes lack.[125] Referring to the thylacine as a dingo, however, accords the animal a nominal subjectivity. Sometimes those who sought their destruction refused even this; they did not hunt 'dingoes' or 'hyenas' or 'tigers' or 'wolves' but 'things'. Eric Guiler, for instance, cites the smallholder H. Pearce (possibly Herbert Eugene Pearce of Ouse) as referring to thylacines as 'them bloody useless things'.[126] Pearce, who lived in the Black Bobs-Strickland-Derwent Bridge District, stated to Guiler that he had claimed numerous tiger bounties. The bushman Bob Wainwright, son of George Wainwright who was hired by the VDLC to kill thylacines, similarly declared to Col Bailey of tigers: 'I caught dozens of those things back in the old days.'[127] The reduction to thingness formed the ne plus ultra of lexical violence and acted as license for extermination. Conceiving tigers as 'things' rendered them as unessential, inconsequential objects. In this sense, as I discuss in Chapter 1, language such as Pearce's also sustains the genocidal gaze, a way of looking that reduced the thylacine to vermin, a purposeless irritant in need of destruction.

Of Wolves

Many regions of the world have viewed wolves ambiguously. In Japan, for example, the wolf was historically revered, featuring prominently in art and literature. They were viewed as a 'good' animal.[128] In a historical novel of the life of Chinzei Hachirō (also Minamoto no) Tametomo, a renowned archer, a wolf bravely aids his master in battles against a monstrous snake (Figure 0.1).[129] Only with the introduction of rabies to Japan in the late seventeenth century and changing in farming practices in the early nineteenth century did the wolf begin to be viewed negatively, and even then not overarchingly so.[130] By the early twentieth century, however, both species of Japanese wolf, the Honshū wolf (*Canis lupus hodophilax*) and Hokkaido wolf (*Canis lupus hattai*), had been hunted to extinction.[131] The Hokkaido wolf fell victim to mass poisonings instigated by Edwin Dun, an Ohioan rancher invited to Japan as an agricultural advisor and tasked with transforming Hokkaido's agrarian system, and to bounty hunting.[132] With the hiring of men such as Dun, Japan imported not just their expertise but also their attitudes to animals such as the wolf. In the case of Dun, he brought his hate of all things lupine with him from the United States.

Wolves have often been perceived negatively in the European and European settler colonial imaginary, associated with 'ideas of wickedness, rapaciousness and danger'.[133] In *The Order of Wolves*, Richard Fiennes emphasises their symbolic potency; they are associated with 'savagery, ferocity and courage'.[134] As the reference to 'courage' demonstrates, they have historically provoked mixed emotions but predominant attitudes have, until recently, been adverse. Theodore Roosevelt begins his chapter on wolves and wolfhounds in *The Wilderness Hunter* with the observation that '[t]he wolf is the archetype of ravin, the beast of waste and desolation'. He adds that in the United States the wolf 'has everywhere retreated from the advance of civilization', constructing the species as antithetical to progress.[135] The 'brutish' nature of wolves is reiterated by Benjamin Corbin in his guide to wolf-killing where he writes: 'The wolf is the enemy of civilization and I want to exterminate him.'[136]

The wolf howl is often cited in unfavourable terms, with Harriet Ritvo referring to its 'eerie menace'.[137] Stanley Young quotes an anonymous stockman describing the howl as 'perhaps the most dismal sound ever heard by human ear'.[138] In 'The Wolf Chase', an anonymous New Brunswick author styles the howl as if 'a fiend from hell had blown a blast from an infernal trumpet'.[139] The wolf is also linked to hell through imagery. This occurs, for example, in the calendar included in the 1885 *Almanach de la France illustrée* (Figure 0.2). The image that accompanies January–March (winter) is of a pack of wolves racing along a forest path in a snow-

Figure 0.1 Wolf, Chinzei Hachiro Tametomo's *Chronicles*: *Volume 1*, Utagawa school woodcut on Japan paper, 1880. Collection of the author.

Figure 0.2. Wolves and a wayside crucifix. *Almanach de la France illustrée*. 1885. Collection of the author.

covered landscape, their tongues lolling and their incisors exposed. They head straight towards the reader, yet succour is close at hand. To their left there is a wayside crucifix. It looms over the pack, its verticality contrasting with their horizontality. The crucifix that overlooks the pack encourages reading the wolves as devilish in aspect.

The negative connotations that Eurasian wolves (*Canis lupus lupus*) possessed stuck to the 'marsupial wolf' and, as David Hunt has noted, were carried to Turtle Island by settler colonists and thus also shaped how the Newfoundland wolf was perceived. Hunt suggests that a possible result of the negative view of wolves may manifest in 'the attempt to exterminate the wolf in the USA by descendants of European immigrants'.[140] Andrea Smalley links the Turtle Island campaign of extermination to British 'wolf-hating traditions' and discusses the successful efforts to eradicate wolves from England and then Scotland.[141] This understanding of the conditions that made wolf extermination acceptable (with past actions used to solve similar 'obstacles' shaping current behaviour) echoes Ervin Staub's recognition that 'cultures carry blueprints for dealing with problems', in this case a past history of genocidal aggression towards wolves.[142] Staub argues that a history of the use of aggression in specific situations makes it more 'available' as a future response in comparable contexts.[143]

Wolf bounties were introduced in Europe in classical times. Solon of Athens, for instance, introduced a wolf bounty in the sixth century BCE, offering five drachmas for each adult carcass and one drachma per dead cub.[144] Garry Marvin details how in Anglo-Saxon Britain in the ninth century CE, dead wolves were used as tribute in an effort to control their numbers, with King Edgar demanding three hundred wolf skins a year from his Welsh subjects.[145] Later, wolf bounties were offered in England, a price was put on the head of the wolf, 'and they doubled, tripled, decupled and centupled the reward as the species became rarer'.[146] Other European countries also offered bounties. France, for example, extended a reward of thirty livres for each wolf head.[147] This sum was also paid in New France, demonstrating how European practices accompanied colonisation.[148] Wolf bounties introduced in Turtle Island are discussed at length in Chapter 1, the first recorded instance being in Massachusetts in 1630.[149] Bounties continue to be offered in the United States of America. Idaho, for example, offers various levels of payment including $1,000 per wolf killed in the north of the state.[150] The bounties offered in Idaho are in response to stock losses. This has long been used as a justification for genocidal policies enacted against the animal.

In *The Companion Species Manifesto*, Haraway gives national figures for wolf predation in the United States in 2002, noting that two hundred cattle, five hundred sheep, seven llamas, one horse and forty-three dogs were identified as killed by wolves, before asking 'who were those 43 dogs?'[151] In the same period, 'over 125 errant wolves' were killed. Errancy can refer to roving but also means misbehaving, straying from accepted standards of comportment. It is a loaded term. That even someone as sensitive to nonhuman animal welfare as Haraway chooses to use it demonstrates the deep-seated, perhaps instinctive misgivings that exist regarding wolves. Haraway does not discuss the number of livestock killed by feral dogs, by dogs that have turned away from companionship. She does, however, acknowledge that in the 1970s 'loose dogs killed lots of sheep', noting that livestock guardian dogs emerged as a counter-measure to curtail this predation.[152] In 2000 (no figures are available for 2002), 23.8 per cent of cattle losses to carnivores in the United States were due to feral dogs (only coyotes caused more losses; after dogs come mountain lions and then the category of 'other predators', which includes bears, lynx, wolves, ravens, eagles, vultures and others).[153] More recently, data from the US Department of Agriculture – Animal and Plant Health Inspection Service (USDA) covering 2015 again shows that stock losses to wolves are minimal. This data uses unverified sources and therefore potentially inflates wolf predation. Figures provided by the US Fish and Wildlife Service (FWS) are lower.[154]

Pressure on the wolf population in Turtle Island has sometimes come not only from bounty schemes but from demand for pelts. As Marvin explains, after a lull in demand by the mid-nineteenth century 'the price for wolfskins began to increase ... and it became worthwhile for hunters to spend time trapping them'.[155] Poisoning using strychnine was the preferred method of killing.[156] Wolf pelts sold for $1.25 each in the early 1860s.[157] Calculating the value of wolf pelts post-1870 is complicated by the decision of the Hudson Bay Company in 1870 to consolidate the returns of coyote, ordinary wolf and timber wolf pelts under a single category: 'wolf'.[158] For a long time the value fluctuated between $1.50 and $3.50.[159] In the 1880s prices for wolf hides varied from fifty cents to $2.50. By the early twentieth century most of the wolf skins sold in Turtle Island derived from Canada or Alaska.[160] Despite the decline in wolf numbers being noted in many regions, trapping continued, and trappers as well as bounty hunters are therefore culpable of animal genocide.

It is not new to describe the wholesale killing of wolves in genocidal terms. Writing in 1925, Ernest Thompson Seton decries the 'bitter, exterminatory war' that modern humankind has waged against wolves.[161] Farley Mowat would later refer to 'the strictly human crime ... of biocide' perpetrated against wolves.[162] In 'A Race of Wolves', Carla Freccero argues that '[f]or a long time humans have intended genocide for wolves'.[163] In the context of a discussion of the campaign of extermination committed by European settler colonists against wolves, Andrea Smalley traces how wolves became subject to legislation because of their perceived transgressive behaviour, refusing to respect the neat (if imaginary) boundaries the colonists had established between the wild and their 'civilised' enclaves.[164] In Newfoundland, this behaviour would ultimately result in the extinction of the island's indigenous wolf population.

Of Newfoundland Wolves

One of the earliest settler colonial references to the Newfoundland wolf occurs in 1620 when Richard Whitbourne states of them that they 'did something affright us with their howlings'.[165] The wolves, however, 'did not hurt us: for we had dogs, fire and sword to welcome them'.[166] Whitbourne is puzzled by the wolves as they seem to befriend his mastiff: 'the one began to fawne and play with the other, and so went together into the Woods.'[167] The dog went with the wolves for nine or ten days at a time, returning 'without any hurt'.[168] The behaviour of the wolves thus challenged the preconceptions he held of them:

> Hereof I am no way superstitious: yet it is something strange to mee, that the wilde beasts, being followed by a sterne Mastiffe-dogge, should grow to familiarity with him, feeling their natures are repugnant: surely much

> rather the people, by our discreete and gentle usage, may be brought to society, being already naturally inclined thereunto.[169]

Here he seems to believe sorcery of some kind may be behind the unexpected behaviour of the animals he thinks abhorrent.

Wolves were clearly relatively numerous on the island. In his *Geography of Newfoundland*, James Howley describes the wolf as 'tolerably abundant in the interior' and as frequently making their way 'to the seashore'.[170] The many place names in Newfoundland that refer to wolves are another indicator of their one-time omnipresence on the island. Howley's 'Report of Geological Explorations on the Gambo and Gander Rivers' names a river that runs into the Southwest Gander River 'Dead Wolf River'.[171] The toponym derives from him discovering a wolf skeleton on its banks. Other places historically named after lupines include Wolf Point (East Coast, St. Mein Bay), Wolf Head (Notre Dame Bay), Wolf Harbor (a former settlement on the south-east coast), Wolf Cove (now Upper Amherst Cove), Wolf Pond (near to Chapel Arm) and Wolf Mountain (close to the Meelpaeg Reservoir). As the European population increased, wolves were increasingly blamed for livestock losses. Because of these losses, on 14 September 1839, 'An Act to Encourage the Killing of Wolves in this Colony' was passed by the Newfoundland legislature. The second paragraph of the Act reads:

> *Be it therefore enacted*, by the Governor, Council and Assembly of Newfoundland, in General Assembly convened, and by the authority of the same, that from and after the passing of this Act, any Person or Persons who shall produce, or cause to be produced, before any Justice of the Peace in this Island, the Skin of a Wolf recently Killed within this Colony, and shall make and subscribe, or cause to be made and subscribed, a declaration in manner and form as in Schedule A, hereunto annexed, that the same was Killed by or for such Person or Persons, such Person or Persons shall be paid a reward of *Five Pounds*; and if any declaration so made shall be false or untrue, the Person wilfully making such false declaration shall be deemed Guilty of Misdemeanour.[172]

The bounty was suspended in 1863, which may have been a financial or political consideration. It was then reintroduced in 1872.

For a time, reports of wolf attacks on livestock were common. The naval captain D. Miller states of the small community living at Ship Cove in 1873 that '[s]ome of their cattle were killed by wolves not long ago'.[173] He also notes of Indian Point that wolves roamed the area all year round 'and must be a great detriment to cattle or sheep farming, besides the loads of hungry dogs generally kept'.[174] As this last remark shows, it was common knowledge that dogs were also responsible for livestock losses. In fact, dogs were identified as the primary cause of livestock predation. Richard Henry Bonnycastle writes that Newfoundland dogs (a breed of working dog, often black or white and black in colour) had the propensity 'of worrying cattle and sheep'.[175] He explains:

> [Newfoundland dogs] have been known to make an entrance under an outhouse where sheep have been stalled, and to commit serious devastation. In short, half the mischief said to be performed by wolves, which are very rare near St. John's, is done by their starved congeners.[176]

There was increasing legislation in the 1860s designed to curb against dog attacks. As with the thylacine, the wolf was likely blamed for many attacks that were the result of wild dogs or uncontrolled domestic dogs.

In the speech that the Governor of Newfoundland, Alexander Bannerman, gave at the opening of the First Session of the Seventh General Assembly of Newfoundland in 1860, he acknowledged 'that great complaints continue to be made of the ravages committed by dogs on sheep &c.' He adds: 'It is an evil to which my attention has been called by many industrious farmers, and I know that they expect the Legislature will provide a remedy for so great a nuisance.'[177] The Legislature did indeed provide a solution, passing the 'Act to Provide against the Destruction of Sheep and Cattle, and Injuries to the Person, by Dogs' on 14 May 1860. This act granted the right to shoot any dog without an owner or without adequate identification of their owner and any dog killing, maiming or worrying livestock. Provisos were also made for compensation to be paid to livestock owners for the loss of animals killed by dogs. The Act seemed aimed primarily at working dogs as the identification required, which was to be fastened to the animal's neck, was 'a clog or piece of wood not less than five pounds weight, and not less than eighteen inches in length, with the name of the owner stamped or marked thereon at full length'.[178] Pointers, setters, spaniels and terriers were explicitly excepted from this rule and only required to 'wear a collar of leather or brass, with the name of the owner marked thereon'.

In a 'Statement of Expenditure on Account of Unforeseen Contingencies' provided by the Financial Secretary's Office for 1860, there are also several references to dogs. On 6 July 1860, Charles Cozens was granted £3 9s 4d for buying guns to kill dogs. On 10 August, Timothy Mitchell was paid £2 18s 6d for killing dogs. Further payments for killing dogs were made to John Hayward (23 October, £17 8s 6d) and R. J. Pinsent Jnr (29 October, £8 0s 4d). This does not equal the £50 in bounties made for killing wolves over this period but it still represents a substantial sum and signals the probable role of dogs in livestock destruction. The actions enabled by the Act were, however, clearly seen as insufficient by the farmers. In 1865, they demanded further action from the legislature. That year saw petitioning from the Agricultural Society of Newfoundland who provided statistics for killings attributed to dogs in various locales around the island. The figures of sheep killed are quite considerable: 4,136 in total. In addition, 228 goats, eighty-two cows and seven horses are listed as victims. As in Lutruwita, some of these deaths may have been wrongly attributed to predation. It is also possible that a few of these attacks were perpetrated by wolves and mistakenly attributed to dogs, yet it is clear that dogs were recognised to be a major threat to livestock. The tallies kept of destroyed dogs in later years demonstrate that the problem was ongoing and unresolved.

Another act was passed on 7 April 1865: 'An Act to make further provision against the Destruction of Sheep and Cattle, and Injuries to the Person, by Dogs'.[179] This Act was far more detailed. The weight of the identifying clog or piece of wood a dog should wear was increased to seven pounds but the dog could now also simply wear a muzzle. Stipendiary magistrates were also required 'once in each year, to prepare a statement, from information furnished on the Oaths of the Owners, of all Sheep, Lambs, Cattle, Horses and Goats, destroyed in their respective districts by Dogs'.[180] A dog licence introduced for dogs such as pointers and setters permitting them 'to be at large' cost $4 for each dog.[181] Police Constables or other people destroying dogs under the provisions of the Act were entitled to receive the sum of fifty cents for each dog destroyed.[182] In 1864, the year before the dollar was adopted as currency on the island, the wolf bounty stood at £10. The first bounty paid (in 1873) after the change to the dollar was for $10. The bounty for a wolf was therefore twenty times the amount allocated for the destruction of a dog. This disparity is revealing. It speaks to the symbolic status of the wolf as threat but, in hard economic terms, also to a recognition of the carnivore's rarity. Offering a $10 payment for the killing of a dog was probably unaffordable to the treasury. The persistently high wolf bounty is therefore a damning indictment of the government, demonstrating their knowledge of the animal's scarcity.

In 1879 it is noted that the 'chief impediment to sheep raising is *dogs*'.[183] It is tragic that despite it being common knowledge dogs were predominantly responsible for attacks against farm animals, the wolf bounty was reinstated and there were numerous payouts in the 1870s and 1880s (all detailed in Chapter 6). The numbers of dogs being destroyed was large. In 1865, for instance, J. Wilcox reported a total of 174 dogs destroyed.[184] In 1872, 617 dogs were destroyed, costing $308.50.[185] These totals vastly outnumber the amount of wolf bounties being claimed in the same period. In 1873, 457 dogs were destroyed at a cost of $228.50.[186] In that same year, four wolves were killed at a cost of $40. Wolves, however, were easier targets than dogs in terms of vote-winning. For this reason, as David Mech has argued, wolf bounties in general continue to be popular among politicians irrespective of 'whether or not the bounty is necessary or useful'.[187] In this sense, the Newfoundland politicians such as William V. Whiteway who knowingly advocated unnecessary wolf bounties aimed at eradicating the island's wolf population were culpable of genocide. Perversely, while the wolf bounty was in force and wolves were regularly being killed efforts to protect domestic animals from violence were underway. On 16 April 1878 'An Act respecting Cruelty to Animals' was passed by the island's legislature. The animals named in this act are, however, all domestic animals.

As with the thylacine, wolves were culturally framed as negative. An account of a Newfoundland shipwreck by the geologist Joseph Jukes is revealing in this context. He writes of the wreck of the 'Onondago' (*sic*):

> She appeared to have been a fine ship of 700 or 800 tons; her hull seemed not to have sustained much injury, but her masts were now gone, and she was of course stove in below by sharp rocks on which she lay. Two small sloops, two or three fishing-boats, and several punts lay alongside, and their crews were on shore getting out her cargo, piling the deals and other timber of which it consisted, and loading their own vessels with all imaginable coolness and deliberation. She was evidently considered fair game, and 'first come first served' seemed the order of the day. The dirty little vessels around her put me in mind of a parcel of wolves preying on the carcase of a race-horse.[188]

Here something of the European mindset regarding wolves reveals itself through the description. The hunting expression 'fair game' is noteworthy. Jukes sees the salvage in terms of a hunter or predator and their prey. The extract demonstrates how hunting, which was a regular pastime, fed into everyday perceptions and sense-making activities. In Jukes's picturing of events, dirty boats have the character of wolves, animals that are, by association, also to be viewed as unclean. The image of a pack of wolves preying on a dead racehorse is incongruous but the noble racehorse (the fine ship) is clearly meant to be contrasted with the ignoble wolf (the dirty boats). The use of 'wolves' as a 'dirty word' or insult is recorded in testimony by the butcher John Eales relating to the November 1836 elections for the Newfoundland House of Assembly (in the build-up to which he served as a special constable).[189] In everyday speech then, wolves signified negatively. The bad language that surrounded the species was deleterious, facilitating the extermination of the island's wolf population because people were unlikely to mourn the passing of what they perceived as such a cruel and depraved creature. Their perception can be seen to embody a 'genocidal gaze'.

The Newfoundland wolf likely became extinct in the 1920s. Writing in 1923, Gower Rabbits, the Secretary of the Game and Inland Fisheries Board in Newfoundland, states that a relatively big pack of wolves (eight animals) was spotted over twenty years previously near Gaff Topsails. Two wolves were seen in the vicinity of St. George's on the south-west coast in 1921 and two were seen in 1922 crossing over Birchy Lakes.[190] Additionally in 1922 the tracks of two wolves were found at Doctor Hills in the Great Northern Peninsula.[191] The bounty scheme probably contributed to the wolf's disappearance. John Maunder suggests that a decline in the island's caribou population in the period 1915–25 was also a major contributing factor.[192] Such a decline would certainly have had a detrimental impact on the wolf population of the island. In a study conducted in Alaska of wolves whose primary prey was caribou, Stephenson and James found that over 95 per cent of the biomass consumed by two wolf packs they studied was from caribou. To an extent, however, wolves can switch prey. They can turn to moose, for instance, when caribou are less numerous.[193] When ungulates are sparse, other prey such as beavers, birds and fish offer alternatives. Fluctuations in caribou populations in Newfoundland have historically probably been recurrent albeit infrequent. Severe winters, for instance, can cause population declines. Caribou decline alone would therefore probably not have caused the extinction of the wolf.

Chapter and Verse

The chapters that follow offer occasional correctives and additions to the factual record, some highlighted and others not. Reading outwards from my chosen case studies, I provide a broad history of the thylacine and the Newfoundland wolf from their first contact with Europeans to their ultimate disappearance. In relation to the thylacine, recent research by Stephen Sleightholme and Cameron Campbell that has identified and catalogued films and photographs has been of significant help. In 2005, Sleightholme also established the International Thylacine Specimen Database, which is crucial when seeking information regarding physical remains of the animal. The sixth revised version of the database was released in 2017. In 1999, Campbell founded the online Thylacine Museum, a venture that has subsequently been expanded and is also a key resource when researching tigers.

The thylacine is a polarising research topic as current robust discussions relating to the last captive thylacine have demonstrated.[194] Threading the needle of thylacine research to avoid conflict is virtually impossible and no doubt my own efforts here (with their indebtedness to critical theory and clear ethical and political engagement) will 'raise hackles' in some quarters. My approach has been to try to build on the broad body of scholarship that already exists regarding the thylacine by signalling occasional disagreements but never descending into vituperation. There are many scholars I am indebted to. Important early work was performed by scientists such as the zoologists Eric Guiler and Heinz Moeller.[195] Robert Paddle is another major figure whose monograph *The Last Tasmanian Tiger* has been an important reference point. The recent publication of *Tasmanian Tiger: Precious Little Remains*, co-authored by David Maynard and Tammy Gordon, which provides an inspiring combination of scientific research and cultural history, is also noteworthy. In relation to cultural history, Hannah Stark's work relating to physical remains of tigers has been inspirational, as has the groundbreaking research by Carol Freeman and Nic Haygarth who have focussed, respectively, on how visual and textual representations historically framed ways in which the thylacine was perceived and, by extension, treated.[196] To my mind, Freeman's *Paper Tiger* stands as the benchmark for visual cultural research on the thylacine.[197] Her efforts are sometimes unjustly sidelined.[198] Col Bailey's writings have also provided valuable insights into how settler colonists (such as trappers and farmers) perceived the tiger. Bailey is one of the few to make direct links between the fate of the thylacine and that of the Newfoundland wolf.[199] Regarding the Newfoundland wolf, the literature is minimal but I am beholden to the pioneering research of Leslie Tuck and, more recently, John Maunder.

The concept of genocide is central to my analyses. It was developed by Lemkin in response to atrocities perpetrated by the Axis powers during the Second World War. He used the term in his 1944 book *Axis Rule in Occupied Europe* to describe coordinated efforts to destroy a nation or an ethnic group.[200] Lemkin conceived the term as denoting an 'old practice in its modern development'.[201] He only considered human genocide but the term began to be used in the context of the annihilation of nonhuman animals in the 1960s. In Chapter 1, I trace the history of 'animal genocide' and the different ways the concept has been understood and employed. Drawing on the work of Elizabeth Baer and Griselda Pollock, in particular, the chapter also discusses the role of the 'genocidal gaze' (a specific way of seeing) in fostering the conditions that made animal genocide possible. Building on Chapter 1, the ensuing chapters each, in their different ways, examine how images, remains and other traces of the Newfoundland wolf or the thylacine incarnate narratives of animal genocide which close readings can bring to visibility.

In Chapter 2, I examine how the physical remains of the thylacine register the genocidal violence that was perpetrated against them. In this sense, the remains constitute a kind of 'material testimony' to genocide, archiving attitudes and emotions. Through close analyses of remains derived from collections in Australia and the United States, I identify how they preserve evidence of histories of violence. The remains include scientific 'specimens' and also artefacts such as thylacine skin rugs and a pincushion made from a young thylacine's mandible. In the same way that artworks, newspapers and photographs embody outlooks and opinions, the pincushion and the rugs record the effects of the genocidal gaze as it was historically directed towards the thylacine. Reading these artefacts in purely aesthetic terms risks overlooking the politics of cruelty they record. In the chapter, I draw on insights derived from cultural studies but also forensic anthropology and veterinary forensics as a means to unpack the varied histories of thylacine remains in museum collections. This transdisciplinary methodological approach provides me with a vocabulary adequate to describing how physical remains document a history of violence. Two of the thylacines I study probably died in the 1950s. This is not interpreted as evidence of the continuing survival of the species but as drawing attention to the arbitrariness of declaring 1936 as the year the species went extinct.

Most of the remains discussed in Chapter 2 are rarely displayed by the museums where they are held. Chapter 3 focusses on the thylacine-related materials that are exhibited. Through close readings, I consider the ways in which exhibition strategies frame a visitor's reception and understanding of the thylacine in the Tasmanian Museum and Art Gallery (TMAG) (Nipaluna/Hobart), the Queen Victoria Museum and Art Gallery (QVMAG) (Launceston), the Australian Museum (Gadigal/Sydney), the Melbourne Museum, the Tiegs Zoology Museum (Melbourne) and the Smithsonian National Museum of Natural History (Washington, DC). While reading the kinds of histories offered by the museum displays, I also sometimes move outwards from them to consider the broader history of the war of extermination of the thylacine in its direct and indirect guises. In this sense, readers who have never seen the exhibitions may find some of the historical information in this chapter of interest. The most extensive analyses are given over to the major exhibitions devoted to the thylacine by TMAG and QVMAG, with the implications of similarities and differences between the two displays examined. The readings of these two exhibitions help to demonstrate the varied forms animal genocide took in Lutruwita, with animal traffickers, bounty hunters and collectors found culpable. My analyses of the museums are informed by cultural semiotics and by the idea of affective encounters. In this way, I can examine not only the kinds of information museums disseminate (or fail to) but also the sorts of feelings they encourage (or discourage) in visitors.

In Chapter 4, I engage at length with Erle Wilson's 1953 novel *Coorinna* – which gives a fictional account of the life and death of a young male thylacine – positioning the novel within Wilson's broader corpus of works. My approach to the novel involves close reading, focussing in detail on form and figurative elements the better to grasp the complexities of Wilson's palimpsestic attitude to history, one in which human and nonhuman animal genocide are often imperfectly overlayed. *Coorinna* memorialises the thylacine and their historical persecution but also draws parallels with the fate of the palawa. As such, in the same way that Brian Johnson has identified Canadian animal stories as haunted by human history, *Coorinna* is preoccupied by human genocide in Lutruwita.[202] The thylacine is therefore frequently reduced to a cipher through which human genocide is obliquely acknowledged and processed. Coorinna, the central character of the novel, is an instrument for addressing human concerns and Wilson imbues him with human qualities. This limits the novel's effectiveness at engaging with animal genocide but does not entirely negate it. *Coorinna* also raises important questions about the representation of nonhuman animals, which are developed more fully in Chapter 5.

Chapter 5 focusses on the film *The Hunter*, an adaptation of Julia Leigh's novel of the same name, which tells the story of a hunter on a mission to locate a thylacine in present-day Lutruwita. The film combines representational and non-representational modes of engaging with the tiger and, drawing on the psychoanalytic theories of Bracha Ettinger, I examine the aesthetic potential of dimensions of the film (notably its use of 'absence' as a way of imaging the thylacine) for resisting anthropomorphism. A straightforward narrative analysis would be incapable of registering and understanding the film's aesthetic complexities and affective dimensions. I also explore the use of archival film footage in *The Hunter* and in other narrative films featuring thylacines. The footage I concentrate on is David Fleay's 1933 film of 'Benjamin' (probably the last thylacine held in captivity by Hobart Zoo in Nipaluna/Hobart), assessing what this footage reveals about cultural attitudes towards the tiger.[203] I also address how these attitudes carry into the 2021 colourised version of Fleay's film and the CGI tiger in *The Hunter*.[204] The colourised version is intended to restore 'Benjamin' to how he 'originally looked' but I demonstrate that this comes at the expense of traumatic histories embodied in the black and white footage. Colourisation as a practice, one that involves technical processes underpinned by aesthetic choices, is therefore revealed to be an ethical issue, one that, in this instance, works to mute a particular genocidal history.

In Chapter 6, I offer an object biography of the pelt and skull of a wolf shot in Newfoundland in the late nineteenth or early twentieth century by William Whiteway. Object biographies track the changing perceptions and significances regarding an object through time, attending to the varied histories it accumulates. Through this biography, I hope that something of the objecthood of the object is diminished and the wolf becomes a subject for the reader, a singular individual rather than a type.[205] The wolf skull and skin are powerful testaments to the violence wolves were subject to in Newfoundland. The chapter also builds on Leslie Tuck's pathbreaking research into the bounty scheme to provide what is, perhaps, the most comprehensive account to date of its scope.[206]

I conclude *Animal Genocide and its Aftermath* by considering a series of photographs taken of the last known 'wild' thylacine, examining what they reveal about the genocidal gaze as it was operative in Lutruwita. These photographs pose profound ethical questions regarding the representation of nonhuman animals. The thylacine in question was shot and killed in 1933 by Wilfred 'Wilf" Batty at a time when the marsupial's rarity was common knowledge. The very taking of the photographs attests to the 'event' that an encounter with a thylacine constituted. Batty knew the financial value of the tiger as he tried to capture him alive. When the tiger was on the cusp of escaping, however, the smallholder shot him.[207] This killing was an act of genocide. Retrospectively, Batty attributed the cause of the species' extinction to canine distemper, downplaying the impact of actions such as his own. The death of the tiger Batty shot was not quick; he lingered for roughly twenty minutes, becoming a local spectacle: people 'came from all about to look'.[208]

Notes

1. The palawa (First Nations Tasmanian) name of Lutruwita is used to refer to Tasmania throughout this book. Lutruwita and other palawa place names are capitalised in response to the April 2024 decision by the Tasmanian Aboriginal Centre to adopt this format (it was not previously standard practice to capitalise palawa places names).
2. In this book the terms thylacine and tiger are used interchangeably to refer to *Thylacinus cynocephalus*. The term Tasmanian tiger, while common, is misleading as it attributes feline characteristics to the thylacine. I nonetheless reluctantly choose to employ it for reasons of literary style to reduce repetition. As I go on to discuss, many palawa names for the tiger are known and there are strong arguments for using these in certain contexts. Discussions of the thylacine in this book, however, are primarily about European settler colonist attitudes towards the species and their actions against it. European terminology reflects and is bound up with those attitudes and actions, which is why I retain it. For a discussion of how European-derived classification systems in natural history erase Indigenous recognition of some species, see Bienvenue and Chare, 'Introduction', 11–12. See also Gillman and Wright, 'Restoring Indigenous Names'. When referring to thylacines and wolves as a species or when the gender of a given animal is uncertain, 'they', 'them' and 'their' pronouns are used. When the gender

of a given animal is established, male or female pronouns are adopted. My choice of pronouns aims to resist objectifying the animals that are being discussed, although it inevitably and problematically involves imposing human binary conceptions of gender on them. For a recent discussion of inclusive language in relation to nonhuman animals, see Chau and Jacobs, 'Applied Linguistics'. See also the discussion of pronouns in Hoquet, 'Animal Individuals'.

3. The term palawa is written in lowercase throughout this book. This is in deference to the decision by the Tasmanian Aborigine Centre that, with the exception of place names, palawa kani only be written in lowercase. palawa kani is a composite Lutruwitan language constructed from recorded vocabulary from various palawa peoples. Breen, 'Extermination, Extinction, Genocide', 81.
4. Meston, *The Van Diemen's Land Company*, 53.
5. For a discussion of the passage of the bounty legislation, see Chapter 6 of Paddle's *The Last Tasmanian Tiger*.
6. Rothfels, 'Immersed with Animals', 207.
7. 'Pickled Devils', *The Mercury*, 16 July 1874, 2. For a long time the Jamrach family, based in Germany then England, were major traffickers in exotic animals and birds for the European market. See Rothfels, *Savages and Beasts*, 47.
8. The thylacine was not the only animal in Lutruwita to become extinct due to European invasion. The Tasmanian emu (*Dromaius novaehollandiae diemenensis*) disappeared in the 1850s and the King Island emu (*Dromaius novaehollandiae minor*) slightly earlier, the last known birds dying in captivity in France in 1822. These birds were both subspecies.
9. Baker and Dickman, *Secret Lives of Carnivorous Marsupials*, 75.
10. Ibid., 83.
11. Ibid.
12. See Brook et al., 'Resolving When (and Where) the Thylacine Went Extinct'.
13. Gould, *The Mammals of Australia*, unpaginated [56].
14. Ibid.
15. Newfoundland has been inhabited for about nine thousand years. The major Indigenous population at the time of European invasion in the seventeenth century were the Beothuk. After invasion, the island was ruled by a succession of naval commanders until 1832 when government by colonial assembly commenced. The island was a self-governing colony from 1855 until 1907 when it became a dominion. It continued to be self-governing until 1934 when it began to be governed by royal commission. Newfoundland became a province of Canada in 1949.
16. Col Bailey suggests the bounty legislation was only repealed in 1963. See Bailey, *Lure of the Thylacine*, 140.
17. As discussed in Chapter 6, a wolf that had crossed from the mainland using sea ice in the Strait of Belle Isle was shot on the Bonavista Peninsula in 2012. DNA testing confirmed he came from the wolf population of Labrador. See Barb Sweet, 'It's a Labrador Wolf: DNA Tests', *Telegram*, 24 May 2012, https://www.saltwire.com/newfoundland-labrador/its-a-labrador-wolf-dna-tests-126023 (retrieved 6 February 2025).
18. Haraway, *The Companion Species Manifesto*, 12, 30. Haraway, however, emphasises in *When Species Meet* that companion species are not restricted to companion animals such as dogs but encompass any species that there is an ongoing 'becoming with'. Haraway, *When Species Meet*, 16.
19. Haraway, *When Species Meet*, 35.
20. Haraway, *The Companion Species Manifesto*, 49.
21. Ibid., 6.

22. Haygarth, 'The "Walking Museum"', 35–36. I refer to these 'dealers' as traffickers as although their actions were not illegal they were plainly unethical. My understanding of trafficking is inspired by Sollund's definition in *The Crimes of Wildlife Trafficking*, which extends trafficking to the *legal* wildlife trade. She defines trafficking as the 'legal or illegal trade, smuggling, abduction, theriocide and collection of [free-born] animals pertaining to (often endangered) animal species, either alive or by turning them into products' (3).
23. I am by no means the first to link their fates. Col Bailey, for instance, includes a discussion (Chapter 32) of the Newfoundland wolf in *Lure of the Thylacine*. Others, however, have not examined their histories through the prism of animal genocide.
24. The term 'nonhuman animals' is used on occasion throughout this book to signal that humans are also animals and that there is continuity between humans and other animals. The term has been critiqued for implying a qualitative variance with human animals possessing traits and capacities other animals lack, but I believe that the way it signals continuity and difference between human animals and other animals justifies its continued usage. This opens the question as to why I do not employ the term 'nonhuman animal genocide' instead of 'animal genocide' or use 'animal genocide' to cover all instances of genocide (human and nonhuman). These are certainly possibilities, but I have opted to retain 'genocide' and 'animal genocide' because these are established terms with specific histories that I value.
25. Brantlinger, *Dark Vanishings*, 73.
26. Anonymous, *Report of the Parliamentary Select Committee on Aboriginal Tribes (British Settlements)*, 10–11.
27. Ibid., 5.
28. Goldie suggests that the extermination of the Beothuk was part intentional and part a result of White settler colonists interfering 'with the delicate ecology of an inhospitable environment'. See Goldie, *Fear and Temptation*, 154.
29. Lemkin, *Lemkin on Genocide*, 18–19.
30. Kuper, *Genocide*, 40.
31. Anonymous, 'The Black Natives', *Colonial Advocate*, Thursday 1 May 1828, 29.
32. Ibid., 30.
33. See Archives and Special Collections, Memorial University, St. John's, NL, COLL-262 Howley Family Papers, 6.0 William Epps Cormack Original Documents, 1818–1829: 6.03.005. In another fragment (6.04.013), Cormack writes, 'Civilization and barbarity might shake hands in Nfland [Newfoundland]'.
34. Ibid., 6.03.005.
35. Ibid., 6.03.007.
36. Cartwright, 'An Account of the Red Indians of Newfoundland', 32. Unpublished manuscript held by the Newfoundland and Labrador Collection, Provincial Resource Library, St. John's NL: 971.6 C24. Extracts from Cartwright's account are published in Howley's *The Beothucks or Red Indians*.
37. Cartwright, 'An Account of the Red Indians of Newfoundland', 32–33.
38. As discussed at length later in this Introduction, the thylacine was frequently referred to as a 'hyena' or 'wolf' because of their assumed behaviour. There were therefore clear similarities in how the palawa and the thylacine were perceived by settler colonists. Anonymous, 'Untitled', *Hobart Town Courier*, Saturday 13 November 1830, 3.
39. Lemkin, *Axis Rule in Occupied Europe*, 90.
40. Staub, *The Roots of Evil*, 52.
41. Ibid., 61.
42. Ibid., 88.

43. The evidence for the extinction date of the thylacine on the mainland is largely based on the dating of a mummified 'specimen'. The notion that the near-to-last thylacine was one that became mummified in exceptional circumstances is questionable. Samuel Turvey lists the calibrated radiometric of the last occurrence of the thylacine on the mainland as 3816–3336 BP but the last historically recorded occurrence as in the 1830s–40s. This would suggest relatively late survival but is not based on the dating of fossil or physical remains but on oral historical accounts. Turvey, 'Holocene Mammal Extinctions', 42. See also Turvey and Cooper, 'The Past is Another Country', 196.
44. See White et al., 'Ancient Mitochondrial Genomes', 9.
45. Harris, 'Description of Two New Species of Didelphis from Van Diemen's Land'. For a discussion of nineteenth-century ideas about speciation and classification see Berlocher, 'Origins', 4–5.
46. Allen and Barbour, 'The Newfoundland Wolf'.
47. See John Gittleman and Matthew Gompper. 'Plight of Predators', 371.
48. Ibid.
49. Turvey and Cooper, 'The Past is Another Country', 196.
50. Baker and Dickman nonetheless suggest in *Secret Lives of Carnivorous Marsupials* that there remains some convergence between thylacines and wolves (57). Freeman also discusses convergence in *Paper Tiger*, emphasising its lack (92).
51. 'Description of a Tasmanian Tiger Received by Banks from William Paterson, 30th March 1805', SAFE/Banks Papers/Series 27.33. Sir Joseph Banks Papers, State Library of New South Wales. The Yorkton settlement was soon abandoned in favour of the site that would subsequently become Launceston.
52. Anonymous, 'Untitled', *Sydney Gazette and New South Wales Advertiser*, 21 April 1805, 3.
53. Ibid.
54. On 4 November 1804, one horse, three sheep, four cows and fifteen pigs arrived at Port Dalrymple. I do not know whether any of these animals were still alive in 1805. For a discussion of the history of livestock at Port Dalrymple, see Walker, 'The Discovery and Occupation of Port Dalrymple'.
55. See Philip King's 30 April letter to the Earl of Camden reproduced in Bladen, *Historical Records of New South Wales*, 607–8. King notes that 288 of the cattle being transported by the *Sydney* died on route to Port Dalrymple.
56. The measurements of the female are as precise as the male. She has nineteen stripes, for instance, to his twenty. Unless guesswork, this necessitated the study either of a captive animal or of a cadaver.
57. Anonymous, 'Natives', *Sydney Gazette and New South Wales Advertiser*, 21 April 1805, 2.
58. For an extended account of these resistance activities (and of massacres perpetrated against the Darug by Europeans), see Ryan, 'Untangling Aboriginal Resistance'.
59. Anonymous, 'Natives', 2. Emphasis in the original. All emphases are present in the original unless otherwise stated.
60. Ibid.
61. Meston, *The Van Diemen's Land Company*, 51.
62. Cited in Ellis, *Trucanini*, 44.
63. Wolfe, 'Settler Colonialism and the Elimination of the Native', 133.
64. For a discussion of Newfoundland's mining history, see Martin, *Once Upon a Mine*, particularly Chapter 1. The growth in mining led to increased demand for lumber. Forestry activities therefore expanded at the same time as mining. See Cadigan, *Newfoundland and Labrador*, 162.

65. Ibid., 104–5; Mackinnon, 'Farming the Rock'.
66. The case of the Beothuk is a different story as their territory was regularly encroached on, if not 'legally' annexed, by fishers and trappers. In their context, Wolfe's identification of the settler need for land as a potential contributing to factor to genocide is a major consideration.
67. Harvey, *Newfoundland in 1900*, 132.
68. Wolfe, 'Settler Colonialism and the Elimination of the Native', 141.
69. Jones et al., 'Carnivore Concerns', 423.
70. Thylacines are still sometimes blamed for sheep losses. In the comic strip by François Craenhals, *(Les) 4 as et le Loup de Tasmanie* [*The 4 Aces and the Tasmanian Wolf*], for example, thylacines with their 'wolfish ways' are said to have hunted in packs and to have caused huge damage to farmers (38).
71. Paddle discusses theft as an issue in *The Last Tasmanian Tiger*, 163.
72. Edward Curr, for instance, mentions the poor sheep management skills of Alexander Goldie, the superintendent of the VDLC holdings at Ningherner/Parteenno (known to him as Hampshire Hills) and Wulaninka or Woollunnengar (known to him as Surrey Hills). See Meston, *The Van Diemen's Land Company*, 49.
73. Ibid., 28.
74. Ibid., 38.
75. Dallas, 'Conclusion', 59.
76. See the discussion of stock losses in Stokes, *The Settlement and Development of the Van Diemen's Land Company's Grants*.
77. See Thomas Walker's advertisement warning trespassers on his landholdings in Rhodes that they would be prosecuted. Walker observed that now that he would actively prosecute any intruding splitters, sawyers and stockmen, he expected '[t]he hyenas who are reported to have killed my sheep' to 'disappear'. See Walker, 'Caution', *Launceston Advertiser*, Thursday 26 January 1837, 2.
78. The shepherd Charlie Williams at Mount Cameron, for example, was paid ten shillings for each dog or tiger he caught. See Anonymous, 'Through Tasmania No. 35', *The Mercury*, Saturday 26 April 1884, 2.
79. See Robinson, *Friendly Mission*, 858–59; 860–66. Robinson believed the depredations caused by dogs were exaggerated (865).
80. See ibid., 865.
81. Breton, *Excursions in New South Wales, Western Australia and Van Dieman's Land*, 359.
82. West, *The History of Tasmania: Volume 1*, 323.
83. Ibid., 133.
84. Ibid.
85. Anonymous, 'Parliament of Tasmania: House of Assembly, Tuesday November 2', *The Mercury*, Wednesday 3 November 1886, 3.
86. Stephen Burns, 'Epidemic Dogs the West: Danger Lurking in the Mulga Scrub', *The Land*, Thursday April 25 2019, 3.
87. Ibid.
88. Anonymous, 'Parliament: House of Assembly Thursday, October 7', *Launceston Examiner*, Friday 8 October 1886, 3.
89. In the abstract of this petition (numbered 44) these ravages are specifically identified as against sheep. The petition was issued as Paper No. 165. See *Journals and Printed Papers of the Parliament of Tasmania 1884*, Volume 1 (Hobart: Govt. Printer, 1884).
90. Paddle, *The Last Tasmanian Tiger*, 153.
91. 'Country' encompasses land but also culture and environment. It refers to a physical but also spiritual connection. palawa are connected to Country by ancestry and family.

92. Anonymous, 'Wynyard', *Advocate*, Saturday 19 July 1919, 2.
93. Anonymous, 'House of Assembly,' *The Mercury*, Friday 5 November 1886, 4. Braddon may have used the pun more than once. See Anonymous, 'House of Assembly', *Launceston Examiner*, Saturday 6 November 1886, 3.
94. Chinese tin miners came to greatly outnumber Europeans for a brief period in the late nineteenth century. See Vivian, *Tasmania's Chinese Heritage*.
95. Anonymous, 'House of Assembly', *The Mercury*, Friday 5 November 1886, 4.
96. For 'stickiness' and its relation to attributing negative qualities to specific groups, see Ahmed, *The Cultural Politics of Emotion*. Ahmed is discussed further in Chapter 2.
97. Mitchell, *Picture Theory*, 333.
98. Dickman et al., 'The Dingo Dilemma', 299; Van Eeden et al., '"The Dingo Menace"', 247.
99. See Chapter 3 of Stuart, 'The Dingo in the Colonial Imagination'.
100. See Stuart, 'The Dingo in the Colonial Imagination', 1, 10, 13, 16.
101. Ibid., 19–20.
102. Bewick, *A General History of Quadrupeds*, 319.
103. Anonymous, 'House of Assembly, Friday, August 26', *Devon Herald*, Tuesday 30 August 1887, 3.
104. John Lyne correspondence, 23 January 1888, LMSS 583/1/1, Libraries Tasmania Launceston Manuscript Collection.
105. Krefft, *Mammals of Australia*, 6.
106. Brinsmead, *Echo in the Wilderness*, 150.
107. In 1830 a tiger entered a cottage and sought to seize a child. See Anonymous, 'The Courier', *Hobart Town Courier*, Saturday 17 April 1830, 2. A second example is discussed in my Conclusion.
108. Lumholtz, *Among Cannibals*, 380.
109. Ibid.
110. Oscar, 'The Native Tiger', *The Mercury*, Tuesday 19 September 1882, 3.
111. Freeman, *Paper Tiger*, 97.
112. Ritvo, *The Platypus and the Mermaid*, 8.
113. In Julia Leigh's *The Hunter*, the main character M refers to the names 'striped wolf' and 'marsupial wolf' as evidence of the taxonomic confusion the thylacine caused early settler colonists. Leigh, *The Hunter*, 16. The tiger was not just difficult to name, they were also hard to classify. Of animals extant today, the numbat is close genetically to thylacines although markedly different in appearance and behaviour. For a brief overview of numbats, see Friend and Thomas, 'Conservation of the Numbat'. The thylacine is usually grouped with the Tasmanian devil and the quoll, which, like the numbat, are of the order Dasyuromorphia, an order including most carnivorous marsupials. The devil and quoll, however, are of the Dasyuridae family while the thylacine was of the Thylacinidae family and the numbat is of the Myrmecobiidae. See the discussion in Baker and Dickman, *Secret Lives of Carnivorous Marsupials*, 65–66.
114. Freeman, *Paper Tiger*, 102.
115. References to the 'hyena tiger', for example, occur in Anonymous, 'Hobart Town', *Hobart Town Gazette and Van Diemen's Land Advertiser*, Saturday 2 August 1823, 2; Anonymous, 'Port Phillip', *Bent's News and the Tasmanian Three-Penny Register*, Saturday 9 April 1836, 3.
116. Swainson, 'Australasia: Natural Geography'. In a similar vein, in the 1930s, some Australian newspapers referred to the thylacine as the 'tiger-wolf'. See H. Stuart Dove, 'Remarkable Rocks: Former Haunt of the Hyena', *Advocate*, Tuesday 8 April

1930, 2; Crosbie Morrison, 'It Walks by Night: Tiger-Wolf is Elusive Beast', *Herald*, Tuesday 1 November 1938, 6.

117. Anonymous, 'Untitled', *Hobart Town Gazette and Southern Reporter*, Saturday 24 June 1820, 2.
118. Breton, *Excursions in New South Wales, Western Australia and Van Dieman's Land*, 357.
119. Halsey, *Deleuze and Environmental Damage*, 91.
120. Ibid., 1.
121. Ibid.
122. In *Paper Tiger*, Freeman also attends to the power of words to shape opinions, observing that historically the term 'tiger' 'epitomised what [Europeans] feared in regard to the animal kingdom' (83).
123. Here, images and words are viewed as operating more symbiotically. Images do not simply reflect the connotations of words regarding the objects they depict but rather possess the potential to extend or contest as well as to buttress the meanings of particular names for things. I am grateful to students from my course ART3001 *Écrire l'histoire de l'art* for our enriching conversations regarding word–image relations.
124. Anonymous, 'Parliament', *Daily Telegraph*, Friday 1 October 1886, 3.
125. Storey discusses the mystique surrounding lions and tigers in 'Big Cats and Imperialism', 138.
126. Guiler, *Thylacine*, 115. Herbert Eugene Pearce lived from 1904 to 1973 so certainly could have been interviewed by Guiler in circa 1953. He is too young, however, to have claimed bounties. The man Guiler interviewed claimed to have received fifty-three payments indicating a birthdate in the 1870s or 1880s. The Pearces were renowned trappers and various members are mentioned in histories of the thylacine. Tracing their family history is, however, difficult as records are incomplete. See Chick, 'Tasmanian Family and Community Reconstitution', 319. Chick notes the existence of alternative spellings of the family surname including Pierce and Pears (577).
127. Bailey, *Lure of the Thylacine*, 240.
128. Knight, 'On the Extinction of the Japanese Wolf', 137–43.
129. Tametomo is reputed to have adopted two wolves.
130. In the 1970s a terrorist cell of the East Asia Anti-Japan Armed Front was named *Ōkami* (Wolf) by its members because of the animal's perceived pride and independence. The 2006 video game *Ōkami* designed by Hiroshi Shibata, which features a white wolf, demonstrates Japanese culture's enduring fascination with the animal.
131. The zoologist Coenraad Jacob Temminck identified the Japanese wolf as a distinct species but it is now accepted that there were two subspecies of grey wolf in Japan. Temminck was also the scientist who, building on the work of Harris, created the genus *Thylacinus*. See the section 'Genre Thylacine – *Thylacinus*' (60–65) in Temminck, *Monographies de mammalogie*.
132. For a discussion of Dun's impact on Japanese perceptions of the wolf and a summary of his genocidal conduct towards the species, see Walker, 'Meiji Modernization'.
133. Hunt, 'The Face of the Wolf is Blessed, or is It?', 321.
134. Fiennes, *The Order of Wolves*, 11.
135. Roosevelt, *The Wilderness Hunter*, 188.
136. Corbin, *Corbin's Advice*, 6.
137. Ritvo, 'Calling the Wild', 105.
138. Young and Goldman, *The Wolves of North America: Volume I*, 76.
139. Anonymous, 'The Wolf Chase', 163.
140. Hunt, 'The Face of the Wolf is Blessed, or is It?', 321.

141. Smalley, *Wild by Nature*. One of Scotland's last wolves was stabbed to death in Sutherland in the Highlands by a hunter called Polson in about 1700. There is a stone monument at Glen Loth near Helmsdale commemorating the event (I am grateful to Peter Chare for providing me with a photograph of the monument). The monument fails to acknowledge that the wolf, a female, had cubs that were also killed. For an account of the death of these wolves, see Scrope, *The Art of Deer-Stalking*, 377–80.
142. Staub, *The Roots of Evil*, 54.
143. Ibid.
144. Waters, 'Solon's "Price-Equalisation"', 186.
145. Marvin, *Wolf*, 81.
146. My translation. All translations are my own unless otherwise stated. Rozier, 'Loup, Louve', 298.
147. Ibid.
148. Ball, 'Grim Commerce', 92–93.
149. Ibid., 4n14.
150. Associated Press, 'Idaho Reaches Deal to Reimburse Hunters Who Kill Wolves', *US News*, 11 October 2021, https://www.usnews.com/news/politics/articles/2021-10-11/idaho-reaches-deal-to-reimburse-hunters-who-kill-wolves (retrieved 15 July 2022). Montana currently has a de facto bounty system. See Ted Williams, 'America's New War on Wolves and Why It Must Be Stopped', *Yale Environment 360*, 17 February 2022, https://e360.yale.edu/features/americas-new-war-on-wolves-and-why-it-must-be-stopped (retrieved 15 July 2022).
151. Haraway, *The Companion Species Manifesto*, 78.
152. Ibid., 71.
153. US Department of Agriculture, *Cattle and Calves Predator Death Loss in the United States, 2005*, 26. In 2015, coyotes (40.5 per cent) were still the major cause of cattle predation with unknown predators second and feral dogs (11.3 per cent) third. Wolves (4.9 per cent) were sixth on the list. US Department of Agriculture, *Death Loss in U.S. Cattle and Calves Due to Predator and Nonpredator Causes, 2015*, 53. The next study is due in 2025.
154. For a comparison of the data see The Humane Society of the United States, 'Government Data Confirm that Wolves Have a Negligible Effect on U.S. Cattle and Sheep Industries', 6 March 2019, https://www.humanesociety.org/sites/default/files/docs/HSUS-Wolf-Livestock-6.Mar_.19Final.pdf (retrieved 15 July 2022). The data does show that in the state of Idaho (where a wolf bounty starting at $1,000 is currently offered), wolves are a major source of stock losses alongside coyotes. Often establishing what predator is responsible for killing livestock or whether predation occurred post-mortem is difficult and cost-prohibitive. The unverified data from Idaho therefore needs to be treated with considerable caution.
155. Marvin, *Wolf*, 98.
156. Ibid., 100.
157. Young and Goldman, *The Wolves of North America: Volume I*, 170.
158. Ray, *The Canadian Fur Trade in the Industrial Age*, 24–25.
159. Young and Goldman, *The Wolves of North America: Volume I*, 170.
160. Ibid., 170.
161. Seton, *Lives of Game Animals Volume 1*, 319.
162. Mowat, *Never Cry Wolf*, viii. This preface is not included in the first edition of the book. The epilogue to *Never Cry Wolf* details efforts to eradicate the wolves of the Keewatin barrens using cyanide and strychnine. See *Never Cry Wolf*, 247.

163. Freccero, 'A Race of Wolves', 116.
164. Smalley, *Wild by Nature.*
165. Richard Whitbourne, *A Discourse and Discovery of New-found-land*, unpaginated.
166. Ibid.
167. Ibid.
168. Ibid.
169. Ibid.
170. Howley, *Geography of Newfoundland*, 51.
171. Howley, 'Report of Geological Explorations on the Gambo and Gander Rivers', 437.
172. See 'An Act to Encourage the Killing of Wolves in the Colony [passed 14 September 1839]', published in the 1839 *Acts of the General Assembly of Newfoundland* (St. John's: Ryan & Withers, 1839), unpaginated.
173. Miller, 'Report of the Newfoundland and Labrador Fisheries', 765.
174. Ibid.
175. Bonnycastle, *Newfoundland in 1842*, 24.
176. Ibid.
177. *Journal of the House of Assembly of Newfoundland* (St. John's: House of Assembly of Newfoundland, 1860), 11.
178. 'An Act to Provide against the Destruction of Sheep and Cattle, and Injuries to the Person, by Dogs', *Statutes of Newfoundland* (St. John's: J.C. Withers, 1860), 49–50, 49.
179. 'An Act to make further provision against the Destruction of Sheep and Cattle, and Injuries to the Person, by Dogs', *Statutes of Newfoundland* (St. John's: J.C. Withers, 1865), 53–56.
180. Ibid., 53.
181. Ibid., 53. Newfoundland changed its currency from pounds to dollars in 1865.
182. Ibid., 56. The payment was two shillings and sixpence in sterling.
183. Appendix of the *Journal of the House of Assembly of Newfoundland* (St. John's: E.W. Bowden, 1879), A-647.
184. Appendix of the *Journal of the House of Assembly of Newfoundland* (St. John's: James Seaton, 1866), 837.
185. Appendix of the *Journal of the House of Assembly of Newfoundland* (St. John's: Morning Chronicle, 1873), 104.
186. Appendix of the *Journal of the House of Assembly of Newfoundland* (St. John's: James Seton, 1874), 139–40.
187. Mech, *The Wolf*, 333.
188. Jukes, *Excursions In and About Newfoundland*, vol. 1, 179. The ship was likely the *Onondaga*, a first-class ship built in Québec in 1829 and weighing 568 tons. See *Lloyd's Register of British and Foreign Shipping: From 1st July 1838 to the 30th June 1839*, unpaginated. The *Onondaga* is listed as abandoned in the *Lloyd's Register of British and Foreign Shipping* of 1839. See *Lloyd's Register of British and Foreign Shipping: From 1st July 1839 to the 30th June 1840*, unpaginated.
189. 'Mad dogs' is another animal-inspired insult recorded by Eales. John Eales, 'No 2 – John Eales, Butcher', *Minutes of Evidence Taken Before the Select Committee to Inquire into the State of the Colony of Newfoundland* (London, 1841), 50.
190. Birchy Lake is near the Gaff Topsails.
191. Cited in Seton, *Lives of Game Animals: Volume 1*, 263.
192. In his 1982 article 'The Newfoundland Wolf', Maunder blamed the extinction of the wolf firmly on settler colonists and the bounty scheme they instituted (36). He has subsequently revised his opinion and now attributes extinction to a sparsity of car-

ibou. https://web.archive.org/web/20240421051959/https://www.therooms.ca/the-newfoundland-wolf-0 (last accessed 27 March 2025).

193. See Spaulding et al., 'Summer Diet of Gray Wolves'.
194. Loretta Lohberger and James Dunlevie, 'New Thylacine Research Project Casts Doubt on Last Captive Tasmanian Tiger Assertions', *ABC News*, 8 November 2023, https://amp.abc.net.au/article/103065256 (retrieved 1 February 2024).
195. Despite some factual errors, Guiler's research is remarkable given that he worked from paper archives, painstakingly reconstructing the fates of many tigers and assessing the scope and impact of the bounty scheme.
196. In *Paper Tiger*, Freeman observes that 'wolf-like illustrations of the thylacine predominated in natural history literature' (91). She suggests that common names for the thylacine (such as 'marsupial wolf') 'reflected animosity towards the species and had the capacity to produce and reinforce destructive attitudes' (30). Haygarth notes that the thylacine was often referred to as a 'hyena' by Europeans but gradually the name 'tiger' became dominant in local newspapers. Lupine-inspired nomenclature was infrequent, which opens the possibility that zoological visions of the thylacine differed markedly from popular perceptions or that visual descriptors were at odds with verbal ones. See Nic Haygarth, 'Theophilus Jones and the Thylacine', 29 December 2016, http://nichaygarth.com/index.php/2016/12/29/theophilus-jones-and-the-thylacine-or-the-case-for-the-prosecution/#_ftn6 (retrieved 14 July 2021).
197. My decision not to engage in depth with paintings and drawings of the thylacine in this book is due to the extensive and insightful treatment they have already been accorded by Freeman. I do address photographs in the Conclusion.
198. For example, a recently published chapter by Sleightholme and Campbell, 'The Thylacine: A Wolf in Name Only', that seeks to foreground how different the thylacine was to wolves, makes no mention of Freeman's analyses of the influence of visual culture on lupine perceptions of *Thylacinus cynocephalus*. Freeman, however, helps us understand important elements of the process that led to the production of the 'marsupial wolf'.
199. See Chapter 31 of Bailey's *Lure of the Thylacine*.
200. Lemkin, *Axis Rule in Occupied Europe*, 79.
201. Ibid.
202. Johnson, 'National Species', 340. Johnson suggests that in some Canadian stories nonhuman animals readily stand in for the human because of perceived kinship between so called 'lower races' and animals (342).
203. Hobart Zoo, which opened in 1923, replaced Beaumaris Zoo. The latter, which opened in 1895, was housed on a site deemed not conducive to animal welfare. I am grateful to Robert Paddle for nuancing my understanding of the histories of these zoos. The name 'Benjamin' first emerged in 1968 and seems an invention and not an appellation ever actually used for the last thylacine. See James Dunlevie, 'Stop Calling the Last Thylacine Benjamin, Tasmanian Tiger Researcher Says', *ABC News*, 5 December 2022, https://www.abc.net.au/news/2022-12-06/benjamin-thylacine-tasmanian-tiger-naming-myth-persists/101734442 (retrieved 1 December 2024).
204. The name 'Benjamin' is given in scare quotes throughout this book as, while welcome as a means of attributing retrospective specificity to a particular thylacine (breaking with type), it is a human name imposed on the nonhuman animal. When some nonhuman animals differentiate among their kind (as they are clearly capable of doing) they may be employing qualities such as smell to engage in this process. Smell, in this sense, would potentially function in ways comparable to aspects of naming.

205. Here, although she does not refer to it as object biography, I am influenced by Valérie Bienvenue's inspiring research relating to A544, a quagga that was part of a menagerie kept by Louis XVI of France. See Bienvenue, 'From the General to the Particular'.
206. Due to incomplete records, however, the total number of claims detailed is likely too low. At times information about the names of claimants, where they lived or where a given wolf was killed is also absent.
207. Brass, 'The $55,000 Search to Find a Tasmanian Tiger', 40.
208. Ibid.

Chapter 1

Animal Genocide

Beginnings: Raphael Lemkin

Animal genocide is not referenced in Adam Jones's vast *Genocide: A Comprehensive Introduction*. There, Jones outlines various conceptualisations of genocide and examines diverse historical examples (including the colonial genocide of the palawa of Lutruwita and of Newfoundland's Beothuk) but clearly believes only human groups can be subject to the crime.[1] His thinking is shared by most scholars of genocide. Raphael Lemkin, who invented the term, also intended genocide to apply to crimes directed exclusively at humans. He first referred to genocide in his 1944 work *Axis Rule in Occupied Europe*, where he defined it as 'the destruction of a nation or of an ethnic group'.[2] Lemkin conceived the crime in broad terms so that it encompassed not solely killings but also 'the destruction of essential foundations of the life of national groups, with the aim of annihilating the groups themselves'.[3] These essential foundations include political and social institutions, culture and language, and, on an individual level, liberty, health and dignity. Genocide impacts individuals but does so specifically because of their membership of a given group.

Genocide is sometimes equated with the total annihilation of a specific group. In instances of genocide, however, 'the action of killing denoted by the suffix "cide" need not be completed'.[4] As Berel Lang observes, 'the term [genocide] applies even if the intended killing of a "kind" is no more than partly realized'.[5] The new word was coined 'to denote an old practice in its modern development' and 'is made from the ancient Greek word *genos* (race, tribe) and the Latin *cide* (killing)'.[6] In a footnote, Lemkin offers 'ethnocide' as an alternative term.[7] He perceived the circumstances (Nazi atrocities) in 1944 to require expanded terminology as a means to adequately describe and account for them. In his *Introduction to the Study of Genocide* (unpublished in his lifetime) he states that although 'words can be created spontaneously, like poetry they are essentially the reply of man [*sic*] to a social need', and goes on to note that as 'these needs emerge and become crystallized the necessity arises to describe them as shortly and as poignantly as possible'.[8] Through the use of the neologism 'genocide', Lemkin sought to introduce the *genos* (group) to the study of history as what he perceived to be 'a source of cultural creation, tension, and conflict'.[9]

Genocide was intended as 'a comprehensive term for the most violent manifestations of intergroup conflict'.[10] It came to replace 'extermination' or 'extirpation' as a description of the destruction of groups.[11] The social groups that concern Lemkin – racial, religious, national, linguistic or political in nature – were bound by a collective identity.[12] He states that these groups 'are exposed to genocide only when they constitute a minority or subjected majority within the community or sphere of control in which they are destroyed'.[13] Patrick Wolfe summarises Lemkin's conception as referring to 'a denominate group with a membership that persists through time'.[14] Lemkin suggests that all the groups other than those that are political or religious 'may be broadly classified as ethnic groups'.[15] They are groups 'tied by common bonds' that 'retain a separate identity'.[16]

Lemkin himself came from one such group. He was from a Jewish tenant farming family in Poland and had to flee the country in 1939 when the Nazis invaded. Many members of his family were killed in the Shoah. Lemkin's idea of justice as a basic element of reason was cemented for him in childhood. He learnt about justice through fables featuring animals 'who personified the virtues or vices of men'.[17] For Lemkin, the fact that these lessons about justice were delivered by animals amplified their power as they appealed more directly to his youthful imagination. He was able to put his idea of justice 'into practice in [his] own little world of dogs, birds, horses and playmates'.[18] Lemkin's fight against persecution and discrimination therefore began in the context of nonhuman animals. He writes that in childhood, he 'did not differentiate at that time among human beings, animals and birds'.[19]

Animals feature again later in his memoir, when he recounts his escape from Poland and recalls eating breakfast at a farm. In the presence of the farmer, his wife, their horses, chickens, a cow, a dog and a songbird, he finds considerable 'basic logic and peaceful interdependence in all these lives placed within one framework of yesterday, today and tomorrow'.[20] The brief stay at the farm reminds him of his own rural upbringing and, painfully, of his present disconnection and uprootedness. With the exception of the songbird, the animals that Lemkin relates to are all domesticated, their linkage with humans born of historical processes of resource management that led to their gradual domination. The peaceability that Lemkin identifies therefore has its origins in the exploitation of nonhuman animals for utilitarian purposes. It is also noteworthy in this context that he suggests the songbird stops their chant because they 'didn't want to disturb' the farmers. Here, wild animals are conceived as out of place, as intruding on this bucolic framework of quiet coexistence. Despite its shortcomings, Lemkin's idea of a harmonious human–nonhuman animal interrelatedness, of different 'groups' of animals quietly coexisting, is striking.

Farm animals, specifically poultry, feature in an anecdote about Lemkin's time as a student of law. In 1921, the assassination of Talaat Pasha by Soghomon Tehlirian brought what is now known as the Armenian genocide to Lemkin's attention. In June 1921, Soghomon was tried and acquitted of Talaat's murder.[21] Months later, Lemkin would ask one of his law professors at Jan Kazimierz University (now the Ivan Franko National University of Lviv), Julius Makarewicz, about the case.[22] Lemkin questioned why Tehlirian had not sought legal redress. Makarewicz is reputed to have replied that there was no law by which Talaat, who was by then living in exile in Germany after being sentenced to death in absentia by a Turkish court for his role in perpetrating atrocities against Armenians, could be put on trial for the massacres carried out by the Ottoman Empire:

> 'But Talaat was responsible for the death of those people,' Raphael retorted.
>
> 'Consider the case of a farmer who owns a flock of chickens. He kills them and this is his business. If you interfere, you are trespassing.'
>
> 'But the Armenians are not chickens.'[23]

In this exchange, Lemkin baulks at Makarewicz's clumsy analogy between poultry farming and mass murder. Makarewicz seeks to find an everyday equivalence for national sovereignty in the idea of private property. Yet sovereignty, for Lemkin, cannot be used as an excuse to justify massacres. He clearly also recognises the killing of chickens as qualitatively different from the slaughter of a people.

Makarewicz's analogy implies that national sovereignty and individual sovereignty are, at some level, similar. The people of a nation are nonetheless also the source of its sovereignty whereas a farmer's livestock are, in this instance, only a means by which the farmer can express their individual sovereignty.[24] A farmer's chickens are their personal property, farm chattels or agrichattels, and not readily equated to citizens representing popular sovereignty.[25] Lemkin can disagree with Makarewicz because, at the level of sovereignty, the law professor offers a false analogy. He also clearly opposes the professor's comparison on anthropocentric grounds. The life of a chicken is not of the same order as the life of a person. The analogy is also unworkable on historical grounds. In the Ottoman world, the Armenians were one group of people living alongside other groups of people, rendered distinct through their culture. They were constructed negatively by the Young Turk movement, labelled as profiteers and usurers, made into an out-group.[26] The farmer who kills their chickens, by contrast, is not virulently nationalistic and engaged in scapegoating.

Viewed through the prism of animal genocide, Makarewicz's analogy nevertheless remains interesting as the flock of chickens is implied to somehow correspond to an ethnic group such as the Armenians. The flock (or brood or clutch) is viewed as like a people.[27] Is it ever possible to extend the crime of genocide, of violence towards the *genos* or group, to groups of nonhuman animals?[28] Nonhuman animal groupings take several forms. Animals often live as groups. Many are gregarious (or can be so described from a human perspective). Chickens, for example, live in social groups.[29] The group dynamics of domestic chickens have been described as despotic, with a clear peck order in operation that influences flock behaviour.[30] Social network analysis has been employed as a means to study such collective animal behaviour.[31] Some groups of animals also display what humans understand as distinct cultural characteristics.[32] The form of grouping that has informed much of the extant literature on animal genocide refers to taxonomic groupings imposed on animals by humans. In zoology, animals are grouped, for example, into genera, species and subspecies.[33] The term speciecide, which refers to the systematic extinction of an entire species, is dependent upon taxonomy but not restricted to animals.[34] Non-scientific-based grouping, such as the description of chickens living in flocks or wolves living in packs, is also commonplace.

In an effort to move beyond a reliance on groups and address contemporary mass extinctions, the term ecocide has been proposed to denote the 'large-scale destruction, in whole or in part, of ecosystems within a given territory'.[35] In *Eradicating Ecocide*, Polly Higgins proposes differentiating between non-ascertainable (where the cause of destruction to territory cannot be specifically pinpointed) and ascertainable ecocide (where liability for destruction can be determined).[36] She suggests that a legal framework and principles similar to those underpinning the Genocide Convention are now necessary to protect against ecocide.[37] Higgins conceives of the occupants of an ecosystem as humans and nonhuman animals but also minerals, water and fertile land.[38] Although James Lovelock refers to them as 'genocidal acts', the crimes of 'suffocation by greenhouse gases and the clearance of the rainforests' that he has identified seem close to Higgins's conception of ecocide.[39] Scholars including Audra Mitchell question the usefulness of ecocide because the term applies only to specific events and requires the identification of individual culprits, which means that 'it cannot address the massive spatio-temporal scales and complex, distributed causality of a global extinction crisis'.[40] The wider crimes against the environment that ecocide addresses are beyond the scope of this book.

Ecocide and its synonym, geocide, are useful terms for broad crimes against the environment but are unhelpful in contexts such as the targeted extermination of a specific group of animals such as, for example, the Newfoundland wolf or the thylacine.[41] The impact on the ecosystem of the persecution of both these animals, apex predators, has not been negligible but the system was changed by their extinction rather than destroyed outright. Animal genocide, which is less encompassing than ecocide, therefore seems to have a place in efforts to describe local crimes of this kind. It is a legal concept that resonates with some of the concerns of the emerging field of green criminology, particularly as it relates to animal rights and species justice. In green criminology, the rights of an animal not to suffer harm at the hands of institutions or individuals is foregrounded. Green criminology also engages with ecological justice and the reality that human beings live within ecosystems rather than somehow standing apart from and above them. This emphasis on imbrication fosters responsibility towards the environment.

Animal Genocide: The History of a Term

The earliest direct reference to animal genocide that I have been able to identify is from 1968 and relates to the indiscriminate destruction of game animals in sub-Saharan Africa as part of efforts to reduce the tsetse fly population and control the spread of sleeping sickness.[42] David Weinman describes this destruction as animal genocide and equates genocide with blanket mass killing. This runs counter to the idea of genocide as referring to the destruction of a specific group. Later references use the term in ways that are closer to Lemkin's understanding of genocide as the targeted annihilation of a specific group. In an anonymous review of Cy Martin's *The Saga of the Buffalo*, a book about the history of the American bison (*Bison bison*), the author refers, for instance, to the 'inexcusable tragedy' of 'this vast animal genocide'.[43] Martin does not use the term genocide in *The Saga of the Buffalo* but does detail the wholesale slaughter of the bison by Europeans, which he refers to variously as annihilation, extermination and extinction.[44] The motivation for this mass killing shifted through time with bison shot for their hide and, sometimes, for meat. When slain for their hide, most of the carcass of a bison was left to rot with only the tongue, viewed as a culinary delicacy, also removed.[45] Additionally, Martin repeats the disputed claim that the slaughter of the bison was, in part, motivated by the desire to harm Native American Indians by depriving them of a vital source of sustenance: '[t]he relentless killing-off of the Plains Indian's cattle would force him [*sic*] onto the reservations, even more effectively than would all the efforts of the United States Army.'[46]

The slaughter of American bison is also referred to as genocide by Tasha Hubbard.[47] Writing from a Cree perspective, Hubbard argues that 'the slaughter of the buffalo constitutes an act of genocide'.[48] For Hubbard, although usually associated with human crimes, genocide can be extended to other animals because the human–animal divide associated with Cartesian thought does not reflect many Indigenous cultures. Hubbard emphasises that in an Indigenous paradigm the concept of people can include 'other-than-human animals' and that 'animals-as-people' are found throughout Indigenous epistemologies.[49] Hubbard contends that as 'humans do not hold exclusive title to personhood', they also do not hold exclusive title to genocide.[50] Although she predominantly engages with the fate of the bison, Hubbard also identifies wolves as subject to genocide.[51] Kim TallBear similarly affirms the broader applicability of genocide in some Indigenous cultures, observing that:

> When one speaks of genocide in the Americas it cannot be understood in relation to the European holocaust, for example, that is seen as having a beginning and an end, and which is focused on humans alone. Our genocide in the Americas included and continues to include our other-than-human relatives.[52]

The parallels that exist between the destruction of the bison and the destruction of human communities are explored in a Canadian Métis context by charlie peters.[53]

One of the earliest extended reflections on animal genocide is Philippe Roger's 1978 essay 'Hallali'. 'Hallali' takes its title from a French hunting cry used when quarry is about to be taken and killed.[54] Roger investigates what makes the mass destruction of animals possible. He identifies Charles Darwin and Sigmund Freud as the godfathers of animal genocide. For him, both thinkers contributed to producing an intellectual climate in which nonhuman animals were no longer valued. Darwin's ideas foregrounded continuity between human and nonhuman animals but with humans representing the apotheosis of evolution and thus all other animals cast as inferior.[55] Freud overarchingly interpreted dreams and fantasies about animals in symbolic terms; to him animals 'always stand for something else or for someone else'.[56] When combined, Darwin's and Freud's thinking leaves animals as 'subhuman' and without voice, 'their tongue cut out'.[57] Because of Darwin's and Freud's devaluation and negation of the nonhuman animal, circumstances now exist in which 'entire species of animal are dying'.[58] These extinctions go unchallenged because 'in our heads, the animals are already dead'.[59] Animal genocide seems to be understood loosely by Roger as human activities that cause mass animal death rather than as the intentional singling out of a specific group of animals for destruction.

Roger begins his essay with a description of a cartoon by Reiser (Jean-Marc Roussillon) from the weekly satirical magazine *Hara-Kiri Hebdo* in which a university student discovers a cockroach in his feed ('*pitance*'), causing a dinner lady to remark, 'And? It's upsetting enough that there aren't storks anymore . . .'.[60] Here, rather than apologise for the contaminated food, the server frames the event in terms of the limits of a capacity to mourn, suggesting that the invertebrate is less worthy of grieving than the endangered bird.[61] Roger goes on to suggest that through works such as *La Vie au grand air* (Outdoor Life), which he calls a modern bestiary, Reiser is 'perhaps the last venerator [*le dernier chantre*] of animality', respecting something of their radical alterity.[62]

Cartoonists such as Reiser, however, also help enable animal genocide through using animals as symbols and figures of fun.[63] Namechecking Reiser again, Roger asks if the news that the Parisian botanic garden the Jardin des Plantes will provide shelter to hedgehogs (*Erinaceus europaeus*) is likely to please the hedgehogs that are subject to a weekly holocaust in satirical magazines.[64] Here, Roger makes a direct link between a practice of representation and the conditions that facilitate genocide. He also identifies a historical shift from demands for animal liberation (the priority no longer being to accord '*habeus corpus*' to animals) to calls to save various species.[65] For Roger, saving animals as an aim can be questionable if it equates to keeping them captive, to a kind of death-in-life. The willingness to endorse captivity (think, for instance, of the phenomenon of the 'captive' breeding programme) is symptomatic of a broader problem concerning the way animals are perceived.[66] For Roger, once their strangeness is rendered familiar, they are reduced to a decorative backdrop against which human affairs unfold.

Writing in 1983, Don Allen identifies animal genocide as one of several obstacles impeding nonhuman animals from attaining what he refers to as 'being dignity'. 'Being dignity' describes the right of all beings to share in 'well-being, respect and affection'.[67] Like Roger, he clearly recognises how culture shapes the perception and treatment of nonhuman animals. Allen never explicitly defines animal genocide but the examples he provides (such as the overhunting of the passenger pigeon [*Ectopistes migratorius*]) suggest that he equates it primarily with hunting to excess or overharvesting. Such activities are, nevertheless, not intentionally aimed at exterminating a given species and do not straightforwardly conform to the understanding of genocide being developed here. Hunting and harvesting would only become genocidal if they were continued once a clear knowledge had emerged that a species was critically endangered. This was the case, for example, with the great auk (*Pinguinus impennis*) in Newfoundland.[68] The auk was initially subject to mass killing but with no comprehension that this would cause their annihilation. Once it became clear, however, that the auk was disappearing and proclamations prohibiting their killing were broadcast, the continuing slaughter became genocidal. In the case of the passenger pigeon, legislation was introduced to protect the birds in 1897 in Michigan but it was too late to be effective.[69] The bird's endangerment was not recognised until they were beyond saving and their overhunting was therefore not genocidal. The intent was not to eradicate them and there was no sustained effort to kill them once their imminent disappearance was recognised. Only if a notion of 'accidental genocide' is adopted could the destruction of the passenger pigeon be described as genocidal.[70]

'Accidental genocide' should be a non sequitur given that genocide is a crime of intent. The term is, however, still sometimes seriously employed. Bob Brissenden, for instance, describes the massacre of Lutruwita's palawa as 'a process of partly deliberate and partly accidental genocide'.[71] Nick Brodie also refers to the fantasy that the violence wrought upon the palawa was an 'accidental genocide'.[72] Additionally, Berel Lang discusses the possibility that 'a group and its members might be attacked or annihilated by chance'.[73] In the context of groups of nonhuman animals, annihilation by accident rather than design is a relatively common occurrence and would apply to incidents such as the unintentional introduction of alien predators or diseases. Some collecting practices that involve indiscriminate killing (such as the use of an ichthyocide [a fish-killing substance] by marine scientists to secure fish 'specimens' for study) could also conceivably involve accidental eradication of a species.[74] Intention is crucial to genocide as I understand it and the term would therefore not apply to such events. This means that an 'extinction icon' such as the dodo (*Raphus cucullatus*), for example, cannot be considered a victim of animal genocide – despite their being subject to habitat destruction and at least occasional hunting and abductions (with captive birds sent to Europe, India and Japan) – unless proof exists that the bird continued to be killed once their disappearance was recognised as a possibility.[75]

A more recent engagement with animal genocide occurs as part of Kirstin Marie Waldkoenig's extended consideration of nonhuman genocide.[76] Waldkoenig draws on Claudia Card's idea of social death, linking it to animal communities (including those of wolves) and examining how processes comparable to the destruction of human communal relations, social vitality and identity can also impact nonhuman animals. Carol Adams's important essay 'The War on Compassion', offers a sustained consideration of the social conditions that make animal genocide possible. Adams's discussion of genocide draws parallels between the reduction of a group of individuals in human genocide to a *people*, the subsumption of the particular into the general, and the incorporation of individual animals under general terms such as 'meat': '[w]hen humans turn a nonhuman into "meat," someone who has a very particular situated life, a unique being, is converted into something that has no individuality, no uniqueness, no specificity.'[77] Adams's argument might be extended to the classing of animals into species, which also involves the negation of the particular by the general. In the absence of individuality, Adams suggests that compassion is foreclosed. The generalising logic she identifies renders nonhuman animals into animals that do not matter, the deaths of whom pass unremarked and ungrieved.[78] Adams sees continuity between this violent generalising and the animalising language often directed at human peoples singled out for genocidal acts.[79] This language 'transforms people into false mass terms, since animals already exist in that linguistic no-man's-land of lacking a recognizable individuality'.[80] Leo Kuper has similarly drawn attention to the way practices of dehumanisation directed towards hunter-gatherer peoples often involve promoting ideologies that equate them with animals and then 'hunting them down in the same way as animals'.[81] Donna Haraway also notes that when humans become 'merely killable', reduced to the status of animals, it is 'the height of moral outrage' and 'the definition of genocide'.[82] She observes that the scapegoat is never the human, always the animal.[83] Humans or animals who are scapegoated, who are sacrificed, are reduced to the status of things. For Haraway, human exceptionalism is the issue here. There is a need to recognise that '[k]illing sentient animals is killing someone not something; knowing this is not the end but the beginning of serious accountability inside worldly complexities.'[84]

Haraway's stress on acknowledging subjectivity, like Adams's emphasis on recognition, shares common ground with Judith Butler's arguments about grievability in *Precarious Life*. Butler observes that a life that is ungrievable 'is not quite a life'.[85] In Western culture, Butler suggests that obituaries can perform the role of rendering a life grievable through signalling a life's singularity, making the stranger both unique and familiar, rendering them recognisable. Narrative forms such as the obituary therefore potentially have an important role to play in countering the homogenising violence of mass terms. Through narrative, singularity is affirmed and compassion is encouraged. Josephine Donovan explores this potential of narrative in *The Aesthetics of Care*. Donovan's own use of the term genocide in relation to animals is, to me, excessively broad. She describes the breeding of a transgenic rabbit as 'a fascist genocidal operation'. The rabbit in question was exhibited by the artist Eduardo Kac in 2000 as a work named *Alba*. The aim of the French laboratory that produced the rabbit, the Institut national de la recherche agronomique (INRA), was not to destroy rabbits as a species or even a local population of rabbits. The actions may therefore be ethically questionable, potentially criminal, but they are not genocidal in my understanding of the term.

While most early references to animal genocide understand it as a crime perpetrated by humans against animals, occasionally it is discussed as a potential crime enacted by nonhuman animals against other species. William Ellos has this understanding when he writes that there 'are few, if any, documented examples of animal genocide'.[86] Ellos views nonhuman animals as more accommodating towards each other with symbiotic working relationships cultivated that 'enhance surviving and thriving'.[87] Kuper notes that although 'animals . . . engage in intra-species killing, genocide is essentially a human crime'.[88] Intra-species killing often takes the form of infanticide and may be motivated by limited resources, social pathology or predation.[89] Some carnivores such as hyaenas and wolves carry out surplus killing, slaying many more prey than they can eat. Hans Kruuk describes such events as massacres and mass killings.[90] He suggests, among other reasons, that surplus killing may aid a carnivore to refine their hunting techniques. Surplus killings are not interpreted as intentional efforts to exterminate a prey species and Kruuk is therefore right not to refer to such activities as genocide.

Nonhuman animals can also become weaponised and used to assist the perpetration of human genocide. Writing in an Australian context, Tony Barta argues that although 'sheep may seem unlikely instruments of genocide . . . together with the cattle that tramped edible plants and fouled water holes, they were the innocent embodiment of historical pressures which wrought massive and irremediable destruction'.[91] Dogs were involved in maintaining discipline in Nazi concentration and extermination camps and sometimes killed prisoners. Robert Tindol suggests the dogs performed a 'police function' and furthered 'the Nazi program of psychological terror'.[92] They were 'helpless animals' used to amplify human cruelty.[93] In these examples, animals are instrumentalised by humans to aid genocidal endeavours. The animals do not perpetrate genocide, they are *used* to support it.

Jacques Derrida and Animal Genocide

Jacques Derrida's discussion of animal genocide is the most well-known engagement with the topic. His treatment occurs as part of a broader consideration of animal–human relations. Derrida describes a kind of cruelty at work in the very concept of the animal, which violently yokes together heterogeneous forms of life. He also considers how the animal is employed to raise up the human. As Susan McHugh explains, he reveals 'the logic of substitution through which the animal's sacrificiality (their real and representational consumption) supports the human'.[94] In this context, Donna Haraway points out in *When Species Meet* that '[w]ithin the logic of sacrifice, only human beings can be murdered'.[95] Derrida works to trouble the human–animal binary, stressing that the human is a kind of animal rather than in opposition to the animal. As humans, for instance, we know what animal suffering is because we are animals.[96] In a sensitive reading of Derrida's conceptualisation of animals, Matthew Calarco explains that the philosopher 'is keenly aware of and intent on problematizing the anthropocentric underpinnings and orientation of philosophy and associated discourses'.[97] Calarco suggests that Derrida 'is at pains to argue that binary oppositions between human beings and animals are not only empirically inaccurate, but also overlook the various differences we find between and among human beings themselves and animals themselves'.[98]

Derrida's reference to animal genocide in his 2006 book *L'animal que donc je suis* [*The Animal That Therefore I Am/The Animal That I Am Therefore Following*] is well known.[99] He had, however, already discussed the use of the term 'animal genocide' as part of a conversation with Élisabeth Roudinesco published in 2001 as *De quoi demain. . . Dialogue*, subsequently translated as *For What Tomorrow. . . An Exchange*.[100] In *For What Tomorrow. . .*, Derrida is hesitant about using the term animal genocide, aware that to do so risks causing upset to victims of human genocide. Carol Adams similarly recognises the term's potential to cause offence.[101] Derrida acknowledges his hesitation in referring to 'animal genocide' because he does not wish 'to take unfair advantage of the inevitable associations' of that term.[102] Those associations are clearly to the Shoah, as shortly afterwards Derrida recalls using the word at a law school of a Jewish university to describe the daily slaughter of animals on a commercial scale at abattoirs and receiving an indignant response, with one person stating: '"We know what genocide is."'[103] He therefore suggests taking back the word: '*retirons le mot*'.[104] This withdrawing of the term is conditional, as after suggesting it be retracted, he adds, 'but you clearly see what I want to talk about', thereby intimating the word is apposite.[105]

As a Sephardic Jew, Derrida directly experienced antisemitism as a youth in Algeria, suffering at the hands of the Vichy regime acting on behalf of the Nazis.[106] He was expelled from his lycée of Ben Aknoun in 1942 when quotas on Jewish students were introduced. The experience marked him for life.[107] He would clearly never use a term such as genocide without careful reflection. When he was chastised – 'We know what genocide is' – the 'we' of the Jewish questioner includes the philosopher. The Shoah, like the animal, is a topic that haunts much, perhaps all, of Derrida's work, coming to the forefront in texts such as the 1982 essay 'Feu la cendre' ['Cinders'].[108] Derrida's belief that the term genocide *is* appropriate in the context of nonhuman animals therefore merits careful consideration. He seems to use it to refer to the mass slaughter of livestock, speaking to Roudinesco of 'the genocidal torture that we inflict on [animals] in a way that is often fundamentally perverse, which is to say through the hyper-industrialized mass breeding of herds so that they can then be killed for supposed human needs'.[109] Here quantity seems to be a primary concern; the mass production of domestic livestock is seen by Derrida as aberrant. James Waller suggests that mass killing is distinct from genocide in that there is no clear group that is being targeted or the intention is not to eliminate the whole group.[110] Abattoirs in this definition are sites of mass killing but not of genocide.

Another important, if secondary, connection is also made between animal genocide and animal extinction, with Derrida mentioning 'the hundreds of species that disappear from the face of the earth each year at the hands of humans, who, when they don't kill them in sufficient numbers, leave them to die'.[111] Here genocide applies not just to actively slaughtering animals but also to not intervening to ensure their protection or conservation. Passivity, standing idle while a species goes extinct, is also identified as a form of culpability. Derrida's linking of the intentional destruction of species (either through intentional slaughter or active inaction) with animal genocide is much closer to how the term is understood throughout this book. For Lemkin, genocide is the comprehensive term for 'the most violent manifestation of intergroup conflict'.[112] Derrida extends the term to additionally cover *inter-species* conflict. Animal genocide would nonetheless remain exclusionary in this context and might justifiably potentially be supplemented by other kinds of inter-species genocide such as 'plant genocide', 'algae genocide' and 'fungi genocide'.[113]

Derrida's seeming condemnation of bystanders as well as direct perpetrators of genocide is problematic. A possible historical example of such guilt by inaction would be the conduct of David Lyall. Lyall was part of the initial team of lighthouse keepers employed on Takapourewa/Stephens Island in Cook Strait in Aotearoa/New Zealand. He brought the existence of the Stephens Island wren (*Traversia lyalli*), also known as Lyall's wren, to the attention of the naturalist Walter Buller in 1894.[114] A few years after the bird was identified it became extinct, exterminated by a family of feral cats.[115] Lyall was aware that the cat predation was annihilating the wren population. He noted: 'The rock wrens are very hard to get, and in a short time there will be none left.'[116] Despite knowing the wren was at risk of disappearing, Lyall did nothing to reduce the number of cats. He idly allowed the birds to die out. Drawing on Derrida's definition, this would constitute an act of animal genocide on the part of the lighthouse keeper.

Derrida affirms that there is not really a reliable difference between killing and letting die, yet if letting die is taken to refer to being a bystander, not all bystanders are homogeneous in their comportment. As Giorgia Doná suggests in relation to the Rwandan genocide, some onlookers, to whom she refers as 'situated bystanders', can be understood to 'stand back from the pressure of participating in genocidal violence'.[117] Doná views the situational bystander as engaging in passive resistance.[118] Lyall is not a bystander of this kind but there are historical examples of bystanders who refuse to adopt a genocidal outlook. In late March or early April of 1902, 39-year-old Priscilla Murray, who was visiting one of her mother Sarah's smallholdings in the village of Springfield, was attacked by a thylacine who seized her by the arm.[119] Murray was able to fight off the 'man tiger' by pressing on his windpipe but the arm was left 'lacerated in two places'.[120] After this experience, which I discuss in more detail in the Conclusion to this book, Murray might be expected to have fully embraced the demonisation of the thylacine, berating the species and contributing to the climate of opprobrium that enabled tigers to be killed largely unopposed. Instead, as Edward Alexander Bell notes in a draft of his unpublished book *Thylacine*, Murray showed 'tolerance' towards her aggressor. She 'said it seemed possible that the one-eyed creature, on rounding the corner of the house, mistook her moving arm for a small bird or mammal'.[121] Her empathy, seeking to put herself in the tiger's position, to see and react as he did, is an active response. Murray concludes that: 'There was probably no actual intention to attack the human, whom it normally would avoid.'[122] Through her reaction, she did nothing to stop the persecution of the thylacine, but to hold her morally accountable for their destruction seems perverse. In fact, her empathy for the tiger and her willingness to go against popular opinion and not demonise the animal is courageous. 'Letting die' is therefore a dangerous catch-all that elides the nuances of behaviour of people who have not actively killed endangered species. Derrida globalises guilt for species extinction. This approach erases people (often from Indigenous populations) who have never endorsed a genocidal perspective and are sympathetic to persecuted species.

To understand why Derrida chose to employ the term animal genocide in his conversation with Roudinesco, it is necessary to examine an earlier exchange between the two, one concerning vegetarianism. Derrida suggests that meat is not just consumed for nutrition but for symbolic reasons.[123] He identifies a sacrificial structure at work. Mulling over how this structure might change qualitatively in the future, Derrida suggests that humankind will come to find the 'spectacle' of animal cruelty intolerable. He asks Roudinesco: 'If you were placed daily before the spectacle of . . . industrial slaughter, what would you do?'[124] Roudinesco admits she would cease eating meat but questions the power of seeing to provide real insight into animal suffering: 'I don't think that the visibility of a situation allows one to know it better. Knowing is not the same as looking.'[125] Derrida responds by way of an imagistic assault, seeking to prove the potency of the spectacle by way of performing that power: 'But if, every day, a truck filled with calves leaving the cowshed on its way to the slaughterhouse passed before your eyes, slowly, without giving you time to be distracted, would you still be able to eat veal?'[126] Here Derrida casts Roudinesco as bystander and paints a slow-motion scene of suffering for her to witness. He is insistent on the potential for sights of suffering to change behaviour. This is hardly a new position, yet it does help explain his later reference to genocide, with the images of horror it conjures. Roudinesco, for her part, still queries what kind of knowledge the spectacle of horror can provide: 'to have the necessary detachment and really understand a situation, it's best not to be an eyewitness to it.'[127] This is also a commonplace assertion about objectivity and the need to be dispassionate.

What is key in the exchange is the importance Derrida accords to the visual as an agent of change. When Derrida subsequently uses the word genocide to describe mass butchery of livestock, it is, in part, the imagery invoked by the word that he finds apt. After suggesting that he will withdraw the word, Derrida says to Roudinesco, 'But you can see [*vous voyez bien*] what I'm talking about'. Here he is referring to seeing in the sense of understanding, but it is also possible to hear an indicator of the capacity of the word 'genocide' to *show* the horror of slaughterhouses. Derrida goes on to say that animal suffering of the kind present at the abattoir must cease if only because of 'the image that it reflects back to man of himself'.[128] He therefore seems to be encouraging links similar to those made by Siegfried Kracauer in *Nature of Film*, namely between the abattoir and the concentration camp. Inspired by Georges Franju's *Le Sang des bêtes* (dir. Georges Franju, France, 1949), Kracauer observes that in 'experiencing the row of calves' heads or the litter of tortured human bodies in the films made of the Nazi concentration camps, we redeem horror from its invisibility behind the veils of panic and imagination'.[129] For Kracauer, seeing the reality of suffering becomes an act of courage akin to Perseus overcoming his fears to look at the reflection of Medusa in the shield. It is clear from his conversation with Roudinesco that, for Derrida, seeing horror has a twofold effect; it is both a way to encourage changes in behaviour and a potential threat to well-being.

The visual returns as part of the discussion of animal genocide in *The Animal That Therefore I Am*.[130] There, Derrida is focussed particularly on animal genocide as the death-in-life of factory farming, what he calls 'an artificial, infernal, virtually interminable survival'.[131] He makes a direct link with the Holocaust, suggesting that it is as if 'instead of throwing a people into crematoria and gas chambers (let's say Nazis) doctors and geneticists had decided to organize the overproduction and overgeneration of Jews, gypsies [*sic*], and homosexuals by means of artificial insemination, so that, being continually more numerous and better fed, they could be destined in always increasing numbers for the same hell, that of the imposition of genetic experimentation, or extermination by gas or by fire. In the same abattoirs.'[132] Here, Derrida's thinking is at its weakest because it instrumentalises particular instances of suffering (of the Jews, Romani, Sinti and Gay community under the Nazis), sidestepping historical specifics, to make a point about brutality towards animals. This violent collapsing together of the experiences of the Gay community, Jews, and Romani and Sinti necessitates eliding the singular dimensions to their experiences and treatment during the Holocaust.[133]

Derrida also fails to distinguish between different kinds of genocide and animal genocide. Factory farming could arguably be viewed as similar to cultural genocide. Lemkin conceived culture as 'a functional, structural force that integrated individuals into social groups'.[134] For him, the destruction of culture could contribute to genocide. Nowadays, cultural genocide is conceived as a form of genocide that can occur independently of the physical destruction of a people.[135] Factory-farmed animals are transformed behaviourally by their living conditions, their comportment differing from that exhibited by them in alternative farming environments. Behaviours have been linked to efforts to theorise animal cultures.[136] Pär Segerdahl cautions against conflating behaviour with culture (and then seeking to distinguish natural from cultural behaviours) but goes on to suggest that the idea of animal culture is useful for recognising how diverse farming environments influence animal comportment.[137] Using this understanding, the rise of factory farming could be said to have inhibited animal cultures associated with less intensive farming practices. Turning to cultural genocide to explain Derrida's condemnation of factory farming as genocidal, however, might be less useful than simply accepting that the term is inapplicable when used to describe industrial livestock production.

Such farming is ethically unacceptable but not genocidal. Derrida's vision of animals who are purpose-bred for research and for factory farming (although he does not call it as such) is also unnuanced.[138]

Susan McHugh has noted the offensiveness of linking concentration camps with industrial slaughterhouses, but suggests that there are historically grounded alignments between the treatment of human and nonhuman populations by modern institutions of biopower.[139] Calarco has also considered comparisons made by some animal rights groups between 'current animal rearing and slaughtering methods' and the Holocaust.[140] He finds these 'false analogies' because 'there are significant and irreducible differences between factory farms and concentration camps and between the historical, social, and economic forces behind the abuse and murder of human beings and violence towards animals'.[141] Although he exercises caution in his use of the term genocide, Derrida arguably does not sufficiently foreground these differences. Other analogies, for instance, between factory farming and human incarceration might be more apposite.[142] Calarco nonetheless goes on to caution against viewing 'the comparison of violence toward human beings and animals as scandalously inadequate *simply* because it compares human to non-human life'.[143] He suggests that an anthropocentric value hierarchy 'that places human life always and everywhere in a higher rank than animal life' is 'presupposed by nearly every major thinker in recent Continental thought'.[144] He is therefore reluctant to rule out any comparison between human and animal genocide as objectionable. This is because to do so is to keep in place value hierarchies that will always operate to the detriment of nonhuman animals. Calarco argues that '[t]he very difficult task for thought here is to bear the burden of thinking through [animal and human suffering] in their respective singularity *and* to notice relevant similarities and parallel logics at work where they exist'.[145] Derrida may be over-hasty to identify similarities, but this does not render his assertion that 'animal genocides' take place invalid.

After inviting his readers to imagine a Holocaust embodied by purpose breeding and factory farming, Derrida turns once more to reflect on the issue of representation. His position in *The Animal That Therefore I Am* differs from *For What Tomorrow...* in that rather than using imagery to try and provoke a specific response, as he does with Roudinesco, he consciously refrains from employing troubling portrayals of suffering. Derrida does nonetheless foreground the power of images, stating: 'Everybody knows what terrifying and intolerable pictures a realist painting could give to the industrial, mechanical, chemical, hormonal, and genetic violence to which humankind has been submitting animal life for the past two centuries.'[146] Here a particular style, realism, is privileged. Realism is conventionally associated with truthfulness, objectivity and impartiality.[147] Derrida views it as engendering particular effects. He seems to have precise artworks in mind (he talks of potentially awakening memories of the images in his audience) although the abattoir was not a common subject for French realist painters.[148] After staking a claim for the power of such images, Derrida explains that he will not show them to his audience as this would be both too easy to do and an endless task. Nevertheless, he calls attention to how such images arouse pity and prompt reflection on pathos and the nature of compassion. Here again, then, the agency of images is foregrounded. The compassion such imagery engenders can potentially change attitudes to the human. In both his engagements with animal genocide, Derrida's ideas about the topic are bound up with reflections on the nature of images and their effects. This relationship between animal genocide and the image, the capacity of representations to facilitate or discourage atrocity, is one that is examined repeatedly in forthcoming chapters.

Animal Genocide: Looking Back to the Future

Having provided an overview of the different ways animal genocide has historically been conceived, I will now move on to examine some past examples in order to outline my own understanding of the term. These examples often involve concerted efforts through initiatives such as bounty schemes to eradicate a particular species from a given area. A bounty is a financial incentive to destroy a specific animal.[149] Early efforts at such destruction occurred in Europe in relation to the wolf. In the sixth century BCE, a wolf bounty existed in ancient Greece.[150] There are records from Genoa from the fourteenth century detailing payments for the elimination of wolves. The earliest of these dates to May 1377.[151] In Sassello in Savona, a sixteenth-century document shows that bears were also subject to a bounty.[152] Bounties on wolves in many parts of Europe from the mediaeval period to the twentieth century led to their decline across the continent and contributed to their disappearance from some countries.[153] The last known wolf in Scotland, for instance, was stabbed to death in the county of Sutherland in the Highlands by a hunter called Polson in about 1700.[154] In Ireland, the wolf bounty introduced by Oliver Cromwell's government in 1653 (and published on 1 July 1656) led to a sharp increase in the killing of wolves.[155] The last well-documented killing of a wolf in Ireland was in 1786 near Stua Laighean (Mount Leinster) in the south-east of the country.[156]

Bounty schemes were also employed by European settler colonists to encourage the extermination of native species. A wolf bounty was introduced in Massachusetts in 1630 (by 1840 wolves had disappeared from the state).[157] Other states soon followed. Paul Barclay notes that wolf bounties in the 1700s accounted for 10–20 per cent of the budget of some towns and counties.[158] Early wolf bounties varied depending on the ethnicity of the claimant, with Native Americans discriminated against.[159] The persecution of wolves continued through the centuries. Bounties were often paid upon delivery of the head of the wolf. To avoid the head being reused for subsequent claims, sometimes the ears were cut off or the tongue cut out.[160] The existence of bounty schemes in particular states and not others and of differing bounty payments between states created dishonest practices. In Wisconsin in the early twentieth century, for example, wolves were submitted for bounty payments that were actually killed in Minnesota.[161] Wolf bounties have persisted, although the term 'reimbursement' is now sometimes used, a euphemism designed to hide the reality of hunting for monetary reward.[162] In Idaho a reimbursement programme adopted in 2021 paid $2,500 per wolf for wolves killed in areas where Idaho Fish and Game (IDFG) stated wolves were 'chronically preying on livestock'.[163] The culling of wolves in Idaho is referred to as 'harvesting'.[164] To claim their 'reimbursement', bounty hunters must present the hide and skull of a wolf to IDFG within ten days of 'harvesting' them.

Other animals of Turtle Island were also subject to settler colonist bounty schemes including the black bear (*Ursus Americanus*), the cougar (*Puma concolor*), coyotes (*Canis latrans*), crows (*Corvus brachyrhynchos*), prairie dogs (genus *Cynomys*), squirrels (genus *Sciurus*) and the lynx (*Lynx lynx*) (Figure 1.1). In the 1790s, for instance, a bounty scheme was instituted against the black bear on Prince Edward Island. This ran until 1861 when the last recorded bounty payments were made. Stan Vass estimates that a thousand bears were slaughtered on the island in this period.[165] Efforts to reinstate the bounty in 1874 were rejected, with the Liberal politician William Welsh observing that he did not believe many bears remained on the island.[166] The last known bear on Prince Edward Island was killed on 7 February 1927 but warnings about the animal's endangered status began in the early nineteenth century.[167] Any bears killed once their scarcity was cause for concern would be victims of animal genocide. Prince Edward Island also offered a bounty for the lynx or *loup-cervier* from 1825 onwards. This bounty was motivated by fears (probably unfounded) concerning sheep predation. The lynx had disappeared from the island by the 1890s. Wolves were present on Prince Edward Island when Europeans first settled there but were last sighted in 1721.[168] They were quite possibly also subject to genocide, but with no bounty records or mention of their being hunted despite a knowledge of their scarcity their maltreatment must remain conjectural. Many of the bounties offered for the killing of native fauna in Turtle Island ran until relatively recently.

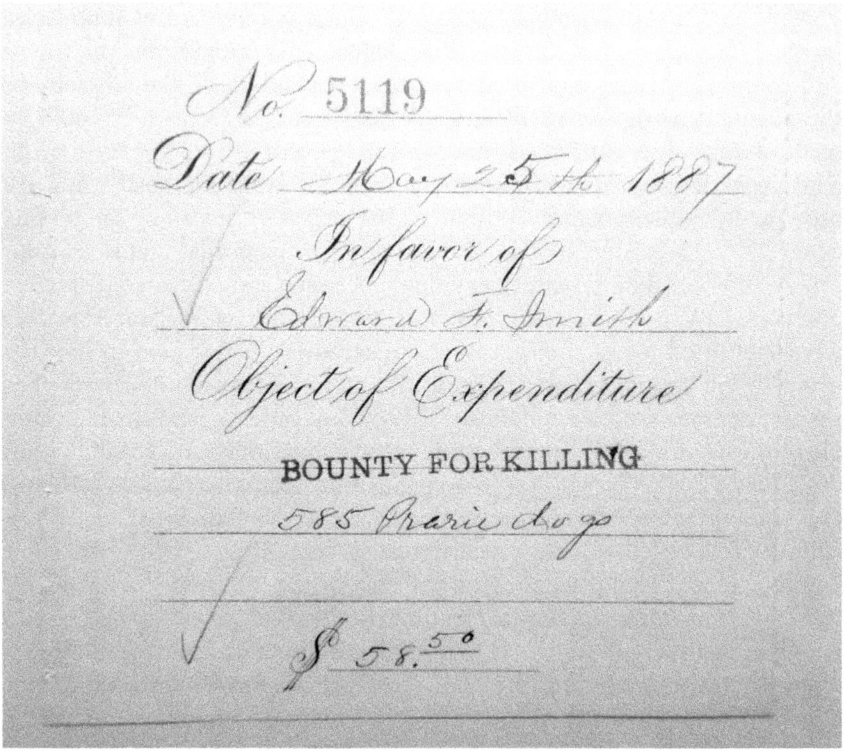

No. 5119
Date May 25th 1887
In favor of
Edward F. Smith
Object of Expenditure
BOUNTY FOR KILLING
585 Prairie dogs
$ 58.50

Figure 1.1. Prairie dog bounty receipt. Collection of the author.

In the United States, bounties were paid for the 'scalps' of specific species such as coyotes. Scalp bounties were recognised as ripe for corruption. Writing in 1909, Vernon Bailey noted that the scalps of dogs, foxes, racoons, badgers and cats were sometimes presented as wolves or coyotes. He suggested discontinuing accepting scalps or ears and accepting skins as proof instead (which could then be slit in such a way as to prevent their being presented again but not render them unusable).[169] The use of the term scalp in the context of Turtle Island cannot be dissociated from the Native American practice of scalping, which was rapidly copied by settler colonists. In *Corbin's Advice, or, The Wolf Hunter's Guide*, Benjamin Corbin moves in short order from a discussion of the fourteen human scalps taken by his settler colonist father to mentioning wolf bounties, clearly viewing these acts of killing as on a continuum.[170]

Scalp bounties for humans as well as animals were instituted by colonies and states. A 1749 (sometimes erroneously reported as 1756) proclamation was issued by Edward Cornwallis in Nova Scotia, for instance, that offered a bounty of ten guineas (later increased to fifty guineas) for the scalp of any member of the Mi'kmaq people.[171] The proclamation stated that the bounty would be paid 'upon producing such savages taken or his scalp (as is the custom of America)'.[172] Anne-Christine Hornborg links the scalp bounty with genocide.[173] In the United States, states such as Massachusetts and Pennsylvania instituted scalp bounties for Native Americans. In 1703, a scalp bounty of £20 per individual was instituted in Massachusetts. Samuel Drake observes that '[t]his was treating Indians and wolves alike'.[174] Pennsylvania instituted scalp bounties on three occasions, in 1756, 1764 and 1780. One of the scalps presented for the 1780 bounty was subsequently exhibited in Pierre-Eugène de Simitière's American Museum.[175] In Pennsylvania the bounty was controversial, with colonists recognising that it caused mistrust with 'friendly Indians' (even if, as was the case with the Iroquois, they were exempted) and encouraged atrocities.[176]

Animal bounty schemes were not limited to Europe and Turtle Island. In Australia, in addition to the bounties instigated against the thylacine in Lutruwita, many other animals were also subject to persecution on the mainland. In New South Wales (NSW), for example, the state government identified species groups including bandicoots (order *Peramelemorphia*), bilbies (*Macrotis lagotis*), kangaroos (family *Macropodidae*), rat-kangaroos (family *Potoroidae*), wallabies (genus *Notamacropus*) and wombats (family *Vombatidae*) as pests with bounties offered for their eradication.[177] Over 280,000 bounty payments were made in NSW for the killing of dingoes in the period 1883–1930.[178] Rabbits, an invasive species, were also subject to a bounty from 1880 to 1888.[179] These animals were not all persecuted to extinction, but as their eradication was the aim, they were subject to animal genocide. Some species of rat-kangaroo did completely disappear. In the period 1883–1920 roughly three million rat-kangaroos were slaughtered. The vanishment of three species of rat-kangaroo is attributed to factors such as drought and predation by foxes.[180] While the campaign of destruction wrought against rat-kangaroos may not have caused the direct extinction of any species it still conforms to the definition of animal genocide being advanced here.

The wedge-tailed eagle (*Aquila audax*) was subject to state-sponsored bounties in Queensland and Western Australia. Between 1958 and 1967 in those two states, an average of thirteen thousand bounties were paid a year.[181] The eagle was perceived as a threat to the pastoral industry, a killer of lambs. They were therefore viewed as vermin and was ruthlessly targeted. Like the thylacine, however, the extent of the eagle's ovine predation is questionable.[182] Eagles will eat carrion and may therefore feed on lambs that have died of natural causes or been killed by other predators. In the absence of eyewitnesses, identifying when a lamb has been killed by an eagle is not straightforward. Various private and state bounty initiatives targeted the eagles in the past. In the period 1958–67, Queensland and Western Australia paid out for an average of 13,056 'scalps' per year between them.[183] In Western Australia, the bounty scheme had little effect on the overall eagle population but, as the aim was the eagle's destruction, this failure does not preclude the enterprise being genocidal.[184]

Another bird species subject to a bounty was the kea (*Nestor notabilis*), an alpine species of parrot, in Aotearoa. The kea was also a victim of animal genocide. The bounty was in effect from 1884 to 1886 and then 1898 to 1971 and approximately 150,000 kea beaks were presented to claim payment.[185] Graham Renshaw links the bird with the Tasmanian tiger, observing that '[a]n undesirable trait in the character of the thylacine was the ghastly readiness with which it took a fancy to live mutton like the kea parrot of New Zealand'.[186] The kea was persecuted because the bird predated sheep, but stock losses were probably exaggerated. A celebrated photograph from 1938 of a kea 'caught in the act' of attacking a sheep is also ambiguous. It is unclear if the sheep is alive or if the bird is feeding from a carcass.[187] As occurred in Lutruwita, wild dogs (*Canis lupus familiaris*) were responsible for destroying a significant number of young sheep.[188] In Aotearoa, wild pigs also killed and ate newborn lambs.[189] Additionally, sheep sometimes consumed tutu (*Coriaria*), a genus of poisonous shrub. Moreover, sheep farming suffered considerably from the damage to grazing pastures caused by the European-introduced rabbit. Kea are now classed as endangered and have fully protected status but are still occasionally shot by farmers.[190] In 2009 and 2011, the Department of Conservation issued permits for the killing of kea, demonstrating that the bird's protected status is not absolute.[191] Given the well-known precarity of the kea population, these actions are potentially genocidal in intent.

Privately Funded Genocide

The examples I have so far discussed involve government-instituted bounties (often preceded by private bounty schemes). Instances of privately initiated efforts to eradicate species also exist. In an arrangement overseen by Captain William Moriarty, Lady Jane Franklin offered one shilling per snake head in Lutruwita in an effort to eradicate all the species of the reptile from the island.[192] James Erskine Calder refers to Franklin's efforts to 'exterminate, or at least greatly diminish, the snakes in the settled districts'.[193] Calder notes that Franklin's 'hatred, or perhaps fear of [snakes], was intense' with 'her dislike of them bordering on absolute horror'.[194] She seems to have suffered from ophidiophobia. Franklin, however, requested that George Augustus Robinson furnish her with examples of the island's different kinds of snakes, showing that her repulsion was tinged with fascination. She arrived in Lutruwita in 1837 and must have started offering her bounty soon after.[195] Bounty payments were already being paid early in 1838.[196] On 26 January 1838, Robinson mentions the one shilling premium in his journal and also that Franklin encouraged him to share details of the scheme with the palawa on Flinders Island.[197] On 2 March 1838, Robinson records the European names of six palawa who had killed snakes or found them and then had others kill them.[198] Franklin paid out over £600 in bounties before abandoning the scheme.[199] It seems bounties were offered for about a year as the scheme ceased in August 1838.[200]

In a note dated 2 July 1838 that details money (£6, 8 shillings) received by Longford Police Office from Matthew Forster, money expended on 'account of Lady Franklin for snakes destroyed', it is made clear that this was a private scheme aided by the colonial administration and its infrastructure.[201] Some members of the police appear to have objected to their involvement. In a conciliatory letter from Jane Franklin to William Moriarty dated 24 August 1838, she rues the lack of an official policy to destroy snakes but explains that in the absence of such a policy, she undertook in her 'quiet way' to use her 'private means' to 'remedy the evil'.[202] She discusses having mulled over the idea of setting up a permanent fund for the 'continued destruction of snakes after [her] departure from the colony', this because if hostilities towards snakes were to cease, they would increase in numbers again unless 'pursued to extermination'.[203] Towards the end of the letter, however, Franklin urges 'unhesitatingly, *to discontinue henceforth at your office, all further rewards for snakes*' (emphasis in the original). This call to cease the bounty seems motivated by a wish not to further trouble the police with a private venture (Franklin refers to it as her 'hobby') rather than because of any emerging concern for the welfare of snakes.[204] Franklin closes by assuring Moriarty that she will make funds available again in future should he feel the need to reactivate the bounty. Franklin's actions, motivated by a deep-seated enmity of the ophidian, provide a rare example of a specific individual assuming a clearly identifiable central role in the perpetration of animal genocide.

Franklin's actions conform to what is known as an intentionalist paradigm of genocide. Dirk Moses has examined 'intentionalist' and 'structuralist' genocides, which he identifies as rival paradigms of genocide, assessing their capacity to account for genocides that have occurred in colonial contexts such as that of Australia.[205] Intentionalist understandings of genocide emphasise perpetrator agency and the role of official policies in supporting murderous aims towards specific groups. Structuralist understandings by contrast highlight the role of anonymous processes in producing the enabling conditions for genocide. Moses finds the former approach wanting in colonial histories as 'perpetrator agency is often difficult to identify' in such contexts.[206] In colonies, settlers 'often outstripped the regulatory capacity of the metropolitan colony'.[207] For this reason, he urges a more capacious understanding of intention that enables scholars to look for genocidal intentions 'not in explicit, prior statements of settlers or governments, but in the gradual evolution of European attitudes and policies as they were pushed in an exterminatory direction by the confluence of their underlying ideological assumptions, the acute fear of Aboriginal attack, the demands of the colonial and international economy, their plans for their land, and the resistance to these plans by the Indigenous peoples'.[208]

Often, like genocides against Indigenous peoples, animal genocide is best understood not from an intentionalist but from a structuralist perspective. As Patrick Wolfe observes, settler colonial efforts at elimination of Indigenous populations for the purposes of usurping their lands, while genocidal, have historically often not depended on 'the presence or absence of formal state institutions or functionaries'.[209] In the context of genocide specifically, Wolfe asserts that 'the extent to which settler colonialism conduces to genocide [is] not a matter of the presence or absence of the formal apparatus of the state'.[210] In this context, Barta has usefully urged a conception of genocide that focusses on '*relations* of destruction' rather than emphasising 'policy and intention'.[211] He recognises the risk inherent in this position of seeming to abrogate individual responsibility for acts of genocide but suggests that encounters between individuals are varied in character such that, by implication, an explanation of genocide in simple terms of an us versus them logic lacks granularity, failing to adequately address the complex lived experiences characterising local encounters between settlers and Indigenous groups. Attention therefore should rather focus not on individual interactions but on the 'larger encounter' in which the '*inescapable relationship* was between totally incompatible forms of economy and society'.[212]

Drawing on the insights of Barta and Wolfe, it is clear that formal legislation inviting the destruction of animals such as the thylacine or the Newfoundland wolf so as to protect land is not needed for animal genocide to occur. Working within a structuralist framework, the private bounty offered for the destruction of wedge-tailed eagles in Lutruwita could be interpreted as an incitement to genocide despite an absence of state sponsorship.[213] In the context of Lutruwita, intentionalist policies have been increasingly downplayed as contributing factors for the extinction of the thylacine. Following Robert Paddle's lead, Nic Haygarth, for instance, argues that the thylacine was a victim of collateral damage caused by graziers rather than of systemic persecution.[214] A structuralist account of the genocide perpetrated against the thylacine is therefore important as both intentional and structural factors contributed to the tiger's extinction.

Quiet Genocide

There are historical examples of individuals who, while not openly advocating their destruction, were knowingly financing the obliteration of specific groups of animals. Many collectors of the nineteenth and early twentieth centuries paid for 'specimens' in the full knowledge that their actions were encouraging the killing of critically endangered species. The Stephens Island wren discussed earlier was, for example, recognised as extremely rare yet Walter Rothschild purchased multiple examples. Rothschild knew the birds were scarce from the inflated prices he paid yet he bought at least eight of them. In an essay, 'On Extinct and Vanishing Birds', he laid the blame for the disappearance of the wren on a single cat 'which has completely exterminated the species'.[215] In this same essay he identifies an indirect cause of extinction by humans as being the 'introduction of animals of prey' to their habitat.[216] Rothschild's assertion that the bird went extinct simultaneously with its discovery makes it appear that his eight 'specimens' were all secured at the same time. This was not the case. The feline procured the birds that Rothschild would purchase over a period of several months. An initial eight birds were killed (either by a cat or by Lyall after 'rescuing' the bird from the cat) in the period between February and October 1894. A further four to seven birds came onto the market after that time. As Ross Galbreath and Derek Brown acknowledge, '[t]he latter specimens may have been collected some time earlier, but even allowing for this uncertainty it seems more likely that *Traversia lyalli* declined to extinction over several years at least, rather than being exterminated promptly on its discovery in 1894 as has generally been stated'.[217] If action had immediately been taken upon the detection of the bird to eradicate cats from the island, the wren may have survived. Cats were subsequently eliminated with none recorded there after 1925.

Rothschild should have added the actions of collectors to his list of direct causes of extinction. For instance, once the killing of great auks for their feathers was no longer commercially viable because of their greatly reduced numbers, the demand for birds and their eggs from collectors meant that they continued to be persecuted. Allan Eckert's novel *The Great Auk* captures something of the collecting frenzy that led to the killing of the last of the auks. At one point, a naturalist working on behalf of a museum locates nine birds, the remnants of an auk colony. Responding to advice that he should leave the small group alone he observes:

> We've been scouring the shorelines for these birds for months and certainly we're not simply going to look at them now that we've found them. Remember, the museum is sponsoring this trip and they expect results. They have only one mounted specimen and one skin and both, unfortunately, are in rather disreputable condition. It'll be a distinct feather in our caps to bring back specimens like these.[218]

Here the birds are viewed as supplies for the museum's collection. Faced with criticism, the unnamed museum representative seeks to differentiate his actions from those who 'were wantonly slaughtering [the auks] for their feathers or their meat like those club-wielding killers we've heard so much about'.[219] To demonstrate his peaceability he suggests just killing four of the birds, two adults and two chicks.

The naturalist's use of the expression 'feather in your cap' is noteworthy. It may derive from the practice of wearing feathers to signify exploits in battle.[220] Alternatively, it may originate in the practice of the sportsperson who kills the first woodcock on a hunt, plucking out the pin feathers (developing feathers sometimes called 'blood feathers' as they have a blood supply) and sticking them in their cap.[221] Either way, the use by the man of an expression that derives from the killing of birds seems designed to undercut his protestations that his actions are qualitatively different from those of the crews of the feather companies who 'harvested' auks for down.

Like Eckert's fictional naturalist, Rothschild does not see himself as akin to the exterminators he refers to in his essay on extinction, the men who cause a species to vanish through overkilling them for food or engaging in wanton slaughter.[222] In reality, Rothschild also contributed to the decline of critically endangered species through his drive to amass 'specimens'. Similarly, Charles Haskins Townsend's efforts to source the skeletons and skins of the northern elephant seal (*Mirounga angustirostris*) should be viewed as genocidal. Well aware of the seal's scarcity, he viewed himself as in a race with sealers to secure the last of these marine mammals. In October 1884, he located three juveniles (later shooting one of them) at Bahia San Cristobal (known colloquially as Elephant Beach) in Mexico. He returned there in December to find fifteen seals (twelve females, a male and two pups), which he observed for a few hours before having them all killed. The skins and skeletons were shipped back to the United States National Museum. Townsend's account of this expedition ends downbeat; he closes with an account of the many seal boneyards he had encountered and of the nameless and lonely graves of sealers he witnessed.[223] Something of the tragedy of the seal's persecution and of his own guilt potentially registers affectively in these sombre reflections.

In 1892 at Isla Guadalupe in Mexico, he found a further group of eight seals, of which seven were killed (but four of the bodies were subsequently abandoned because of rough weather), and, for a time, believed these animals represented 'about the last of this exceedingly rare species'.[224] It is telling that one female was deliberately left alive. It seems that Townsend did not want to be responsible for killing the very last of the species. In fact, other seals survived and there were sporadic visits to the island and killings by later expeditions. In 1911, upon receiving reports of further seals at the island, Townsend travelled there again and found approximately 125 of the animals. He killed four (a female and three males) and captured six juveniles, reducing the colony by roughly a twelfth. By his own admission, between 1892 and 1911, thirty-five seals were killed at Isla Guadalupe for 'museum purposes'.[225] Each expedition to the island by Townsend and others operated in the belief that the seals they were killing were the last remnant of the species. It was only the intervention of the Mexican government that put a stop to such collecting expeditions. In 1922, they made Isla Guadalupe a reserve and the seals a protected species.

Townsend was clearly aware of the dangers of overkilling, as he wrote of the disappearance of the 'Guadalupe flicker' or northern flicker (*Colaptes auratus*) that it was 'believed to have been exterminated by former collectors'.[226] Despite this he repeatedly returned to Isla Guadalupe to slaughter seals for research purposes. Retrospectively, there have been efforts to champion Townsend as a conservationist, yet his actions demonstrate that in the context of the elephant seal he was an architect of genocide.[227] Something of the measure of the man can be gleaned from how he sourced 'action photographs' of the seals in 1911, having sailors kick them in the ribs and, in one instance, jump on their back.[228] His 1912 *Zoologica* article includes some of the photographs from this trip, including Figure 61, featuring an adult female in distress captioned: 'Under excitement the nose is projected into a pointed tip.' This 'excitement' indexes abuse encouraged by Townsend. For him, the seals were all simply data sources, morphologies and behaviours to be recorded, potential 'specimens' to secure, study and display. This scientific gaze potentially enabled his genocidal practices.

The actions of private collectors and of museum and zoo employees contributed directly to the disappearance of a number of species in the late nineteenth and early twentieth centuries. As they were aware that many of the species were threatened but carried on their accumulating regardless, parts of their collections were built on animal genocide. A justification offered for collecting critically endangered species such as the Stephens Island wren or the thylacine is that they are doomed to extinction anyway, a vision of natural history equivalent to the anthropological idea of doomed races. The notion of 'doomed races' was advanced by nineteenth-century anthropologists as a 'self-evident truth'.[229] It was believed that certain groups of humans such as First Nations Australians faced imminent extinction. They were perceived as 'living fossils', evolutionary throwbacks who were unfit for the modern world and possessed no long-term prospects for survival.[230] Echoes of this logic can be heard in the identification of some species as 'living dead' or 'zombies'. David Woodruff, for example, wrote in 2001 of the failure of conservationists to recognise that some species were 'already members of the doomed, living dead'.[231] Here, he picks up on Daniel Janzen's idea of the 'living dead', of organisms that are alive but have no reproductive future. Janzen identifies a member of the living dead as being '[a]n individual stripped of the ecological circumstances that allow it to be a reproductive member of its population, but which is living out its physiological life'.[232] For Janzen, an organism does not have to be the last of their kind to be one of the living dead, but is isolated from others of their species and unable to reproduce. Some organisms are 'living extinct species', visible yet 'vanished'.[233] Janzen views the 'living dead' as organisms that 'show us what once was', as natural historical artefacts, residues of an ecological past that persist in the present.[234] Living dead status is not bound to existing animals though. If a given animal has been sighted in a region within living memory, the idea that the species is alive there may continue long after any population has disappeared.[235] Janzen's living dead therefore applies to living creatures that are dead and to dead creatures that are thought to be living.

The concept of living dead species has been interpreted more broadly by others to encompass small populations of species that seem to have no long-term viability.[236] Some argue that species loss is inevitable and that a triage system should therefore be introduced.[237] Moral arguments regarding the notion of taxon-oriented conservation prioritisation and the formulae used to calculate rankings exceed the scope of this book.[238] Identifying some species as living dead, however, does potentially enable moral qualms about abandoning efforts to conserve them to be sidestepped. It is not possible to be responsible for the death of a species that has already been declared 'dead'. Whatever the moral implications of triage, it would not constitute animal genocide as species that are deemed expendable are not actively killed. Not acting might be akin to 'letting die' as Derrida conceives it, but there is no concerted effort to eradicate the species through activities such as bounty schemes, collecting and overexploitation.

Sometimes conservationists must also eradicate one species to protect another. Such an approach, if adopted sooner, might have saved the Stephens Island wren. There are plans to eliminate mice (*Mus musculus*) from the Farallon Islands where the mouse population threatens native plant species.[239] The mice also attract owls, which feed on the endangered ashy storm petrel (*Oceanodroma homochroa*). Additionally, the mice are suspected of attacking seabird chicks.[240] This century, in an effort to safeguard endemic species, rat populations have been exterminated on Hawadax Island and South Georgia Island among other isles and efforts are ongoing to eradicate them from the entirety of Aotearoa.[241] Despite the rats being identified as an invasive species, their killing would potentially constitute genocide as it has been defined here, as it involves the targeted killing of an entire group, a group that is potentially distinct in aspects of its behaviour to other groups of members of the same species. Whether island eradications of this kind are genocides depends upon the distinctiveness of the group, something that can be difficult to establish and gauge.[242] The call by Don Colgan to exterminate foxes (*Vulpes vulpes*) in Lutruwita, a non-native species, would not constitute genocide as, if they were indeed present, the vulpine population would not have had sufficient time to develop characteristics distinct from their mainland relatives.[243] The case of the rat population on Hawadax is potentially more complicated as Norwegian rats (*Rattus norvegicus*) form social groups and territories.[244] Although the eradication of non-native rats on islands forms a local extinction, an extirpation, rather than a global extinction, it is potentially still a genocide dependent upon group distinctiveness.[245]

On Hawadax Island (previously named Rat Island) in the Aleutian Archipelago, the introduction of Arctic foxes (*Alopex lagopus*) and Norwegian rats in the eighteenth century led to the diminishment of nesting birds, predominantly seabirds. Foxes were eliminated on Hawadax in 1984 when a total of 175 foxes were killed using various methods.[246] In 2008, the rats were eradicated using aerial drops of the rodenticide Brodifacoum, a highly toxic anticoagulant. Although some birds were expected to die as a result of eating poisoned rats, mortality rates among avifauna were higher than expected.[247] It is nonetheless clear that in the long term the killing of the rats aided the recovery of endemic species.[248] This is not always the case. The extirpation of cats (*Felis catus*) on Macquarie Island (Lutruwita), which began in 1998 and was accomplished by 2001, may have caused other non-native species (rabbits [*Oryctolagus cuniculus*], rats [*Rattus rattus*] and mice) to increase, harming vegetation and seabirds.[249] On occasion, native species have been eradicated to preserve introduced species. This was the case, for example, with the Miller Lake lamprey (*Lampetra minima*), a unique species of parasitic lamprey that was the smallest known. To protect non-native trout, the population was destroyed in the 1950s using ichthyocides. The intentional destruction of this distinct species is an example of animal genocide.[250]

Genetics and Genocide

Efforts to target invasive species in the future may increasingly involve the use of gene editing, specifically the technique of suppression-drives that artificially suppress the 'fecundity and viability' of a given species.[251] Genetic pest management has existed for many years. Recent developments involve using synthetic DNA sequences (transgenes) that are engineered into the genome of species identified as pests. Although directed initially at insects, the possibility of using transgenes to target vertebrate invasive species is now being explored. Sometimes transgenes remove traits that have been identified as harmful to humans or agriculture, rendering a species benign but not causing eradication. Other transgenes can be used to exterminate a local population of a species, an act that could potentially be genocidal. Island ecosystems that are suffering disruption from invasive species have been identified as 'ideal locations' for trials relating to transgene approaches to pest control.[252]

The idea of entirely eradicating some species has been given serious discussion in recent years. In his 2016 book *Half-Earth*, E. O. Wilson observes that '[t]he biosphere would not mourn the loss of mosquitoes of the African *Anopheles gambiae* group of species, specialists on human blood, expert hiders in native dwellings and the principal carriers of malaria'.[253] He also suggests that not even a dedicated conservationist would lament 'the complete elimination of the African guinea worm . . . the most ghastly of human pathogens'.[254] All told, Wilson estimates the number of species 'worthy of extinction' as 'fewer than a thousand'.[255] In a subsequent interview, Wilson restated more forcefully his desire to eradicate such species, advocating getting the *Anopheles gambiae* 'off the face of the Earth'.[256] Wilson justifies this annihilation because such mosquitoes are highly harmful to the human population that coexists with them. He suggests that although he would 'gladly throw the switch and be the executioner', specimens of the mosquito should be intentionally preserved in liquid nitrogen. This would represent genocide and cause extinction (as the mosquitoes in stasis could not be said to be living), but of an unusual kind (with deliberate efforts made to render extermination potentially reversible). Through his use of the metaphor of the electric chair, Wilson, tellingly, frames the genocide as capital punishment, with the mosquitoes condemned to death for their implied 'crime'. This crime is presumably mass murder as *Anopheles gambiae* is a major cause of malaria in sub-Saharan Africa, where the disease kills approximately four hundred thousand people a year.[257]

For Wilson, the species of mosquito is criminal and their eradication is a form of justice.[258] Insects such as flies are not perceived sympathetically. This is clear from one of the earliest references to animal genocide, which occurs in a discussion of methods to control tsetse flies (*Glossina austeni*, *Glossina brevipalpis*, *Glossina morsitans*, *Glossina pallipides* and others) and prevent the spread of sleeping sickness.[259] Weinman describes wholesale game killing as a means to control tsetse flies as a procedure that 'is not without repugnance, for maximal killing is animal genocide'. Pointedly, he is here concerned by the killing of vertebrates and not by the indirect extermination of flies that their destruction leads to.[260] He avows no sympathy for the fate of the insects.

In this context, it is revealing that one of the few examples of animal genocide portrayed in children's literature features the extermination of ants. In Tove Jansson's novel *Moominpappa at Sea*, a colony of ants in a thicket on a remote island is discovered by one of the characters, Moomintroll, entirely destroyed using paraffin: 'And there they were, masses of them, but all dead every one. They were all shrivelled up. Horror of horrors! An appalling [*förfärlig*] massacre had taken place and not a single ant had survived!'[261] The perpetrator, a character called Little My, justifies her actions as 'ants are like mosquitoes' so it's 'a good thing to get rid of them!'[262] My's analogy, her assumption (in the English translation) that ants are similar to mosquitoes, is used to rationalise her genocidal act.[263] She maps negative qualities associated with mosquitoes (presumably their irritating bites) onto ants.[264] Like Wilson, she views mosquitoes as abominable although her dislike is general rather than directed at specific species. Ants are, in many ways, highly dissimilar to mosquitoes; descending from ancient wasps, they are eusocial while flies are not social insects. Mosquitoes live independently of each other. Ants (who can be scavengers, predators or indirect herbivores) live in colonies, usually in a caste society.[265] Wilson has written of ants that we need to take care not to tread on them: 'Be careful of little lives.'[266]

Because of her negative perception of ants, My is unperturbed by their annihilation. The thicket also holds evidence of guilt at animal death; 'rows of tiny little crosses' mark the graves of birds killed by striking a nearby, now abandoned, lighthouse.[267] Deaths of birds due to collisions with lighthouse lanterns have long been recognised.[268] Moomintroll speculates that the island's absentee lighthouse keeper may have forsaken the lighthouse because of the continual avian casualties caused by the lantern. Such a course of action would mean putting the safeguarding of birdlife above protecting human life. The burial plot with its makeshift crosses (broken sticks bound together with twine) expresses concern for the post-mortem fate of the birds.[269] The site of mourning, with its tiny grave markers, contrasts starkly with the unmarked mass grave of the ants. Neither Little My nor Moomintroll contemplate erecting a comparable monument for the insects.

As well as calling for the killing of certain species of insect outright, Wilson also advocates the targeted killing of mosquito populations elsewhere but only as an informed decision with the impact on local ecosystems taken into account.[270] Here, he is calling for mass killing but not total eradication. It is killing of this kind that he refers to when asked if there are precedents for 'intentionally eradicating an entire animal species'. Without providing specifics, Wilson gives the example of the 'flesh fly' (of which there are roughly 2,500 species) and notes that 'it made sense to take it out in the main areas where it was causing damage to animals, to the economy'. He is probably referring to the New World screw-worm fly (*Cochliomyia hominivorax*), a species eradicated from the United States in the 1960s using the sterile insect technique (SIT).[271] This technique conforms to the process of breeding and releasing sterile male flies outlined by Wilson. Crucially, however, an entire species was not exterminated nor was that the intent. The SIT campaign of the 1960s was not animal genocide. A more obvious example of a precedent for intentional eradication would be the thylacine, a species purposely and systematically targeted for destruction. Indeed, in *Half-Earth* the thylacine is depicted on the page facing the discussion of mosquito eradication, but with no discussion in the main text as to how the species became extinct.[272] The idea that mammals might historically have been targeted for annihilation seems impossible for Wilson to contemplate.

Perception and Representation

The differing perceptions of species of animals, with insects generally viewed more negatively than mammals, and, within the category of mammals, with rodents viewed more negatively than, for example, equines, has tremendous implications for their likelihood to be targeted for genocide.[273] Wilson offers a considered rationale for the eradication of the African *Anopheles gambiae* group of mosquito species, but the common dislike for mosquitoes more broadly helps his cause.[274] A proposal to eradicate the cabbage white butterfly (*Pieris rapae*), an agricultural pest, would probably be viewed less sympathetically not solely because the insect does not transmit deadly diseases but also because butterflies are seen as charismatic microfauna.[275] Robert Holt suggests that '[a] person must have a heart of stone to glimpse a butterfly flickering in flight across a sun-dappled field, and not feel a simple joy at the sight'.[276] The emotional responses to different animals, feelings such as hatred, disgust, terror, admiration or love, influence willingness to inflict injury on them.

Lemkin indicates that 'it is the psychology of emotions which can contribute most to the etiology of individual participation in genocide'.[277] He writes of 'deeply rooted antipathy and fear' leading one group to attack another. Negative emotions certainly accrued around the figure of the thylacine, who was identified as a sheep killer. The Newfoundland wolf was framed as a child killer, with the disappearance of a youngster from the vicinity of St. John's blamed on them.[278] For Lemkin, persecution is predominantly rooted in emotion rather than simply derived from rational calculation. Genocide emerges from a complex interplay of emotion and rationalism:

> Vested interest groups often foster or actually supervise the carrying out of genocide for reasons of expediency. They may want to distract the people on whom their power depends from certain grievances or gain other advantages from the destruction of the victim group. In this case genocide is a premediated act, a means to an end, and this is rational behaviour. However, even those that plan mass murder in cold blood are rarely pure cynics. They may develop an emotional attitude toward the victim group which is endangering their power, or they may single out a particular harmless group for attack because of their deeply rooted antipathy and fear of that group. Thus it is the psychology of emotions which can contribute most to the etiology of individual participation in genocide.[279]

These insights assume considerable importance in the context of instances of animal genocide such as the bounty against the thylacine or the kea. Both these animals were blamed for sheep predation. They were exploited to explain livestock losses caused by multiple factors.

Holding the thylacine responsible for stock losses (and resultant economic hardship) generated a visceral hatred for the animal. In a clear identification of their radical objectification, they were referred to by the trapper Bob Wainwright as 'those things'.[280] In 1886 when John Lyne was seeking finances for a government bounty on the tiger, Edward Braddon suggested 'a bill be brought in for the purpose of exterminating the native tiger'.[281] Ervin Staub notes in the context of genocide that '[t]he authorities have great power to define reality and shape the people's perception of the victims'.[282] Often, however, the shaping of perception is more insidious, difficult to trace back to a specific group. In Newfoundland and Lutruwita, negative perceptions of wolves and thylacines respectively emerged in and through everyday discourse more than through official pronouncements. A diffuse negativity fuelled by rumour operated and manifested itself culturally.

Staub also emphasises the role of culture in constructing 'shared explanations and images of the world, shared values and goals, a shared symbolic environment'.[283] He suggests that sources such as art and literature contribute to creating the cultural characteristics that make group violence possible.[284] Some of the ways of imaging and imagining the thylacine discussed by Carol Freeman in *Paper Tiger* contributed to the production of a symbolic environment in which the species was viewed adversely. Freeman notes how from the mid-nineteenth century, anxieties about tigers became 'distilled into loathing for the animal' that manifested in representations of them. She rightly observes that such 'images encouraged the hunting and killing of the species and justified the first bounty and the implementation of the second'.[285] The emergence of denigrating depictions and vocabularies, destroying the image of the animal alongside the animal itself, forms the first step in a 'continuum of destruction', to use Staub's term.[286]

Writing in 1871, Thomas Potts contributes to the production of a vilifying image of the kea, giving a graphic description of the shift in nature of the bird from 'the mild gentleness of a honey-eater, luxuriating against fragrant blossoms' to a daring, thieving 'render and tearer of flesh', regularly attacking sheep.[287] Potts blames this perceived changed behaviour upon the corrupting influence of Europeans. Prior to colonisation, the kea is portrayed as living in a prelapsarian paradise. The 'meat-gallows of the backcountry squatters' drew the attention of the birds, which made 'many a hearty meal torn from the dangling carcass as it swung in the frosty air'.[288] Potts suggests that the Paneka has broadened the eating habits of the bird. Paneka is a corruption of the Māori word *pākehā* referring to Europeans and other non-Māori. Potts therefore blames his fellow Europeans for their own misfortune. His construction of the kea is, however, highly negative. He states that the Paneka 'has been destined to supply the enterprising Kea with a dainty only equalled perhaps by that which the epicurean African cuts warm from his bovine victim – our *educated* bird now tears his food from the back of the living sheep'.[289] Potts's reference to epicureanism should be understood here in the popular sense as indicating a love of good food and therefore read as sarcasm. The 'African' and the kea, both consumers of raw meat, are portrayed unfavourably.[290] In Europe, the consumption of raw meat has long been used to signal being uncivilised.[291] Jean de Joinville's circa 1309 *Histoire de Saint Louis*, for instance, describes the Mongols (he refers to them as 'Tatares' [Tatars]) negatively as consuming raw meat and rotten meat.[292] Potts goes on to suggest that the kea possesses the capacity for 'bold thievery of some of the Corvidae' and 'an amount of daring akin to the savage fierceness of the raptorial'.[293] These combined traits lead him to argue that the bird defies categorisation while simultaneously contributing to a negative image of the kea as lawless and brutish.[294] Demeaning the bird in this way gives validation to their destruction. The example from children's literature referred to earlier in which ants are likened to mosquitoes and therefore tainted by association (like Potts's association of the kea with the crow family) also demonstrates how vilification can be used to justify genocidal actions.[295]

Waller identifies three disengagement practices that are necessary for perpetrators of human genocide and mass killing to make their conduct appear acceptable to themselves and not morally rebarbative: moral justification, dehumanisation of the victims and euphemistic labelling of evil actions.[296] These practices were employed to differing degrees in the context of many animals subject to genocide including the kea, the thylacine and the Newfoundland wolf. Dehumanisation is unnecessary when a victim is an animal, but degrading of the animal's appeal and reputation is a commonplace. Moral justification in relation to the genocide practised against species such as the kea and thylacine took the form of protecting domestic animals from acts of savagery and safeguarding human livelihoods. Evil deeds are here masked by a sense of moral right. Euphemisms for large-scale killing such as 'pest control' contribute to disengagement.[297] Waller suggests that perpetrators of genocide become able to morally disengage from their actions through the use of language that camouflages 'their evil in innocuous or sanitizing jargon', causing the evil to lose 'much of its moral repugnancy'.[298] Bounty schemes, financial incentives for countering criminality, contribute to such sanitising through casting 'pest' species as outlaws.[299] The thylacine trappers in Lutruwita, for example, as bounty hunters, had the law on their side.

The use of emotive language serves to promote more than disengagement, cultivating enmity. In the case of the thylacine, this has applied not solely to adult literature but also to publications for children. Reginald Horsley's fictional story 'Boy Life in Australia', which appeared in *Boy's Own*, describes the thylacine and Tasmanian devil as taking 'first class rank as vermin', hence the efforts to exterminate them.[300] Ernest Bryant's 'factual' chapter on 'The Pouched Mammals' in Arthur Mee's *Popular Science* refers to the 'hideous thylacine' and calls the thylacine and the Tasmanian devil (*Sarcophillus harrisii*) 'savage carnivores'.[301] Mee goes on to label the thylacine fierce and stupid and suggests 'its rapacity is not simulated' and that they are 'the worst foe that the sheep-breeder has to fear'.[302] The use of the term 'stupid' implies a kind of intelligence deficit on behalf of the marsupial, opening the possibility of blaming the thylacine for their own demise because of their lack of guile.

Thylacines were also sometimes described as gutless. William Ogilby, for instance, hypothesised that their extinction (and that of the Tasmanian devil) in mainland Australia was caused by the introduction of 'a more powerful adversary' such as the dingo. Ogilby calls the Tasmanian devil and the thylacine 'the slowest, most cowardly' animals in the country.[303] Framing the thylacine as pusillanimous encourages viewing them as a 'wimp' species. 'Wimp' species are the species most vulnerable to human-caused extinction, not fit enough to survive in the long term.[304] Their 'wimpish-ness' is not related to a lack of courage but to a particular vulnerability to human pressures.[305] There is nonetheless semantic crossover between being 'chicken', a 'mouse' or a 'scaredy cat', being a coward, and being a 'wimp', a species susceptible to extinction. Like a 'living dead' or 'zombie species', the idea of a 'wimp species' bears some similarity in its effect to the idea of the 'doomed race'.

As well as fostering an image of spiritless weaklings, calling an animal such as the thylacine cowardly invites viewing them as a 'wimp species' in the zoological sense. Being designated a 'wimp species' brings with it ideas of inevitable rapid disappearance. The wimp species is nominally marked for abrupt extinction. Species constructed as 'wimps' can be seen to have 'had it coming'. Their vanishment, be it by animal genocide, habitat loss, niche displacement or other factors, is understood as unavoidable. Michael Jeffries discusses how the dodo was perceived in wimpish terms, as a 'doomed, dumpy convenience food', when, in reality, they were 'a lithe ground dweller' who were likely well suited to their island habitat.[306] The bird's short-term extinction was not preordained (most animals go extinct eventually) but the product of human callousness. As Stuart Pimm notes, the 'dodo did not go extinct. Humanity bludgeoned it into oblivion.'[307] Here the very term extinction appears euphemistic, offering a sanitised vision of the dodo's fate as naturally determined rather than the result of human action. The dodo was subject to mass killing, usually conceived as a crime against humanity but here perpetrated against a nonhuman animal. Pimm's blunt language inserts human agency into the narrative of the dodo's disappearance. His linguistic reframing of what happened enables us to ask if the dodo's fate could have been different based on alternative actions and attitudes towards them.

Pimm subsequently explained that his blunt description of the dodo's fate was motivated by the Oxford English Dictionary definition of dodo, which he summarises as: 'The dodo went extinct.'[308] Lera Boroditsky observes that when used in English, non-agentive language (such as the kind used by the OED to describe the fate of the dodo) signals evasiveness. Using the example of accounts of Dick Cheney's shooting of Harry Whittington in a hunting accident, she shows how language of this kind can be used to circumvent admitting guilt for a given action.[309] As will become clear in subsequent chapters, issues of language and responsibility assume considerable importance in efforts to attest to (or suppress) acts of animal genocide. Boroditsky's analysis of how people spoke about Cheney's mishap explores how various verbal formulations insert increasing distance between perpetrator (Cheney) and victim (Whittington). Comparable acts of distancing in human–nonhuman animal relations facilitate violence. This distancing can take the form of a denial of any commonality between humans and another species. This is a regular component of practices of scapegoating.

In Staub's exploration of the psychological motivations for genocide, he draws attention to the role of scapegoating:

> Blaming others, scapegoating, diminishes our own responsibility. By pointing to a cause of the problems, it offers understanding, which, although false, has great psychological usefulness. It promises a solution to problems by action against the scapegoat. And it allows people to feel connected as they join to scapegoat others.[310]

Staub discusses scapegoating in the context of an 'us'–'them' differentiation akin to that which Leo Kuper had earlier recognised as enabling social distancing, a facilitator of the perpetuation of atrocity.[311] Killing, as Waller discusses, is rendered easier through distancing, be it physical or on social and psychological levels.[312] Groups who are scapegoated, rendered as 'them' and placed at a distance, are identified by Staub as 'the enemy of the dominant group's well-being, safety, and even survival, or as an obstacle to the realization of its ideological blueprint'.[313] Through scapegoating, a cause is found for hardships and 'life problems become comprehensible'.[314] Scapegoating coupled with devaluation of the 'them' renders retaliation easier: 'People who are judged mean or vicious or worthless deserve to suffer.'[315] Those who scapegoat also 'become an ingroup whose members feel less alone'.[316] Kuper identifies scapegoating as a particular type of genocide, describing how it usually involves projecting 'hostile stereotypes' onto the victims and directing 'vilifying propaganda' at them.[317] He suggests that vulnerability is a key characteristic of victims of scapegoating, with scapegoated groups forming 'easy prey'.[318] Scapegoating as a process forms a stage in a continuum that can ultimately result in death.

In the context of a discussion of *animal* genocide, it is important to note the origins of the concept of the scapegoat. Donal Dorr provides a useful summary:

> Originally, the scapegoat was a real goat. Once a year the Jewish high priest presided over a solemn ritual in which the sins of the community were symbolically loaded on to a goat. The animal was then driven out into the wilderness, to carry the evil away and leave the people free to resume their everyday lives with 'a clean sheet'.[319]

Scapegoats were therefore initially 'beasts of psychological burden'. Laden with the negative affects of a given community, they were expelled but not killed. More recently, many animals have served as scapegoats with fatal consequences for them, their negative impact grossly exaggerated to explain different kinds of failure. The thylacine, for example, was 'an easy scapegoat for livestock losses and the associated loss in income'.[320] They were routinely blamed for sheep deaths that were usually the result of poor farming practices and attacks by wild dogs. Failing farms were the outcome of a bad understanding of the local climate and habitat and, concomitantly, of ovine needs. As Edward Richall Richardson reported, the tiger was made into a patsy by shepherds, who, when questioned by their employers about a deficit in the flock, took it as a rule 'to "blame the tiger"'. Richardson goes on to note that the 'native tiger is a kind of scape-goat' akin to the dingo on the mainland or the cat in domestic situations: '"It was the cat ma'am, as broke the sugar basin" and "It was the tiger, sir, who ate the sheep".'[321]

In Newfoundland, wild dogs were also a major nuisance. William Robert Kennedy suggests that much destruction blamed on the island's wolves was caused by dogs, 'packs of half-wild, half-starved curs which infest the country in the neighbourhood of the settlements'.[322] Kennedy advocates a bounty for the dogs as 'they are, without doubt, the curse of the country, no farmer being able to keep sheep or cattle for fear of them'.[323] As mentioned in the Introduction, Richard Bonnycastle also writes of Newfoundland's dogs that they have the propensity 'of worrying cattle and sheep'.[324] He suggests that 'half the mischief said to be performed by wolves . . . is done by their starved congeners'.[325] During his discussion of hardships as a motivating factor in genocide, Staub suggests that: 'In the face of persistent frustrations or threats, an important motive is gaining *hope* for control over events and renewed faith in the future goodness or benevolence of life itself.'[326] He suggests that ideology can provide this hope. In such circumstances, scapegoating offers 'the illusion of understanding and control that arises from identifying the cause of one's problem'.[327] As a mechanism for justifying violence, scapegoating makes it possible for people to rationalise their aggressive actions through the belief that 'victims have earned their suffering by their actions and character'.[328] Once the thylacine was identified as the cause of sheep farming difficulties, the possibility of eradicating the marsupial offered hope for an end to troubles for graziers.

Staub suggests that in-group–out-group differentiation of the kind employed, for example, in processes of scapegoating can be enabled by 'fear as a common human response to the unusual, unknown, and different'.[329] In this context, the perceived or expressed foreignness of an animal such as the thylacine counted against them. As discussed in the Introduction, the thylacine refused to readily conform to extant animal templates such as the dog, the hyena, the tiger or the wolf, the varied nomenclature used to describe them attesting to their strangeness. Blaming Indigenous wildlife or 'First wildlife' also distracted from the damage wrought by feral dogs, domesticated animals gone rogue.[330] Wild dogs continue to pose a problem to Australia's pastoral industry today, killing livestock and spreading disease.[331] The threat these dogs pose to livestock is a theme in the film *Mystery Road* (Dir. Ivan Sen, Australia, 2014), in which their poisoning (through a shot of a dead dog beside a fence warning of the use of DDT) and shooting is referenced.[332] Domestic dogs, however, are often associated with values such as faithfulness, reliability and docility.[333] They are frequently perceived positively. Perception is a key issue when seeking to understand the conditions that facilitate animal genocide.

The Genocidal Gaze

In her book *The Genocidal Gaze*, Elizabeth Baer argues that a particular mode of perception that she terms the 'genocidal gaze' acted as a license for extermination in German South West Africa.[334] Baer focusses on continuities in relation to perception of the Other across the genocide of the Herero and Nama and the Shoah. She carefully distinguishes between the 'imperial gaze' and the 'genocidal gaze'. The former 'describes the dominating look of the imperialist who assigns an inferior identity to the colonized'.[335] It is a gaze that 'has as its aim the control or even enslavement of the colonized'.[336] The genocidal gaze, a gaze that the imperial gaze may eventually morph into, 'has as its aim extermination'.[337] In German South West Africa, 'the genocidal gaze cast the indigenous people in the position of being subhuman, of being *expendable*, a perspective that in turn permitted the Germans to achieve their goal of domination and exclusive possession of the land'.[338] The gaze was communicated, learned, by way of cultural outputs such as literature and photography.

For Baer, the genocidal gaze as a way of seeing gave German settler colonists the moral justification to engage in acts of killing. The gaze functioned as an unconscious salve for their consciences as they brutalised the Herero and Nama. Frequently, Baer shows, the salve operated by seeing oppressed groups as less than human, as 'animal'. The genocidal gaze as embodied in works of early twentieth-century German literature and visual culture relating to the peoples of South West Africa frequently served to affirm German humanity as against the perceived animality of the Herero and Nama.

Griselda Pollock also refers to the genocidal gaze in her article 'Thinking the Feminine', in which she explores the artist Bracha Ettinger's engagement with photographs of a punishment massacre of Jews from the Mizocz (also Mizoch) ghetto. For Pollock, the gaze is embodied in the viewpoint of the photographer – 'who took this photograph must always be a question when dealing with this archive' – who shows the women just prior to their massacre.[339] It is the gaze of a perpetrator. In looking at the photograph, we become aligned with this gaze. The gaze of the photographer manifests by way of another gaze, that of the camera. In her multiple discussions of atrocity photographs, Pollock always admits that the camera possesses a 'gaze' that is its own, linked with but not analogous to the person depressing the shutter button.[340] The camera provides an 'inhuman gaze' that has become 'deeply involved in war, in violence, in distancing the act of violence from the vulnerable bodies upon which industrialized warfare is inflicted'.[341] The camera gaze is discussed in Chapter 5 as part of my analysis of the film *The Hunter*.

Written and visual cultures have functioned to demonise nonhuman animals including the thylacine and the Newfoundland wolf, who have been contrasted with other, 'more innocent', animals such as chickens and lambs. This demonisation, I would suggest, also constitutes a genocidal gaze, albeit one in which dehumanising is unnecessary. The vilification of the thylacine and the wolf led to a comparable perceptual enabling for genocide. The conditions for animal genocide to occur were created through particular ways of seeing that were adopted by science and culture. The thylacine and the wolf were viewed as animals of a particular kind, perceived as vermin and as vicious. This negative outlook regarding wolves is still present today.

In preparation for the reintroduction of wolves to Yellowstone in the United States, Alistair Bath examined public attitudes towards the lupine and found negative perceptions still extant because knowledge of wolf behaviour was lacking. Many of the people Bath surveyed, for example, believed wolves severed the hamstring muscles of their ungulate prey, whereas there is no scientific evidence for this practice. Bath speculates that '[b]elief in such a tale has nurtured the image of the wolf as a vicious killer, and has maintained public fear of the animal'.[342] Bath also found there was still uncertainty about the propensity of wolves to attack people (wolf attacks on people are extremely rare and are usually committed by rabid animals) and that if the reality of the paucity of such attacks could be communicated to people, 'negative attitudes to fear of the wolf would probably be reduced'.[343] Additionally, Bath found that popular cultural forms such as cinema shape public perception of wolves. As an example of the impact of 'multi-media coverage' on perception, he provides the case of the film *Never Cry Wolf* (dir. Carroll Ballard, USA, 1983), which led some people to believe, erroneously, that wolves survive on a diet of mice through winter.[344] Films such as *The Grey* (dir. Joe Carnahan, USA, 2011), released twenty years after Bath's study, are far more nefarious, reinforcing the idea that wolves actively prey on humans.[345]

The cover artwork for the first volume (the executive summary) of the *Wolves for Yellowstone?* report to the United States Congress is a painting (seemingly a watercolour) by Dave Cowan of a lone wolf, their body in profile but with their head turned to look directly at the reader-viewer.[346] The head is low to the ground, as unthreatening as possible. The wolf's eyes, when coupled with the report's title – *Wolves for Yellowstone?* – which appears just below the animal's muzzle, invite being read as imploring or sad. The wolf has a crafted gaze designed to support the recommendation of the report, that wolves be reintroduced. Here an artwork is used to contest visions of wolves as 'blood-thirsty marauders'.[347] In an article co-authored with Doug Blandy, Cowan discusses the varied attitudes that exist towards the wolf, viewed by some as a negative symbol and by others as a source of reverence.[348] Cowan's own image is sympathetic, attributing a gaze to the animal, suggesting an emotional state. The wolf is portrayed as a subject not an object. The artwork is not shaped by a genocidal gaze.

In the chapters that follow, as well as providing examples of the genocidal gaze, I seek to foreground moments of resistance to that gaze. Baer asserts that '"looking" is . . . a two way process and equally important as the gaze of the colonizer is the gaze of the colonized'.[349] Through the act of looking back, the colonised can resist their objectification and recognition of agency, and its restoration, can take place.[350] This return of the gaze comes into being through cultural productions such as literature. Baer's case studies are the writings of the Nama revolutionary Hendrik Witbooi (!Nanseb Gâbemab) and the work of contemporary Ghanaian author Ama Ata Aidoo. She demonstrates how different ways of seeing, oppressive and resistant gazes, register through cultural forms such as correspondence, diaries and novels. Nonhuman animals do not have access to these cultural forms but human modes of representing them are not uniform and some modes of depiction are clearly more ethically desirable than others. Although human representations such as Cowan's painting constantly risk anthropomorphism, I will argue that, at times, something of the nonhuman animal's opaque singularity catches the eye, is glimpsed.

Singularity here refers to the unique characteristics of any given animal, their refusal to be typecast, their lived refusal of typology. Singularity, like individuality, refers to distinguishing qualities, attributes that confer uniqueness. Individuality, however, has accrued considerable negative connotations and is seen by Carla Freccero (working from a Deleuzian perspective), for example, as an embodiment of patriarchal ideology. She refers to the 'oedipalized individual' and 'masculinist heroic or demonic individualism', which she contrasts with collectivities such as the wolf pack.[351] In this conception, individuality is bound up with individualism, which can manifest as self-centredness and selfishness, as insularity and a lack of acknowledgement of interconnection with others. Individuality also implies stasis, an accomplished identity, rather than a dynamic, ongoing becoming.[352] I use the term singularity in an effort to accord specific animals their distinctiveness yet without thereby negating discrete multiplicities, without overlooking their collective existences and continual becoming.[353] Singularity in the context of the genocidal gaze becomes a means to puncture the deprivation of identity that characterises processes of scapegoating. In such processes, categories become all-defining, groups homogenised. Ways in which scapegoating can be challenged through emphasising singularity and difference feature repeatedly in the chapters that follow.

Notes

1. See Jones, *Genocide*, 164, 188.
2. Lemkin, *Axis Rule in Occupied Europe*, 79.
3. Ibid.
4. Lang, 'The Concept of Genocide', 119.
5. Ibid.
6. Lemkin, *Axis Rule in Occupied Europe*, 79.
7. Ibid., 79fn1.
8. Lemkin, *Lemkin on Genocide*, 21.
9. Ibid., 9. The term in English is not bound to the human but, as Kirstin Marie Waldkoenig notes in her thesis 'Can Nonhumans be Victims of Genocide?', sometimes the term is translated in ways that 'denote only humans can be victims of genocide' (6).
10. Ibid., 35.
11. Moses, 'Genocide and Settler Society', 152.
12. Lemkin, *Lemkin on Genocide*, 34.
13. Ibid.
14. Wolfe, 'Settler Colonialism and the Elimination of the Native', 144.
15. Lemkin, *Lemkin on Genocide*, 34.
16. Ibid.
17. Lemkin, *Totally Unofficial*, 17.
18. Ibid., 18.
19. Ibid.
20. Ibid., 31.
21. The assassination occurred on 15 March 1921 and Tehlirian's trial ran from 2 to 3 June. Lemkin enrolled at the Jan Kazimierz University (University of Lwów) in October 1921.
22. In *East West Street*, Sands suggests that Makarewicz was probably the professor that Lemkin had this exchange with (154–55). Makarewicz clearly admired the younger man's scholarship as he agreed to write a preface to Lemkin's co-authored (with Tadeusz Kochanowicz) book *Kodeks Karny Republik Sowieckich* [Criminal Code of the Soviet Republics].
23. Bartlett, *They Stand Invincible*, 97.

24. For an exploration of the nature of popular sovereignty in relation to the nation state, see Chapter 4 of René Grotenhuis's *Nation-Building as Necessary Effort in Fragile States*.
25. For a discussion of farm chattels, see Jensen, 'Agrichattels'.
26. Kévorkian, *The Armenian Genocide*, 25.
27. This is another aspect of Makarewicz's analogy that is highly questionable, as a flock of domestic chickens is artificially engineered by the farmer and not like an ethnic or religious group with a shared culture and history.
28. Piers Beirne proposes the term 'theriocide' for the killing of individual animals. See Beirne, 'Theriocide'.
29. See Chapter 6 of Nicol, *The Behavioural Biology of Chickens*. Domestic chickens exhibit gregarious nesting behaviour. This manifestation of gregariousness appears to be a result of domestication. See Riber, 'Gregarious Nesting'.
30. Gottier, 'The Dominance-Submission Hierarchy in the Social Behavior of the Domestic Chicken'.
31. See Wey et al., 'Social Network Analysis of Animal Behaviour'.
32. See Whitehead, *Sperm Whales*.
33. In *Genocide*, Lang notes that the idea of the species as a distinct group is seldom contested. He observes that a criticism levelled at the concept of genocide is that groups are 'only the assembly of a number . . . of individuals' and that groups cannot be distinguished from the individuals that constitute them (29). Lang rebuts such an assessment by pointing out that groups form a 'collective "self"', the capacities (such as language) and inclinations of which would be foreign to a solitary individual (29). Groups attain and retain their identity through shared actions. He then goes on to observe that although the idea of human groups sometimes has to be justified, 'political movements in defense of animal species threatened with extinction have rarely been impeded by the question of justification, even as concerns the character of the particular species being protected' (30). Implicitly here Lang is making some kind of equivalence between human genocide and animal extinction.
34. The restricted scope of speciecide, which cannot be applied to elimination efforts directed at local groupings of a given species or at the deliberate eradication of a subspecies, is a significant drawback. Ross MacPhee notes that 'substantial loss of sub-species' reduces the genetic endowment of a species, 'which in the long run is an invitation to disaster'. Attacks on subspecies therefore potentially contribute in the long term to speciecide. See MacPhee, 'Remember the Islands', 143.
35. Higgins, *Eradicating Ecocide*, 62. For further discussion of the origins of the tern ecocide, see Waldkoenig, 'Can Nonhumans Be Victims of Genocide?', 18–19.
36. Ibid., 63.
37. Ibid., 67.
38. Ibid. The reference to fertile land is problematic as 'barren' lands such as deserts would therefore potentially be excluded from ecocide as territory that is unproductive and hence 'uninhabited'.
39. James Lovelock, 'Beware: Gaia May Destroy Humans before We Destroy the Earth', *Guardian*, Tuesday 2 November 2021, https://www.theguardian.com/commentisfree/2021/nov/02/beware-gaia-theory-climate-crisis-earth (retrieved 7 January 2025).
40. Mitchell, 'Is IR Going Extinct?', 10.
41. For a discussion of geocide (the killing of the earth), see Berat, 'Defending the Right to a Healthy Environment'.
42. Weinman, 'The Human Trypanosomiases', 133.
43. Anonymous, '*The Saga of the Buffalo*, by Cy Martin'.

44. See, for example, Martin, *The Saga of the Buffalo*, 136–38. Martin's use of the term extermination follows that of Hornaday in *The Extermination of the American Bison*. Writing in 1887, Hornaday describes the 'remorseless slaughter' of the bison (464). He attributes the decimation of the species to factors including human greed, 'wanton destructiveness', failure to legislate to protect the animal and the animal's 'stupidity' and 'lack of courage' (464–65).
45. The tongue as a delicacy is referenced in *True Grit* (dir. Joel and Ethan Coen, USA, 2010), with Marshall Reuben J. 'Rooster' Cogburn observing that he would give $3 for a pickled buffalo tongue. The historical slaughter of bison purely for their hides was given recent cinematic expression in *Prey* (dir. Dan Trachtenberg, USA, 2022). The wanton massacre of the bison causes anguish to the Comanche healer Naru.
46. Martin, *The Saga of the Buffalo*, 66. In 1915, Oliver Farrington suggested that 'the practical disappearance of the Indian from the North American continent was not wholly due to his forcible segregation by the white man, but in part at least the destruction of the buffalo, which constituted his [*sic*] normal food supply'. Farrington, 'The Rise of Natural History Museums', 207. In *Nature's Mirror*, Andrei writes that there is no historical evidence to support the idea that the decimation of the bison was a racist US government policy (79).
47. For a helpful discussion of Hubbard's understanding of genocide, see Mitchell, 'Revitalizing Laws', 917–18.
48. Hubbard, 'Buffalo Genocide in Nineteenth-Century North America', 293.
49. Ibid., 294.
50. Ibid., 295.
51. Ibid., 302.
52. Kim TallBear, 'Failed Settler Kinship, Truth and Reconciliation, and Science', *Indigenous STS*, 16 March 2016, https://indigenoussts.com/failed-settler-kinship-truth-and-reconciliation-and-science/ (retrieved 18 May 2021).
53. peters also uses the term animal genocide to describe the fate of the bison. See peters, 'Girls, Animals, Fear, and the Iterative Force of the National Pack'.
54. For an extended discussion of the term, see Pairault, *Nouveau dictionnaire des chasses*, 200–1. The English equivalents of 'halloo' and 'tally ho' usually refer simply to sighting quarry. See Blaine, *An Encyclopaedia of Rural Sports*, 488.
55. Roger, 'Hallali', 835.
56. Ibid., 836. I explore the issue of reducing animals to symbols in Chapters 4 and 5.
57. Ibid., 837.
58. Ibid., 830.
59. Ibid.
60. The magazine was subtitled *Journal bête et méchant* (Beastly and Mean Newspaper).
61. In the 1970s the white stork (*Ciconia ciconia*) population in France was in decline, although it has now recovered. In Alsace in 1974, only nine breeding pairs were found. The dinner lady in the cartoon wears a 'coiffe alsacienne' (Alsatian headdress) so seems to be from that region. See Merle, 'Les Espèces en phase de reconquête'. The black stork (*Ciconia nigra*), which is probably not being referenced in the cartoon, remains rare in France.
62. Roger, 'Hallali', 829. Reiser's *La Vie au grand air* features mainly African wildlife and their interactions with humans (who are portrayed as racist caricatures). Some of Reiser's animal cartoons (often with a sexual theme) appear in Yves Frémion's *Reiser*.
63. In *Vive les Vacances* (published after Roger wrote 'Hallali'), Reiser uses some hedgehogs and a cat to discuss sex (71).
64. Roger may be referring to *Le Hérisson* [*The Hedgehog*], a weekly satirical magazine.

65. Roger, 'Hallali', 829.
66. Some of the problems that accompany captive breeding (such as domestication and a reduction in the urgency of habitat preservation) are taken up by Gordon Sayre in his essay 'Rare Books and Rare Birds'.
67. Allen, 'Notes', 377n2.
68. See Chare, 'The Beothuk, the Great Auk and the Newfoundland Wolf'.
69. See Barrow, 'Teetering on the Brink of Extinction', 54.
70. Menno Schilthuizen seems to imply that accidental speciecide is possible. He uses the example of the accidental eradication of the condor-specific feather louse, the California condor louse (*Colphocephalum californici*), as part of a discussion of speciecide. Specicide is contrasted with 'wilful, intentional genocide waged against biodiversity in the name of the human appetite for wild species'. By implication, speciecide for Schilthuizen is inadvertent and unintentional. See Schilthuizen, *The Loom of Life*, 116.
71. Brissenden, 'On the Edge of the Empire', 146.
72. Brodie, *The Vandemonian War*, 381. For a critical review of Brodie's history of the Lutruwitan genocide, see Russell, 'Remembering and Forgetting'.
73. Lang, *Genocide*, 98.
74. I am grateful to Tony Gill for sharing his knowledge regarding ichthyocide. Adam Kerezsy has noted the danger of employing ichthyocide in physically isolated ecosystems inhabited by critically endangered species. Kerezsy is referring to the use of ichthyocide to kill invasive fish species rather than for collecting purposes. Kerezsy, *Gambusia Control in Spring Wetlands*, 18.
75. For a discussion of captive dodos and their travels, see Winters and Hume, 'The Dodo'.
76. Waldkoenig, 'Can Nonhumans Be Victims of Genocide?'
77. Adams, 'The War on Compassion', 23.
78. Ibid., 25.
79. Josephine Donovan makes a similar point in *The Aesthetics of Care*, 41–42.
80. Adams, 'The War on Compassion', 26–27. For an important discussion of the link between the term 'animals' as a classificatory tool and instances of slaughter, see Hoquet, 'Animal Individuals', 73.
81. Kuper, *Genocide*, 54. The attentiveness given by Kuper to the power of metaphor in aiding and abetting ideologies of dehumanisation is particularly important.
82. Haraway, *When Species Meet*, 78.
83. Ibid., 79.
84. Ibid., 106.
85. Butler, *Precarious Life*, 34.
86. Ellos, 'Ethical Similarities in Human and Animal Social Structures', 214.
87. Ibid., 213.
88. Kuper, *Genocide*, 52.
89. See the discussion of infanticide in Mörner et al., 'Diseases and Mortality', 301.
90. Kruuk, 'Surplus Killing by Carnivores'.
91. Barta, 'Relations of Genocide', 240–41. In 'Settler Colonialism and the Elimination of the Native', Wolfe suggests that underpinning settler colonialism is 'the driving engine of international market forces', forces that link, for instance, Australian wool to Yorkshire mills (394). The growing links between Australia and the United Kingdom by way of the wool trade are catalogued by Thomas Southey, who traces a 326¼ per cent increase in imports between 1829 (1,838,642 lbs) and 1838 (7,837,423 lbs) from Australia (Gadigal/Sydney and Lutruwita) to the UK. Southey, *A Treatise on Sheep*, 25.
92. See Tindol, 'The Best Friend of the Murderers', 116.
93. Ibid., 120.

94. McHugh, *Animal Stories*, 9.
95. Haraway, *When Species Meet*, 78.
96. Derrida and Roudinesco, *For What Tomorrow. . .*, 70 (118). Page numbers given in brackets refer to the original French edition of this text.
97. Calarco, 'Thinking through Animals', 2.
98. Ibid., 3.
99. The book draws together ideas that were first articulated in a conference paper delivered in 1997 and published in the proceedings as 'L'animal que donc je suis'. See Marie-Louis Mallet, 'Avant-propos'.
100. I am grateful to Tom Tyler for drawing this text to my attention.
101. Adams, 'The War on Compassion', 22. Adams goes on to cite Derrida's 'The Animal That Therefore I Am' in her text (32).
102. Derrida and Roudinesco, *For What Tomorrow. . .*, 73 (122). Translation amended.
103. Ibid.
104. Ibid.
105. Ibid. Translation amended.
106. For a brief discussion of the incident, see Rabaté, *Les guerres de Jacques Derrida*, 87.
107. Peeters, *Derrida*, 31–32.
108. Derrida, 'Feu la cendre'. For an extended discussion of the significance of this text in relation to Derrida's writings more broadly, see Parent-Thivierge, 'Relire *Feu la cendre* de Jacques Derrida'.
109. Derrida and Roudinesco, *For What Tomorrow. . .*, 67 (112). Translation amended.
110. Waller, *Becoming Evil*, 14.
111. Derrida and Roudinesco, *For What Tomorrow. . .*, 67 (112).
112. Lemkin, *Lemkin on Genocide*, 35.
113. The intentional destruction of a species of plant, algae or fungi can be envisaged through habitat destruction such as deforestation if it is carried out in the knowledge that it will lead to the annihilation of a given species of these organisms.
114. A construction worker, F. W. Ingram, noted the presence of two species of wren on the island in 1892 but did not capture any of the birds. Buller wrote the first description of the bird, which he named *Xenicus insularis*. His paper was, however, published after Walter Rothschild's so the latter's name of *Traversia lyalli* was adopted. For a discussion of Rothschild's ungentlemanly conduct, see Galbreath and Brown, 'The Tale of the Lighthouse Keeper's Cat', 195.
115. This may have been as early as 1895 and was almost certainly before 1900. See Galbreath and Brown, 'The Tale of the Lighthouse Keeper's Cat'.
116. Cited in ibid., 194.
117. Donà, '"Situated Bystandership" during and after the Rwandan Genocide', 8.
118. For a discussion of passive resistance in the context of the Shoah, see Chare and Williams, 'Introduction'.
119. Anonymous, 'Attacked by a Tiger or Hyena', *Examiner*, Tuesday 8 April 1902, 7. Bailey summarises this attack in *Lure of the Thylacine*, 91–92.
120. Ibid. Murray was left physically scarred by the attack.
121. E. A. Bell, *Thylacine*, 87. Tasmanian Archives, Edward Alexander Bell Papers, NS463/1/2.
122. Ibid.
123. Derrida and Roudinesco, *For What Tomorrow. . .*, 71 (119).
124. Ibid.
125. Ibid.

126. Ibid. Translation amended.
127. Ibid., 72 (120). Translation amended.
128. Ibid., 73 (122). Translation amended.
129. Kracauer, *Nature of Film*, 306.
130. Derrida, *The Animal That Therefore I Am*, 26 (46–47). Page numbers given in brackets refer to the original French edition of this text.
131. Ibid. (47). Karen Davis also links industrial farming practices with animal genocide, employing the chicken as a case study. See Davis, 'Chicken-Human Relationships'.
132. Ibid. Translation slightly amended.
133. Ian Hancock, for instance, notes that Romani, victims of the *Baro Porrajmos*, were solely persecuted for race-based reasons while Jews were singled out for racial persecution but were also feared economically and politically. Hancock also notes that many massacres of Romani did not occur in concentration and extermination camps (many Jewish victims were also murdered elsewhere than in the camps). See Hancock, 'Romanies and the Holocaust'. Gay people, mainly men, were interned and some were executed. See Grau, *Hidden Holocaust?* Schoppmann details specific examples of lesbians interned at concentration camps in 'National Socialist Policies Towards Female Homosexuality'.
134. Irvin-Erickson, *Raphaël Lemkin and the Concept of Genocide*, 222.
135. See Novic, *The Concept of Cultural Genocide*, 4.
136. Segerdahl, 'Can Natural Behaviour Be Cultivated?', 183.
137. Ibid., 184. For Segerdahl, culture and biology are not readily dissociated (184–85).
138. Some animals bred for research purposes, for instance, are now granted 'retirement'. This is not something that occurs in factory farming. For a discussion of 'retirement' of monkeys from research facilities and of surmounting some of the barriers to this, see Kerwin, 'Overcoming the Barriers'.
139. McHugh, *Love in a Time of Slaughters*, 15.
140. Calarco, 'Thinking through Animals', 8.
141. Ibid.
142. See, for instance, Morin, 'Carceral Spaces'.
143. Emphasis in the original. Calarco, 'Thinking through Animals', 8.
144. Ibid. Jane Caputi notes that acknowledging animal genocide challenges human exceptionalism and, hence, hierarchisation. See Caputi, '*Jaws* as Patriarchal – and Ecocidal – Myth', 13.
145. Ibid., 10.
146. Derrida, *The Animal That Therefore I Am*, 26 (47). Although Derrida writes 'un peinture réaliste' it seems possible he means 'un peintre réaliste' (a realist painter).
147. Nochlin, *Realism*, 13.
148. The German painter Lovis Corinth painted slaughterhouses, but his style with its loose facture is better described as expressionistic. For a discussion of his paintings of butchery and slaughter with a strong emphasis on their chromatics, see Zimmermann, 'Corinth et la chair de la peinture'. Derrida may have more contemporary artworks in mind such as Sue Coe's project *Dead Meat*, but her sketches are tenebrous, more baroque than realist. Cary Wolfe interprets them as melodramatic. See Wolfe, 'From *Dead Meat* to Glow in the Dark Bunnies'. As Johanne Lamoureux has explored, early twentieth-century depictions of slaughterhouses in film and literature often served the dual purpose of showing the becoming-animal of the worker and the reduction of the worker to an appendage of flesh of the machine. At times, Coe's works seem to perform a similar function. See Lamoureux, 'Le travail de la viande'.

149. For a discussion of various bounty schemes, most of which were ineffectual in terms of the total eradication of a given species, see Whitehouse, 'Bounty Systems in Vermin Control'.
150. Buxton, *Myths and Tragedies in their Ancient Greek Contexts*, 35.
151. See Hearn et al., 'The Wolf in the Landscape', 11.
152. Ibid., 13.
153. As Fiennes notes, habitat destruction was also a major factor in the extirpation of wolves in England and Scotland. See Fiennes, *The Order of Wolves*, 169–70.
154. As mentioned in my Introduction, there is a stone monument at Glen Loth near Helmsdale, erected in 1924, that commemorates the event. The monument fails to acknowledge that the she-wolf had cubs that were also killed. For an account of the murder of these wolves, see Scrope, *The Art of Deer-Stalking*, 377–80.
155. Hickey, 'A Geographical Perspective', 195.
156. Ibid., 193.
157. The Massachusetts Bay Colony introduced the bounty in 1630. See Foster et al., 'Wildlife Dynamics in the Changing New England Landscape', 1347.
158. Barclay, 'A "Curious and Grim Testimony to a Persistent Human Blindness"', 25.
159. Ibid., 31.
160. Barclay, 'A "Curious and Grim Testimony to a Persistent Human Blindness"', 32.
161. See Anonymous, 'Traffic in Wolf Pelts', *Duluth Evening Herald*, 11 April 1906, 9.
162. See Jennifer Sherry's discussion of a 2019 effort to reintroduce a wolf bounty in Montana. 'Montana's Wolf Bounty Will Remain a "Fossil"', *NRDC*, 5 April 2019, https://www.nrdc.org/experts/jennifer-sherry/montanas-wolf-bounty-will-remain-fossil (retrieved 15 September 2022).
163. For further specifics on the bounty programme, see Associated Press journalist Keith Ridler's syndicated article 'Idaho Reaches Deal to Reimburse Hunters Who Kill Wolves', https://halifax.citynews.ca/2021/10/11/idaho-reaches-deal-to-reimburse-hunters-who-kill-wolves/ (retrieved 17 March 2025).
164. See the references to harvesting on the Idaho Fish and Game website: 'Wolf – Hunting and Trapping Information', https://idfg.idaho.gov/hunt/big-game/wolf (retrieved 15 September 2022).
165. Vass, 'The Black Bear on Prince Edward Island', 9.
166. Hornby, 'Bear Facts: Part 1', 9.
167. Hornby, 'Bear Facts: Part 2', 27–28.
168. See Douglas Sobey's impressively researched account of historical records of mammals native to Prince Edward Island. Sobey, 'An Analysis of the Historical Records'. Paul Paquet and Ludwig Carbyn wrongly assert that the grey wolf was absent from Prince Edward Island. See Paquet and Carbyn, 'Gray Wolf', 483.
169. Bailey, *Key to Animals*, 1. When scalps were presented, they included ears and one of these would usually be removed to prevent the scalp being presented again. See the entry 'Scalp Bounties' in the State of Oregon's 1899 General and Special Laws, 131.
170. Corbin, *Corbin's Advice*, 12–13.
171. Anonymous, 'Two Hundred Year-Old Scalp Law Still on Books in Nova Scotia', *CBC News*, 4 January 2000, https://www.cbc.ca/news/canada/two-hundred-year-old-scalp-law-still-on-books-in-nova-scotia-1.230906 (retrieved 6 June 2022). See also Landry, 'Pokemouche Mi'kmaq and the Colonial Regimes', 41.
172. Cited in Hornborg, *Mi'kmaq Landscapes*, 8.
173. Ibid., 7.
174. Drake, *The Border Wars of New England*, 166.

175. Young, 'A Note on Scalp Bounties in Pennsylvania', 217. For a general discussion of the museum, see Levey, 'The First American Museum of Natural History'.
176. Young, 'A Note on Scalp Bounties in Pennsylvania', 208.
177. Other states ran similar bounty schemes. Queensland, for example, also offered a bounty for kangaroos under the 1877 Queensland Marsupials Act.
178. See Glen and Short, 'The Control of Dingoes in New South Wales'.
179. In 1890–91, when the bounty was no longer in force, twenty-five million rabbits were destroyed. See Short, 'The Extinction of Rat-Kangaroos', 373.
180. See Short, 'The Extinction of Rat-Kangaroos'.
181. See Olsen, 'Wedge-Tailed Eagle, Australia', 62.
182. Using data derived from nesting sites in the Australian Capital Territory and New South Wales, Leopold and Wolfe report that lamb remains constituted 7 per cent of the eagle's diet but that some of this could be carrion. See Leopold and Wolfe, 'Food Habits of Nesting Wedge-Tailed Eagles'.
183. Ibid., 1.
184. For an analysis of the effect of the bounty on Western Australia's eagle population, see Ridpath and Brooker, 'Age, Movements and the Management of the Wedge-Tailed Eagle', 255.
185. Clio Elizabeth Reid also discusses private bounty schemes initiated from 1868 onwards. She gives a lower estimate of 116,869 kea killed for bounties but acknowledges that as further records become available her estimate will likely prove conservative (30). See Chapter 2 of Reid, *Understanding Attacks by Kea.*
186. Renshaw, *More Natural History Essays*, 221.
187. Kea do attack live sheep but also consume carrion and may therefore sometimes be unfairly blamed for sheep deaths simply because they were seen in close proximity to a corpse. The photograph of the kea atop a sheep appears as part of a photo spread in *The Press*, 31 August 1938, 16.
188. The Māori brought dogs with them to Aotearoa. Archaeological evidence suggests that these dogs were dependent on the Māori and unable to survive in the wild. See Coutts and Jurisich, 'Canine Passengers in Maori Canoes', 82. The number and variety of dogs increased greatly with the arrival of European settler colonists. Soon there were packs of wild dogs roaming both islands. William Colenso makes reference to packs of dogs predating the flocks of early settler colonists. See Colenso, 'Notes, Chiefly Historical, on the Ancient Dog of the New Zealanders', 186–87. In *The Invasion of New Zealand by People, Plants and Animals*, Andrew Clark acknowledges that dogs and pigs predated lambs (206). He suggests, however, that dogs, pigs and the kea were only ever a 'minor nuisance' (181).
189. Belich, *Making Peoples*, 344.
190. Reid, *Understanding Attacks by Kea*, 34.
191. See Will Harvie, 'Department of Conservation Killed Kea Attacking Sheep', *Stuff*, 26 January 2020, https://www.stuff.co.nz/science/118782768/department-of-conservation-killed-kea-attacking-sheep (retrieved 16 September 2022).
192. The three native species are the lowland copperhead (*Austrelaps superbus*), tiger snake (*Notechis scutatus*) and white-lipped snake (*Drysdalia coronoides*).
193. James Erskine Calder, Papers, 1797–1882. MS F BOX 89/4. State Library of Victoria. Unpaginated.
194. Ibid. Louisa Meredith possessed similar sentiments, writing of her 'horror of the whole fearful tribe' and of shrinking aghast when walking if her foot touched a stick or she heard 'rustling in the dead leaves'. See Meredith, *My Home in Tasmania*, 174. Anxieties about snakes are given powerful literary expression in Bram Stoker's no-

vella 'The Lair of the White Worm', in which the malevolent character Lady Arabella March is a serpentine creature able to assume human form. At one point, the possibility that Lady Arabella is a snake is directly voiced but then refused. See Stoker, 'The Lair of the White Worm', 204. In Stoker, snakes seem psychologically linked with anxieties about aggressive femininity and Lady Arabella can be interpreted as a phallic woman.

195. John Franklin was Governor of Tasmania from 1837 to 1843.
196. In the Calder archive the earliest letter mentioning bounty payments dates from 14 February, in which the police from Launceston detail bounty expenditures. Another letter from the electoral district of Morven dated 28 February 1828 mentions a payment of £2 and 14 shillings for fifty-four snakes. There are several other documents including lists. The last letter to refer to payments dates from 3 August 1838. These documents are pasted into a scrapbook created by Calder located in Box 4 of his papers. James Erskine Calder, Papers, 1797–1882. MS F BOX 89/4. State Library of Victoria.
197. Robinson, 'The Flinders Island Journal of George Augustus Robinson', 530. The palawa were exiled to Flinders Island beginning in 1830 and ending in 1847.
198. Ibid., 538.
199. See Organ, *The Illawarra Diary of Lady Jane Franklin*, 44n25. See also George Augustus Robinson's journal entry for 19 November 1838. Robinson, 'The Flinders Island Journal of George Augustus Robinson', 604.
200. Longman observes that in the first year the bounty was offered, over 1,900 snake heads were tendered, implying the scheme lasted longer. Heber Longman, 'Nature's Ways: Snake Bounty Might Rid State of the Killers', *Courier-Mail* (Brisbane), 9 December 1950, 2.
201. Ibid.
202. Ibid. The letter is numbered 417 in the scrapbook.
203. Here Franklin seems to accept that the scheme was limited in scope (probably because large tracts of Lutruwita were visited rarely, so many snakes were beyond the reach of bounty hunters) but the idea of 'total extermination' does not perturb her. It is simply an ideal that is not currently achievable.
204. Boyer suggests that some of the opprobrium directed at Franklin regarding her efforts to eradicate snakes was motivated by sexism. Her campaign was, Boyer implies, viewed as an unladylike pursuit. Boyer, 'Leaders and Helpers', 62.
205. Moses, 'Genocide and Settler Society in Australia', 158.
206. Ibid., 159.
207. Ibid.
208. Ibid., 167.
209. Wolfe, 'Settler Colonialism and the Elimination of the Native', 393.
210. Ibid.
211. Barta, 'Relations of Genocide', 238.
212. Ibid., 247.
213. The bounties offered on wedge-tailed eagles in Western Australia, by contrast, would comprise state-sponsored genocide. For a discussion of these schemes, see Knobel, 'The Conservation Status of the Wedge-Tailed Eagle', 294–95.
214. Haygarth, 'The Myth of the Dedicated Thylacine Hunter', 29.
215. Rothschild, 'On Extinct and Vanishing Birds', 200.
216. Ibid., 191.
217. Galbreath and Brown, 'The Tale of the Lighthouse Keeper's Cat', 197.
218. Eckert, *The Great Auk*, 161.

219. Ibid., 161–62.
220. This is practised by the Hidatsa and Dakota peoples among others. The Hidatsa wear eagle feathers to signify courage, skill and success in war. For a discussion of contemporary use of eagle feathers in Native American Indian cultures of North Dakota, see Murray, 'Feathers, Fasting and the Eagle Complex', 148–49.
221. See the entry for 'Woodcock' in Maddams, *Game*.
222. Rothschild, 'On Extinct and Vanishing Birds', 191.
223. Townsend, 'An Account of Recent Captures of the California Sea Elephant', 93.
224. Townsend, *Pelagic Sealing*, 267; Townsend, 'The Northern Elephant Seal', 161.
225. Townsend, 'The Northern Elephant Seal', 171.
226. Townsend, *The USS 'Albatross' in Lower Californian Seas*, 405.
227. Andrei, *Nature's Mirror*, 201. Andrei makes a more convincing case for Hornaday's conservation credentials.
228. Townsend, 'The Northern Elephant Seal', 165.
229. McGregor, 'The Doomed Race', 14.
230. As McGregor notes, the idea that various 'races' were dwindling away was usually not based on empirical evidence such as population numbers. See McGregor, 'The Doomed Race'.
231. Woodruff, 'Declines of Biomes and Biotas and the Future of Evolution', 5474.
232. Janzen, 'Latent Extinction', 590.
233. Janzen, 'The Future of Tropical Ecology', 308.
234. Ibid., 307.
235. In 'Latent Extinction', Janzen offers the example of the giant anteater (*Myrmecophaga tridacytyla*) in Costa Rica. In Costa Rica, the animal is not officially classified as extinct but there are currently no protected areas of sufficient size to maintain a genetically viable population for long-term species fitness. If there are still giant anteaters in Costa Rica, they are living dead. See Vaughan, 'Change in Dense Forest Habitat', 112–13.
236. The cut-off below which the time span of survival is not sufficient to justify conservation efforts could potentially be measured in tens, hundreds or thousands of years.
237. Wayt Gibbs, 'On the Termination of Species', 49.
238. For an analysis of site- and taxon-oriented prioritisation in conservation, see Rondinini et al., 'Reconciling Global Mammal Prioritization Schemes into a Strategy'.
239. At the time of writing, the project was in the final stages of the permit/approval process. The California Coastal Commission had approved the eradication in principle but projected costs had increased by 43 per cent in the period 2019–23. See US Department of the Interior, 'Budget Justifications and Performance Information: Fiscal Year 2025', https://www.fws.gov/sites/default/files/documents/2024-03/fy2025-508-fws-greenbook.pdf (retrieved 28 November 2024).
240. For the most up-to-date discussion of mouse attacks on seabirds, see Connan et al., 'First Evidence of Mouse Predation'.
241. For a discussion of some of these efforts, see Chapter 11 of Stolzenburg's *Rat Island*.
242. The killing of the group of cattle known as the 'scrubbers' in Frank Dalby Davison's novel *Man-Shy* would seem to be genocide as the bovines are constructed as a distinct group who are psychologically qualitatively different from domesticated kine.
243. Colgan, 'Thylacine Cloning is a Long Shot', 15. Colgan was writing in 2002 and may have had in mind the June 1998 incident in which a fox scarpered from a container ship in Burnie. The fox was sighted repeatedly in the ensuing months. See Kirsty Eade, 'Fox Seen Near Burnie', *Examiner*, 29 September 1998, http://www.examiner.com.au/story/641145/fox-seen-near-burnie/ (retrieved 25 June 2021).

244. See the summary of John Calhoun's experiments with a population of Norwegian rats in Lawson, 'The Social and Psychological Issues of High-Density City Space', 285. Calhoun discovered that in artificially created dense populations, behavioural repertories broke down and aberrant behaviours arose. Calhoun, 'Population Density and Human Pathology', 148.
245. Field research might be able to establish the presence or absence of group distinctiveness in an animal population.
246. Bailey, *Introduction of Foxes to Alaskan Islands*, 41.
247. In a survey conducted in May–June 2009, 328 bird carcasses were observed including forty-three bald eagles and 251 glaucous-winged gulls. Most of the eagles and some of the gulls probably died from Brodifacoum exposure having consumed poisoned rats. For further information, see the report of the investigation into 'Rat Island Mortality' conducted by the Office of Law Enforcement of the US Fish and Wildlife Service, which was published in 2009 (Case Number 2009703127). The name of the author of the report is redacted. A rat extermination campaign on Palmyra atoll, one of the Northern Line Islands in the Pacific, also caused significant 'non-target' casualties or collateral damage. See Pitt et al., 'Non-Target Species Mortality'.
248. For a discussion of the positive impact of the eradication of rats on Hawadax island, see Croll et al., 'Passive Recovery of an Island Bird Community'.
249. In *Rat Island*, Stolzenburg suggests that the extermination of the cats caused 'ecological whiplash' (207). Keith Springer is sceptical of this interpretation of the effects of eradication. See Springer, 'Eradication of Invasive Species', 15.
250. For a discussion of the killing of the Miller Lake lamprey, see Miller et al., 'Extinctions of North American Fishes', 23.
251. Thresher, 'When Extinction is Warranted', 134.
252. See Harvey-Samuel et al., 'Towards the Genetic Control of Invasive Species', 1696. Harvey-Samuel et al. provide a useful overview of different genetic approaches to controlling invasive and pest species.
253. Wilson, *Half-Earth*, 53.
254. Ibid.
255. Ibid.
256. Steve Paulson, 'Why a Famous Biologist Wants to Eradicate Killer Mosquitos', *The World*, 4 April 2016, https://theworld.org/stories/2016/04/04/why-famous-biologist-wants-eradicate-killer-mosquitoes (retrieved 4 May 2021).
257. Roughly 380,000 people in sub-Saharan Africa died of malaria in 2018. See Lindsay et al., 'Recommendations for Building Out Mosquito-Transmitted Diseases'.
258. For a discussion of the ethics of mosquito eradication using transgenes, see Pugh, 'Driven to Extinction?'
259. Weinman, 'The Human Trypanosomiases', 133.
260. Tsetse flies feed on various mammals (and birds and reptiles) but with differences in tendency across species of the fly and with seeming regional preferences (perhaps based on availability). Indiscriminate killing of game animals to achieve starvation-dependent mortality of the fly therefore risked causing the destruction of non-likely host species such as, for instance, impala (*Aepyceros melampus*) and zebra (*Equus quagga*). It also potentially drove the flies to seek out humans for sustenance in the absence of alternatives. For a study that informed Weinman's thinking, see Weitz, 'Feeding Habits of Bloodsucking Anthropods', 74–77. For a more recent study, see Clausen et al., 'Host Preferences of Tsetse'. Efforts to control the tsetse fly in the early twentieth century often ignored Indigenous knowledge that would have helped considerably refine the methods adopted. See Mavhunga, 'Big Game Hunters'.

261. Translation amended. Jansson's use of the word *förfärlig* potentially implies a moral dimension to Moomintroll's response to the massacre; it is reprehensible to him. Jansson, *Moominpappa at Sea*, 96 (Jansson, *Pappan och havet*, 99).
262. Jansson, *Moominpappa at Sea*, 97.
263. The Swedish term *myggor* that Kingsley Hart translates as 'mosquitos' can also refer to 'midges' (midges are smaller blood-sucking flies that swarm [unlike the solitary mosquito], making the analogy with ants work better).
264. For a summary of how analogy functions as a process, see Cheetham, 'Analogous Histories?'
265. For a good short summary of the behaviour and diversity of ants, see Ward, 'Ants'.
266. Wilson, 'Ants', 13.
267. Jansson, *Moominpappa at Sea*, 125.
268. Barrington discusses various species of birds killed by Irish lighthouses and lightships, speculating on the role played by migration in collisions and on why some species of bird seem more prone to strike the lanterns and others less. Barrington affirms that fixed lights cause more deaths than revolving ones (*The Migration of Birds*, 265). His research findings derived from over two thousand birds killed by lanterns and then sent to him by lighthouse keepers in the period 1881 to 1887. Munro conducted a study of bird fatalities caused by lighthouses in British Columbia in the early 1920s. His findings were published in 1924 in two instalments in *The Canadian Field-Naturalist*. Montevecchi discusses the potential impact of lighthouses and other forms of artificial lighting on endangered species such as Barau's petrel (*Pterodroma baraui*), the Mascarene petrel (*Psuedobulweria aterrima*) and Newell's shearwater (*Puffinus newelli*) in 'Influences of Artificial Light on Marine Birds'.
269. The boneyard is not a pet cemetery as the dead that are interred there are wild birds.
270. Janet Fang has argued that outside of an Arctic context, the eradication of mosquito populations may have minimal impact on ecosystems: '[u]ltimately, there seem to be few things that mosquitos do that other organisms can't do as well.' Fang, 'A World Without Mosquitos', 434.
271. For a historical overview of this eradication, see Gutierrez et al., 'Deconstructing the Eradication of New World Screwworm'.
272. Wilson, *Half-Earth*, 52.
273. There are always exceptions to these hierarchies. In the United Kingdom, red squirrels (*Sciurus vulgaris*), despite being rodents, are viewed positively. The use of a fictional red squirrel (Tufty Fluffytail) to front a road safety campaign aimed at children from the 1950s to the 1980s may have contributed to this. Butterflies and bees are also often well perceived.
274. Although mosquitoes are major spreaders of diseases such as malaria, their capacity as virus carriers is sometimes exaggerated, as is evinced by the myth that AIDS could be transmitted via mosquito bites. See Cline and Johnson, 'Mosquitoes, Doorknobs and Sneezing', 275. Mosquitoes are potentially not feared enough in some countries and the dangers they pose underestimated. See Widmar et al., 'Public Perceptions of Threats from Mosquitoes in the U.S.'.
275. For an analysis of the cabbage white butterfly as an agricultural pest (particularly in relation to cruciferous crops [members of the cabbage family]), see Ryan et al., 'Global Invasion History'. Daniels, Magdich and Tolson describe butterflies and pollinators as 'particularly charismatic groups'. Unfortunately, they do not reflect on what generates and sustains this perceived charisma. See Daniels et al., 'Butterfly Recovery Planning', 2. Lorimer attributes butterfly charisma to the importance of the visual in their differentiation and orientation (917). To me, aesthetic charisma (one

of three categories of charisma proposed by Lorimer) would also seem important. Lorimer, 'Nonhuman Charisma'.

276. Holt, 'Charismatic Mesofauna', 1. Holt also later notes how popular butterflies are (3).
277. Lemkin, *Lemkin on Genocide*, 27.
278. Tocque, *Wandering Thoughts*, 112.
279. Lemkin, *Lemkin on Genocide*, 26–27.
280. Wainwright was the son of George Wainwright, the 'tiger man' employed by the Van Diemen's Land Company at Woolnorth to kill thylacines. Bailey, *Lure of the Thylacine*, 240.
281. Anonymous, 'Parliament: House of Assembly Thursday, October 7', *Launceston Examiner*, Friday 8 October 1886, 3.
282. Staub, *The Roots of Evil*, 19.
283. Ibid., 51.
284. Ibid., 52.
285. Freeman, *Paper Tiger*, 47.
286. Staub, 'The Psychology of Bystanders'.
287. Potts, 'The Kea', 489.
288. Ibid.
289. Ibid.
290. Given the reference to beef, Potts may be thinking of Ethiopian cuisine. Dishes such as *gored gored* or *tere siga* use raw meat. Ethiopian cooking is negatively portrayed in William Barker's *Narrative of a Journey to Shoa* where he observes that Ethiopia 'is a nation who generally live on raw meat, and it cannot be supposed that they have made great advancement in their cuisine' (25). Barker is clearly unfamiliar with the French fondness for steak haché. *Narrative of a Journey to Shoa* is sometimes misattributed to James Barker. That William Barker was a member of the party sent by William Cornwallis Harris to the royal court of Shoa (Shewa) to meet the Negus in 1841 is confirmed in Burton, *First Footsteps in East Africa*, 221.
291. Steak tartare became popular in France in the early twentieth century, so after Potts was writing. See Piatti-Farnell, *Beef*, 62.
292. Joinville, *Histoire de Saint Louis*, 268.
293. Potts, 'The Kea', 489.
294. Potts carefully constructs an unbalanced image of the bird. The kea is rendered as beautiful and ugly, as bold and daring yet also thieving and savage, refusing ready classification. This cultivated ambiguity that renders the bird a source of disorder can also be read as enabling violence against the kea. Ambivalence is a source of discomfort and threat. See Bauman, *Modernity and Ambivalence*.
295. Slander and calumny have been identified as argumentative strategies employed by Little My to undercut the validity of another's point of view. See Mikkonen, '"Believe You Me"', 99–100.
296. Waller, *Becoming Evil*, 202.
297. Marks discusses the use of euphemism in the context of the killing of Australian feral cats in 'Killing Schrödinger's Feral Cat', noting that sodium fluoroacetate (colloquially known as '1080') is referred to as a 'tool' rather than a poison by some bureaucrats (61). Marks's essay resists detachment, acknowledging the ethical difficulties he, as a scientist working on vertebrate pest research, feels when administering poison to feral cats.
298. Waller, *Becoming Evil*, 211.

299. The term 'bounty' has its etymological roots in the Latin *bonitātem* meaning goodness or virtue.
300. Horsley, 'Boy Life in Australia', 394.
301. Bryant, 'The Pouched Mammals', 2009–10.
302. Ibid., 2014.
303. Ogilby, 'Notice of Certain Australian Quadrupeds', 122.
304. See Huston, 'Pieces of the Sky', 192; Pimm, Moulton and Justice, 'Bird Extinction in the Central Pacific', 84.
305. Jeffries, *Biodiversity and Conservation*, 129.
306. Ibid., 132.
307. Pimm, 'The Dodo Went Extinct (And Other Ecological Myths)', 196. Pimm would later revise this to 'It doesn't say "The Dutch bludgeoned the poor bloody thing to oblivion"'. See Pimm, 'It's a New Century', 7. The change is important as it avoids attributing universal guilt to humankind for the fate of the dodo and instead directs blame squarely at Dutch colonists. Pimm provides further discussion of the dodo and reflects on the idea of 'wimp species' in '"Putting Nature Back Together Again"', 103–4.
308. Pimm, 'It's a New Century', 7. *The Oxford English Dictionary* (1971) gives the definition 'An extinct bird, *Didus ineptus*, belonging to the order *Columbidae*, formerly inhabiting the island of Mauritius; it had a massive clumsy body and small wings of no use to flight.' *The Chambers Dictionary* (1998) gives the definition 'A clumsy flightless bird about the size of a turkey [native of Mauritius] which became extinct about the end of the 17c.'
309. Boroditsky also traces how different conventions across languages influence the ways particular events are perceived and remembered. Boroditsky, 'How Language Shapes Thought', 64.
310. Staub, *The Roots of Evil*, 17. Scapegoating here differs from understandings such as that of René Girard in which violence against scapegoats is ritualistic and imitates the foundational violence upon which a given community is founded. See Girard, *Violence and the Sacred*, 93–103.
311. Kuper, *Genocide*, 86.
312. Waller, *Becoming Evil*, 197.
313. Staub, *The Roots of Evil*, 19.
314. Ibid., 48.
315. Ibid., 49.
316. Ibid.
317. Kuper, *Genocide*, 43.
318. Ibid.
319. Dorr, 'The Scapegoat', 92.
320. Fletcher, *Mendel's Ark*, 74. In her article 'Bring 'Em Back Alive', Amy Lynn Fletcher also refers to the Tasmanian tiger as 'a scapegoat for the mass slaughter of sheep' (195).
321. E. Richall Richardson, 'Tour through Tasmania Letter No. 89: Cattle Branding', *Tribune*, Thursday 14 March 1878, 3.
322. Kennedy, *Sport, Travel and Adventure in Newfoundland and the West Indies*, 121.
323. Ibid.
324. Bonnycastle, *Newfoundland in 1842*, 24.
325. Ibid.
326. Staub, *The Roots of Evil*, 41.

327. Ibid., 72.
328. Ibid., 79.
329. Ibid., 59.
330. I have chosen to capitalise Indigenous here as the use of lowercase implies a lesser status relating to nonhuman animals based on their lack of 'peoplehood'. The lowercase also obscures the close spiritual relationships between First Nations Australians and native fauna. 'First wildlife' as a term helps to affirm this relationship.
331. See the survey (which excludes ACT and Tasmania) conducted by Wicks et al., *An Integrated Assessment of the Impact of Wild Dogs in Australia*. Wicks et al. identify tapeworm (*Echinococcus granulosus*) as one such disease. For a good summary of the range of impacts (economic, environmental, social) wild dogs can have, see Howard et al., 'Wild Dog Management in Australia', 244.
332. Dichlorodiphenyltrichloroethane (DDT) is an insecticide the use of which is now banned in most of the world because of its negative environmental impact and toxicity to humans.
333. See Moesnner, 'Dog – Man's Best Friend', 207.
334. Baer, *The Genocidal Gaze*, 3.
335. Ibid., 6.
336. Ibid.
337. Ibid.
338. Ibid., 6–7.
339. Pollock, 'Thinking the Feminine', 16.
340. The 'camera's gaze' is referenced in Pollock, 'Aesthetic Wit(h)nessing in the Era of Trauma', 850.
341. Pollock, *Art in the Time-Space of Memory and Migration*, 14.
342. Bath, 'Identification and Documentation of Public Attitudes toward Wolf Reintroduction', 2-19.
343. Ibid., 2-18.
344. Ibid., 2-18–2-19.
345. This film is discussed in more detail in Chapter 5.
346. Yellowstone National Park et al., *Wolves for Yellowstone?*
347. This is how Lett describes the grey wolf, which lived in 'all the remote and sparsely settled portions of Canada, Newfoundland and Cape Breton' (455). Lett, 'The Wolf'.
348. Blandy and Cowan, 'Imagine! Yellowstone', 45.
349. Baer, *The Genocidal Gaze*, 7.
350. Ibid., 8.
351. Freccero, '"A Race of Wolves"', 122. I am, however, wary of engaging in the kind of binary logic Freccero briefly embraces, in which the pack is privileged over the individual. Research into wolf packs shows that they are a complex formation of distinct voices; they act as both many and one.
352. For a discussion of becoming versus fixity, see Davies, 'Difference and Differentiation', 20.
353. Braidotti's notion of the subject as a 'singular configuration' that is 'intensive and dynamic' is useful here. See Braidotti, 'Posthuman, All Too Human', 201.

Chapter 2

Material Witnesses

Thylacine Remains and Animal Histories

Introduction: Data of the Dead

The International Thylacine Specimen Database (ITSD), which is being continually revised and updated, is a remarkable initiative that catalogues almost all known physical remains of the thylacine that are held in museums and private collections. These include fluid-preserved remains or 'wet specimens' (whole animals or body parts suspended in alcohol or formalin), skins, osteological remains (skulls and skeletons or parts thereof) and taxidermy mounts. The majority of the remains, 64 per cent, are osteological.[1] The materials are mainly located in Europe (just over 50 per cent), Australasia (just under 40 per cent) and Turtle Island/North America (about 9 per cent) but 'specimens' in India and Japan are also recorded.[2] The ITSD includes selective information such as the sex and provenance of each of the remains. There are also photographs, usually including scale bars, of most of the materials. These were incorporated 'to complement the data component' and 'to significantly enhance its educational worth'.[3] The photographs are perceived to grant virtual access to the materials, to prevent the need for handling (hence helping preserve materials), to aid with security and to keep a record of the current state of preservation of the materials.[4]

The database is clearly aimed at an academic, primarily scientific, research audience as it is not open access and can only be consulted from a limited number of university libraries and museums. In their text accompanying the fifth revision of the database, Stephen Sleightholme and Nicholas Ayliffe clearly recognise that in addition to their scientific value, the material remains are historical documents. They are not simply data resources to be mined for anatomical and physiological insights. As Sleightholme and Ayliffe note: 'These physical remains trace the history of the thylacine's untimely demise and teach us the price of complacency and failure to act.'[5] In this reading, thylacine remains constitute both a history lesson and an ethical imperative.

As the name of the ITSD makes clear, the enduring remains are nonetheless regarded foremost as 'specimens'. A 'specimen' is an example of a species that is regarded as typical or representative of that species. Writing in the context of anthropology, Hannah Turner explains that artefacts become specimens through a process of documentation.[6] An artefact without data (about its acquisition, identity and physical characteristics, for example) cannot become a specimen.[7] In such an understanding, a thylacine remain is not innately a 'specimen' but must submit to encoding procedures to become one. The ITSD both records physical remains that are already accepted as 'specimens' and, to a lesser extent, engages in practices of encoding that assign 'specimen' status (this is particularly the case for the few remains held in private collections). The term 'specimen' is frequently used in scientific contexts to denote material for study or testing.

The 'specimens' collated in the ITSD are viewed as embodying data about the thylacine, although, as Caitlin Donahue Wylie points out in a discussion of specimens more broadly, they should really be described only as '*potential* data' until they are deployed to argue for a knowledge claim.[8] Some of the physical remains recorded in the ITSD have probably never been utilised in this way because of their poor condition and lack of provenance (i.e. no information on the place and date of collection). Thylacine remains assume the status of 'specimens' once installed or put to work in a scientific context. This is something Sleightholme and Campbell recognise, describing how thylacines were bought from farmers and trappers by local agents who then resold the animals 'to museums and universities as specimens'.[9] The agents bought *native tigers* from farmers and trappers who viewed them as *vermin* and then sold them to scientific institutions who perceived them differently as objects of knowledge, as *specimens*. Here the lack of fixity of the nature of the remains is foregrounded. Just as remains can shift from being those of vermin to being those of a specimen, so a specimen can potentially cease to be seen solely as such, can also become something else. In the light of this, in this chapter, alternative ways of looking at the physical remains will be considered. My methodological approach is narrative-based, seeking to coax the remains into sharing aspects of their stories.

The status of the materials recorded by the ITSD is bound up with the database itself and how it is structured. Sleightholme and Campbell contribute to the construction of the remains as 'specimens' through naming them so and through the kinds of information they gather about them and how they organise that information in Excel spreadsheets.[10] Spreadsheets are a mode of organising data in cells that appear in rows and columns. They involve pre-programmed ways of presenting information that reflect the preconceptions of their authors. The spreadsheets and the database of which they are a part, like any such enterprise, are a repository of assumed values and senses as much as they are a source of information. Geoffrey Swinney's analysis of the role that museum registers play in constructing the artefacts they catalogue is pertinent here. Swinney notes that documentary practices, of which a database would be an example, 'are part of a technology through which collections are made and delineated'.[11] Thylacine remains, in this understanding, emerge as 'specimens' through a complex and ongoing process of classification and interpretation. The act of transforming physical remains into 'specimens' is one of objectification, reducing the relics to the status of things.

In *Negative Dialectics*, Theodor W. Adorno writes of loss of life in the Nazi concentration camps that 'it was no longer an individual who died, but a specimen [*Exemplar*]'.[12] Humans reduced to specimens are not singular but infinitely substitutable: 'fungible and replaceable'.[13] Viewing physical remains as 'specimens', a way of seeing that can be cultivated, renders them interchangeable and enables formal detachment (as research on human dissection in medical contexts, for example, has shown).[14] Detachment suppresses emotional responsiveness and moral reflection.[15] It enables distancing to take place. Sue Black observes of human dissection in a teaching context that the transformation of an individual into an anonymous number, an abstraction, negates the possibility of 'crippling empathy' emerging.[16] As discussed in the preceding chapter, psychological distancing of this kind can also facilitate violence.

Here, I will go on to explore how distancing has led to a failure to engage adequately with traces of the aftermath of violence, with legacies of animal genocide, as they are embodied in the thylacine remains. As a meaning suppressant, distancing has caused semantic violence. Framing the thylacine remains as 'specimens' encourages specific modes of perceiving and interpreting them and discourages others. Through the information architecture of the ITSD, with its investment in particular kinds of knowledge, the remains are designated as potential sources of data about, for instance, morphology and behaviour.[17] The theory of knowledge and, by extension, of the thylacine that guides the ITSD also significantly informs many museum collections holding remains of the animal. These institutions understand themselves as repositories of facts.[18] Their mammalogical holdings form data storehouses for aiding zoological research, for contributing to practices of classification and investigations into physiology. As this chapter demonstrates, natural history museums are admirably open to having their collections read and used in ways that do not conform to common museum-based scientific research (such as the study of morphology or genome structure), but their methods of cataloguing and storing and the kinds of equipment they provide nevertheless continue to encourage particular ways of receiving and understanding their collections and support the production of certain kinds of knowledge (such as bioinformatics or insights into evolutionary processes). The 'truth' of an animal such as the tiger is seen to be their skull morphology or their DNA. The thylacine is thus read through and reduced to limiting discursive frames. Mark Halsey therefore rightly suggests that biological science should always be subject to critique rather than 'blind acceptance' because it is 'constituted through all manner of rhetorical manoeuvres and assumptions about the limits of knowledge and the operation of power'.[19]

Material Remains

In this chapter, thylacine remains are not seen solely as 'specimens' but also viewed as forms of material testimony. They are still sources of information, but information regarding the history of attitudes of settler colonists towards the thylacine. The remains therefore contribute to the production of historical narratives in addition to scientific ones. A conscious effort is made not to refer to the remains as 'specimens', as representations of an animal type. Eschewing the term 'specimen' potentially contributes to a process of reducing detachment, encouraging greater recognition of human entanglement with these nonhuman animals and their fate. The remains that are examined derive from collections in Australia and the United States, namely from the Tasmanian Museum and Art Gallery (TMAG) in Nipaluna/Hobart, the Queen Victoria Museum and Art Gallery (QVMAG) in Launceston, the Melbourne Museum and the Tiegs Zoology Museum, both in Melbourne, the Australian Museum and the Chau Chak Wing Museum in Gadigal/Sydney, and the Smithsonian National Museum of Natural History in Washington, DC.[20]

The majority of the remains in these museums derive from thylacines that were either trapped and killed or were caught and sold to zoos before dying in captivity. They consist predominantly of osteological materials, of skins and mounts, and of wet specimens (pups preserved in alcohol). Traditionally, skulls and skins were the most valued remains as these were key to species identification and description. They predominate in the collections that were studied. Space is also always at a premium in natural history museums and has influenced what remains have been preserved and how they have been stored. At one museum, a staffer spoke of playing 'Tiger Tetris' when trying to rehouse several mounts on storage shelves. With issues of space in mind, Charles Darwin advocated keeping skins rather than mounts for most 'specimens'.[21] At the Smithsonian, thylacine skins are carefully stored folded in metal drawers. The creases caused by this index the need to maximise the use of space and embody how economic concerns intersect with scientific research. At TMAG, thylacine skins used to be housed folded in discarded ammunition boxes.[22] They are now stored flat on shelves.

Many of the remains discussed in this chapter are rarely exhibited, if ever, by the museums where they are preserved.[23] Some, such as the study skins, were prepared exclusively for scientific research and not for exhibition. Study skins are quick to prepare compared to a mount (a skin mounted over an internal support and presented in a lifelike pose) and have long been appreciated for their space-saving attributes. They are also viewed as less fragile than mounts.[24] Joseph Grinnell distinguished between the exhibition museum (open to the public) and the research museum (the preserve of the specialist) as spaces that coexist in the same institution.[25] The study skin is intended for the latter museum. They are conceived to act as sources of potential data and do not possess the complexity of a mount that has been created to serve as informative spectacle.

Study skins were originally filled with materials such as cotton, newspaper, oakum and tow.[26] Those I handled at the Smithsonian seemed to be filled with wood wool (excelsior). The study thylacines form depressed creatures, flat and lifeless. Strange to handle, they crinkle and rumple, crumple, their straw-filled bodies indenting even when gently touched. These sounds betray the illusionary solidity of the animals. Their seeming fleshliness is undone by telltale acoustic indications of hollowness and by the excessive give of the skin, which discloses a supple inner architecture. Study thylacines are skins without bones. They are frequently roughly stitched as aesthetics were not a primary concern when they were prepared. The acts of suturing suggest those made by a surgeon closing a wound except, unlike in many medical cases, the stitches are not interrupted (separately tied). The stitched mouths of these thylacines remind me of asylum seekers who have sewn their lips together. For asylum seekers, lip-sewing can be a form of protest and an expression of political agency.[27] Depriving oneself of the capacity of speech in such a way speaks louder than words. The tigers, however, do not appear to possess a comparable agency. In their case the stitched-up mouths seem to figure a lack of voice. They are consigned to invisibility and silence.[28]

Other remains are withheld from exhibition because as representations of the thylacine, as 'specimens' of it, they are found somehow lacking, their condition less than ideal. In the case of the thylacine because of the paucity of specimens, some that would otherwise likely be relegated to a drawer or disposed of altogether are put on public display.[29] The Tiegs Museum, for instance, displays a broken thylacine skull. Usually remains damaged to such an extent would not be on show. 'Below stairs' thylacine remains (they are often located on lower levels or in basements) of this kind, forever unseen by the museum-going public, are nonetheless retained for research purposes, occasionally employed in the service of scientific inquiry. There are, for example, a number of damaged skulls that were retrieved from 'retired' mounts prior to their destruction. These are retained in case they have scientific value but they are not exhibited.

Relic

While I studied thylacine remains at TMAG and QVMAG, scientific research was being carried out in proximity to me. At TMAG eagles were being necropsied on an adjacent workbench and at QVMAG some reptiles were being dissected down the corridor. I overheard one staff member at QVMAG asking a colleague: 'Are you into reptile guts?', the visceral dimension to the research clearly registering and impressing itself on them. As a non-scientist, at such moments, I could feel my outsider status and a tangible expectation that I should be repulsed by the messy, fleshy reality of this research into recently living things.[30] Such instances of scientific investigation brought home the vitality of the collections. Much of the research being conducted concerns the still extant (working with the recently deceased rather than the extinct) and relates to contemporary concerns rather than historical events. It reminded me that animal death (recent or ancient) is integrated into the everyday of natural history collections, routinised and rendered mundane.

The detachment towards animals that forms part of this everyday shares some similarities with the violence-enabling disinterestedness discussed in the preceding chapter but is also qualitatively different. The man necropsying eagles and taking samples, for example, clearly cared deeply about the raptors as a species even as he was unmoved by individuals. Each bird was a 'specimen', a set of internal organs, a liver to be tested, to be studied for signs of toxicity.[31] The overall research clearly aimed at the conservation of the eagle population and the 'specimens' were collected, already dead, by volunteers rather than deliberately killed. Lead poisoning is a recognised issue for Lutruwita's raptor population and even non-acute poisoning may have detrimental effects on health, including reproductive capacity.[32] Reducing the ingestion of lead through prohibiting lead-based ammunition might aid the raptor population. Data derived from the necropsies would potentially help build a case for such a ban. For this researcher acting as a pathologist, detachment therefore intersected with care. Compassion was a central component of the scientist's endeavours. Studying death in a detached manner to try and prevent it is very different from killing in a detached manner with the aim of extermination. Both forms of placing at a remove nonetheless involve the denial of singularity.

The disavowal of singularity (of each animal's difference from another of their kind) occurs regularly in natural history collections.[33] In most collections, each 'specimen' is individually catalogued, given a unique alphanumeric identifier rather than referred to by name. Most 'specimens' will only possess this identifier because personal names (rather than zoonyms and binomials) are reserved for nonhuman animals that are kept as pets.[34] Some animals appear to individuate within their species, to know and differentiate one another, but in museum collections within a given species the singular cedes to the general. Numbering is an administrative imperative that makes vast amounts of 'specimens' easily identifiable and manageable. The number is often an accession number, part of an ever-ascending series arranged according to the order of arrival of each 'specimen'. Every number designates a particular entity but, as a number, also operates to transform the organism into a 'thing', an object without a specific name. Classified by a few reference letters and digits, the organism is consigned to an abstract anonymity. Numbered identifiers have been linked with dehumanisation in authoritarian settings such as the military and the prison system.[35] For nonhuman animals, numbering doubles down their objectification.[36] Some 'specimens' may accrue a local celebrity, be granted nicknames (often not shared with the public) by museum staff, but most remain anonymous ciphers.

'Specimens', each carefully numbered and catalogued, are cautiously stored, usually in cabinets and drawers. The reference letters and numbers become each thylacine's posthumous identity and, out of necessity, are retained here.[37] One of the most well-known of thylacine remains, an adult female preserved in ethanol that has been held by the Swedish Museum of Natural History in Stockholm since the 1970s, is referred to as NRM 566599. The identifier offers no suggestion as to how remarkable these remains are. This thylacine – which was sourced from London by Stockholm University, and was exhibited at London Zoo between 1884 and 1893, having been purchased by the Zoological Society from Arthur Bingham Crowther of Launceston – is the best-preserved mature animal in existence.[38] Yet as an identifier, NRM 566599 is no more revealing than

NRM 533079, the catalogue identifier for a wet specimen of a juvenile common opossum (*Didelphis marsupialis*) in the same collection.

In life, NRM 566599 was estimated to have weighed approximately 14.4 kg. She has been described by Douglass Rovinsky as a 'very emaciated' and a 'very beautiful, but very sad specimen'.[39] Rovinsky was part of a team of researchers who assessed the body mass of this thylacine and of others and, based on their results, postulated an average weight of 13.7 kg for females (and 19.7 kg for males).[40] This average is lower than historical calculations and indicates that the thylacine as a species was probably not heavy enough to predate sheep, preferring smaller prey.[41] The scientific paper in which the findings were initially published does not reflect on the aesthetic qualities of the physical remains or the emotional responses they generate but Rovinsky's musings on social media give insight into how scientists sometimes look past the 'specimen' as simply data, negotiating various feelings and thoughts, not all of them straightforwardly 'scientific'.

Catalogue numbers are either inscribed on labels attached to physical remains or written directly onto the remains themselves. The labels can become indirect catalogues of different staff members. At QVMAG, for example, a label for some mandibles that is in the hand of the curator and first director of the museum, Herbert Hedley Scott – 'Two mandibular elements of Native Tiger Thylacinus Gunn's Estate / Feb 1921 no crania came with these HHS' – is supplemented by a more recent effort: 'From R.C. Gunn's estate, acquired Feb 1921. See collection guide by T. Gordon 2006.' The differing hands of the two labels demonstrate how artefacts can become places through which museum employees inscribe themselves into the histories of specific objects. The numbers on skulls and bones are usually written in black India ink. Sometimes more than one number will be written if a cataloguing system has changed or if the remains have been transferred from one collection to another.[42] In databases and ledgers or card catalogues, catalogue numbers are usually recorded alongside information about when and where a given 'specimen' was collected and therefore contribute to establishing provenance for remains. Provenance is central to increasing the usefulness of zoological collections for studying biotic change.[43] It also contributes to the 'life history' of the collected material. Some physical remains become interesting through who collected them rather than for what they are in themselves.[44] This phenomenon is referred to as 'positive contamination'.[45]

The need to carefully preserve remains and to keep them in good order has long been recognised. Writing in 1926, Hartley Jackson noted that 'whether common or rare [each] should be treated as sacred'.[46] Jackson's reference to sacrality suggests he conceives of 'specimens' as akin to relics, materials that are themselves often corporeal (the skulls, bones, blood or hair of saints, for instance). Relics are usually associated with religion and subject to veneration. As Helen Hills has observed, as a sacred trace, relics possess aura.[47] In the context of devotional objects, aura refers to what Anthony Hughes has described as a 'nimbus of awe', a powerful psychological effect that enacts distancing between beholders and artefacts.[48] In this sense, Jackson's sacralising of 'specimens', like the term 'specimen' itself, also encourages a kind of detachment. The 'specimen' is, in some sense, other-worldly, a thing to be revered rather than simply regarded as of scientific significance. Sometimes the way 'specimens' are stored encourages their being viewed as objects of veneration. A humerus at QVMAG is now preserved in a padded Styrofoam container, the receptacle encouraging reading the forelimb as akin to a relic. This solitary bone, which is now accorded the catalogue number QVM 2016: GFV: 0001, formed part of a drawer of humeri of various species used for comparative purposes by Scott. The preservation of casts of pawprints is also in keeping with the idea of the mammal collection as a kind of reliquary; 'miraculous footprints' have a long history as relics.[49]

Relics, as Hills notes, are all 'potentially miraculous'.[50] Jackson does not mention miracles but does note that 'often a particular specimen is of unexpected value for some particular purpose'.[51] Although he is not discussing them specifically, his remark has subsequently been borne out in relation to thylacine remains. With the emergence and development of molecular biology, there has been growing interest in the information that DNA extracted from remains can provide about extinct species. In the context of the thylacine, this led to a notorious attempt to clone a thylacine, which was spearheaded by the Australian Museum. The project was launched in 1999 and abandoned in 2005, although there have subsequently been other efforts to source a viable genetic blueprint of the animal, including a later effort that made use of NRM 566599.[52] The cloning endeavour began by extracting DNA from a wet thylacine pup (P762) and two dry remains (M822 [a skin and skull of a male thylacine] and M27836 [a skull]).[53] It was the thylacine preserved in alcohol that possessed a tangible aura and became iconic of the project even though it ultimately did not provide the most useful DNA.[54] As Amy Lynn Fletcher notes, the animal in the jar offered an 'evocative image' that helped generate publicity for the project.[55] The enclosed thylacine pup 'vividly embodied the dream that propels any project that seeks to resurrect an extinct species: the hope that a dusty old jar in a museum basement will turn out to contain an elixir of life that will allow us to bring back all those species whose loss now haunts us'.[56] It was anticipated that the remains, which dated from 1866, might perform the 'miracle' of resurrection.[57] In this sense, the pup was treated very much like a Christian religious relic, offering 'resistance to the devouring power of death'.[58]

Viewing the museum remains as holding the promise of renewed life is a means of sacralising them, of granting them a redemptive potential. Reading them as offering an important historical lesson about the price of complacency, as Sleightholme and Campbell and others do, also risks according them a sacred character. The deaths of the animals are imbued with meaning, incarnating a moral lesson. Similarly, David Maynard and Tammy Gordon describe how information gleaned from the remains can be 'used in management and recovery plans designed to save some of today's threatened species'.[59] Focussing on positive outcomes generated by studying the remains, granting them useful functions, lends meaning to their destruction. It unintentionally gives a sense of purpose to the genocide perpetrated against the thylacine. Here the aim is to look beyond the sacred aura that the remains have accrued and the redemptive narratives that have emerged regarding them, foregrounding instead the often-calamitous histories of human–nonhuman animal relations that they index, the traumatic encounters they embody and the senseless loss they register. Matthew Chrulew has referred to an 'unextinguished legacy of violence' that haunts zoo exhibits.[60] A comparable legacy haunts the museum collections considered here.

Arrested Development

Looking beyond the physical remains as storehouses of data, Hannah Stark has examined how thylacine remains can form repositories of emotions. She focusses on one example in particular, the female thylacine pup catalogued as P762 that is held in the Australian Museum. The pup was about twelve weeks old when she was killed and was probably in the process of transitioning out of the pouch.[61] She is preserved in ethanol, a method of safeguarding that can be remarkably effective but has dangers. If oxygen enters a storage container it can cause ethanol to oxidise into acetic acid, damaging the wet remains that the alcohol is intended to preserve. Lipids including fats also dissolve in an alcohol such as ethanol. The compound therefore both protects and imperils the pup.

P762 is the pup that became iconic for the cloning project. Drawing on Judith Butler's *Precarious Life* and her exploration of the limits that are set on what losses can be avowed and what loss must remain unacknowledged and therefore ungrieved and ungrievable, Stark considers how the thylacine can be made to matter, to be mourned.[62] Working from photographs by Heinz Moeller that form part of the record of P762 in the ITSD, she draws attention to the poignancy and beauty of the images of the pup. In reality, P762 has an ugly wound the length of her belly where she was cut open to enable her innards to be studied. The injury, however, is only visible in one of the photographs Stark reproduces, and then only partially.

Stark finds Moeller's photographs arresting because the pup's 'remains are a poignant rem(a)inder of a species that was lost as a by-product of Western imperial expansion'.[63] She regrets the way the pup has become bound up with debates about de-extinction, with revivification, a reality that distracts from the loss embodied in this stilled life. Stark suggests that to 'mourn for P762 is to have an affective connection tied up with the uncanny rupture of an animal who lived in a different historical time'.[64] This act of mourning, she contends, can potentially invite recognition of human culpability in the thylacine's fate, fostering a sense of responsibility and also prompting action in relation to contemporary extinctions.[65] Here too then, the loss of the thylacine assumes a positive meaning, serving to reshape perceptions of the environment.

Stark's valuable insights into the affective potency of the pup can potentially be related to all thylacine remains. Her decision to focus on wet remains, specifically on P762, therefore merits reflection. Part of Butler's argument in *Precarious Life*, as Stark appreciates, is that grievability is afforded to some lives but not to others. Butler indicates that it would be possible to enumerate a 'hierarchy of grief'.[66] In this sense, why is one tiger privileged over another among the many recorded in the ITSD? Stark's choice was, she explains, influenced by the nature of the photographs, which possess a beauty and intimacy that is absent from most of the other images.[67] Some of Moeller's photographs still manifest scientific intent through the neutral backdrop and the inclusion of a measuring tape, but the presence of the pup grants them additional aesthetic and emotive dimensions. Stark observes: 'She is a baby creature; she could almost be a sleeping puppy.'[68] This life to be mourned is a life hardly lived, an infant's life.

One museum staffer at the Australian Museum said to me that it 'feels like [I'm on] a maternity ward' when they wheeled a trolley with a pup to the mammalogy study space. The thylacine pups therefore seem to possess a broad evocative power that transcends their 'specimen' status. Their lives cut short bespeak potential unfulfilled. Infancy is also associated with innocence, a pup such as P762 offering a symbolic shorthand for blamelessness. Physically immature, the creature cannot be confused with the savage, sheep-ravaging thylacine of myth. The gender of the animal is evidently also important to Stark. She contrasts the female pup with the male scientist who extracts DNA from her, interpreting the act of the masked man probing her interior cavities as 'invasive', a violation.[69] The 'babe-in-alcohol' in Stark's reading becomes powerfully emblematic of vulnerability and guiltlessness. The museum pup P762 was already celebrated through media uptake regarding the cloning project before Stark's reading, but the latter reaffirms her extraordinariness and her symbolic potency.

Encountering these wet thylacines in their fleshy physicality, their remarkable preservation, is certainly an extraordinary experience and initially seems to grant a kind of proximity to the animal. At the Melbourne Museum, when C5756 was removed from the jar, despite her physical flaws, she looked as if asleep, as if, at any moment, she might move, her chest heave and her legs spasm, her lungs cough and breathe.[70] Laid gently on a plastic tray that has been scratched and scored by repeated use, the young centenarian seemed still animate. Yet, grimly motionless, she gradually assumed her lifelessness. Pale and limp, stark in the artificial light, the pup moved me through her remaining unmoving. None of the other physical remains possess this strange inert agency. This explains why the wet thylacines have always attracted considerable attention. The skulls and bones, the skins and the mounts, offer different, less affecting visions of the animal. As such, the wet remains risk overshadowing these other vestiges and the stories they hold.

At TMAG, for a long time there were thought to be five wet thylacines, but after research the museum conducted in tandem with the University of Melbourne it was revealed that two of the pouch young (A932; A934) possessed epipubic bones and were likely to be Tasmanian devils (*Sacrophilus harrisii*).[71] The thylacine has no epipubic bones.[72] The researchers used a non-invasive procedure, CT scans, to produce 3D digital models of each of the pups. Of the three animals preserved in ethanol that were confirmed as thylacines, two are pouch embryos (A931; A933) and one is a pouch young (A930), two are male (A930; A931) and one (A933) is of undetermined sex. The embryos are both catalogued as dating from prior to 1910 and from 'Pearson's collection'. Joseph Pearson, a zoologist, was director of TMAG from 1934 to 1952. He published several articles on marsupial anatomy.[73] In the 1940s, Pearson worked with other pouch specimens loaned from the Australian Museum and the National Museum (Melbourne), using dissection to study their anatomy. This led to a co-authored article detailing the urogenital system of thylacines published in 1953.[74] The dissected preparations from the Australian Museum's pouch young were rediscovered at TMAG in 2005 when the Vertebrate Zoology collections were moved from the city centre site to a new collection storage facility at Rosny on the outskirts of Nipaluna/Hobart.[75] These were returned to Gadigal/Sydney.

The two pouch embryos at TMAG seem a world apart from the thylacine as they are popularly known, the striped animal with the broad gape.

Their bodies are the colour of alabaster, the mouths show no easily discernible teeth. The white colour lends the animals a sculptural quality; they resemble a couple of finely carved ivories, the slight wrinkles around the neck of A931 like folds in drapery.[76] These animals are alien in comparison with popular imagery of the tiger, foreign curiosities. Pouch embryo A933 lies curled up at the bottom of the jar (although from photographs they appear as if they were formerly in a more upright position reminiscent of TMAG's two other wet thylacines). The tail of the creature curls towards its tip, still pliant when they died. A931 has pronounced dimpling along his back, numerous small craters in the surface of the skin. A933 also has these around the base of the tail but they are difficult to see because of how the animal lies in the jar and are evidently less pronounced. The two embryos have small incisions in the dermis atop their crania and the skin in this area is pulled into the shape of a bow. These little holes in their heads form traces of a time when the animals were both strung up by hooks from the top of their jars, hung in 'suspended animation'. The tiny perforations now serve as reminders of this previous mode of display.

The biggest of the three pouch thylacines, the pouch young, has fine hairs on the top of his head and slight downy fur across his wrinkled body. He possesses whiskers but these are not the dark, hard bristles of an adult, they are slender and silvery. A930 seems to be just beginning to find hints of his adult colouring, there are the beginnings of reddish-brown hair near the ears.[77] This larger tiger also has bones that stick out, pressing up through the creased skin: the humeri protrude and the vertebrae of the spine are evident. In raking light, the ribs are also clearly visible. The obtruding bones suggest at once a body yet to be grown into, one waiting to be filled out, and a body in an advanced state of emaciation, one wasting away. The creases of skin similarly evoke both youth and old age; this dichotomy is fitting in that this animal is a juvenile yet also over a hundred years old, an ageing infant, ancient before his time, an oxymoron incarnate. A930 was brought to TMAG along with his mother and female sibling in 1902. They were caught near Campbell Town in 1902 with the aim of claiming bounty on them.[78] A930 is therefore a victim of genocide.

The circular jars in which all the wet young are kept impact a beholder's perception, each animal shrinking and expanding as their jar is turned.[79] The head of A930, the male pouch young at TMAG, measures approximately 6.5 cm when seen from the side of the jar where his head currently touches the glass, whereas its perceived size almost doubles (to 11 cm) when seen from the jar's opposite side. The liquid in the jar acts as a magnifying glass, seemingly enlarging the head, engorging it. It distends and contracts, calling to mind Jacques Lacan's discussion of anamorphosis and, particularly, of the skull portrayed in anamorphic perspective by Hans

Holbein in *The Ambassadors* (1533).[80] Holbein's painting shows two French ambassadors, depicted life-size, flanking shelving that displays diverse objects. In front of the two men, in the bottom foreground of the picture is a peculiar shape, a carefully modelled stain, the skull. Jurgis Baltrušaitis, one of Lacan's points of reference, calls the anamorphic skull '[a] strange object, like a cuttle-fish bone'.[81] He makes sense of the distorted projection through the prism of natural history, seeking to render the strange more familiar by using the simile of the cuttlebone, an internal shell found in cuttlefish.

Drawing parallels between Lacan's remarks on anamorphosis and the jars seems apposite because the skull in *The Ambassadors* forms a memento mori and constitutes part of a broader iconography within the painting that is read by some as linked to death and resurrection.[82] Skulls have traditionally been viewed as intermediaries between the living and the dead.[83] For Lacan, the skull also provides the imaged embodiment of castration, gesturing towards symbolic lack.[84] In the context of the thylacine, the pups are sometimes read not as memento mori but as symbols of potential resurrection, as the cloning effort featuring P762 attests. Such endeavours seek to make good the lack of the thylacine, the animal's poignant absence, through reinstantiation. The thylacine, as Fletcher notes, is 'a potent symbol of conservation failure and colonial hubris'.[85] The cloning project, which viewed a pup such as P762 as nascent data waiting for extraction, promised to undo the extermination of the species. As there were no thylacines left to engage in sexual reproduction, the species would be reproduced through cloning.

Reproduction

Patricia Woolley used one of the testicles of A930 for an experiment as part of research into marsupial sperm. She also removed a testis from a thylacine held in the National Museum of Australia's collection.[86] This experiment was seemingly unsuccessful as no results were published. Pearson's and Woolley's experiments are examples of how thylacine remains continue to be sampled for research. Maynard and Gordon write of the dilemma posed by requests to museums by scientists who wish to engage in 'destructive sampling', noting the 'trade-off between supporting new and novel research and preserving finite resources for the future'.[87] The advent of new technologies has led to an increase in destructive sampling in some contexts. Such sampling involves chipping, cutting, drilling or scraping portions of physical remains. Many of the thylacine remains I studied carried traces of said procedures, particularly drilling. These index knowledge that has been gleaned through damage, through hurt. Even minimal sampling can cause unforeseen harm. Some experiments using samples also fail to yield results. The efforts to extract DNA from the liver of P762, for instance, were less successful than had been hoped as it was discovered the sample was contaminated.[88] Similarly, hair samples from a mount of an adult male at the Swedish Museum of Natural History, NRM 592206, failed to yield DNA.[89]

There are also three wet pouch young (C5755, C5756, C5757) held in the collection of the Melbourne Museum. The age of these pups has been estimated at thirty-nine days postpartum.[90] The pups are all offspring of C5752, an adult thylacine who was shot in the wild and whose corpse and those of her progeny were then acquired by William M. McGowan, the Superintendent of Launceston City Park Zoo.[91] The pups therefore died as a direct result of genocidal violence. A fourth pup, the sibling of the other three, was cut up and made into slides by Milan Klima (an expert in embryology) at the Institute of Anatomy at the Johann Wolfgang Goethe University in Frankfurt in 1994. The 'histological specimen' (C5754), his biological tissue stained to grant him greater visual legibility and make him easier to study, is designed to be viewed by microscope. The slides are housed in a case in a storeroom and appear to be seldom used.[92] Pouched young are extremely rare with only fourteen known to exist so the decision in the 1990s to transform one of the young into slides, to slice and dice him, is scientifically and ethically questionable.[93] The slides show damage as do the sections and their scientific value is therefore compromised.[94]

The three remaining wet thylacines in Melbourne are semi-upright in their jars, a pose that anthropomorphises them: their forelegs become arms, their paws hands. Each thylacine is curled inwards, in a foetal position. The Melbourne pouch young also have the hole in the skin of their heads where they were strung up, hung as morbid mobiles. The spirit in the jars is filled with particles, with very small fragments, perhaps, in part, traces of dissolved fat. Some of the particles are probably not from the thylacines but from curators and researchers. When the wet female thylacine in Sweden, NRM 566599, was sampled, 1,037,369 reads were generated of DNA fragments but of these, 44,493 originated from the human nuclear genome and 136 from the human mitochondrial genome.[95] Because the human DNA was quite degraded, of similar length to the marsupial's, the team behind the project speculated that it was probably of similar age to the thylacine. Human residues cling to the nonhuman animal, clouding her genetic profile. As the criminologist Edmond Locard is said to have observed: 'Every contact leaves a trace.'[96] Efforts were made to decontaminate the hair shafts employed for the sequencing efforts using bleaching but without great success. The researchers speculate that the ethanol 'has helped the contaminant DNA to penetrate much deeper into the shaft than conventional sources of contamination'.[97] The effects of preservation are therefore to amplify contamination and fuse human and thylacine more forcefully together.

The effort to clone a tiger has been framed as an exercise in blue sky thinking, enhancing knowledge of the utility of ancient DNA for research purposes and aiding conservation efforts (through encouraging the preservation of suitable habitat in preparation for the eventual return of tigers to Lutruwita). As Fletcher observes, however, the project was also interpreted as an effort 'to erase past mistakes'.[98] She notes that the 'Janus-faced themes of guilt and potential redemption through science' were prominent in debates about the initial cloning effort.[99] Cloning therefore offered not just the potential resurrection of the tiger but a simultaneous redemption of science. The bounty scheme is regularly acknowledged as contributing to the thylacine's extinction, but scientific efforts to secure 'specimens' were also a factor in their disappearance.

P762, the pup at the centre of cloning efforts, was among a significant quantity of thylacine remains collected by George Masters, then Assistant Curator at the Australian Museum, in 1866. Writing in 1868, Gerard Krefft makes reference to a nine-month trip to Lutruwita by Masters.[100] He also describes Masters as having sourced not only the skins of native animals but 'a fine series of skeletons, and young from the pouch'.[101] Some of the thylacines were obtained in the Central Highlands and the pup may also come from there.[102] By the 1860s the decline of the thylacine had already been noted, so Masters's collecting can be viewed as complicit with the private bounty campaign then being conducted to extirpate the tiger.[103] The other thylacine remains used for the initial cloning project became part of the museum collection in 1893 (skin and skull) and 1922 (skull). The skin and skull from 1893 were formerly part of the TMAG collection and the skull from 1922 was given to the collection by William G. Lyne and derived from an animal from the Great Western Tiers.[104] The circumstances in which these thylacines died are unclear but M822 seems to have been shot. It is highly possible that all these remains employed with the aim of resurrecting tigers were victims of genocide. Cloning is sometimes justified from a deontological perspective as it can right the wrong of a human-induced extinction of a species.[105] In the case of the thylacine, however, while cloning may negate extinction it cannot undo genocide, only potentially work to obscure it (as having living animals present again would likely lead to less engagement with the past and the dead). In this sense, the thylacine remains exploited to support cloning initiatives may one day serve to efface the harrowing histories they embody.

The Way of All Flesh

Skulls and skeletons are sometimes accorded a complex significance. As signalled by the earlier discussion of Holbein, human skulls in Western culture have traditionally signified the impermanence of life. In art, the genre of the *vanitas* employed skulls to serve as reminders of the transience of earthly achievements and pleasures and the need to guard against excessive self-opinion. Michael Kearl notes in his cultural history of skulls, however, that their connotations changed in the twentieth century when they became increasingly politicised, symbolising 'needless brutality and injustices served by political regimes'.[106] Skulls were associated with genocides in Cambodia (enacted under the Khmer Rouge led by Pol Pot) and, more recently, Rwanda.[107] Nonhuman animal skulls have never attained the same symbolic potency as those of their human counterparts.[108] Jonathan Koestle-Cate suggests that the 'human skull has a metaphysical puissance in the way that, say, an animal skull does not'.[109] Nonhuman animal skulls have generally been valued, if at all, as sources of scientific data or as aesthetic curios, although as remnants of dead creatures they still potentially recall mortality.

Krefft used skulls collected by Masters to mistakenly argue for the existence of two distinct species of thylacine. Skulls have traditionally held a particular importance in natural history and the study of cranial morphology including measurements and physical characteristics was, for a long time, of central importance for species identification.[110] In some natural history museums, post-cranials (the rest of the skeleton apart from skull) were not traditionally kept. Skulls and skins were retained, the animal decapitated, flayed, with the flesh and most of the bones discarded. Sleightholme and Campbell claim that of adult thylacine skulls, 'the most important from an historical perspective' are two skulls in the Natural History Museum in Leiden, RMNH39000 and RMNH39001 as these were described by Coenraad Jacob Temminck in 1824.[111]

The ITSD authors privilege these skulls because Temminck is credited with separating the thylacine into their own genus of *Thylacinus*.[112] He identified morphological differences that distinguish the skull of the thylacine from that of a dog and from the genus of *Dasyurus*.[113] Temminck noted of the skull that he primarily worked from (he studied four thylacines in total, three adults and a juvenile, all male) that it was definitely that of a fully developed animal and that all the teeth were worn.[114] Here the condition of the teeth is read by Temminck as evidence for the tiger's maturity. The abrasion patterns index a life lived into adulthood and are used by the zoologist to support the precision and veracity of his findings.[115] Temminck believed that when studying the thylacine, George Harris had worked with a younger, less fully developed animal, the back upper molars of which had not yet grown.[116] This rendered his description imperfect. The remains of the male thylacine that Harris described as *Didelphis cynocephala*, of the holotype, are no longer extant.[117] In the absence of this type specimen, the skulls connected to Temminck therefore have a special status.[118]

Like the preserved pup as it is understood by Stark, thylacine skulls and other bones are not solely scientific resources. The osseous remains provoke feelings that are often qualitatively different from their wet counterparts. Skeletal remains are usually unfleshed or very nearly so. They seem uncompromisingly inanimate in contrast with fluid-preserved animals. This inanimateness is powerfully communicated by the skull QVM2013. GFV.1. The skull, retrieved from Flowery Gully in Northern Tasmania in 1951 by Eric Oswald Gale Scott and Thomas Edgar Burns, is part of the QVMAG collection, catalogued as a fossil.[119] It gives a profound sense of the ancientness of the thylacine, of the species' remote past. Limestone has grown around the skull, encasing it. A ventral view reveals that the mineral carapace has something of the appearance of florets, of flowering stems, organic outgrowths made of lifeless stone. When holding it upright, peering within the cavities of the rock-skull, it is also still possible to see thin, white matter, the enduring, petrified bone. Here, geological processes that were potentially at work over several millennia help figure something of the tiger's ancient history.

The fossil holdings in QVMAG are not thought to contain genetic material and are therefore often overlooked in favour of the organic specimens.[120] QVM2013.GFV.1, however, gives a sense of the thylacine as an animal having a remote as well as a recent past. The skull, with its evocative stalagmitic film, does not derive from an animal that was being harried to extinction. The cranium is very old but its age is insignificant compared to how ancient some of the fossil remains of branches of the thylacinid family are. Certain members of the family, such as *Badjcinus turnbulli*, *Thylacinus macknessi* and *Nimbacinus richi*, may date back to the late Oligocene, while most derive from the Miocene.[121] Thylacinidae therefore existed for roughly twenty-five million years. In geological terms, twenty-five million years ago is recent, but the time span still seems 'far into the abyss of time'.[122] Fossil remains appear remote and otherworldly, deriving from a period readily affirmed but less easily imagined. *Thylacinus cynocephalus* first occurs in the fossil record in the early Pliocene, about five million years ago. The calcium carbonate-encrusted skull is probably only a few thousand years old.[123] Despite its relative 'youthfulness', the skull gestures towards this much longer history of the thylacinid family. The extinction of *Thylacinus cynocephalus*, caused at least in part, I would argue, by animal genocide, brought an end not solely to a species and a genus but to an entire family of marsupial carnivores with an ancient past.

Stories in and around Skulls

Works of fiction about the thylacine have made multifaceted uses of skulls and skeletons as imagery. The novels *Coorinna* and *The Hunter*, for example, both include skulls as figures and symbols. In Julia Leigh's novel *The Hunter*, the first bones to appear in the narrative are human. The protagonist, known simply as M, finds the physical remains (minus the skull) of a missing man, Jarrah Armstrong, and carefully reassembles the skeleton. Engaging in an improvised osteoarchaeological investigation, at one point he measures a rib against the side of his own chest to gauge the probable gender of the dead person. The remains here are treated as an osseous archive that preserves information and potentially records a man's fate: 'Each bone he studies closely, looking for any tell-tale chips, the mark of a bullet.'[124] The relative cleanness of the remains leads M to guess there is an ants' nest in the vicinity. He measures, he looks, he touches, he extrapolates, and from the bare bones comes the flesh of insight, the emergence of a fragmentary narrative.

Later in *The Hunter*, M discovers a cache of bones that lead him to the lair of the thylacine he is hunting. In the lair, he finds the skeleton of a tiger pup, a treasure that is so 'alarmingly beautiful' that 'he touches it as he would the Holy Grail or his own first child'.[125] Here bones are venerated like a religious treasure and appreciated aesthetically. M reveres the skeleton because it confirms that he is close to attaining his goal of finding a living thylacine, a Lazarus.[126] He 'trails a finger over the curved lumpy spine' and then 'lies down on the ground in a mirror position, eye to eye with the skull, and imagines for a second that he, too, will rot in this cave'.[127] The skull in this moment is read conventionally by M, understood as a reminder of the inevitability of death. With an eye to the empty orbit of the skull, he fantasies his own demise. It is noteworthy that during his contemplation of the bones, M caresses the vertebrae. Touching bones is an important aspect of osteoarchaeology but here the haptic is not employed to assess physical characteristics.[128] The trailing finger, the lingering handling, implies wonderment and a measure of tenderness. The bones grant M a sense of physical connection with the thylacine as a species. M's touch here is qualitatively different to the instrumental handling he engaged in to examine Armstrong's skeleton. The pup is not an object of analysis but of adoration. The encounter with the pup's bones demonstrates that there are multiple possible ways for humans to interact with such remains.

As well as thinking about the significance of the bones for himself, M speculates on the solace that the last tiger derives from them; he ponders that they may offer her a kind of companionship, of kinship: 'lonesome, is this where she comes for company?'[129] The skeleton in its substantiality potentially provides the solitary thylacine with a tangible reminder of another.[130] Erle Wilson's novel *Coorinna* also explores how a tiger might potentially relate with bones. In *Coorinna*, a thylacine pup becomes fascinated by 'the sun-bleached skull of an old ewe', adopting it as a plaything that he tosses and buffets and worries.[131] The skull here is not understood symbolically by the young thylacine, it is not a prop against solitude or a prompt to meditation on mortality, but registered as a physically stimulating thing, a spur to actions and reactions. The young thylacine in *Coorinna* is a creature of instinct. The adult female in M's imaginings in *The Hunter* is gifted an emotional palette, the capacity for longing and lonesomeness. Each of these visions of the thylacine is articulated by way of an engagement with bones. The novels give a sense of their storytelling and affective potentialities.

Skulls and other bones, fossilised or otherwise, possess distinctive traits, hold unique life stories. QVM2013.GFV.1 stands out from other fossil thylacine materials at QVMAG because the crystallisation grants it a singular appearance. The non-fossil skulls at QVMAG and elsewhere nevertheless each possess their own 'personality'. Particular skulls manifest their individuality not solely through differences in scale but also by way of varied patinas and states of preservation. Adult skulls unsurprisingly have more weight to them. The skulls of pups are almost insubstantial, arrested densities. Inside adult skulls there is also significant cartilage. Most skulls have accrued some kind of damage, possessing distinctive cracks and fractures. Some of these are caused by the actions of 'museum beetles' and their larvae.[132] Beetles embody the natural world invading and consuming a part of its history. Preventative measures, the laying of traps, form an effort to denaturalise the collections as much as possible, to shield them from the decay cycle and the actions of detritivores and scavengers. In this context, the ITSD, the digital archive, the most unnatural of settings, is ideal as a repository. The zeros and ones keep the virtual remains safe from burrowing and gnawing, from the mess that is decomposition.

Other breaks and flaws in skulls index repeated handling and moving or more violent events, the rough use of calipers or a moment of clumsiness, a fumble and drop.[133] The teeth are noticeably loose in some mandibles, necessitating that they be kept upright. Several skulls and mandibles are stored alongside Ziplock bags holding dislodged teeth and cartilage fragments. The bags indicate ongoing efforts to hold on to all the pieces, the bits, of decaying remains, to shore against their ruin. Damage is occasionally also deliberate. In the Smithsonian, 49724, a yellowy-red-coloured thylacine skull, shows evidence of the performance of a necropsy. The top of the cranium has been sawed open and can now be removed and replaced at will, the action not unlike lifting and replacing the lid of a teapot.[134] The value of skulls that retain dry tissue is now appreciated (in the past great efforts were made to clean skulls, removing all trace of the fleshy matter that once encased them), making them stand apart. The residual flesh is now increasingly referred to as osteocrust.[135] QVM1962.1.55 has flesh adhering to the cranium and some whiskers attached.[136] The mandible is still associated, held together by dried flesh. The skull has clearly not been cleaned.[137] The tiger may have been discovered like this in the bush, dead in a snare or poisoned, rotted to bones. The back of the skull is intact, suggesting that the thylacine was probably discovered post-mortem. The full description of this 'specimen' in *Precious Little Remains* reads: 'There is dried flesh present on parts of this skull and mandibles, including part of the nose. A few whiskers are still present.'[138] The text is revealing; it shows the importance accorded to the flesh and whiskers, they are foregrounded as the skull's distinguishing features, differentiating it from other skulls in the QVMAG collection.[139] The description interprets the skull, puts meat on its bones.

Skulls are always labelled, conceptualised, accompanied by words and ideas that guide their reception and interpretation. The labelling sometimes records histories of exchange, of the passage of remains from one collection to another. In the Melbourne Museum, a skull (C5750.2) of a female thylacine currently has R2154 written in black ink on it and another identifier that clear efforts have been made to scratch off.[140] There are abrasions on the skull, scoring caused by a sharp implement, and the bone has been discoloured, lightened, by the scraping in that area. The identifier has been rendered illegible but traces of ink remain. A possible explanation for this act of damage is that the wrong identifier was initially inscribed and could not easily be corrected. The scratching indexes entitlement, the curator affirming their ownership of the remains. The rough handling shows that this skull was initially not conceived as something to be honoured or revered. C5750 was listed as part of an accumulated lot from QVMAG sourced in an exchange.[141] A note written to accompany the skeleton observes that the thylacine 'came in the flesh' from Launceston on Monday 1 June 1903 and was 'stripped' on the Saturday (6 June).[142] The paper used to cushion the remains of the skeleton (excluding the skull and mandibles) of C5750 (R2155) is black (in most of the boxes it is white) and, like a display case in the Tiegs Museum discussed in the next chapter, evokes mourning. It is an accidental expression of sorrow.

As well as text written directly on skulls and skeletons or on labels stored with them or in ledgers, sometimes far more expansive writings are inspired by osteological remains. The skull and mandibles numbered 1129 at the Tiegs Museum in Melbourne, for example, have a back story that has attracted significant attention.[143] Ada Lambert donated the remains to the collection in the late 1890s or early 1900s along with those of a common ringtail possum.[144] Lambert studied at the Biology Department at the University of Melbourne and subsequently taught there. She was an expert in Australian leeches, giving papers on the subject in the late 1890s that were published in the *Proceedings of the Royal Society of Victoria*.[145] Lambert was appointed assistant lecturer in biology at the University of Melbourne in 1901.[146] She may have been the first woman appointed to lecture there.[147] Lambert would, however, abandon her university career shortly after marrying Thomas Archibald à Beckett in 1903. Ada à Beckett resumed teaching after the birth of her third child, mainly working in schools. The skull is linked with a time in her life before marriage. It is central to an article by Rohan Long, a former collection manager of the Tiegs Museum, entitled 'Miss Lambert's Thylacine Skull', which is about pioneering women zoologists in Australia.[148]

Handled by Lambert over a century ago, the skull's importance as an object was clearly recognised by her, its scientific value known. Later this same skull would be employed to draw attention to a carefully constructed narrative designed to give the impression that in the early twentieth century the University of Melbourne was a forward-thinking institution regarding the equality of the sexes. The women mentioned in the article all donated specimens to what is now named the Tiegs Museum. As well as Lambert, Georgina Sweet and Gwynneth Buchanan are discussed.[149] Georgina Sweet was a student and then teacher and researcher in biology at Melbourne. She is referred to in the article as 'Australia's first female professor'.[150] Gwynneth Buchanan was a zoologist who worked at the University of Western Australia and the University of Melbourne. She wrote *Elements of Animal Morphology*, which was a standard biology textbook in schools for many years.[151]

Given that Lambert's expertise was in relation to leeches, the thylacine skull is hardly representative of her research interests. It is here exploited because of its cultural cachet. In the article itself, there is little discussion of the skull and what potential insights it might hold, although a photograph of it is included and Long remarks that the thylacine is extinct and that 'specimens of this animal are incredibly valuable'. He suggests the skull was likely collected south of Cressy in the northern midlands of Lutruwita.[152] Lambert may also have carried out fieldwork in Kooparoona Niara/the Great Western Tiers as she makes reference to leeches sourced from there.[153] The skull could therefore alternatively derive from that area. At the time of writing, 'Miss. Lambert's Thylacine Skull' was in storage, enfolded in bubble wrap taped up with yellow Scotch tape. The large skull (25 cm in length) of an adult animal, discoloured a dirty sepia in parts, has gathered a variety of accretions and stains during its time in the collection (and, perhaps, beforehand). There is brown discolouration where the vibrissae used to be. Examining the remains, the cartilage in the turbinals draws the eye, its thin, fragile layering resembling *pâte feuilletée* or puff pastry. There are no obvious signs of cranial trauma.

Although Long uses a possessive that grants Lambert posthumous ownership of the skull, in reality it is clear that she has no hold over it. Long takes possession of the bones, using them as a promotional prop for his narrative about the gender history of zoology at Melbourne. The skull and mandible in themselves are of little interest to him. They are instrumentalised, adapted to a specific purpose, used and manipulated to historical ends. Any narrative necessarily involves a measure of instrumentalising but Long's employment of the bones makes no effort to give voice to them in their singularity or to attend to their affective potency. There is no sense of what it is like to encounter the skull Lambert donated; the bone matters little. Part of the issue here, of course, is provenance. Lambert's acquisition of the vestiges – where she got them from, who she got them from, how the thylacine died – involves many unknowns. Nonetheless, the skull is there in its obstinate, grimy materiality. There is a distinctive 'something' behind the words, a singular materiality that is not easily communicable or readily owned but that demands consideration. Through studying thylacine remains, through noting differences in size and appearance, it is possible to break with the typecasting that was enabling for genocide, the reduction of the thylacine to sheep-killers, to brutes, to 'them things'. It is also possible to move beyond scientific notions of typology.

Beyond Biomolecular Analysis

In the context of human skulls, John Harries has noted of the skulls of the Beothuk couple Nonosabasut and Demasduit when they were held by National Museums Scotland that they often existed only as 'a vague outline of presence' in discussions about them.[154] The idea of the skulls is of greater importance than the remains themselves. In this understanding, skulls can act as catalysts to reflections on the past and efforts to describe it, but the bones themselves are resistant to interpretation, muted. When looking at the skulls of Nonosabasut and Demasduit, Harries reflects on his not seeing a great deal in them. The Beothuk are 'not made present in their skulls'.[155] This is a sensible reaction. Harries clearly refuses to fantasise some mystical connection with the Beothuk or figurative reanimation enabled by their bones. For him, the skulls matter as prompts to reveal the limits of narration, their usefulness being to keep the historian grounded. The value in a historian (rather than, for instance, a forensic anthropologist) examining human physical remains exists solely in their being reminded of the limits of history-writing, of the presence of unknowability in knowledge of the past. Bones in this conception cannot attain the status of historical sources. They are not documents which can be compared to the eyewitness testimonies to the abduction of Demasduit and the murder of Nonosabasut that exist.[156] Harries's description of the skulls is minimal. For him, they can be read into but are not in themselves readable.

Writing in the context of archaeology, Joanna Sofaer has noted that 'the skeletal body is rarely used explicitly for interpretation'.[157] This observation is borne out by Harries's negligible engagement with the matter of the skulls themselves. The bones are also peripheral to the research findings about Beothuk ancestry and diet derived from DNA extracted from teeth from the two skulls. In the paper summarising this research, some description is provided for how DNA was obtained from each tooth, with drilling (Demasduit) and fracturing (Nonosabasut) used to source dental materials that were then converted into demineralised solutions.[158] The reader learns in passing that Nonosabasut was suffering from a dental abscess, an ailment differentiating him from Demasduit, but the focus is on how DNA and isotopes were secured. Teeth are means to ends, interesting for the genetic material they hold but viewed as generic in themselves. The DNA, once sequenced, takes on a life of its own, divorced from the teeth and bones that held it.[159] The paper begins by naming Demasduit and Nonosabasut but by the end they have become 'one male and one female individual of the Beothuk', types rather than personalities.[160] The skulls were initially sourced as 'specimens' of a 'disappearing race' and something of their 'specimen' status persists in the DNA research.

Sometimes specimen logic also seems to reduce remains such as skulls to data repositories. The genocidal violence perpetrated against Demasduit (abducted) and Nonosabasut (murdered) recedes from view under such a logic. In the DNA and isotope research conducted with input from the McMaster Ancient DNA Centre, settler colonial enmity is implied by the acknowledged loss of Beothuk access to coastal resources. This harm, however, is not one that can be deduced from the carbon-isotope data derived from the skulls, which 'suggests a significant marine dietary source'.[161] In the case of the Demasduit and Nonosabasut, the DNA and isotopes do not register land theft and violent aggressions. Similarly, research on thylacine DNA focusses on ancestry and cannot speak to the animal's persecution, to the tiger's modern history.[162]

The biomolecular analysis carried out on the skulls of Demasduit and Nonosabasut is not the only way of conducting osteoarchaeological investigation. Other approaches, like that performed by M in the novel *The Hunter*, are based on close visual examination. Such a method is used in combination with other techniques by Sue Black, Ingeborg Marshall and Andrew Kitchener in relation to the skulls. Although Harries conceives of the two skulls as enabling yet not inhabiting the stories that are told about them, perceiving language and materiality as at odds rather than imbricated, Black and her co-authors gesture towards an alternative relation between words and bones, one that resonates with the readings of thylacine remains offered here. They close-read the Beothuk remains, noting, for example, the writing on the skulls. Nonosabasut's skull bears the numeric identifier '1827-35' and high on the frontal bone the words 'Male Red Indian Newfoundland', 'Chief', 'Mr. Cormack' and more text that is now too faded to decipher. Additionally, the word 'Mongolian' was written prominently on the frontal bone and then subsequently crossed out.[163] The authors also remark the reddish colour of the skull, which they attribute to the Beothuk burial practice of overlaying remains with red ochre. Nonosabasut's skull retains nine teeth with a tenth a cast of the tooth that was pulled and used to source DNA. The remaining teeth show 'advanced occlusal attrition', which suggests a diet including abrasive foods. Nonosabasut's mandible showed traces of an injury that had healed and was probably the result of an attack. The analysis therefore brings back something of his biography, providing a vague and imperfect but singular narrative.

Demasduit's skull, twice referred to as 'relatively gracile', is also described in depth. Gracile is a term used in forensic anthropology in relation to crania and mandibles to express slenderness. A study published in 2009, for example, suggests that in the past century in the United States, the human mandible has been 'transforming into a longer, narrower, more gracile bone'.[164] In the context of a discussion of Demasduit, however, it seems possible that her gender and the description of her that survives has influenced how her skull is interpreted. Hercules Robinson calls Demasduit 'tall and rather stout-bodied, limbs very small and delicate, particularly her arms; her hands and feet were very small, and beautifully formed, and of these she was very proud.'[165] Although referred to as stout rather than slender, Demasduit's delicate limbs as noted by Robinson lend themselves to the adjective gracile. Black and her co-authors also discuss at length the evidence of blunt-force trauma present in the cranium.[166] The injury was perimortem and 'most certainly not a *post mortem* artefact'.[167] Perimortem injuries occur at or near the time of death, opening the possibility that Demasduit was, like Nonosabasut, murdered.[168] George Busk may have noted the injury when he examined the skull in 1876, as he states that

Demasduit 'died, or was killed, in the year 1823'.[169] Demasduit's skull lacks the sanguine tinge of Nonosabasut's, which probably indicates a degree of post-mortem social ostracism caused by her contact with Europeans. Both the skulls are encouraged to tell their stories by Black and her co-authors. Through this storytelling as it intersects with the historical record, something of the singular lives of Demasduit and Nonosabasut is brought to the fore. The skulls are looked upon not as 'specimens' or types, or, at least, not solely as these, but also as records of the lives of two specific people.[170] Such an approach is helped considerably by knowing in advance the names of the victims and something of their lives. It is nonetheless possible to ascertain biographical information from bones themselves in the absence of other sources.

Bone Biographies

Often in forensic anthropology much effort is expended in establishing an 'antemortem skeletal biography'.[171] This biography may aid the positive identification of the remains. Michael Portillo states of forensic anthropology that 'establishing an identity helps to define a crime'.[172] Portillo is here writing of the ability to put a name to physical remains. Identification in forensic anthropology often begins with establishing the age, sex and stature of a person. Identity therefore operates on a continuum with group characteristics ascertained before individual ones. The study of remains may be a gradual process of moving from the general to the singular. Basic information such as the gender of victims can aid in determining what kind of crime has occurred even in the absence of further particulars.[173] Forensic anthropologists sometimes work with veterinarians, studying nonhuman animal remains for signs of non-accidental injury, looking for evidence, for example, of trauma arising from physical abuse.[174] Veterinary forensics involves the application of forensic science techniques to legal investigations where nonhuman animals are the victims of criminal offences.[175] Forensics offers a language for describing bones that permits the sometimes violent histories embedded in them to be spoken. Some of the thylacine remains held in museums show clear indications of deliberately inflicted injuries.

Sleightholme and Campbell have noted that 'a number of [thylacine] skulls reveal evidence of direct trauma i.e. fractures of the cranium resulting from the method of dispatch (clubbing)'.[176] In 1858, Charles Shum Henty refers ambiguously to breaking the heads of snared thylacines.[177] Additionally, Campbell has indicated that some thylacine skins show 'evidence of scarring', of injuries to the head or legs, or exhibit bullet holes.[178]

He also observes that the rump of certain skins is marred by 'puncture wounds suggestive of dog bites or claw marks'.[179] Some thylacine remains may also index snaring itself. The mount A1299 (a female juvenile) in the TMAG collection, for example, has damage to the skin consistent with injury by a snare.[180] Several thylacine skulls that I examined did show evidence of trauma. A skull in the collection of the Melbourne Museum, C2164.1, has a hole in the cranium with jagged edges and fractured bone that may be consistent with clubbing.[181] At TMAG, A1243 has a smashed cranium that probably resulted from clubbing.[182] Sometimes, however, injuries manifested by the skulls are not consistent with blunt-force trauma of the kind produced by a club but with damage more consistent with penetrating or sharp-force trauma caused either by gunshots or blows with a spiked implement.[183] Clubbing, as blunt-force trauma, would potentially cause greater fracturing.[184] Similarly, a trap hoe retained by the descendants of James Millar Kaine and a trapper's hatchet on display at TMAG are designed in such a way that they would probably cause greater damage to a thylacine's skull if used for killing than many of the injured crania display.[185] The skull S788 in the Australian Museum shows evidence of injury that looks like a gunshot wound. The skin and skull of the thylacine M822 at the Australian Museum manifest the traces of a ballistic injury, of violence that carries across skin and bone. The skin has a neat hole where it was punctured but the skull has a rougher fissure where the bullet impacted hard matter. As mentioned earlier, these remains were used in cloning efforts, genetic material sourced from a victim of atrocity. The thylacine's physical remains are therefore occasionally marked by the genocidal violence perpetrated against the animal.

Eric Guiler believed that due to the cost of bullets, snared thylacines would usually have been despatched with a hatchet by trappers rather than shot. His view is replicated in the exhibition 'Tasmanian Tiger: Precious Little Remains' (discussed in depth in the next chapter), in which a label accompanying a thylacine skull with a '"v"-shaped cut . . . probably made by a snarer's hatchet or tomahawk' explains that 'Thylacines caught in snares or traps were killed by a blow to the head' as bullets 'were expensive and considered to be wasted on Thylacines'.[186] The fracture morphologies (the forms of the holes in the skulls), which differ across thylacine skulls displaying cranial trauma, strongly indicate that both methods of killing, bludgeoning and shooting, were employed.[187] The decision on how to kill the animal was likely made based on practical and psychological reasons. There would be slightly more risk of injury to the trapper if a hatchet were used because it would require bringing the hand and forearm close to the animal.[188] The killing distance when using a gun is marginally greater.[189] This extra distance is physical but also has a psychological dimension.[190]

Whichever method was used, the skulls bearing these traumatic injuries form graphic testaments to genocidal violence.[191]

The Windows of the Skull

One of the commonly identified difficulties when writing a history of a nonhuman animal has been the absence of written documentation offered by animals themselves: 'the only documents available to the historian in any field are documents written, or spoken, by humans.'[192] Animals, as Margaret Atwood observes, 'do not speak a human language; nor do they write stories'.[193] This lack of story-telling ability, as Erica Fudge notes, means that any history of nonhuman animals is necessarily 'the history of human attitudes toward animals'.[194] Fudge suggests that working from human representations of animals that appear in historical records runs the risk of disappearing the animal. Nonhuman animals become *figures* through which to articulate human concerns.[195] She identifies three (sometimes interrelated) modes of writing animal history, intellectual, humane and holistic.[196] The first two modes focus on what the study of animals in the past reveals about human ideas and attitudes, whereas the third holistic mode also foregrounds how the status of the human itself is bound up with the animal.[197] These three modes are, however, each allied in that they all work from documents by humans. Their differences manifest through the use and interpretation of those documents.

The study of animal remains such as skulls, however, permits the emergence of documents that are not in a human hand, or not entirely so. The bones communicate 'brute' facts and hold nascent feelings and meanings. They are documents of a kind, texts that are inscribed with human attitudes but not completely overwritten by them. These documents can be studied and read. Atwood suggests of nonhuman animals that it is 'impossible to get the real inside story, from the horse's mouth so to speak', yet through studying the mouths of the skeletal remains of horses, David Anthony and Dorcas Brown have been able to trace the history of early equine domestication.[198] Anthony and Brown view their work as combining cultural and zoological insights, merging histories natural and not so natural.[199] The have examined the teeth of horses dating from 2200 to 1600 BCE that lived in the environs of the ancient city of Anshan (located in what is now western Iran), looking for damage caused by the use of bits. In a sense, their research and other comparable efforts confirm Atwood's contention that it is impossible to access nonhuman animal perspectives. Through studying the dentition of the horses, Anthony and Brown extract information about the impact of the human upon another animal. They do not get inside the horse's head. Yet their study does index physical damage, bevelled and fractured teeth, and indicates the pain the bitted horses must have endured.[200] In this sense, something of the inside story, the equine affective experience (if not the cognised one), is communicated. A more recent study of the teeth of horse remains from the seventeenth century excavated in Turtle Island has also found evidence of the use of bits. This study shows that Indigenous peoples from Turtle Island had spread horses through the American West by the first half of the seventeenth century, so prior to European contact in that region.[201] Through studying dentition and employing osteoarchaeology, research of this kind enables horses to speak without speaking, to articulate something of their lives.

Garry Marvin notes of animal histories that they are always histories of relationships between humans and other animals.[202] The fractures that Anthony and Brown locate and interpret materialise an instance of such relating, a painful one. Marvin differentiates between natural histories (written by scientists) and cultural histories (written by humanities scholars such as himself) of nonhuman animals. For Marvin, humanities scholars cannot write about biological animals (his specific example is wolves) but only about how animals have been imagined and constructed culturally. Scientists, by contrast, have a qualitatively different experience of animals. He notes of scientific studies that 'they do attempt to foreground how animals are for themselves, and they do emerge, in the main, from a direct encounter with the embodied – sometimes living, sometimes dead – animals'.[203] Although scientists produce cultural representations of the animals they study, these representations benefit from a proximity that is lacking in other histories. When scientists work from nonhuman animal remains, Marvin notes appreciatively that they can provide evolutionary narratives through tracing morphological changes and can potentially offers insights into a creature's behaviour, social life and interspecies interactions.[204] Marvin, however, does not appear to view physical remains as a potential resource for the writing of cultural history or to accord great emphasis to how the questions natural historians ask of nonhuman animal remains are shaped by culture. Anthony and Brown demonstrate that nonhuman animal bones are sometimes directly inscribed by human culture and have a potentially important role to play in the writing of cultural histories. Bones held in museum collections are always, visibly or invisibly, culturally inscribed. This goes even for those of animals that died of natural causes in the wild. Such is the case, for instance, of A1317, a cranium and mandible along with some vertebrae and small bones that were acquired by TMAG in 1984.

A1317's remains (see Figure 2.1) were found by D. J. Rae (a doctor at Repatriation General Hospital in Hobart) in sand dunes at St. Helen's Point in north-east Lutruwita, in nature, in 1972.[205] A note in the box in

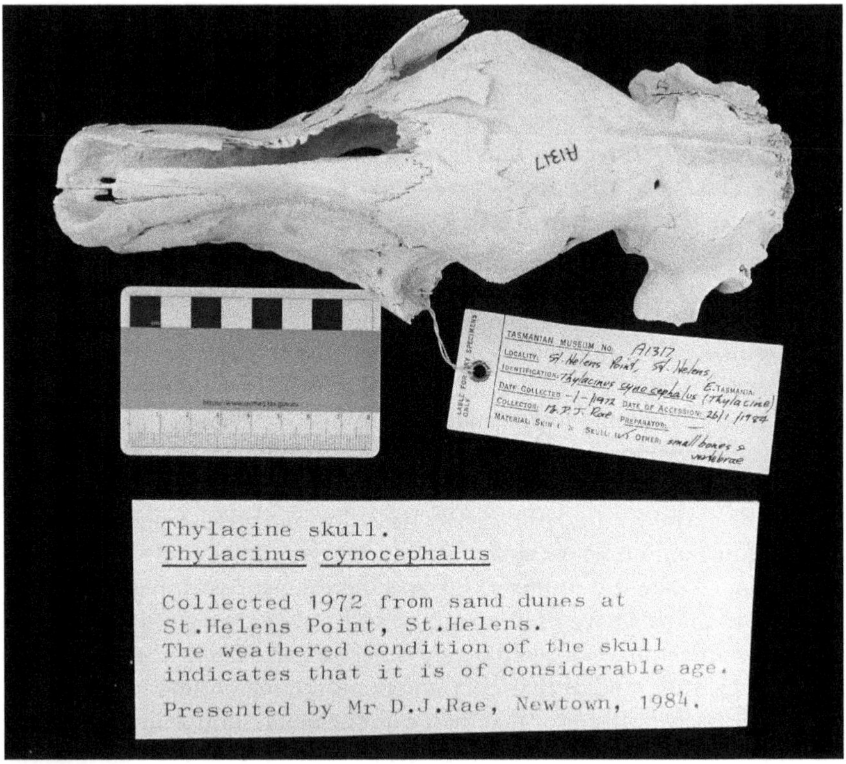

Figure 2.1. A1317. Tasmanian Museum and Art Gallery, Nipaluna/Hobart. Photo taken by Nicholas Chare. Reproduced with permission.

which they are now stored observes that 'the weathered condition of the skull indicates that it is of considerable age'. The skull is certainly decomposed, the joins of the cranium are coming apart and a white substance, a glue or paste, has been used to hold them together. This paste is also applied to the interior of the right side of the mandible, a mandible including cracked and split teeth. The right and left sides of the mandible have been glued together. Sand is still present in parts of the bones. The conditions in which the skull was found, whether it was buried or exposed to the elements, are not recorded.[206] From the weathering, it seems the skull was lying on its side, semi-exposed. If the skull and the bones were entirely buried in the sand, then they could indeed be of considerable age. If, however, they were part-exposed (as seems likely, given their discovery and condition) they may only have been in situ for twenty or thirty years, perhaps even less.

There are considerable variables in terms of the decomposition process of skeletal remains in different climates; in the tropics thirty years is about the longest time it will take for exposed human bones to completely lose their integrity but in more temperate climates it can take longer.[207] Bone weathering varies dependent on the size of a species. In her discussion of bone weathering, Anna K. Behrensmeyer explains that most nonhuman animal bones decompose beyond recognition in ten to fifteen years.[208] The skull of the thylacine has not yet lost its structural integrity but it is not far from it. The remains are somewhere between what are referred to as stages 3 and 4 of bone weathering.[209] It is therefore possible, probable even, that this thylacine died after 1936, perhaps in the late 1950s. Like plaster casts of thylacine tracks that were made during an expedition in 1938, they may provide evidence of the species' survival for a period of time after the death in 1936 of the last captive animal.[210] A1317 is also important as they may have met their fate naturally rather than in a snare, for instance, and died in the wild rather than in captivity.

Another thylacine that seems to have avoided a violent end at human hands, one whose remains were also found in 1972, is A890. This tiger is listed as having been discovered in Zulu Cave in the Florentine Valley on 16 July 1972.[211] The skull and skeleton were recovered in two separate trips on 16 and 23 July.[212] Writing that same year in an essay about these remains, the vertebrate zoologist Phil Andrews who worked at TMAG suggested that the very good state of preservation of the remains indicated that the thylacine had lain at the bottom of the cave for, at most, thirty to forty years (placing the tiger's demise in the 1930s or 1940s). Andrews seems to be striving here to make the remains fit with the species disappearing in 1936.[213] Reading between the lines, he clearly feels the tiger died more recently.[214] He notes that the skeleton appears of 'fairly recent origin' and is 'in a truly remarkable state of preservation'.[215] Andrews also emphasises that the year-round wet conditions of the cave are not conducive to the preservation of osteological remains.[216]

The remains of A890 and of A1317 are important because they are of thylacines that potentially died of natural causes in the wild and also may have lived after the official extinction date of 7 September 1936. That their deaths seemingly occurred post-1936 should not be treated as evidence for the continuing survival of the thylacine in the present. The finds do, however, foreground the arbitrary decision to treat extinction as having occurred with the death of the last captive animal. The identification of the last captive thylacine as an endling registers a desire to 'own' the death of the animal, one that will be explored in more detail in Chapter 4. A study conducted by Colin Carlson, Alexander Bond and Kevin Burgio estimated the extinction of the thylacine to have occurred in 1940, with their most optimistic modelling suggesting 1956 as when the tiger disappeared.[217] The two skulls just discussed could plausibly date from the early 1950s. There is also the possibility the modelling is flawed as more recent, alternative modelling efforts have proposed the 1990s or 2000s as when extinction probably occurred or even offer the outside possibility that the tiger is still extant.[218] Hans Naarding, a Parks and Wildlife Services employee, claimed to have sighted a thylacine in north-west Lutruwita in 1982. Naarding's sighting was in an area where thylacines were known to still occur in the 1930s and, although it does not fit with some modelling, it is difficult to dismiss.[219] The naturalist and former wildlife officer Nick Mooney also believes the tiger might have persisted into the 1980s.[220] If the thylacine did live on for fifty or more years after the death of the last captive animal, the extinction date of 1936 is arbitrary and inflicts a figurative violence against the species, reducing the relict population that endured to a twilight existence, alive yet chronicled as already dead. A890 and A1317 materially challenge the narrative that 1936 marked the end of the tiger in Lutruwita.

Dead Tigers Tell No Tales

In their entry for the thylacine in *A Synopsis of the Vertebrate Animals of Tasmania*, Clive Lord and Herbert Hedley Scott record: 'As we write these lines there is before us, in the form of a rug on the study floor the skin of a Thylacine which measures 7ft. 9in. from tip to tip, but this is an exceedingly large specimen.'[221] Because of the distinctive banding of the tiger, skins were sometimes transformed into rugs. The significance of individual animals to the meaning attributed to rugs by their owners is noted in an 1891 article on the use of fur rugs in modern interiors. The author suggests furs are not only noteworthy 'from a mere decorative point of view' but that 'the animal whose skin is thus utilized has a history of its own, which imparts a rare charm to the rug itself'.[222] Rugs, whether placed on the floor or hung on the wall, register a particular attitude towards the animal that they are made from.[223] They were usually created as 'mementoes or trophies'.[224] Rugs manufactured by (or for) private individuals have historically been cherished for the associations they held.[225] In the case of a thylacine rug, these associations may have been of elation at the slaying of a 'vicious sheep-killer'. Two hearth rugs are known to survive, as well as a buggy or carriage rug, a rug used to keep warm and to protect clothing when riding in horse-drawn vehicles such as traps, sulkies or phaetons.[226]

The buggy rug, measuring 118 x 108 cm (roughly 1.1 x 1.2 metres), is composed of eight well-matched thylacine skins, stitched together and backed by a red-coloured wool baize with a scalloped border.[227] Some small holes in the pelts may derive from accidental damage to them when the tigers were skinned and/or from when they were sanded or shaved to thin them. There is a circular brass fitting at one end of the rug near the main seam.[228] The rug was made circa 1903, reputedly by the landowner Robert David Stevenson.[229] The skins are skilfully tanned, their dark colour evidence of a job well done.[230] Arranged in two groups of four with each skin oriented so that the thicker bands of the tiger's rump face the same direction as that of the others, the rug has been criticised for not mirroring the skin pattern along the horizontal seam to produce a 'symmetrical arrangement'.[231] Symmetry was nonetheless clearly a design concern. A mirroring arrangement would work well for a floor rug (where it would be clearly visible), but as this was for a buggy, translating (repeating) the first quartet of skins rather than reflecting it may have been judged preferable.

Although a large artefact, the rug is quite small when compared with other buggy rugs held in museums and private collections. The relatively modest scale of the rug may index a paucity of pelts (the addition of two further tiger skins would have brought the size into line with other small-scale buggy rugs). By 1903 individual tiger skins were hard to come by so the rug's modest scale is offset by the use of several tiger skins together, which would have given it exclusivity.[232] It is reported to have been sold at a clearing sale at Stevenson's farm of Aplico in 1945 for £3 5s.[233] At the sale, it was bought by Robert John Leake Foster and it was kept at the Foster family home of Pleasant Banks where, for a time, it adorned a piano stool.[234] The family sold the rug at auction on 7 September 2002 (the death date of the last captive thylacine) for $270,000. It was purchased by TMAG and QVMAG with assistance from the State Government of Tasmania and the Federal Hotels and Resorts group. Now permanently housed at TMAG, the rug is viewed as an example of colonial decorative art and is therefore not stored with most of the other thylacine remains.[235]

It is the only surviving thylacine rug made from the skin of more than one tiger, although the idea of making a rug out of multiple tiger pelts was discussed in the press before Stevenson undertook his project. A person writing to the local newspaper *The Mercury* in 1901 (using the pseudonym Salmo Fario) suggested presenting a rug made from devil or tiger pelts to the Duchess of Cornwall on her visit to Lutruwita because they were typical animals of the isle.[236] Making lap rugs from multiple animals was a common technique in Australia at the time.[237] Buggy rugs were often made from wool (with a striped pattern or checked design), India rubber or the skins of animals such as possums, platypuses and native cats.[238] Lighter rugs made of cotton were used in the summer. Demand for fur rugs in general was significant in Lutruwita, with a writer in the 1870s attributing the high quality of the goods produced by local furriers to the intense competition that existed between them.[239] Furriers regularly advertised for skins, with Francis Gardner running an advert to procure five hundred dozen opossum skins and two hundred dozen native cat skins in the *Cornwall Chronicle* and the *Launceston Examiner* in 1863.

A newspaper article from 1908 that details furs offered for sale by the Tasmanian Fur Depot (a business owned by the furrier George Patterson) lists rugs using multiple pelts made from 'Tasmanian black opossums' (Brush-tailed possums) (*Trichosurus vulpecula*) and platypuses (*Ornithorhynchus anatinus*). The article gives a sense of the qualities that were looked for in a rug using several skins. Describing a rug made of platypuses, the writer observes that 'the skins are very uniform and the fur of exceptional quality'.[240] Another article about the Depot from 1918 points towards the importance of good tanning, remarking that the leather of all the skins prepared by Tom Latham is 'soft and pliable, quite firm, and yet so soft'.[241] Many of the skins were 'tanned in their natural colours'.[242] Gardner, who worked in Launceston in the 1860s, produced opossum skin rugs that stood out for their softness, evenness of appearance and 'joinings [that] are barely perceptible'.[243] The thylacine buggy rug manifests good tanning and relative uniformity. The rug continues to be appreciated both tactilely and visually, with the TMAG report of the purchase noting that the skins are not faded and are 'very thick and soft'.[244]

One possible aim in displaying the rug may therefore have been to testify to Robert Stevenson's skill in taxidermy. Katrina Schlunke offers other proposals, suggesting that the cover may have been 'a calculated bid for status' that showcased prowess at hunting or needlework.[245] In a statement about the rug's significance, TMAG refers to rug-making in general as 'a practical use of available resources'.[246] There is a 1903 report in the *Australian Star* of a (probably wool) buggy rug with a tiger pattern that was recovered from a presumed thief. Making a rug out of thylacine skins enabled the Stevensons to emulate the fashionable tiger stripe pattern (also common in railway rugs) but through using actual animal pelts, locally sourced, rather than engaging in simulation. TMAG notes that '[t]he thylacine skin rug is representative of frontier colonial culture and the tradition of using immediately available resources'. Efforts are made here to frame the rug as an embodiment of pioneer spirit, as evidence of practicality and self-sufficiency. The museum encourages an image of the Stevenson family as adapting to adverse circumstances through creativeness. In reality, by 1903 Robert Stevenson was a wealthy man who employed a private tutor for his five children.[247] He could afford to buy a carriage rug from a commercial furrier so the decision to create one, as Schlunke indicates, clearly potentially had a more complex symbolism that went beyond economy and practical ingenuity.[248]

When seeking to understand what the rug probably meant to Stevenson and his family, however, their specific relation to the thylacine also needs to be considered. At Aplico, which was located in Upper Blessington, Stevenson kept sheep (merinos and crossbreeds) and cattle.[249] He claimed that he had lost between fifty and eighty sheep a year to the 'ravages of tigers' in the 1890s.[250] He also professed to have 'lost as many as 400 sheep in a Winter' because of the 'native pest'.[251] Given his belief that thylacines were sheep killers, it is clear his opinion of the animals was low. A keen amateur engineer known for designing a private bridge at North Esk, Stevenson also put his technical skills to the destruction of tigers, devising a trapline at Aplico that consisted of a wire fence line 'with wings running off it . . . designed to direct wildlife to snares, gin traps, pitfall traps and box traps'.[252] Most of the wings led to pitfall traps, large wood-lined pits. Once the thylacine was extinct and the pits were no longer needed these were either burned or filled in, the architecture of genocide obliterated.[253] The eight juvenile thylacines used to construct the lap rug were killed along the trapline. In all, about sixty tigers were caught at Aplico between 1890 and 1906.[254] To the farmer, the rug therefore probably commemorated his ingenuity and the lethal efficiency of his fencing setup. It was a trophy, a symbol of victory over a reviled adversary. Stevenson possessed another farm, Elverton, where as many as sixteen skulls of thylacines were 'fixed onto old harrow tines wedged into a large tree stump'.[255] The rug should be viewed as on a continuum with this callous display but, if it was used in a buggy or a cart, it would have offered a more public, travelling exhibition of the destruction of tigers.

The rug may have been intended to be worn in a particular direction. The heavier wear visible on one quartet of skins suggests that it may have been used with the central seam kept at a horizontal (probably running across the knees).[256] This would mean that the thicker stripes from the rump of the thylacine pointed upward and the thinner stripes faced groundward. If this was how the rug was worn, it would have created a striking visual effect of downward motion as the stripes (which look like downfacing arrows) grew steadily thinner and smaller towards the earth. People seeing the rug would unconsciously be encouraged to look down, to momentarily lower their eyes. Pierre Bourdieu links this gesture with humility and gives it as an example of submissiveness.[257] In this sense, the rug helped to affirm the elevated status of the Stevenson family, masters of all they locally surveyed. Psychologically, it is also significant that the heads of four tigers, were they still present, would have been located at the feet of those travelling in the vehicle, a fallen adversary. The design therefore potentially works to affirm human dominance over the thylacine. The second owner of the rug, Foster, was also a sheep farmer, a breeder of prize-winning merinos.[258] For him, too, the rug therefore probably represented triumph over a despised enemy.[259] By using it to cushion a piano stool, human domination was affirmed even more strongly than when the rug was employed in a buggy. People in the Foster household sat their bums on tigers, figuratively wiped their arses with them. The dead thylacines were degraded and demeaned.

Now popular perceptions of the thylacine have shifted and the rug is valued differently. It is feted for its fine condition, with emphasis placed on the resourcefulness it embodies. TMAG also observes that it is 'a poignant reminder of the effects of colonisation in transforming the landscape and its ecosystems'.[260] The mention of the transformation of the ecosystem can be taken as an oblique reference to the wholesale slaughter of the thylacine and the apex predator's extinction. The genocidal perspective that led to the rug's creation is readily indexed through its history of use and should not be overlooked. Schlunke observes that 'so many skins, so much death standing in for the ultimate death of all thylacines' is 'sickening'.[261] Her queasy response to the rug points to its contemporary significance as a record of animal genocide.

In a similar way to the psychic violence encoded in how the buggy rug was used in the Foster household to articulate unequal power relations, the hearth rugs also speak to fantasies of domination and humiliation. Humiliation here is understood as being treated with contempt and reduced to lowliness. The word is etymologically rooted in 'soil', in *humus*. To be humiliated is therefore to fall in stature, to be put down, debased. It is recognised as an important dimension of some genocides. Evelin Lindner has explored the role of humiliation (as act, feeling and process) in the history of genocidal killing. She notes that humiliation can be a motor force for genocide (the humiliated seeking revenge) and a process inflicted upon its victims.[262] Usually, humiliation is thought of as a harm inflicted on human individuals.[263] The humiliation of inanimate things, however, also has a long history.[264] Through the creation of artefacts such as rugs, some settler colonists in Lutruwita gave expression to their disdain for thylacines. These artefacts acted to humiliate post-mortem, exploiting the radical passivity of death to produce subordinate bodies, bodies brought low and repeatedly trampled underfoot. Nic Haygarth waggishly suggests that the skin of a female thylacine killed by Clem Penney was used as a doormat.[265] The quip captures the psychological dynamics underpinning the rugs; they figuratively enabled humans to clean their feet on tigers. In one of two photographs of the Penney family gathered around the skin, Clem's mother has her foot on the thylacine.[266] The nonhuman animal is downtrodden, figured as defeated by way of this minimal yet potent gesture.

Cultural practices of humiliation, as Myra Mendible notes, work to constitute and enforce unequal power relations.[267] They serve to shape the popular imaginary. Thylacine hearth rugs potentially contributed to such a shaping through their power to symbolically demean.[268] In Erle Wilson's novel *Coorinna*, a shepherd shoots a thylacine and carries the carcass back to his shack where '[h]is boot spurned the wolf's body contemptuously'.[269] He also unceremoniously drags the dead tiger to an outhouse and throws him inside.[270] Here Wilson catches something of the attitude of farmers to the animal, an attitude that was probably embodied in numerous hearth rugs. The significance of such rugs would have depended on who made and owned them. The value (or lack thereof) accorded to individual animals contributes to the meaning attributed to rugs by their owners, a phenomenon noted in an 1891 article on the use of fur rugs in modern interiors. The author suggests that furs are not only noteworthy 'from a mere decorative point of view' but that 'the animal whose skin is thus utilized has a history of its own, which imparts a rare charm to the rug itself'.[271] For a farmer, a thylacine rug would potentially have levels of significance different to those of a city-dweller.

Farmers such as Foster and Stevenson viewed thylacines primarily as vermin. In such circumstances, the making and use of a thylacine buggy or hearth rug served to embody disdain and perform the tiger's subjugation. Hearth rugs, in particular, would have staged human superiority over the thylacine very effectively. Rugs were, of course, made from other animals and the meaning of a possum skin hearth rug, for example, would have been considerably different to one made from that of a thylacine. Although both animals were viewed as inferior, possums were not seen as a threat to livelihood and 'demonised'. In an article published in *The Examiner* in 1926, H. B. Martin described possums as beautiful gentle, pretty and harmless and railed against their wholesale slaughter.[272] A more critical article written in 1908 acknowledged their natural historical interest and the market value of their skins but saw them as a potential threat to crops.[273] In contrast to a rug manufactured out of a thylacine, a possum skin rug would therefore probably not have acted as a ritual staging of humiliation.

Thylacine skins that were not transformed into rugs also embody cultural values. Some skins, such as those held at the Smithsonian, were explicitly prepared for museum collections. They were viewed as of scientific importance. Others were clearly perceived differently. The Marchwiel thylacine skin (A1957), which is now held at TMAG, was previously in the possession of the Dunbabin family, a farming family who can trace their island heritage back to the arrival on the island of John Dunbabin in the early 1830s.[274] Dunbabin, sent to Lutruwita as a convict, came from Marchwiel in North Wales and when he purchased a property, he named it after the town.[275] The farm, Marchwiel at Bream Creek in south-east Lutruwita, sold sheep and wool.[276] It was kept in the family until 2008. Located near marshland where pademelons are still abundant and red-necked wallabies are also present, the farm was in habitat well suited to thylacines.[277] Approximately twenty tigers were caught at Marchwiel. Although the Dunbabins clearly viewed the tiger with loathing, the potential role of dogs in any sheep losses they may have incurred is indicated by a notice they posted in the Miscellaneous section of *The Mercury* in 1903: 'Trespassers Prosecuted. Poison Laid. Dogs shot. T. Dunbabin and Sons. Marchwiel.'[278]

The Marchwiel skin was donated to TMAG by James Dunbabin under the Cultural Gift programme.[279] Its geographical provenance is known, which makes it a rarity. Only four skins worldwide can be traced back to specific Lutruwitan locations or properties.[280] The Marchwiel skin was nailed to the farm's woolshed.[281] This shed was built very soon after John Dunbabin acquired the property. When Kenneth Graham visited it in 1975, he reported seeing newspapers from 1883 tacked to the walls.[282] The thylacine skin was also presumably still nailed there but unrecognisable, the tiger's telltale bands long since faded. In his *Natural History of Selborne*, Gilbert White writes of finding what transpired to be a peregrine falcon 'nailed up at the end of a barn, which is the countryman's museum'.[283] Unlike the falcon, the thylacine skin at Marchwiel was not a curio but a trophy. David Fleay observed that it used to be common practice in Lutruwita to nail the heads of tigers to barn doors.[284] Similarly, a 1929 newspaper article reminiscing about north-east Lutruwita in the late 1870s also recounts of thylacines that '[a] number of trees near the woolshed at Waterhouse were decorated with heads of these animals as trophies'.[285] Some of the thylacine skulls collected by George Masters for the Australian Museum were probably retrieved by him from fenceposts where they had been displayed as symbols of triumph or as threats. The erosion at the rear of the jaw of the skull PA767 (collected in 1866), for example, indicates it may have been found exhibited in this way. Most of the skulls retrieved by Masters were probably victims of genocide. The chances of him randomly coming across thylacine remains while travelling are negligible. Even those skulls recovered by Masters that bear no outward marks of violence (such as P768) therefore probably met a cruel death at human hands.

The display of the skin should be understood as on a continuum with this custom. Rachel Poliquin calls trophies 'souvenirs of termination'.[286] Like the Stevenson rug, the skin symbolised victory over the thylacine as 'vermin'. Choosing to exhibit the skin at the woolshed was not happenstance. Sheep were key to the farm's success and tigers were perceived to threaten flocks. Farmers have often displayed the carcasses of predators by stringing them up on fences as if advertising their comeuppance would discourage other predators from 'intruding'.[287] Nailing the skin to the woolshed seems qualitatively different to this practice. Skins must be prepared and even if such preparation is crude this involves a measure of effort. If the dead thylacine were intended to simply serve as a 'warning' to other tigers then their body would have been left elsewhere on the farm (such as on a fence), unskinned. Woolsheds are prominent structures, often large and enduring, nicknamed 'cathedrals of the bush'.[288] Andrew Chapman describes them as 'like living museums', combining ancient and modern technologies.[289] Marked by long-term use, their interiors preserve 'the cumulative smell of sheep and the knotches of shearers gone by'.[290] Yet as the reference to shearing suggests, they are also transitional spaces, used for cyclical practices. Sheep are therefore not kept in them for long periods of time. The sheds frequently host seasonal workers including shearers and roustabouts. Other farmers will often tour the woolsheds of neighbouring properties when visiting on sheep business.[291] Woolsheds are also sometimes used for social gatherings such as family reunions and dances.[292] The tiger was therefore positioned in a place where certain visitors to the farm were likely to encounter them.

Thomas Dunbabin, who was born at Bream Creek in 1883 to Thomas and Sarah Ada (née Murdoch) Dunbabin, wrote an article, 'Fossil Wealth', about vanishing wildlife, that was published in the New South Wales newspaper *The Sun*. Dunbabin described Australia as 'a museum of living fossils'.[293] He went on to note that '[s]ince white men burst the bars of the ocean they have been busy turning the living fossils into the more orthodox kind'. He describes this destruction as 'bad business' and calls for their preservation as they have 'a cash as well as a scientific value'. Coming from a family of tiger haters, Dunbabin's conversion is clearly not damascene but motivated by money. Making no reference to his family background, he writes: 'It is not many years since sheepowners in parts of Tasmania used to pay £1 each for the heads of that carnivorous marsupial the Tasmanian tiger (Thylacine) which is confined to the island. Now a good "tiger" is worth £50 and the price is going up.'[294] Like the Stevenson family, the Dunbabins seemed to come to recognise that the financial value of live tigers outweighed the satisfaction they derived from killing them. The rarity of the tigers by the 1920s was, of course, partly attributable to the genocidal actions of farmers such as Thomas Dunbabin Senior.

Like the rugs, the Marchwiel skin archives feelings directed towards the thylacine. Ann Cvetkovich has explored the emotional power possessed by some archives, the ways they act as repositories of feelings as well as storehouses of data. Cvetkovich is particularly interested in absent archives, in histories that have left no records, but her ideas about the emotional power of archives are also useful in the context of unwritten documents such as the rugs and the Marchwiel skin. These artefacts provide material evidence of animosity. They are, however, usually only mentioned in passing, overlooked in favour of textual or visual expressions of cruelty towards the tiger.

The rugs and the skin are nonetheless important as personal manifestations of detestation embodying the effects that those negative feelings had in practice. The genocidal gaze facilitated the killing of the thylacines used to make these artefacts and it lives on in them. Cvetkovich encourages paying attention to 'the strong feelings that get attached to things'.[295] The rugs and the Marchwiel skin are made out of loathing, a loathing that still adheres to these remains. In *The Cultural Politics of Emotion*, Sara Ahmed writes of the way affects can become attached to objects, to bodies, can stick to them. These affects work to shape how bodies are received and understood. The reception of the thylacine's body, dead or alive, was shaped by extremely negative feelings that became 'stuck' to them through time. In the present, different feelings now accrue to the physical remains of the tiger – feelings including guilt, shame, sadness and wonderment – but they do not obscure the historical detestation, a lingering hatred that can still be felt when examining the archive.

Other examples of this loathing have not survived but are known through textual or photographic evidence. One of the last thylacines captured alive, for instance, was transformed into a waistcoat upon her death.[296] In August 1930, the brothers Gordon and Roy Delphin caught the tiger and kept her captive in a pigeon loft (an outdoor shed) where she died shortly afterwards. Their aim was no doubt to peddle her to a zoo as they did with a male tiger caught by Roy and his uncle Dan Delphin in July of that year and sold to Hobart Zoo for £25 in October. Upon her death, the thylacine would still have possessed substantial financial value but Dan reputedly had her skin made into a waistcoat that he then wore with 'pride'.[297] He was well aware the tigers were critically endangered, using their rarity as a marketing ploy when exhibiting one at the Burnie Show in October 1930. An advert he placed in the local newspaper *The Advocate* noted that '[d]ue to the march of civilization [they] are becoming extinct'.[298] The capture was therefore a genocidal act, knowingly removing tigers (that might have otherwise bred) from their habitat to serve as spectacle and a source of financial gain. Dan referred to himself as 'the great West Coast Hunter' and turning the tiger into a waistcoat seems to have been a way to provide direct evidence of his trapping prowess.

Perhaps the most disturbing expression of abhorrence towards tigers is recorded in a pincushion, A1280 (see Figure 2.2), now in the TMAG decorative arts collection. The cushion was manufactured by Emily Harriett (née Nicholls) Ferrar (1857–1940) using the mandible of a juvenile thylacine.[299] It was donated to the museum by Sarah Mitchell (1853–1946) in 1937.[300] Ferrar made it for the Glamorgan Horticultural Society's show, which was held at Albert Morris's 'big room' in his store at Swansea.[301] Morris regularly gave over space in the upper storeys of the three-storey

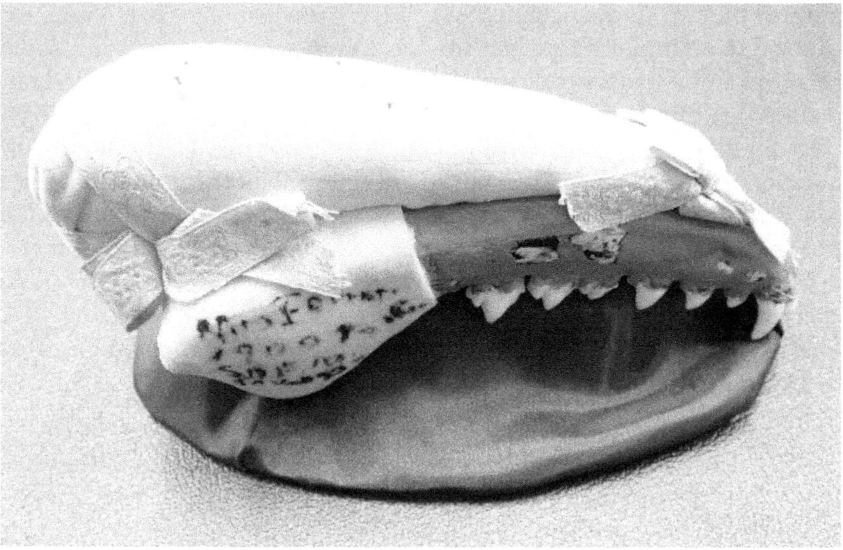

Figure 2.2. A1280. Thylacine pincushion. Tasmanian Museum and Art Gallery, Nipaluna/Hobart. Photo taken by Nicholas Chare. Reproduced with permission.

brick building to local events such as jumble sales and the show. The free space in his shop in the early 1900s registered the population decline of Swansea and the accompanying dwindling market for store goods.[302] Emily Ferrar seems to have been a regular participant in the show with her presence first mentioned in 1901. At the show, which was attended by around two hundred people, flowers, vegetables, baked goods and needlework were exhibited and prizes awarded. The pincushion is reported by TMAG to have garnered second place in the 'fancy needlework' category.[303] A note accompanying the cushion suggests that it was made to win 'most original' but does not specify the category.[304]

Beverly Gordon has noted that in fancywork, natural phenomena were frequently subject to processes of domestication. She gives the example of animal hooves stuffed with velvet.[305] Jerry Jeanmard also discusses how animal parts were viewed as highly decorative by the Victorians but now potentially appear repulsive.[306] In this context, Ferrar's pincushion, while deeply unsettling today, would not have been outlandish. The pincushion is thought to have been submitted to the show in 1900 as it has 1900 written on the fabric in black ink. This information was added later, as it is initialled 'S.E.E.M' in Sarah Elizabeth Emma Mitchell's unsteady hand, and its accuracy is therefore questionable.[307] Mitchell probably inscribed the pincushion at the same time (21 May 1930) as she wrote a note detailing its provenance. There is no mention of 'Mrs Ferrar' in the newspaper reports of the two shows held in 1900.[308] The participation of Mrs W. Ferrar is noted in autumn 1901, when she was commended for exhibiting 'some beautiful and artistic woodcarving'.[309] She is then mentioned regularly in reports about the show thereafter.[310] Mrs Ferrar is referred to as receiving a prize for 'Best Unique Pin Cushion' in 1911.[311] This may be the cushion in question. At the time she was contributing to the horticultural show, her son Fergus Moore Ferrar was running a 5,400-acre farm, Milton, which was located at Cranbrook near Swansea. The Ferrar family raised merinos and crossbreeds and, like the Dunbabin and the Stevenson families, claimed to suffer considerable stock losses caused by thylacines.[312] In a note penned by Sarah Mitchell in 1930, she explains that the jawbone used by Ferrar derived from a thylacine shot at Milton:

> Mr John Lyne M.H.A got a bill Through Parliament For the Ferrar family to make a fenced yard with gaps and spring guns that went off when [a tiger] stepped on the gap many Tigers were shot (17?) This jaw was one, they called him 'Gun Lyne', because of it.[313]

Lyne was a member of another farming family in the Swansea area, living at Apslawn adjacent to Cranbrook. While the elected representative of Tebranuykunna/Glamorgan (a seat he held from 1881 to 1893), he pressured Lutruwita's House of Assembly to adopt a bounty on the thylacine. Spring-guns, shotguns rigged to fire using a triggering device, were illegal in Lutruwita from 1863 (bringing the island into conformity with English law).[314] There was, however, clearly confusion about their legality and they were employed on farms. A legal case was brought in October 1893 against a member of the Ferrar family after a dog belonging to a shepherd in the employ of the grazier Charles Headlam was killed by such a gun. The shotgun had been set by one of the Ferrar brothers 'for the purpose of killing tigers'.[315] The Ferrar family won the case but Lyne does not appear to have been directly involved and by 1893 he had already retired from representing Tebranuykunna/Glamorgan in the House of Assembly. The nickname 'Gun Lyne' could derive from earlier lobbying for spring-guns or from a different event. In 1885, for instance, while Warden of Tebranuykunna/Glamorgan, Lyne sought to form a rifle corps to support the Glamorgan Association for the Suppression of Sheep and Cattle Stealing, and the Extinction of Tigers.[316]

The extended Lyne family participated regularly in the Glamorgan Horticultural Society's show alongside the Ferrar family.[317] Members of the Dilger family were also usually present.[318] Alfred William Dilger, who was born in Swansea but moved to Mathinna, caught a 5 ft thylacine in a snare on his farm of Myrtle Grange in 1912.[319] He sent the dead animal to QVMAG where he is now a prized exhibit. The pincushion was therefore displayed in a milieu where farmers with historic animosity towards the tiger grouped together. Emily Ferrar could therefore expect (but not guarantee) a sympathetic reception for her creation. Like Stevenson's buggy rug, the pincushion can be seen to celebrate killing efficiency. Stevenson employed a complex system of wire netting and pit traps to eradicate tigers, William Moore Ferrar constructed a method using fencing and spring-guns.[320] The pincushion indexed the 'value' of the spring-gun in the same way as the buggy rug did the trapping pit.[321] Both registered family pride in tiger slaying. The rug may have been created by a woman and the pincushion certainly was. This demonstrates that animal genocide was not solely enabled by men. Women also contributed to the cultural climate of permissiveness regarding violence towards thylacines. Sometimes they actively facilitated killing. For example, the thylacine Wilfred Batty killed on his father William Batty's property was first spotted by one of his sisters (probably the youngest, Hilda Batty). It was she who drew the family's attention to the presence of the tiger 'prowling' in the yard.[322]

Pincushions, as fancywork, formed one means by which women in the nineteenth and early twentieth centuries could explore their creativity. They were part of the domestic everyday, mute witnesses to quotidian life. This was recognised in literature through Mary Ann Kilner's *The Adventures of a Pincushion*, which granted the cushion the powers of thought and sense and traced its existence from its being made, through escapades with numerous owners across social classes, to its eventual *fall* beneath a water tub.[323] In *Pins and Pincushions*, Eleanor Longman and Sophy Loch write that if a magic wand could give old pincushions 'the gift of speech, what interesting and romantic tales might be unfolded to us – tales that would carry us back into that golden age we call the "past," over which it is customary to mourn'.[324] Although initially resistant to the idea that objects can 'speak' of their history, they later observe of one cushion, 'when our thoughts turn to the loving hands, long since at rest, that fashioned and made use of it, such memories fill us with "a gentle sense of gloom," and with a passing sigh we tenderly cover up and put away this little treasure.'[325] They acknowledge that pincushions and similar objects are 'treasures touched by a thousand memories'.[326] Although the objects cannot speak and provide specifics of their past, they clearly possess a strong affective dimension, handling them leading Longman and Loch to poignant reflection. The thylacine pincushion possesses a comparable affective charge but one that is deeply unsettling and painful rather than mildly upsetting.

Although it was stored with the osteological remains at TMAG when I first examined it in 2019, the pincushion's importance to mammalogy is marginal. Its significance lies in its encoding of the intense hatred the tiger was subject to in Lutruwita's rural communities. Now, the pincushion is located in the decorative arts collection along with the buggy rug. To grasp the psychic underpinnings of the cushion, it is first necessary to study its form closely. The thylacine mandible is inverted, the ramus is wrapped in velvet, but the bone of the corpus (the thick part that anchors the teeth) is left exposed. The corpus has been painted blood red, but the thylacine's teeth are unpainted except at their necks where pigment applied to the bone has scuffed the enamel. The cushion proper (which is made either of batting, bran or wool) is coated in cream-coloured velvet and rests in the empty space between the body of the mandible. It is located where the floor of the tiger's mouth would have been. The front and rear of the cushion are decorated by floral patterned ribbon tied in bows. This ribbon, edged in salmon-pink, is not handcrafted. As already mentioned, the velvet covering the mandible has been written on by Mitchell. Either she or, much less likely, Ferrar positioned the pins in the cushion so that they are in the shape of an 'S' (presumably for Sarah). TMAG has inscribed the catalogue number A1280 in black ink on a band of unpainted bone adjacent to the inscribed velvet.

Pins have long been associated in European folklore with protecting against malevolent spirits. Historically, pins were used as 'charms for the prevention and cure of various diseases, to keep off bad dreams, to prevent ill-luck, to bring good-luck, besides wounding one's enemies'.[327] Something of these superstitions may feed into the decision to transform the mandible into a pincushion. This encourages regularly enacting violence upon the remains of the 'verminous' animal. In her poetic exploration of the pincushion and its significance, Danielle Wood calls the pins 'thinner sharper cousins perhaps to the bullet'.[328] The metaphorical linkage of pincushions with physical attack is well attested to in the nineteenth century and was potentially still resonant in the early twentieth century. The hymn-writer Henry Fowler, for example, suggested that any church minister who settled in Birmingham must have made up his mind 'to be a lady's pincushion to be pierced through and through'.[329] Eden Lowther's short story 'The Painter and his Pupil' features a young artist, Francesco Ribalta, who fantasises about being attacked by painted figures who gather a forest of thorns and stick them 'into his hapless body just as if he had been a pincushion'.[330] Seen in this light, Ferrar's cushion seems an encouragement to virtual aggression against the thylacine. The gesture of stabbing a pin in the wadding would therefore have figured enmity towards the animal. Her negative conception of the marsupial is clearly signalled by the gory red jaw, the colour symbolising something of the carnage attributed to the thylacine, an animal referred to as 'bloodthirsty' by some.[331] The artificial cruor gives graphic expression to the notion of the tiger as sheep-killer, rendering the cushion a studied exercise in abhorrence. Made to confirm popular stereotypes, the thylacine is framed as an object of hate by Ferrar's genocidal gaze.

Later, when gifted to Mitchell, the meaning of the cushion would change. Mitchell owned Schouten House in Swansea for decades, sometimes renting the property and at other times living there.[332] In the 1930s, the residence was clearly a local attraction, with visitors from elsewhere in Lutruwita dropping in to admire it. Mitchell, who dwelt there then, had treasures and mementoes on display.[333] The house was a small private museum and it is probable that the pincushion was one of its exhibits. The writing she added to the artefact in black ink was therefore likely intended for the benefit of visitors. By 1930, the thylacine was very rare, changing the reception of the pincushion. It was no longer viewed with active hatred but rather seen as an important piece of local history: the tiger was a thing of the past, a museum piece. Because of this transformed reception, Mitchell would, in her turn, gift the artefact to TMAG. In the next chapter, I will consider how the thylacine is presented in museums today. Often in museums the evidence of genocidal hatred indexed by the remains discussed in this chapter continues to be overlooked or ignored.

Notes

1. Sleightholme and Campbell, 'The International Thylacine Specimen Database', 494.
2. Sleightholme and Ayliffe, *The International Thylacine Specimen Database*, 16.
3. Ibid., 14.
4. Ibid.
5. Ibid., 144.
6. Turner, *Cataloguing Culture*, 66.
7. Ibid., 130.
8. Wylie, 'Overcoming the Underdetermination of Specimens', 24.
9. Sleightholme and Campbell, 'The International Thylacine Specimen Database', 499. Sleightholme and Campbell also discuss how thylacine skins used by furriers were sometimes later acquired by museums, in this instance shifting from 'pelt' to 'specimen' (500).
10. I am grateful to Kristine Tanton for sharing her insights into databases and the digital humanities.
11. Swinney, 'What Do We Know about What We Know?', 37.
12. Adorno, *Negative Dialectics*, 362; Adorno, *Negative Dialektik*, 355. The German word 'Exemplar' that is translated as 'specimen' is used in museums in the same way as 'specimen' in English. A type specimen, for example, is a *Typusexemplar*.
13. Ibid.
14. For a discussion of detachment see, for example, Goss et al., 'Not Just a Specimen'; Robbins, *The Medicalized Body and Anesthetic Culture*, 151–65. Interchangeability here refers to the idea that a 'specimen' of a species is viewed as replaceable by another of the

same species and does not imply that differences across specimens of a given species pass unrecognised. In the past, for example, thylacine mounts would 'wear out' and, if possible, be replaced by 'fresher' ones. The old and the new mount were both 'specimens' of the 'thylacine', a type that transcended their specificities. See, for instance, Medlock's discussion of the replacement at the Tasmanian Museum and Art Gallery (TMAG) of the Buckland group (a display of a thylacine family) created in 1884 by a newer group in the 1920s. Medlock, 'Exhibiting Extinction', 396–98.

15. For this reason, in *Should Trees Have Standing?*, Christopher Stone encourages a rejection of the view of nature as 'a collection of senseless objects' and a move towards recognition of our 'mutual interplay' with other organisms in the natural world (49).
16. Black, *All That Remains*, 19.
17. For a discussion of databases as historically and politically invested technologies, see Christie, 'Computer Databases and Aboriginal Knowledge'.
18. Grinnell, 'The Methods and Uses of a Research Museum', 168.
19. Halsey, *Deleuze and Environmental Damage*, 47.
20. The thylacine remains in the collection of the American Museum of Natural History (New York) were also consulted by my research assistant Valérie Bienvenue. Additionally, I studied the thylacine remains held by the Royal Ontario Museum (Toronto) in Canada. These remains, however, do not feature in the analysis offered here.
21. Poliquin, *The Breathless Zoo*, 129.
22. I am grateful to Kathryn Medlock for this information.
23. This situation is not unusual. The majority of the collections held by large natural history museums never go on public display; some materials are kept back because of their fragility and others because it has long been recognised that displaying physical remains that lack perceived aesthetic appeal can deter visitors. Such remains can be of creatures that are deemed uninteresting or repulsive or that betray obvious signs of damage or decay. See Kohlstedt, '"Thoughts in Things"', 597.
24. Anderson, *Methods of Collecting and Preserving Vertebrate Animals*, 8.
25. Grinnell, 'The Methods and Uses of a Research Museum', 163. The research museum is conceived by Grinnell as a place of darkness and the exhibition museum as one of light.
26. Older methods for stuffing study skins are detailed in Anderson, *Methods of Collecting and Preserving Vertebrate Animals*, 15.
27. Rajan, 'What Do Refugees Want?', 529.
28. Katrina Schlunke finds skins and furs the most evocative physical remains of thylacines, bespeaking connection. See Schlunke, 'The Practices of Care', 185. Schlunke's article, which is partly devoted to an analysis of thylacine remains held at the Berlin Museum of Natural History, calls for natural history collections to embrace Indigenous informed practices of care.
29. If duplicate material in better condition exists, imperfect remains of species are still sometimes retained to be used for destructive sampling.
30. Having had a rural upbringing and later worked on a sheep farm, seeing dead animals in various stages of decomposition did not have the impact on me that some staff, possessing their understandable preconceptions, clearly presumed it might. I therefore had the odd sense of somehow letting these staff down, of not reacting 'properly' and adequately performing my disgust. This made me self-conscious regarding my indifference and its implications. I possess a different outsider status in terms of my cultural positioning relating to the thylacine and the Newfoundland wolf, both 'First Animals' who had close links to First Peoples, links to which I, as a Luxembourger from Europe, cannot be privy.

31. Although my visits to TMAG occurred in 2019 and 2022 and fall outside the time frame of its research (May 2017–March 2018), which was part conducted at the museum, the project regarding wedge-tailed eagles carried out by James Pay et al. seems motivated by similar aims to the autopsying I witnessed. See Pay et al., 'High Frequency of Lead Exposure'.
32. See ibid., 228. Mooney has examined the potential negative impact of rodenticides on Lutruwita's raptors. Mooney, 'Risk of Anticoagulant Rodenticides to Tasmanian Raptors'.
33. Museum collections do acknowledge 'particularity' understood as individuation within a category. Different weights and sizes, for instance, are carefully recorded. For a useful discussion of 'singularity' and 'particularity' and their differences that is indebted to Simone de Beauvoir's philosophy, see Parker, 'Singularity in Beauvoir's *The Ethics of Ambiguity*'.
34. Some nonhuman animals such as dolphins seem to use their own system of 'names' for identifying individuals. See Janik et al., 'Signature Whistle Conveys Identity Information to Bottlenose Dolphins'. Many animals recognise individuals of their species not through the use of names but via auditory and visual cues. See, for example, the discussion of individual recognition in horses in Proops et al., 'Cross-Modal Individual Recognition in Domestic Horses'.
35. See Haque and Waytz, 'Dehumanization in Medicine', 181.
36. It was also common in scientific research to refer to nonhuman animals by numeric identifiers rather than names. Jane Goodall was one of the pioneers who accorded the chimpanzees she worked with names rather than numbers. See Goodall, *Through a Window*, 16–17.
37. Here the numbers are used as names and, where possible, thylacines are referred to as 'she' or 'he' to avoid objectifying the animals, as occurs when the inanimate pronoun 'it' is utilised. In the absence of information about the gender of a given 'specimen', 'they' is used. For a discussion of language issues in relation to the representation of nonhuman animals, see Freeman et al., 'Giving Voice to the "Voiceless"', 600–2.
38. Crowther, a medical doctor, was an amateur naturalist. He presented a paper on duck-billed platypuses to the Royal Society of Tasmania on 14 October 1880. Crowther, 'On Some Points of Interest Connected with the Platypus'. One of his brothers sourced thylacine materials for William Henry Flower. See Crowther, 'Appendix', 173. Following in the footsteps of his father (who desecrated William Lanne's grave and stole his skull), Crowther was responsible for desecrating the grave of Truganini and stealing her skull.
39. See Douglass Rovinsky's X (then Twitter) posts regarding his research on NRM 566599. Rovinsky, Douglass S. X Posts. 21st August 2020, 14:34. http://twitter.com/DSRovinsky (Accessed 24th June 2021).
40. Rovinsky et al., 'Did the Thylacine Violate the Costs of Carnivory?'
41. See also Rovinsky et al., 'Functional Ecological Convergence'.
42. In the Smithsonian, for instance, early policy was for skins separated from skulls of the same animal to be each allocated a different number.
43. See Rosenberg and Khoo's discussion of mollusc 'specimens' and provenance in 'Improving Provenance Data in Natural History Collections'.
44. Pedder-Smith, 'The Glow of Significance', 31.
45. Belk et al., 'Collectors and Collecting', 550.
46. Jackson, 'The Care of Museum Specimens of Recent Mammals', 113.
47. Hills, *The Matter of Miracles*, 40.
48. Hughes, 'Authority, Authenticity and Aura', 41.

49. For a considered discussion of 'miraculous footprints' as relics in mediaeval Europe (with a particular emphasis on Castile), see Nickson, *Toledo Cathedral*. In the Melbourne Museum, there is a set of casts of two distinct animals preserved on the same slab of plaster, an impossible conglomeration of prints of two animals that were donated sixty years apart, C5745.3 (R348) (presented in February 1900) and C3149.4 (presented in 1961). These animals, both male, were clearly quite different in size. The four feet from two different dead animals now function as a type, used to represent a thylacine's tracks.
50. Hills, *The Matter of Miracles*, 29.
51. Jackson, 'The Care of Museum Specimens of Recent Mammals', 113.
52. Miller et al., 'The Mitochondrial Genome Sequence of the Tasmanian Tiger'.
53. The gender of the thylacine that is M27836 is not known.
54. Fletcher, *Mendel's Ark*, 80.
55. Ibid., 82.
56. Ibid.
57. The ethical questions raised by the project are beyond the scope of this book. It bears noting, however, that thylacine pups were likely educated in hunting techniques by their mothers. Their behaviour was not entirely instinctive. A cloned thylacine might look like the animal that went extinct, however it would be unlikely to behave like them. Such a tiger would risk being a human-dependent curiosity.
58. Hills, *The Matter of Miracles*, 51.
59. Maynard and Gordon, *Tasmanian Tiger*, 79.
60. Chrulew, 'Managing Love and Death at the Zoo', 142.
61. Citing Mike Archer, Stark suggests that the pup is six months old. By this stage the pup would have fully left the pouch, so this is incorrect. See Stark, 'The Cultural Politics of Mourning in the Era of Mass Extinction', 71.
62. See Butler, *Precarious Life*, 32.
63. Stark, 'The Cultural Politics of Mourning in the Era of Mass Extinction', 74.
64. Ibid., 77.
65. Ibid.
66. Butler, *Precarious Life*, 32.
67. Stark, 'The Cultural Politics of Mourning in the Era of Mass Extinction', 72.
68. Ibid.
69. Ibid., 75.
70. The pup was removed using metal tweezers manufactured by a company called Australian Entomological Supplies. The tweezers were reminiscent of obstetrical forceps. The thylacine was thus figuratively delivered, birthed from the jar-womb that housed and protected her.
71. See the discussion of A934 in Newton et al., 'Letting the "Cat" out of the Bag', 12–13.
72. This has long been recognised. Haswell and Parker, for example, note that in the thylacine the epipubic bones are represented 'only by small unossified fibrocartilages'. See Haswell and Parker, *A Text-Book of Zoology*, 492.
73. See, for instance, Pearson, 'The Female Urogenital System of the Marsupialia'; Pearson, 'The Relationships of the Potoridae to the Macropodidae'.
74. See Pearson and Bavay, 'The Urogenital System of the Dasyurinae and the Thylacininae'.
75. Trustees of the Tasmanian Museum and Art Gallery, *Tasmanian Museum and Art Gallery Annual Report 2005–2006*, 19.
76. This thylacine was one of four studied by Arthur Lyne as part of his exploration of vibrissae in the Marsupialia. A line drawing that refers to him by his old TMAG

catalogue number of 589 is reproduced in Lyne's thesis. See Lyne, 'The Vibrissae and Pelage Hairs in the Marsupialia', 20.

77. In reality, his colouring may have faded because ethanol can cause whitening of fur, as can light bouncing within the storage container. I thank Belinda Bauer for explaining this to me.
78. Newton et al., 'Letting the "Cat" out of the Bag', 6. The fifth iteration of the ITSD has A930 listed as purchased dead from Hobart Zoo for £3 on 3 April 1924.
79. The thylacines are preserved in glass jars. Acrylic containers, a relatively recent invention, reduce distortion. See Asma, *Stuffed Animals and Pickled Heads*, 75.
80. Lacan, *The Four Fundamental Concepts of Psychoanalysis*, 79–90. For a useful summary of Lacan's reading of the painting, see April Cheetham, 'A Veiling of Identity'.
81. The cuttlebone, the internal shell of the marine molluscs known as cuttlefish, is used for buoyancy. Jurgis Baltrušaitis describes the skull as floating about the floor (for Lacan, it is flying), rendering the simile apposite. Baltrušaitis, *Anamorphic Art*, 91.
82. For a discussion of varied interpretations of the work, see Escobedo, 'Holbein's *Mementi Mori*'.
83. Kearl, 'The Proliferation of Skulls in Popular Culture', 2.
84. Lacan, *The Four Fundamental Concepts of Psychoanalysis*, 88–89.
85. Fletcher, 'Bring 'Em Back Alive', 194.
86. Woolley has considerable expertise in marsupial reproductive systems and published her PhD on the 'Reproductive Biology of Antechinus stuartii Macleay (Marsupialia: Dasyuridae)'. Sleightholme and Campbell, 'The International Thylacine Specimen Database', 485n10.
87. Maynard and Gordon, *Tasmanian Tiger*, 79.
88. See Fletcher, 'Genuine Fakes', 54. In efforts to extract ancient human DNA, the petrous bone (a dense portion of the temporal bone at the base of the skull forming part of the inner ear) has been found to provide particularly impressive DNA samples. Petrous bones are possessed by thylacines and would therefore seem rich for sampling but the process is destructive and may prevent future researchers from properly studying the inner ear of the marsupial. For a discussion of the human petrous bone as a DNA source, see Pinhasi et al., 'Optimal Ancient DNA Yields'.
89. Miller et al., 'Supplement', 4. For this same project, failed samples were taken from the skin of the mount of an adult female, USNM 124662, held at the United States Museum of Natural History (USNM), Smithsonian Institution. DNA had successfully been recovered from 'USNM 49723' by Carey Krajewski et al. They sourced the DNA from 0.6 g of dried muscle and cartilage 'removed from a prepared skeleton' (Krajewski et al., 'Phylogenetic Relationships of the Thylacine', 20). According to the ITSD, USNM 49723 is the catalogue number of the skull linked to taxidermy mount USNM 124662. Either Krajewski misremembered a mount for a skeleton or the skeleton in question was that linked to the skull USNM 49724, the skull of the son of USNM 49723. The skin, brain and skeleton of USNM 49724 are jointly preserved by the Smithsonian under the catalogue identifier USNM 125345.
90. Old, 'Immunological Insights into the Life and Times of the Extinct Tasmanian Tiger', 4.
91. See Sleightholme et al., 'Description of Four Newly Discovered Thylacine Pouch Young', 233.
92. They were studied quite recently by Julie Old in relation to the development of the animal's immune tissue. See Old, 'Immunological Insights into the Life and Times of the Extinct Tasmanian Tiger'.

93. For a discussion of the number of wet remains known to exist, see Sleightholme et al., 'Description of Four Newly Discovered Thylacine Pouch Young'.
94. Old, 'Immunological Insights into the Life and Times of the Extinct Tasmanian Tiger', 3–4. Old nonetheless observes that the serially sectioned pup provided 'invaluable histological descriptions of the immune tissues of an extinct marsupial' (7).
95. The material sampled from the wet thylacine became a sequencing library that was sequenced to produce a set of 'reads'. A 'read' in DNA sequencing efforts refers to the detection of a sequence of base pairs (units of double-stranded nucleic acids that form the foundation of the DNA double helix) that correspond to all or part of a single DNA fragment. The reads for the wet thylacine had an average length of 67 bp (base pairs). This was shorter than for the dry thylacine (87.5 bp) and is a poor length compared to what would be expected from modern samples. See Miller et al., 'The Mitochondrial Genome Sequence of the Tasmanian Tiger', 214.
96. Locard seems not to have used these precise words (although they are regularly attributed to him). He expresses a similar sentiment in the chapter on traces (Chapter 3) in his manual on police techniques. There, in developing what has become known as the exchange principle, he writes that 'it is impossible for a criminal to act . . . without leaving traces of their passing'. See Locard, *Manuel de technique policière*, 80.
97. See Miller et al., 'Supplement', 10.
98. Fletcher, 'Bring 'Em Back Alive', 198.
99. Ibid.
100. Krefft mentions twenty-six skulls sourced by Masters. Using some of these skulls as evidence, he mistakenly argued for two species of thylacine, known by the colloquial names of the greyhound and bulldog-tiger, and proposed the name *Thylacinus breviceps* for the latter, smaller of the two. Krefft, 'Description of a New Species of Thylacine'. In the later publication *The Mammals of Australia*, Krefft continued to urge recognising two distinct species: the dog-headed thylacine (*Thylacinus cynocephalus*) and the short-headed thylacine (*Thylacinus breviceps*). In another publication from 1868, *Notes on the Fauna of Tasmania*, Krefft notes that Masters's trip was 'about a year ago' (3). The Australian Museum's annual report for 1867 lists twenty-five thylacine specimens collected by Masters in the period December 1866 to March 1867. See Krefft, 'Appendix 6'. Whitley suggests that Masters was in Lutruwita some time in 1866 and then from March to May of 1867. See Whitley, 'George Masters, Naturalist', 26. There are twenty-eight candidate skulls in the Australian Museum for the twenty-six skulls (or twenty-five, depending on accounts) sourced by Masters so there is uncertainty over precisely which ones he collected. I am grateful to Harry Parnaby for sharing this information. Parnaby also notes that for those skulls with accompanying mandibles it is not known if Masters bound the mandibles and skulls together when transporting them. If he did not, some mandibles may be attributed to the wrong skull.
101. Krefft, *Notes on the Fauna of Tasmania*, 3.
102. Krefft writes in *The Mammals of Australia* of two skulls sourced by Masters on the Ouse River (6).
103. In *More Natural History Essays*, Graham Renshaw writes that in 1868, 'the thylacine, owing to constant persecution, had been extirpated from the more settled districts of Tasmania, though it still abounded in the wilder and less accessible regions' (223).
104. The years in which the remains entered the Australian Museum collection do not necessarily tally with the year when each thylacine was killed. This is clearly the case with the 1893 skull and skin, which were already part of the TMAG collection (TMAG419).

105. See the discussion of mammoths in Cottrell et al., 'Resuscitation and Resurrection', 6–11.
106. Kearl, 'The Proliferation of Skulls in Popular Culture', 6.
107. For Cambodia, see, for example, the discussion of the significance of skulls displayed at the killing field Cheong Ek in Williams, 'Witnessing Genocide'. In the context of Rwanda, see Mirzoeff, 'Invisible Again'.
108. In non-Western cultures this is not the case, with nonhuman animal skulls frequently possessing ritual importance and symbolic force. See Elbroch, *Animal Skulls*, 1.
109. Koestle-Cate, 'Patrice Moor', 19.
110. Skull measurements were also utilised in racial anthropology to distinguish between 'races'. The skulls of palawa, for instance, were plundered and studied by the University of Melbourne anatomist and eugenicist Richard James Arthur Berry who then positioned the palawa in a racial hierarchy. See Cawte, 'Craniometry and Eugenics in Australia', 35–53, 43–44.
111. Sleightholme and Campbell, 'The International Thylacine Specimen Database', 495.
112. Ibid.
113. Writing in 1927, the then director of TMAG, Clive Errol Lord, also distinguished the skull of a thylacine from that of a dog using careful cranial comparisons. He suggested that the former exhibited such traits as an expanded zygoma, contracted parietals and an incurved lower jaw. Lord, 'Existing Tasmanian Marsupials', 21–22.
114. Temminck, *Monographies de mammalogie*, 60fn1.
115. The age of wolves has been estimated through tooth wear but sometimes age is overestimated by one or two years. Tooth wear may vary dependent upon the region in which an animal lives (and the dietary choices concomitant with that region). See Gipson et al., 'Accuracy and Precision of Estimating Age of Gray Wolves by Tooth Wear'.
116. Temminck, *Monographies de mammalogie*, 65.
117. Harris, 'Description of Two New Species of Didelphis from Van Diemen's Land', 174–75.
118. Lord and Scott note that the type specimen used by George Harris to describe the animal is not in existence. Lord and Scott, *A Synopsis of the Vertebrate Animals of Tasmania*, 264.
119. Scott was a former director of QVMAG. The cave containing the remains was found by Kerry Burns, Dereham Scott (Eric Scott's son) and Robert Tanner. As well as the tiger skull, the bones of bandicoots and wallabies were also discovered. See Anonymous, 'Limestone Caves Discovered', *The Mercury*, 4 June 1951, 9.
120. DNA probably survives in the bone of the skull that is QVM2013.GFV.1. The skull is petrified rather than mineralised. For an examination of DNA in relation to ancient bone, see Campos et al., 'DNA in Ancient Bone'.
121. See Muirhead and Wroe, 'A New Genus and Species'; Muirhead and Archer, '*Nimbacinus dicksoni*'; Muirhead, 'A Specialised Thylacinid'.
122. Playfair, 'Biographical Account', 73.
123. Limestone encrustation can be quite rapid. It is therefore conceivable the skull may be more recent.
124. Leigh, *The Hunter*, 114.
125. Ibid., 159.
126. A Lazarus taxon or Lazarus species is a species thought to be extinct that is rediscovered. Gilbert's potoroo (*Potorous gilbertii*) is an example of such a taxon. 'Discovered' by colonists in 1840, it was thought to be extinct by the 1960s but was rediscovered in 1994.

127. Leigh, *The Hunter*, 159–60.
128. See Sofaer, 'Touching the Body'.
129. Leigh, *The Hunter*, 159.
130. Some nonhuman animals may possess notions of mortality. Research on forest and savannah elephants reveals that they visit dead conspecifics and investigate bones, perhaps processing death events. See Hawley et al., 'Conspecific Investigation of a Dead Forest Elephant' and King, *How Animals Grieve*, 56–57. Barbara King also provides other examples of animals (such as bison) interacting with bones, suggesting that these encounters may evince more than curiosity, potentially manifesting emotional responses to death (139).
131. Wilson, *Coorinna*, 155.
132. For a discussion of various insects that infest natural history collections, see Notton, *Identifying Insect Pests in Museums and Heritage Buildings*.
133. A skull (numbered 643) that is currently on display at the Tiegs Museum of Zoology looks to have been dropped since the skull was acquired by the collection. It has cracked in two. In the case of remains of a more common species, such a skull would likely have been thrown away. Occasionally damage is deliberately inflicted by naturalists. Harriet Ritvo notes the scissor marks left by George Shaw of the British Museum on osteological remains of a duck-billed platypus (*Ornithorhynchus anatinus*). Shaw hacked at the bill because he believed the skull was potentially a forgery, the scratches now providing a lasting index of his scepticism. Ritvo, *The Platypus and the Mermaid*, 4.
134. A thylacine skull is formed of a cranium (the top part) and mandible (or lower jaw).
135. See Timm et al., 'Innovations that Changed Mammalogy', 369.
136. The skull was sourced from the estate of the doctor Walter Henry Tofft. He lived in Campbell Town. Tofft died in 1939 and the skull entered the QVMAG collection in 1962.
137. In the early twentieth century skeletons received by museums were usually fleshed, with 'the "meat" being scraped off with a knife', then placed in a macerating tank so any remaining flesh decomposed. The bones were then cleansed to remove any traces of grease. See Anonymous, 'Embalmed Elephant Joins Stuffed Sea Elephant in Weird Catacombs', *Herald*, Tuesday 16 June 1914, 1.
138. Maynard and Gordon, *Precious Little Remains*, 73.
139. Whiskers, as keratinised tissue, can potentially provide tremendous insight into the life history of an animal including diet and habitat. See, for example, the research into foraging ecology conducted by McHuron et al. using the whiskers of northern elephant seals (*Mirounga angustirostris*). McHuron et al., 'What's in a Whisker?'
140. The R identifier is from an older cataloguing system used at the museum. The skull is currently stored with some mandibles that have the identifier INT 1397 written on their interior. These mandibles may have been incorrectly associated with this skull at some point, and the correct mandibles are now to be found with another skull. The R2154 identifier is written across one of two holes in the frontal region of the cranium. Writing across the damage might seem to indicate that it postdates the inscription, but if the effaced identifier was indeed erroneous the decision to locate the second correct identifier where damage already existed may have been an effort not to create more 'mess' on the bone.
141. She seems to have been collected by Robert Hewett (R. H.) Green, the former curator of zoology at QVMAG.
142. The note also explains that the thylacine's clavicles have been placed in a pill box.
143. The skull may be of a male thylacine based on its size.

144. Long suggests that the year of accession was either 1899 or 1900. See Long, 'Five Previously Undescribed Thylacine (*Thylacinus cynocephalus*) Specimens', 464.
145. Lambert, 'The Structure of an Australian Land Leech', 211–35; Lambert, 'Description of Two New Species of Australian Land Leeches', 156–63.
146. See Long, 'Miss Lambert's Thylacine Skull'.
147. Julie Marginson, 'Ada Mary à Beckett (1872–1948)', *Australian Dictionary of Biography (1891–1939)*, Volume 7, https://adb.anu.edu.au/biography/a-beckett-ada-mary-4963 (retrieved 22 June 2023). In an address given in May 1946, Lambert was among several men and women described by the retiring warden of Trinity College (of which Lambert was an alumnus), John Behan, as great names who had contributed to the 'betterment of mankind'. See Behan, 'The Past', 8.
148. Long, 'Miss Lambert's Thylacine Skull'.
149. In reality, these women had to contend with considerable difficulties that are glossed over by the article. Sweet, for instance, suffered from exhaustion as she had to cover for more senior, male professors as they enjoyed research leave.
150. Sweet retired as an associate professor. In 1919, she was passed over for promotion to full professor in favour of Wilfred Agar. She did serve briefly as Acting Professor of Zoology (1916–17) prior to this. See MacCallum, 'Georgina Sweet'. Another leading Melbourne academic, the botanist Ethel Irene McLennan, was also passed over for promotion to full professor in 1937. The geologist Dorothy Hill is often credited as the first woman to become a full professor at an Australian university (University of Queensland) in 1960.
151. Buchanan, *Elements of Animal Morphology*. In the acknowledgements to this book, which is dedicated to her Melbourne students, Buchanan thanks Georgina Sweet.
152. Long, 'Miss Lambert's Thylacine Skull'. Long provides no supporting information for this claim.
153. Lambert, 'The Structure of the Australian Land Leech', 211.
154. Harries, 'Of Bleeding Skulls and the Postcolonial Uncanny', 412. William Eppes (also Epps) Cormack stole the skulls from a Beothuk burial site and took them to Scotland where he donated them to the collection of the Edinburgh University Museum. They were given the catalogue identifiers NMS.Z.1827.25.1 (Nonosabasut) and NMS.Z.1827.25.2 (Demasduit). This collection was transferred to the Edinburgh Museum of Science and Art in 1854. The Edinburgh Museum of Science and Art became part of National Museums Scotland in 1985. The skulls were repatriated to Newfoundland in March 2020. Élise Dubuc describes the repatriation of artefacts from museums as an epiphenomenon of decolonisation. See Dubuc, 'Museum and University Mutations', 505.
155. Harries, 'Of Bleeding Skulls and the Postcolonial Uncanny', 419.
156. For a discussion of these accounts, see Chare, 'In Her Hands', 296–98.
157. Sofaer, *The Body as Material Culture*, 2.
158. Kuch et al., 'A Preliminary Analysis of the DNA and Diet of the Extinct Beothuk', 595.
159. The DNA of Demasduit and Nonosabasut (as well as of sixteen other now nameless Beothuk) is used by Carr to examine the persistence of ancient Beothuk mitochondrial DNA genome lineage in modern Native American and First Nations peoples. See Carr, 'Evidence for the Persistence of Ancient Beothuk and Maritime Archaic Mitochondrial DNA Genome Lineages'. Collier and Carr compare mitogenome sequences from Demasduit and Nonosabasut among other Beothuk with sequences from some members of the Newfoundland Mi'kmaq in their paper 'On the Persistence and Detectability of Ancient Beothuk Mitochondrial DNA Genomes in

Living First Nations Peoples'. They found that the Mi'kmaq lineages they studied were no closer to the Beothuk than the lineages of other First Nations and Native American peoples.

160. Kuch et al., 'A Preliminary Analysis of the DNA and Diet of the Extinct Beothuk', 603.
161. Ibid.
162. Although thylacine DNA does not register the violence perpetrated against the species, studying stable isotopes derived from the remains of thylacines that had lived part of their lives in zoos might show evidence of dietary changes and thereby indirectly reflect the impact that removing the animals from their natural habitat had on them. For a discussion of the insights that stable isotope analysis can potentially provide, see Hopkins and Ferguson, 'Estimating the Diets of Animals'.
163. These writings are summarised in Black et al., 'The Skulls of Chief Nonosabasut and his Wife Demasduit', 662.
164. Two further references are made in the main text to the mandible becoming 'gracile' (704; 705). See Martin and Danforth, 'An Analysis of Secular Change'.
165. Robinson, 'Private Journal Kept on Board HMS Favorite', 216.
166. Black et al., 'The Skulls of Chief Nonosabasut and his Wife Demasduit', 672.
167. Ibid., 673.
168. Black, Marshall and Kitchener find this unlikely as Demasduit was 'highly regarded' (675). They suggest accidental injury or '*post mortem* carelessness' as causes (675).
169. Busk, 'Description of Two Beothuc Skulls', 231.
170. Even now that they are repatriated to Newfoundland, the skulls are still treated as 'specimens', housed in a museum collection. Chief Mi'sel Joe notes upon their return: 'They're almost home. They're not quite home yet. . . . They are in this museum [The Rooms].' From Joe's comments, it is clear that only once the skulls are reburied at Beothuk Lake (formerly known as Red Indian Lake), the location from which they were originally pillaged, will they cease to be 'specimens'.
171. Işcan, 'Concepts in Teaching Forensic Anthropology', 11.
172. Portillo, 'Foreword', viii.
173. Anonymous remains discovered in the environs of Ciudad Juarez, for instance, once they have been identified as belonging to a woman who has been murdered, can be interpreted as another potential example of feminicide. Such an approach is fraught with difficulty and may enact potential post-mortem violence on transgender or non-binary victims of violence. Their bones may be misidentified as of a gender they did not identify with. For a discussion of forensic science and Juarez, see Chare, 'Concrete Loss'.
174. See Smith-Blackmore and Bethard, 'A Multidisciplinary Investigation of Chronic Animal Abuse'.
175. For further information about veterinary forensics see the website of the International Veterinary Forensic Sciences Association, http://www.ivfsa.org/about (retrieved 7 July 2021).
176. Sleightholme and Campbell, 'The International Thylacine Specimen Database', 495. Merck et al. explain in 'Blunt Force Trauma' that the head is a particularly vulnerable part of the body of an animal, with even apparently superficial injuries capable of resulting 'in coup and countercoup lesions that can affect the brain stem and lead to the death of the animal' (98).
177. Cited in Anonymous, 'Royal Society of Tasmania', *The Hobart Town Courier*, Wednesday 16 June 1858, 3.

178. A degree of caution needs to be exercised when identifying bullet holes as there are a number of ways a skin might be 'holed' post-mortem, particularly at the outer edges. Cameron R. Campbell, 'The Specimens', www.naturalworlds.org/thylacine/biology/specimens/specimens_3.htm (retrieved 9 July 2021).
179. Ibid.
180. I am grateful to Belinda Bauer for sharing her insights regarding this mount. The damage may, of course, have a different cause or have occurred post-mortem.
181. Another skull, C2163, has two holes in it; too neat to be made by a hatchet, they are either caused by gunshots or by drilling for a bone sample (neither explanation is satisfactory given that there is more than one hole, but without the benefits of forensic examination, the former cause seems more likely).
182. The nasal cavity of this skull is stuffed with cotton wool, a substance often associated with medical use including the cleaning of woods, but here a remainder (a material reminder) of former storage practices. The skull was shown post-mortem care, swaddled, in contrast to in life. I thank Belinda Bauer and Kathryn Medlock for providing information about TMAG's historical stowing procedures.
183. Lorenzo Ressel, Udo Hetzel and Emanuele Ricci note, however, that '[a] clear demarcation between blunt- and sharp-force trauma may not exist in some circumstances'. Such circumstances would include when an object is not sharp but is acuminated and leaves a puncture wound. Ressel et al., 'Blunt Force Trauma in Veterinary Forensic Pathology', 941. Some skulls also have holes caused by the taking of post-mortem bone samples, but these are usually distinctive in appearance.
184. Ressel et al. include photographs of the cranium of a dog that was subject to blunt-force trauma (the animal was hit with a stick). The impact has produced a large, depressed fracture. Thylacine skulls do not usually display fracturing of this kind (museums would likely not have thought it worth keeping skulls subject to such extensive damage). Ibid., 947. With blunt-force trauma, blows to different parts of the head produce different effects. As Merck et al. observe in 'Blunt Force Trauma', an object with a small surface area (such as a club at point of impact) 'can cause a depressed skull fracture, sometimes with fragmentation' (114). Clubbing as a method of stunning (rather than killing) is nowadays not considered effective for animals over 5 kg in weight. See Aldworth, 'Welfare Aspects of Commercial Hunting and Climate Change', 259.
185. In TMAG, A1277, the skull of a thylacine of unknown gender, appears to have an injury inflicted by a hatchet. A photograph of the trap hoe is reproduced (as Figure 5) in Sleightholme et al., 'The Kaine Capture', 5.
186. See also the book accompanying this exhibition, Maynard and Gordon, *Tasmanian Tiger*, 73.
187. Wounds made by spiked objects (such as an ice pick) can resemble gunshot wounds caused by shotgun pellets or .22-calibre bullets. See Merck et al., 'Sharp Force Injuries', 126.
188. Historical accounts give a sense of the kinds of weapon used to kill tigers. On 19 November 1887, for example, Edward and James Aylett and Ambrose Walsh cornered a male thylacine on Christmas Hill near Elizabeth Town. Edward Aylett caught the tiger by the tail and Walsh killed him with either an 'axe' or a 'tomahawk'. The dead tiger's 'prepared carcase' was exhibited in Launceston on Wednesday 23 November. It was noted that the thylacine's teeth were worn down, which was taken as evidence of him having 'played terrible havoc among the sheep'. As discussed in Chapter 3, Edward Aylett expressed the wish to have the skin stuffed. Anonymous, 'The Native

Tiger', *Daily Telegraph*, Thursday 24 November 1887, 2; Anonymous, 'Local and General', *The Tasmanian*, Saturday 26 November 1887, 24.

189. An animal caught in a snare would not usually be shot from far away so the difference in distance between killer and victim when using a rifle would only be minimally different (rifles are longer) from using a club or a hatchet. Striking a blow, however, requires more energetic involvement than pulling a trigger.
190. Killing remotely eases the psychological burden accompanying acts of killing. See Rutchick et al., 'Technologically Facilitated Remoteness Increases Killing Behavior'. Given the general aversion to close-range killing (of humans), the slightly greater distance a gun potentially allowed a trapper or farmer intent on killing a thylacine may sometimes have been psychologically significant. In their essay 'Why Can't Johnny Kill?' Molloy and Grossman rightly note that 'people who grow up in a largely pastoral society learn from a young age the ability to kill animals without compunction' (197). As landowners such as farmers grow older, however, some experience a growing reluctance to kill animals (domestic or 'pest') themselves. Generalisations about the ease (or lack thereof) with which farmers killed thylacines (or about the methods they adopted to do so) should therefore probably be avoided.
191. The great auk skulls discovered by Frederic Augustus Lucas on Funk Island in Newfoundland that manifested traumatic head injuries, 'the marks of cuts or blows', potentially formed another instance of osteological memorials to animal genocide (Lucas, *The Expedition to the Funk Island*, 513). Damaged skulls were not, however, collected by Lucas. He also found an ulna that seemed to have been broken by a gunshot (512). For a consideration of animal genocide in relation to the great auk, see Chare, 'The Beothuk, the Great Auk and the Newfoundland Wolf'.
192. Fudge, 'A Left-Handed Blow', 5.
193. Atwood, *Survival*, 74–75.
194. Fudge, 'A Left-Handed Blow', 6.
195. Marvin similarly states that animal behaviour as it is discussed in the humanities is always interpreted 'in terms of human concerns'. Marvin, 'Wolves in Sheep's (and Others') Clothing', 65.
196. Fudge, 'A Left-Handed Blow', 8.
197. Ibid., 9–10.
198. Attwood, *Survival*, 75.
199. Anthony and Brown, 'Looking a Gift Horse in the Mouth', 99.
200. The use of a bit can cause exostosis. For a discussion of the discomfort bits can cause, see Cook, 'Pathophysiology of Bit Control in the Horse' and Cook, 'Bit-Induced Pain'.
201. The horses probably derived from some of those brought over by Hernán Cortés when he began his invasion of the Aztec Empire in 1519. They were gradually transported north in the following decades. See Taylor et al., 'Early Dispersal'.
202. Marvin, 'Wolves in Sheep's (and Others') Clothing', 62.
203. Ibid., 65.
204. Ibid., 63–64.
205. Rae sometimes x-rayed marsupial remains for researchers and had some familiarity with zoology. He receives thanks from A. G. Lyne for furnishing radiographs of a wallaby in the paper 'Syndactyly in the Manus of the Marsupial', 32.
206. The skull may have been buried beneath the sand but exposed by a violent storm or excavated by chance through actions undertaken by Rae. Many of the dunes at St. Helen's are transgressive, constantly shifting, which could also have caused the bones to be buried and then subsequently exposed. Alternatively, the part-skeleton may

already have been on the surface but concealed by marram grass or other vegetation. The paucity of information elicited from Rae and recorded about the conditions in which he found the skull is disappointing as it makes it difficult to assess its potential age.

207. See Ross and Cunningham, 'Time-since-Death and Bone Weathering', 132. Ross and Cunningham examine human bones and their degradation in a tropical environment, and they are therefore used with caution here. For a discussion of the bone weathering of nonhuman animals (sheep) in a temperate environment, see Andrews and Armour-Chelu, 'Taphonomic Observations on a Surface Bone Assemblage'. Andrews and Armour-Chelu discuss how bone weathering is uneven across a skeleton, with more robust bones degrading at a slower rate.
208. Behrensmeyer analyses weathering in mammals such as cows, gazelles and zebras and foregrounds the fact that the bones of larger mammals weather at a different rate to smaller ones and often in distinct ways.
209. Behrensmeyer, 'Taphonomic and Ecologic Information from Bone Weathering', 151.
210. For a recent discussion of the 1938 tracks, see Mooney, 'Review of Footprints from the 1938 Jane River Expedition'.
211. Zulu Cave is a misnomer for a rift that is called Zulu Pot, which was discovered near Maydena in 1967. The rift was properly explored on 16 July 1972 when the thylacine remains were located by a party of four cavers on a ledge near the bottom of the cave. The skull was retrieved that day and the rest of the skeleton on 23 July 1972. Steve Harris suggests that the thylacine may have been curled up. They were located in a hollow in the highest and driest part of the cave. Possible claw marks were spotted on the cave wall near the skeleton. It seems the thylacine fell and became trapped in the cave rather than using it as a lair. Other caves including one at Bubs Hill in western Lutruwita called 'Thylacine Lair cave' have been identified as dens. Cockerill, 'Maydena 21/5/67', 5–6; Harris, 'Junee-Zulu Pot 16/7/72', 50; Harris, 'Florentine Valley 23/7/72', 50–51.
212. The bones of other mammals, possibly a ringtail possum and a New Holland mouse, were found nearby.
213. Andrews erroneously states that the last thylacine died in 1933. Andrews, 'Thylacine Remains', 11.
214. In this context, it is interesting to note that thylacines were reportedly sighted in the area in 1986. See Anonymous, 'Walkers Took Photos of Tasmanian Tigers', *Canberra Times*, Sunday 9 February 1986, 3.
215. Andrews, 'Thylacine Remains', 10.
216. Ibid., 11.
217. Carlson et al., 'Estimating the Extinction Date of the Thylacine'.
218. Brook et al., 'Resolving When (and Where) the Thylacine Went Extinct'.
219. See Sleightholme and Campbell, 'A Retrospective Assessment of 20th Century Thylacine Populations', 120.
220. Gmelch and Gmelch, 'Lost Then Loved', 41.
221. Lord and Scott, *A Synopsis of the Vertebrate Animals of Tasmania*, 265.
222. Anonymous, 'Decorative Furs', 84.
223. In *Still Life*, Melissa Milgrom recounts that the taxidermist Bruce Schwendeman was dismayed by people vacuuming bearskin rugs as it was disrespectful and signalled the idea of human dominance over nature (12). I would suggest that even standing on such a rug or admiring it as a decorative object forms a similar display of human supremacy.

224. Farnham, *Home Manufacture of Furs and Skins*, 217. Milgrom notes in relation to mounts that some taxidermists distinguish between trophies and scientific projects, with trophies conceived as offering a 'generalized depiction' and museums providing a vision of a 'specific individual' (*Still Life*, 25). This distinction does not seem universally applicable.
225. Farnham, *Home Manufacture of Furs and Skins*, 218.
226. There is a hearth rug at TMAG and another in a private collection. For a photograph of the rug in private hands, see 'Skin Specimens', www.naturalworlds.org/thylacine/biology/specimens/specimens_3.htm (retrieved 9 July 2021).
227. The probable age of the thylacines is deduced from the size of the skins, but due to variations in the size of adult tigers, to my mind, there is some ambiguity regarding identifying all these animals as juveniles. Of course, claiming that the eight tigers are young certainly helps to amplify the horror of the rug's creation.
228. Buggy rugs often had a plated ring at the side or the centre. The function of the metal dome fitting on the thylacine rug is not known (it might be intended to add weight or to act as part of a fastening mechanism).
229. Schlunke believes that a woman in the Stevenson household made the rug. Other women in the area are known to have made possum-skin rugs and gifted them to neighbours. See Tassell, *Rural Launceston Heritage Study*, 232. If Schlunke is correct, Frances Stevenson (née Whittle), Robert's wife, would be one candidate. The couple had no daughters but did employ a maid. Making rugs from skins was often undertaken by the women in a given household. See ibid., 233. If Frances created it, it would have served to embody her dedication to her household duties. Historically, both men and women are known to have made animal skin rugs in Australia or to have worked collaboratively in their making. If the rug is collaborative, Frances might have pegged and sewn it.
230. I am grateful to Nicole Zehntner for sharing her insights regarding the tanning process.
231. Crane and Wood, *Island Story*, 228.
232. No thylacines were caught at Aplico after 1906 and their increasing scarcity, which was general knowledge, must already have been very evident to Stevenson in 1903.
233. 'Buggy Rug', Tasmanian Colonial Decorative Arts 1803–1930 website, http://static.tmag.tas.gov.au/decorativeart/objects/misc/P2002.63/index.html (retrieved 9 July 2021). By 1945, Stevenson's farm of Aplico was run by his son Alfred. See Anonymous, 'Obituary: Mr. Robert Stevenson', *Examiner*, Saturday 14 June 1947, 2. On Monday 28 May 1945, the Stevenson family did hold a clearing sale (which probably included the rug) but at the property of Evandale. See Anonymous, 'Clearing Sale', *Examiner*, Thursday 31 May 1945, 3. The sale was advertised in *The Mercury* on Saturday 26 May with many items listed but the rug is not one of them. 'Advertising', *The Mercury*, Saturday 26 May 1945, 18.
234. Anonymous, 'Rarest Tiger Skin a Rugged Survivor', *Sydney Morning Herald*, Tuesday 20 August 2002, http://www.smh.com.au/national/rarest-tiger-skin-a-rugged-survivor-20020820-gdfk28.html (retrieved 9 July 2021).
235. After its purchase, the rug was initially displayed at the Thylacine and Endangered Species Information Display Centre at Strahan.
236. See Fario, 'The Presentation Rug', *The Mercury*, Thursday 16 May 1901, 4.
237. In his book *Home Manufacture of Furs and Skins*, Farnham reproduces an image of a lap robe made of eight coyote skins that is similar in design to the thylacine rug (199). A rug made from the skins of thylacines and domestic cats with a border of Tasmanian pademelon (*Thylogale billardierii*) (referred to as 'Tasmanian black wal-

laby') was produced by the furrier S. R. Clark in 1879 and exhibited in Melbourne. See Anonymous, 'The Brunswick Exhibition', *The Age*, Monday 27 October 1879, 3.

238. Some buggy rugs were also made of sealskin or lined with it (imitation sealskin was popular). There is a reference to the theft of a sealskin buggy rug from the Melbourne hardware store James McEwan and Co. in 1906. See Anonymous, 'Ironmonger's Boy', *The Herald*, Wednesday 6 June 1906, 4. There are also reports of fox skin buggy rugs with a reference to one that incorporated the tails of two animals into the design. See Anonymous, 'Victorian Police Gazette', *Bendigo Advertiser*, Monday 12 June 1905, 6. Additionally, the koala (*Phascolarctos cinereus*) or 'native bear' was used to make travel rugs.
239. Anonymous, 'The Town Hall Show', *The Mercury*, Friday 13 November 1874, 2.
240. Anonymous, 'Tasmanian Furs', *Examiner*, Friday 6 March 1908, 7. Tasmanian tiger skins (presumably hearth rugs) are also mentioned in the article, with the animal described as uncommon.
241. Anonymous, 'Tasmanian Fur Depot', *Examiner*, Thursday 17 October 1918, 8.
242. Ibid.
243. Anonymous, 'Town Talk and Table Chat', *Cornwall Chronicle*, Wednesday 5 August 1863, 5.
244. Trustees of the Tasmanian Museum and Art Gallery, *Tasmanian Museum and Art Gallery Annual Report 2002–2003*, 31.
245. Schlunke, 'Thylacine Buggy Rug', 131.
246. 'Buggy Rug', Tasmanian Colonial Decorative Arts 1803–1930 website, http://static.tmag.tas.gov.au/decorativeart/objects/misc/P2002.63/index.html (retrieved 7 June 2023).
247. Tassell, *Rural Launceston Heritage Study*, 197. Aplico had a custom-built schoolroom. Although it took place over forty years after the rug was made, the clearing sale where it was sold also gives a good indication of Stevenson's financial status. He offered several pieces of cedar furniture, three cut glass decanters, and an antique silk shawl among many other items.
248. Contemporary accounts of the theft of buggy rugs attest to their perceived value. In 1912, Thomas Roberts was found guilty of stealing a silver-grey opossum skin rug from a cart, which he then pawned for thirty-two shillings. Anonymous, 'Larceny of a Rug', *The Mercury*, Wednesday 5 June 1912, 7.
249. Stevenson also sold potatoes and wheat to supplement his income.
250. Robert Stevenson, 'In Other Days', *Examiner*, Saturday 30 August 1941, 6. In the winter of 1880, tigers were blamed for the loss of 448 sheep at Aplico. See Tassell, *Rural Launceston Heritage Study*, 232.
251. Robert Stevenson, 'Fox Farming', *The Mercury*, Friday 27 September 1946, 17. His son, George, who took over at Aplico, maintained that between 1865 and 1870 the deaths of 3,700 sheep were caused by thylacines. Quoted in Smith, *The Tasmanian Tiger*, 23.
252. Pemberton, 'The Tasmanian Devil', 17. As late as the 1920s many farms in Lutruwita still had log fences, which were inefficient at keeping livestock in and predators out. See Tassell, *Rural Launceston Heritage Study*, 37, 238. Stevenson used wire netting (rather than wire fencing proper) for his trapline. He erected paling fences at Aplico and Elverton in the 1920s.
253. Tassell, *Rural Launceston Heritage Study*, 232. The remains of some of the pits were located by members of TMAG in 2004. See Trustees of the Tasmanian Museum and Art Gallery, *Tasmanian Museum and Art Gallery Annual Report 2004–2005*, 12. This rare architectural trace of genocide against thylacines merits conservation.
254. Smith, *The Tasmanian Tiger*, 30.

255. Tassell, *Rural Launceston Heritage Study*, 232. The tree is described as a white gum (*Eucalyptus alba*) in the 'Tasmanian Tiger: Precious Little Remains' exhibition at QVMAG, which includes a photograph of the surviving stump. The tree as it features in the exhibition is discussed further in Chapter 3.
256. I am indebted to Peter Hughes for pointing out the variation in wear across the two quartets of skins. The cause of this wear cannot be ascertained with certitude and is possibly unconnected to the rug's use in a buggy.
257. Bourdieu, *The Logic of Practice*, 72.
258. Anonymous, 'Merino Stud Sheep', *The Mercury*, Saturday 17 June 1933, 7.
259. The potential symbolism (conscious or unconscious) of sitting on top of the rug to play the piano, of the triumph of human over thylacine and of culture over nature encoded in the pose, should be recognised.
260. 'Buggy Rug', Tasmanian Colonial Decorative Arts 1803–1930 website, http://static.tmag.tas.gov.au/decorativeart/objects/misc/P2002.63/index.html (retrieved 9 July 2021).
261. Schlunke, 'Thylacine Buggy Rug', 133.
262. Lindner, 'Genocide, Humiliation, and Inferiority'.
263. See Torres and Bergner, 'Humiliation'.
264. See, for example, Geary's discussions in *Furta Sacra* of the humiliations of holy relics that occurred in the mediaeval period.
265. Nic Haygarth, 'Clem Penney's Stripy Doormat', 27 November 2016, https://nichaygarth.com/index.php/2016/11/27/clem-penneys-stripy-doormat/ (retrieved 18 August 2023).
266. These photographs are discussed again, in more depth, in the next chapter.
267. Mendible, 'Visualizing Abjection', 36.
268. The thylacine's bands obviously also made for eye-catching rugs. The tiger was additionally clearly viewed as a distinctly 'Tasmanian' animal and celebrated for that reason. The market for hearth rugs made from readily identifiable native animals was evidently enduring. In the 1860s in Launceston, Henry Archer prepared rugs and other products from animals including 'the tiger and native cat'. B. R. Patey, 'Colonial Industry', *Cornwall Chronicle*, Saturday 21 October 1865, 10.
269. Wilson, *Coorinna*, 49.
270. Ibid., 50.
271. Anonymous, 'Decorative Furs', 84.
272. H. B. Martin, 'The Opossum', *Examiner*, Friday 9 July 1926, 9. In a letter to the Examiner in 1904, the farmer Alex Kidd also expresses sympathy for possums. See Alex Kidd, 'Opossums and Native Cats', *Examiner*, Tuesday 22 November 1904, 3. For a more recent examination of mixed attitudes to possums in Australia, see Milton, 'Possum Magic, Possum Menace', 69–71.
273. Anonymous, 'Game Preservation Bill', *Examiner*, Thursday 29 October 1908, 4.
274. John Dunbabin is sometimes described as arriving in Lutruwita in 1830 but an article about his son Thomas Dunbabin gives the year as 1832. See Anonymous, 'An East Coast Pioneer: The Late Thomas Dunbabin', *The Mercury*, Thursday 21 October 1920, 10.
275. In 1898, Marchwiel had 2,296 acres of agricultural and pastoral land. There was a six-room stone house and a six-room weatherboard cottage on the property.
276. Comebacks and merinos were bred at the farm. Comebacks are crossbreeds between merinos and British Longwool sheep. In 1907, Marchwiel offered '40 Fat Comebacks' for sale. See the Auctioneers Notices in *The Mercury*, Tuesday 10 September 1907, 8. Cattle were also occasionally bred and sold. It is clear that by the early

twentieth century the farm was commercially successful, as in 1908 Mrs T. Dunbabin advertised for a teacher for her three children for a salary of £50. See 'Advertising', *The Mercury*, Wednesday 15 January 1908.

277. See the results of camera trapping conducted at Marchwiel Marsh from November 2019 to January 2020. Prahalad and Aalders, *Marchwiel Marsh, Marion Bay*, 27–28.
278. Thomas Dunbabin, 'Miscellaneous', *The Mercury*, Thursday 7 May 1903, 5.
279. This programme provides tax incentives for the donation of culturally significant artefacts to art galleries, museums and libraries.
280. The annual report of TMAG for 2003–4 suggests that the Marchwiel skin is the only provenanced thylacine material in the museum's collection and one of only three thylacine skins worldwide from a specific location or property, but these figures do not seem to take the Stevenson rug into account. Trustees of the Tasmanian Museum and Art Gallery, *Tasmanian Museum and Art Gallery Annual Report 2003–2004*, 36.
281. The holes around the edge of the skin suggest that it was pinned out. It gradually folded inward as the pins came away and must now be unfolded with caution to expose the faint remains of the stripes.
282. Graham, 'Wool's Disappearing History', 61. Whether these newspapers were affixed because of what they referred to is not known. In 1883, *The Mercury* did carry a mention of Thomas Dunbabin being elected vice-president of the Bream Creek Cricket Club, but the newspapers could equally have detailed sheep sales or wool prices. Anonymous, 'The Bream Creek Cricket Club', *The Mercury*, Wednesday 12 September 1883, 2.
283. White, *Natural History of Selborne*, 33. White later describes studying two albino rooks similarly displayed (44).
284. David Fleay, 'Has the Remarkable Marsupial Wolf Finally Become Extinct?', *Courier-Mail*, Wednesday 1 July 1953, 7.
285. Nomad, 'A Hunter's Paradise: Fifty Years Ago', *Examiner*, Saturday 13 April 1929, 6.
286. Poliquin, *The Breathless Zoo*, 151.
287. Baughman writes of a cowboy in the United States hanging a dead coyote on a barbed-wire fence 'at the end of a line of at least ten . . . dead coyotes'. See Baughman, *Boat*, 110. Fencing often intrudes on the territory of animals such as coyotes and wolves. Hardin traces the emergence of the human idea of property to animal territoriality. See Hardin, 'Foreword', xi–xii. In *Wild Dog Dreaming*, Deborah Bird Rose writes of contemporary graziers hanging dingo corpses from 'gates, trees and signposts' or throwing them across fences (66).
288. Chapman, *Woolsheds*, 1.
289. Ibid.
290. Dominy, *Calling the Station Home*, 96, 189.
291. I am grateful to Angela Mortimer for sharing her insights regarding the multiple functions of shearing or woolsheds historically and today.
292. Dominy, *Calling the Station Home*, 189.
293. Thomas Dunbabin, 'Fossil Wealth', *The Sun*, Sunday 3 June 1923, 3. Herbert Wendt discusses the thylacine in *Ils n'étaient pas dans l'arche* [They were not on the ark], in his chapter 'The Continent of Living Fossils', which is about Oceania. See Wendt, *Ils n'étaient pas dans l'arche*, 241–42.
294. TMAG offered £5 each for dead tigers in 1922. See Anonymous, 'Tasmanian Tigers', *The Sun*, Tuesday 28 March 1922, 3. Arthur Murray sold a live thylacine to Beaumaris Zoo for £25 on 21 July 1925.
295. Cvetkovich, *An Archive of Feelings*, 73.
296. Anonymous, 'Waratah: Rare Catch', *Advocate*, Monday 11 August 1930, 6.

297. See Linnard et al., 'The Parson, the Psychiatrist, the Publican and his Nephews', 17.
298. Anonymous, 'Tasmanian Tiger', *Advocate*, Monday 22 September 1930, 7.
299. For a full description of the pincushion, see 'Pincushion', Tasmanian Colonial Decorative Arts website, http://static.tmag.tas.gov.au/decorativeart/objects/misc/P2008.33/index.html (retrieved 13 July 2021).
300. Mitchell, who was a resident of Swansea, was given the pincushion by Ferrar (probably in 1930). Mitchell clearly had an interest in natural history, which may have motivated her desire to acquire the artefact. See Peregrine, 'To Correspondents', *The Mercury*, Saturday 30 May 1936, 5. Mitchell was suffering from a lengthy illness when she donated the artefact to TMAG in 1937. See Anonymous, 'Swansea', *Examiner*, Friday 26 March 1937, 11.
301. Anonymous, 'Glamorgan Horticultural Society', *The Mercury*, Thursday 25 May 1905, 2. For a period of time the show was clearly held biannually in spring and autumn.
302. The show was held on the second storey, which may have been on the floor formerly used as a wine and spirit department. See Anonymous, 'Swansea Horticultural and Agricultural Show', *The Mercury*, Thursday 18 May 1899, 4. In 1903 approximately 250 people lived in Swansea, down from a high of 600–700. See Anonymous, 'Swansea', *The Mercury*, Saturday 23 May 1903, 6.
303. There was a 'fine needlework' category but not, so far as I can find, a 'fancy needlework' one. See 'Pincushion', Tasmanian Colonial Decorative Arts website, http://static.tmag.tas.gov.au/decorativeart/objects/misc/P2008.33/index.html (retrieved 14 July 2021). Fancywork was, however, a term used to describe the production of objects such as the pincushion. For an important summary of the term, see Gordon, '"Little Trifles"', 139.
304. The note suggests that the cushion came second to an entry of 'two mice running up a bag', which was presumably needlework. In 1911 (the year Ferrar's submission of a pincushion to the show is explicitly mentioned), Miss Mitchelmore submitted a cushion cover that might have featured the mice. Anonymous, 'Glamorgan Spring Show: The Prize List', *The Mercury*, Tuesday 21 November 1911, 8.
305. Gordon, '"Little Trifles"', 140. In 1873, Peter Lund Simmonds exhibited a polished ox-hoof pincushion as part of a display of making use of waste materials. See British Royal Commission, *The British Section at the Vienna Universal Exhibition, 1873*, 137. Pincushions made using animal horns as feet were also manufactured. Images featuring hoof and horn pincushions are reproduced as part of Jeanmard, 'Creature Comforts', 112–13.
306. Ibid., 110.
307. It is possible that Sarah Mitchell wrote the information on the cushion when she received it from Emily Ferrar in 1930.
308. As the cushion is said to have been awarded second prize, it may not have merited mention. Ferrar, however, regularly receives multiple mentions when the show is reported in other years, so it seems plausible she was not involved in 1900.
309. Anonymous, 'Country News', *The Mercury*, Thursday 9 May 1901, 4. Emily was the wife of William Moore Ferrar.
310. Her daughter Gladys May Ferrar also competed and was awarded a prize in 1909 for a 'floral device for children'. Anonymous, 'Glamorgan Horticultural Society', *The Mercury*, Tuesday 23 November 1909, 2.
311. Anonymous, 'Swansea Show', *Daily Post*, Wednesday 22 November 1911. In *The Mercury*, in a typographic error, the prize for the pin cushion was listed as going to Mrs Ferrier. See Anonymous, 'Glamorgan Spring Show', *The Mercury*, Tuesday 21 November 1911.

312. A thylacine skull (M27836) in the Australian Museum collection is listed as deriving from Mr Ferrar who obtained (how is unknown) the remains in 1922 in Ross in the Western Tier. The skull was then in collection of Dr G. M. Parker. Parker practised as a doctor in Swansea before buying his own practice in Bellerive in 1926, and therefore probably knew the Ferrar family. Fergus Moore Ferrar's grandfather, William Moore Ferrar, had established the farm 'Plassy' in Ross, a property that was located roughly two days walk from the Western Tiers.
313. 'Pincushion', Tasmanian Colonial Decorative Arts website, http://static.tmag.tas.gov.au/decorativeart/objects/misc/P2008.33/index.html (retrieved 15 July 2021).
314. Anonymous, 'Spring-Guns and Man-Traps Are Illegal in this Colony', *The Mercury*, Friday 26 October 1894, 2.
315. Anonymous, 'Country News', *The Mercury*, Thursday 26 October 1893, 4. Although illegal, spring-guns were also used on the mainland at the time to kill 'wild dogs'. See Anonymous, 'Intercolonial News Items', *Launceston Examiner*, Wednesday 3 October 1894, 7.
316. Anonymous, 'Swansea', *Daily Telegraph*, Friday 24 April 1885, 3.
317. In the Spring 1911 show, for example, Mavis Lyne, granddaughter of Henry George Leslie Lyne (John Lyne's brother), participated and received a mention for a 'dressed doll'. Numerous other Lynes also entered the 1911 show. Anonymous, 'Glamorgan Spring Show: The Prize List', *The Mercury*, Tuesday 21 November 1911, 8.
318. In the Spring 1911 show, for instance, the state-school children F. Dilger, H. Dilger and Trevor Dilger all received mentions for their writing. Mrs H Dilger, Mrs W. Dilger and Miss M. Dilger are also noted for their baking. Anonymous, 'Glamorgan Spring Show: The Prize List', *The Mercury*, Tuesday 21 November 1911, 8.
319. Anonymous, 'A 5ft. Tiger', *Examiner*, Friday 24 May 1912, 4.
320. As the murder of the dog by a spring-gun demonstrates, these killing grounds were indiscriminate. It is not known how many purinina/Tasmanian devils, feral dogs or wild cats were also slain by the guns at these properties.
321. Ferrar must have mentioned the spring-guns to Mitchell when she gifted her the pincushion, demonstrating how the artefact was linked in her mind with them.
322. Anonymous, 'A Destructive Animal: Hyena Killed at Mawbanna', *Advocate*, Wednesday 14 May 1930, 6.
323. Kilner, *The Adventures of a Pincushion*. Flint calls stories such as Kilner's 'object narratives' and suggests that they are invariably picaresque: 'shifts in plot, subject, and locale emphasize the indiscriminate changes of ownership that dictate the object's market value'. See Flint, 'Speaking Objects', 212.
324. Longman and Loch, *Pins and Pincushions*, 161.
325. Ibid., 178.
326. Ibid., 180.
327. Ibid., 30.
328. Wood, 'Fabrication on a Thylacine Pincushion', 188.
329. Cited in McCure, *Life in England and Australia*, 60.
330. Lowther, 'The Painter and His Pupil', 442.
331. See Anonymous, 'Town Talk and Table Chat', *Cornwall Chronicle*, Saturday 10 September 1864, 5.
332. The house was put on the market by Mitchell in 1924 but then withdrawn from sale.
333. A travelling party from Wynyard visited the two-storey fourteen-room stone house in 1932. They mention admiring curios. See Anonymous, 'Wynyard Motor Trip', *Examiner*, Monday 25 January 1932, 5.

Chapter 3

Displaying the Thylacine

Exhibiting Genocide by Any Other Name

Introduction: Keeping Genocide Under Wraps

A thylacine mount is exhibited behind a semi-transparent cloth within a display case in a section of the Hall of Mammals at the Smithsonian National Museum of Natural History in Washington, DC.[1] She is positioned near to a mount of a dingo (*Canis lupus dingo*). The decision to veil the thylacine is driven by practical reasons (the mount is in relatively poor condition) but exploited for symbolic ends.[2] The cloth is reminiscent of a shroud, of material for wrapping the dead. In this, the covering prefigures the veiling of endangered and extinct species that took place in 2019 in the natural history section of Bristol Museum and Art Gallery in the UK.[3] There, a thylacine was also veiled as part of a broader project to draw attention to the ongoing loss of biodiversity. The screen at the Smithsonian also provides the tiger with a measure of posthumous privacy, as if there were concerns about intrusiveness. Yet if these were real, she would not be put on display. Issues of consent in relation to exhibiting nonhuman animal remains have not yet received serious consideration. The display of human remains is, however, now increasingly viewed as unjustifiable. In 2023, the American Museum of Natural History in New York removed all human remains from public view for this reason.

Most of the other exhibits in the Hall of Mammals at the Smithsonian portray still extant species. These are placed in schematic wilderness settings (minimal rocks and vegetation) to give a rough indication of their natural habitat. Occasionally predators and prey are grouped together in spectacles of violence or its aftermath. The non-extinct species are therefore shown in relation to a habitat or to other animals. The thylacine, by contrast, appears in a plain setting and is on a dais set apart from other wildlife. The nearest creatures are a koala located some distance away and a dingo posed on a lower dais directly in front of the tiger. The koala – climbing a length of plastic tubing, a notional eucalypt – provides a shorthand for Australia. The tiger and the dingo are deliberately counterposed. Stencilled on the wall behind the two carnivores are the words 'Australia: An Extinction Story: What happened when these two predators faced off?'

The dingo and the thylacine are framed as if about to engage in a dogfight. The museum visitor is encouraged to 'place their bets', before reading the information panels and discovering the outcome of this clash. In reality, the disappearance of the tiger from mainland Australia cannot be conclusively linked to the introduction of the dingo. The decision to represent this aspect of the tiger's history is also perplexing as the animal that has been mounted was sourced from Lutruwita, where there are no dingoes. The thylacine, a female, was bought by the National Zoo (by then a part of the Smithsonian) in 1902, where she was then displayed until her death two years later. The Smithsonian's decision to make this Tasmanian tiger substitute for a mainland tiger shows that they view the mount as standing for a species and accord her no singularity. This is in harmony with Garry Marvin's observation that animals exhibited in museums are usually reduced to being 'an example, a type, a token' with the singular uniqueness of each animal suppressed.[4] Stephen Asma has suggested that 'the educational value of any specimen (whether it's a dissected organ or a stuffed animal) lies in its power to extend illumination beyond its own individuality'.[5] For him, 'a true specimen is a species representative rather than an idiosyncratic particular'.[6]

The context constructed by the Smithsonian acts to stifle the kinds of response the tiger on display can generate. In the information panel accompanying the exhibit, the population of thylacines in Lutruwita is referenced and their disappearance is attributed to 'overhunting and disease'. It would have been possible to juxtapose the tiger with details about the exotic animal trade of which she was a victim, a trade that was potentially a contributing factor in the extinction of her species, or to discuss the bounties instituted in Lutruwita, the genocide perpetrated against the animal, but these more difficult narratives are avoided.

In offering a history of the thylacine, albeit a limited one, the Smithsonian conforms to Didier Maleuvre's conception of the museum as an institution that 'gives shape and presence to history'.[7] Maleuvre's book *Museum Memories* focusses on art museums. This may be because science museums are not seen to offer histories but simply to communicate facts, their storytelling dimension downplayed. Serge Chaumier, for example, contrasts science museums with art museums. The science museum offers 'knowledge for its own sake' whereas other kinds of exhibition spaces attest to a heritage or a history.[8] Exhibits such as the thylacine mount at the Smithsonian are, however, hybrid, communicating knowledge while simultaneously commemorating an extinct species and telling a story of that extinction. The exhibit provides a strong example of how ideas and ideologies intersect with 'facts' in the display of nonhuman animals such as the thylacine, rendering their unnatural histories present. Within natural history museums, there are always histories, human and nonhuman animal histories, interwoven with 'nature'. The high profile of the thylacine, their status as a 'celebrity' extinction, explains why the creature features in the Hall of Mammals in proximity to giraffes, a Bengal tiger and a white rhinoceros 'collected' by Theodore Roosevelt.[9] In the United States, the thylacine is a curiosity. In Australia, they are now an extinct part of the native fauna.

Display Practices

Two thylacine skulls displayed in the Tiegs Zoology Museum at the University of Melbourne are housed together in a black box located in a case near the entrance to the museum. As mentioned in the preceding chapter, one of the skulls (numbered 643) is broken. The value of the remains is clearly signalled by the decision to retain and exhibit the damaged cranium. The box the skulls rest within was likely custom built to house them.[10] It may have been constructed expressly to be in keeping with the décor of the museum when it first opened. The museum was initially housed in the Biology building (later the Baldwin Spencer building).[11] There, displays appeared on painted wooden shelves in cabinets.[12] The box containing the skulls is made of wood that has been painted black, of black sheathing tape and of glass. Although probably not intended to be so, it is reminiscent of Victorian mourning stationery, strongly connoting loss. Like the bones it contains, the box is showing its age. Paint has peeled or chipped off in places exposing bare wood beneath. Retaining the ageing container signals something about the old-world character of the Tiegs Museum and of the resources available to it. It is a small museum, employs few staff and has limited space.[13] It differs from the other museums studied here in that it is not public-facing and receives only a relatively small number of visitors each year.

There is little information accompanying the thylacine remains at the Tiegs. The animals are unusual in that they do not openly stand for anything other than themselves. At the Smithsonian, the tiger mount forms an obvious part of a Darwinian narrative of natural selection, of survival of the fittest, with the dingo coming out victorious in a staged 'clash of the carnivores'. Within the broader context of the narrative of the Tiegs Museum, the remains are positioned alongside other mammals and contribute to a narrative of evolution, but they are not exploited to dramatic ends.[14] The evolution narrative begins with the case containing sponges and rhizomes (the 'lowest' or most 'primitive' organisms on display), and continues through cabinets of insects and other invertebrates to culminate with the vertebrates, of which the greatest importance is accorded to the mammals. These are the 'highest' or most 'advanced' creatures based on perceived morphological complexity. They form the end of the evolutionary story. They are also the first animals encountered by a visitor to the museum as they are nearest the entrance. If a person begins their tour with the mammal remains, they must work their way backwards from the 'higher' to the 'lower' organisms. The structure of the display therefore implicitly affirms the notion of evolutionary progress, the idea that evolution is directed, guided by a drive to complexity. It also obscures the reality that many of the organisms on display continue to evolve side by side, living simultaneously.

Adam Dodd, Karen Rader and Liv Emma Thorsen foreground how the way institutions such as natural history museums represent animal remains impacts how animals are conceived and understood.[15] At the Tiegs, mammals are figured as an evolutionary pinnacle, with the thylacine subsumed in the broader class of Mammalia. This subsumption is unusual as often the thylacine is set apart. As Hannah Stark observes, there is a tendency to group the thylacine with other extinct species such as the dodo and the passenger pigeon in 'extinction cabinets with other extinction celebrities'.[16] This is the case, for example, at the Royal Ontario Museum in Toronto where the thylacine is part of a display, 'Gone Forever', that also includes a dodo (*Raphus cucullatus*), a great auk (*Pinguinus impennis*), two ivory-billed woodpeckers (*Campephilus principalis*) and a Xerces blue butterfly (*Glaucopsyche xerces*). The passenger pigeon is mounted on a branch that emerges from the plinth supporting the thylacine remains, artificially joining fauna of Turtle Island and Australia. The thylacine is similarly 'showcased' as an extinct species in displays at the Australian Museum and at the Melbourne Museum. Other displays focus exclusively on the thylacine. This is the case in Lutruwita, for example, where, as Carol Freeman notes, the animal has historically 'linked ideas about the Tasmanian environment with perceptions of place and identity', providing a locus for collective identification.[17] Because the tiger is bound up with Lutruwita's local character, their importance is reflected in two of the island's main museums, the Tasmanian

Museum and Art Gallery (TMAG) in Nipaluna/Hobart and the Queen Victoria Museum and Gallery (QVMAG) in Launceston, both of which devote considerable exhibition space to the thylacine. The varied ways that tigers and their history are exhibited, with the effects and impact of animal genocide (including as it intersects with collecting practices) often sidelined, will now be considered. I employ close reading (rare in the context of analyses of exhibitions) and draw on cultural semiotics (approaching the exhibitions as systems of signs) coupled with an attentiveness to affect (addressing the atmosphere of specific exhibits) to enable me to trace and describe the kinds of meanings and moods that museums promote relating to thylacines.

Tigers on the Mainland

The Melbourne Museum, part of Museum Victoria, has a display featuring the thylacine in the 'Wild' section of their Science and Life Gallery. The display provides an example of the thylacine (C28745) as an 'extinction celebrity', a famous loss to the natural world.[18] In a cabinet recessed in an artificial wall (at a height ensuring most adult visitors will look down into the display) a thylacine mount and a pig-footed bandicoot mount are posed together in a white space. These animals both act as synecdoche for extinct species. The animals themselves are clearly conceived not as individuals but as types. There are separate texts for the bandicoot (titled 'Too many changes') and the thylacine (titled 'Tassie Tiger gone forever'). Visitors learn that the bandicoot (*Chaeropus ecaudatas*) 'started to decline soon after the arrival of Europeans'. The animal's habitat was destroyed and 'new threats' emerged, namely introduced species and diseases. The narrative provided suggests that the bandicoot's failure to adapt to these changes caused their extinction, the animal implicitly held accountable due to a lack of flexibility. The fate of the tiger, by contrast, is unambiguously attributed to European settlers who 'hunted, trapped and poisoned the species to extinction'. As my analyses of other thylacine exhibits will demonstrate, it is rare for such a clear-cut and explicit account of the extermination of the thylacine to be offered by a museum.

The information provided about the thylacine also foregrounds how the animal was misunderstood, gifted fearsome qualities that belied their nature and feared as a threat to livestock when they actually 'preyed on wallabies and small animals'. Here the museum text draws attention to potential disjuncts between popular perception and the reality of animal lives. The vision that the exhibit offers of the thylacine is one that works to support the textual corrective regarding the tiger's mythical ferocity. The thickset tiger raises and turns his head to the right, mouth closed, as if investigating a sight or sound in the far distance. His left forepaw is elevated, he is in seeming motion, interrupted in his journeying. The taxidermist has crafted a curious animal rather than a fierce one, lively but not frightening. The mount the museum has chosen to figure *Thylacinus cynocephalus* as a species is not the only mount in Museum Victoria's collection. There are two others, both in poorer condition, dog-eared, but also in different poses.[19] One, C28746, has his mouth open, red-painted upper palate and canines exposed, appearing more aggressive. A Post-it note kept with this sleeker thylacine notes that he has a 'loose tongue'. By contrast, the other mount in storage, C28744, a female thylacine, head turned to the left, is tight-lipped. She was likely not chosen for the exhibit because a gash is visible on her neck. C28745 is more aesthetically pleasing. C28744, ghoulish, looks as if she has been decapitated and then recapitated. In the past C28744 has appeared in tandem with C28745 in a diorama but with her head turned towards him, away from the visitor. The exhibit featuring C28745 tells a sympathetic story regarding the fate of the Tasmanian tiger, yet does so in combination with an animal that is not native to Lutruwita, the bandicoot. This impossible pairing is united through their shared extinction.

When I visited in May 2019, the Australian Museum in Gadigal/Sydney included references to the thylacine in three separate exhibition spaces, foregrounding the importance accorded the animal.[20] One of these spaces conformed to Stark's cabinet of 'extinction celebrities', comprising a display of numerous animals, many of which were endangered or extinct. The display, titled 'Our Inheritance', placed several different species together, '[those] living as well as [those] long-gone', to show that the 'Tree of Life' enables relationships between species to be traced back through time. Accompanying text explained that life develops across millions of years by way of 'numerous evolutionary "steps"'. Creatures in the cabinet included a fossil ammonite and trilobite, the skull of a Lord Howe Island horned turtle (*Meiolania platyceps*), a platypus (*Ornithorhynchus anatinus*) mount and skeleton, a Galapagos giant tortoise (*Chelonoidis niger*), a pouched lamprey (*Geotria australis*), an Australian lungfish (*Neoceratodus forsteri*), two Lord Howe Island stick insects (*Dryococelus australis*), a rufous hare-wallaby (*Lagorchestes hirsutus*) and a rufous bettong (*Aepyprymnus rufescens*). The conservation status of the creatures that are not from the fossil record was given and their 'evolutionary distinctiveness' was noted. The evolutionary distinctiveness of the rufous bettong, a small marsupial, is rated as high whereas that of the rufus hare-wallaby, another marsupial, is low. The thylacine, represented by a mount and a skeleton from different animals, appeared close to the centre of the display. The mount was positioned alongside the pelvis of a dodo and of a domestic pigeon (*Columbia livia domestica*) and near to another mount, one of a rufous hare-wallaby.

The display had obviously been carefully conceived to foster links between species. The different kinds of creature included (amphibians, birds, fish, insects, mammals, reptiles) helped to demonstrate biodiversity. The many animals from Australia that featured also foregrounded the country's rich and distinctive fauna. The logic behind placing two extinct species so centrally was less easy to fathom. It seems their 'celebrity' was being exploited to draw attention to the display. The thylacine, with their distinctive appearance, is eye-catching. The text about the dodo was printed alongside a reproduction of an engraving of the bird attributed to George Shaw.[21] The image was a familiar one and likely to attract notice (even if the dodo pelvis itself was nondescript). The iconic extinct animals were therefore used as a lure, as figurative clickbait, encouraging visitors to take a closer look at the display and, hopefully, the less well-known animals including another extinct species, the southern gastric brooding frog (*Rheobatrachus silus*).[22] To the far left of the thylacine a Galapagos tortoise was employed to provide a shorthand for Charles Darwin and his theory of natural selection, which was influenced by his trip to the Galapagos in 1835 (although finches were his main interest while he developed his ideas of differential survival). The giant tortoise was adjacent to the very small extinct frog in a carefully engineered scale contrast. To the immediate left of the thylacine were two Lord Howe Island stick insects. By 1920 this species of stick insect was thought to be extinct, yet it was rediscovered in 2001.[23] Placing the thylacine beside a Lazarus species, a rediscovered animal, encouraged imaginings about relict populations of the tiger in Lutruwita.[24]

The display was close to a wall panel titled 'The Missing Millions' and a touchscreen information interface that examined mass extinctions. The interface provided text and images exploring past and contemporary extinctions.[25] The wall panel explained that the species alive today represent less than 1 per cent of the species that have existed on earth. It also noted that while extinctions usually occurred naturally in the past, today they are often anthropogenic. The panel referred to the 'Tree of Life', linking it with the cabinet containing the thylacine remains. The main page of the interface was also titled 'The Missing Millions' and carried an image of a factory, its multiple chimneys belching smoke, encouraging visitors to link the sixth mass extinction with the rise of industrial capitalism and with global heating.[26] This is an important narrative that positions the museum as a 'site of conscience', a potentially transformative space 'where society can "learn" from past mistakes and create a better world'.[27] Ross Wilson suggests that many contemporary natural history museums exert a moral agency in relation to their visitors, fostering environmental consciousness.[28]

The labels for the exhibits in the 'Our Inheritance' cabinet used short captions and provided information about each animal. They reproduced selective information or offered speculation. Labels of this kind are usually described as interpretative and differentiated from their non-interpretative counterparts.[29] Interpretative labels, Sarah Chicone and Richard Kissel explain, are those that 'carry the exhibition's narrative'.[30] Non-interpretative labels are 'object identification labels, credit panels, donor plaques, wayfinding, and prohibitive signage (e.g. please do not touch)'.[31] Labels are one of the components of museum practice identified by the ethnologist Otis Mason as contributing to making 'specimens eloquent'.[32] While his remarks were directed at ethnographic collections, they also hold true for natural history. The labels that link to 'specimens' on display give voice to them.[33]

Chicone and Kissel do not consider how dimensions such as typeface and colour in labels contribute to their interpretative capacity. In 'Our Inheritance', the font colour was not arbitrary. A green typeface was used for the name of each creature, the colour judiciously chosen. Green signals verdancy and is also frequently associated with environmental protection. The label for the dodo explained that they were 'a victim of human hunting, habitat loss and introduced species'. The reasons underpinning the thylacine's extinction were not given but the death day of the last captive animal was provided, 7 September 1936, and the gloss that this date is 'now commemorated as National Threatened Species Day'. The demise of the animal was called a 'tragic event', encouraging linking the thylacine's fate with familiar narratives of heroes who are victim to an awful, sorrow-inducing destiny. Visitors were informed that the southern gastric brooding frog became extinct 'possibly due to disease'. The frog probably succumbed due to the human introduction of a waterborne pathogen into their habitat.[34] In the case of the frog and the thylacine then, the role of humans in their extinction was underplayed even though 'The Missing Millions' seeks to foreground human responsibility for the sixth mass extinction.

The way that the thylacine, the dodo and the frog were exhibited worked against acknowledging the role of human agency in their destruction. All the animals included in the display appeared against a white background. In art galleries, the white cube format is associated with the decontextualisation of works of art, with their being positioned at a remove from the vicissitudes of history and politics. Brian O'Doherty equates this whiteness with timelessness, with art(efacts) existing in a kind of 'eternity of display'.[35] The whitewash used to coat the interior of the cabinet housing the animals served a similar purpose. The tree of life schema was rendered timeless, natural selection shown to be an eternal, unchanging process. The possibility of 'unnatural' selections, of historical interventions such as the genocide perpetrated against the thylacine, capable of pruning the tree of life, was disavowed.[36] The open artificiality of the white environment that the animals were placed within also asserted their status as 'specimens'; they were to be understood as evidence within a scientific discourse. The display was not about them (they were readily substitutable) but about their broader origins and futures, about general categories and lineages. They were part of a demonstration of pure science. White as a colour scheme has historically been associated with clinical settings and cleanliness.[37] O'Doherty suggests that the white cube aesthetic possesses some of 'the mystique of the experimental laboratory'.[38]

In the context of the natural history display, whiteness, with its scientific overtones, helps to emphasise a seemingly value-free objectivity.[39] Human history as it meshes with animals such as the thylacine and manifests in natural history is studiously avoided. The blanched habitat of the exhibits is reminiscent of early natural history museums where objects were presented without any environmental context. Writing in 1915, Oliver Farrington praised the way that the introduction of artfulness to natural history museums brought local colour to this plain mode of display. For Farrington, artistry manifested through creating a sense of the surroundings from which objects derived, with each 'represented, if possible, in its natural setting'.[40] This approach would be impossible for 'Our Inheritance' as the animals are from diverse habitats and geological periods, heterogeneous organisms yoked by violence together to illustrate the existence of evolution, a phenomenon that transcends specific locales and times.

Each of the 'specimens' that is displayed actually possessed a distinct life history bound to a milieu and natural historical moment. The thylacine mount, M1674.001, is a female who was collected for the museum by William Aitcheson Haswell (who, in 1883, was briefly acting curator of the Australian Museum) and registered in the collection in 1902. The circumstances in which she was obtained are unclear. Some of the mount's hair has fallen away leaving bare patches; she is an animal in her prime yet also in her dotage. Her head is poorly executed, dog-like. Haswell, writing with Jefferey Parker and seemingly not working from this mount, described the thylacine as having 'a remarkable resemblance in general shape, as well as size, to a wolf'.[41] The skeleton (S.383.001) is of a male thylacine and was collected by Alexander Morton, who worked as a curator's assistant at the museum from 1877 to 1884.[42] The bones were registered as part of the collection in 1892. They were acquired through exchange from TMAG. The tiger was from the Central Highlands. Haswell and Morton, who briefly worked at the museum at the same time, are here reunited by way of two thylacines they obtained for the mammalogy collection. Details of the provenance of both these thylacines are incomplete. They were likely caught through trapping. Sleightholme and Campbell acknowledge the potential role that trade in thylacines for museums played in their extinction.[43] The difficult questions posed by the procurement of this pair are, however, elided by the exhibit. Stark against the whitewash, these tigers come from nowhere.

Further thylacine remains appeared in the temporary exhibition *200 Treasures of the Australian Museum*. A replica of the thylacine pup PA762.001 obtained by George Masters in 1866 was one of the 'treasures'. The 'treasures' as a whole exploited the 'emotional potential of object-led exhibitions'.[44] The simulation pup was housed in a display featuring a platypus skin rug, some bird eggs, a preserved Sydney funnel-web spider (*Atrax robustus*) and mounts of a koala and a black swan (*Cygnus atratus*). These all appeared against a black background. The simulation pup in alcohol was stored in a jar atop a black plinth.[45] The label for the pup suggested that she was 'a moving reminder of the importance of collecting and conserving our natural history'. That killing her mother denied her a life and potentially the chance to birth her own pups was not acknowledged. The label exemplified what Susan McHugh has called 'heroic narratives of specimen procurements' that mask the negative impacts of collecting.[46] As with the 'Our Inheritance' exhibit, the troubling history that accompanied the sourcing of thylacine remains for museum collections was ignored.

The thylacine featured for a third time in the 'Wild Planet' section, which is adjacent to the 'First Australians' exhibition space, a space that includes discussion of Australia's 'Secret Histories'. These secret histories are described as histories of First Nations Australians' resistance to colonisation, of Indigenous peoples '[fighting] for their land and their lives' (as a label explains). Examples provided of these struggles include genocidal acts perpetrated against First Nations Australians such as the massacres at Myall Creek and Waterloo Creek and efforts, such as the Black Line (an offensive against the palawa), to suppress First Nations resistance movements. A reproduction of Robert Dowling's oil painting *Tasmanian Aborigines* (1856–57) appears beside the information board detailing the massacres.[47] Dowling used Thomas Bock's circa 1837 portraits from life of a number of palawa as the inspiration for his depiction of a group including Mannalargenna, Maulboyheenner, Tunnerminnerwait, Truganini (also Trugernanner), Woureddy (also Wurati) and others.[48] There is also an information panel about Truganini and a photograph of her. The appalling impact of colonialism on Australia's First Nations cultures and on those of Lutruwita, in particular, is therefore acknowledged, although the use of European perspectives such as Dowling's (he portrays the palawa in an arcadian setting) is questionable.

'Wild Planet' includes an exhibit titled 'Thylacines: End of an Ancient Line' that features a model of *Thylacinus potens*, the cast of the upper jaw of *Thylacinus potens* found in the Northern Territory and a partial skull and jaw of *Thylacinus cynocephalus* that was excavated at the Wellington Caves in New South Wales.[49] *Thylacinus potens* was the largest known member of the Thylacinidae family. Dated to the Late Miocene (eleven to five million years ago), they weighed somewhere between 40 and 55 kg (*Thylacinus cynocephalus* is thought to have averaged somewhere between 30 and 35 kg, perhaps significantly less).[50] Here the prehistoric history of the family of Thylacinidae, which may not be familiar to visitors, is acknowledged.[51] The former presence of *Thylacinus cynocephalus* in mainland Australia is affirmed by the fossil remains from Wellington Caves (where the remains of a number of species of prehistoric megafauna have been found) and by the inclusion on a information panel of a reproduction of a rock painting from Kakadu in the Northern Territory of a thylacine.[52] The thylacine in the other exhibits at the Australian Museum is not permitted any relation with Australia's Indigenous human population, only with European colonists. Here, a different kind of relationship between human and thylacine is hinted at.[53]

The exhibit tentatively links the 'end of the ancient line' of the thylacine on the mainland to the introduction of the dingo to the biota. In this it offers a similar narrative to the thylacine display at the Smithsonian. Possible causes for the extinction of the population in Lutruwita are not mentioned, although an indication is provided by the backdrop to the main text. The text is printed over an image of a thylacine in a zoo enclosure (the image appears to be a still from film footage from Nipaluna/Hobart of a captive thylacine). The impact of colonialism on the thylacine is therefore only elliptically broached by way of an insubstantial image that many visitors will probably not notice. The contrast with the museum's significant, if problematic, acknowledgement of the violent consequences of colonialism on Australia's First Nations human population is striking. Approaches to the history and extinction of the thylacine on the mainland such as in the Australian Museum are visibly different from those adopted in Lutruwita.

The Tasmanian Museum and Art Gallery

Initially the collecting of thylacine 'specimens' by museums in Lutruwita probably formed part of a process of what the artistic duo Snæbjörnsdóttir/Wilson have called the claiming of 'ownership or authority' over an environment, in this case the 'wilds' of Lutruwita.[54] The 'native tiger' was 'brought to heel' through being collected, studied and displayed. Nowadays the thylacine remains that attest to this drive to mastery are central to the collections of both TMAG (in Nipaluna/Hobart) and QVMAG (Launceston) and both museums devote considerable exhibition space to the animal.

TMAG has dedicated an entire room, 'The Thylacine Gallery', to the tiger and its histories.[55] There are two possible entry points to the gallery; one is via 'Earth and Life: A Tasmanian Natural History', an exhibition that explores life in the region from ancient times to the present, including exhibits related to the fossil record and displays concerning contemporary flora and fauna.[56] The other entryway is via 'Power of Change', an eclectic exhibition about Lutruwita in the last century. Neither route into 'The Thylacine Gallery' seems consciously favoured by the museum yet approximately two-thirds of visitors enter the thylacine exhibition by way of 'Power of Change'.[57] There are three display cases in the centre of the thylacine exhibition, containing respectively a mount, a skin and two skeletons. The choice of entrance therefore potentially alters the narrative available to visitors.[58] Depending on the visitor's point of arrival, the narrative encouraged by the displays is one of stripping away layers – a body, then a skin, then bones – or adding meat to the bones.

The small room dedicated to the thylacine is dimly lit, the gloom sharply contrasting with the well-illuminated exhibition spaces that precede and follow it. The subdued lighting is, in part, practical, permitting archival footage of captive thylacines to be projected on the wall. Low lighting can also help to conserve exhibits.[59] Yet the obscurity is also clearly intended to foster a particular ambience.[60] Museum lighting as an interpretative and curatorial tool can help 'create mood'.[61] One of the information panels in the room is titled 'A Sad History', encouraging visitors to respond emotionally to the exhibits, to feel despondent at the thylacine's fate. The sombre lighting contributes to the affective tenor of the room, to its pervasive oppressiveness.[62] It is also a lighting in keeping with the crepuscular and nocturnal habits of the tiger, an artificial means to give a feel for light that was 'natural' to the animal. Curators often favour lighting that make objects appear as they would under daylight illumination. Daylight is viewed as 'unbiased', objective light.[63] 'The Thylacine Gallery' opts for something like twilight. Twilight, as well as being a particular quality of light (one that is softer and more diffuse than daylight), is, of course, also figuratively associated with decline and obscurity. In this sense, the lighting may symbolise the dwindling to disappearance of the thylacine as a species.

A thylacine mount, A1298, forms the room's centrepiece. Taxidermied animals, as Rachel Poliquin explores, are unsettling, 'at once lifelike yet dead, both a human-made *representation* of a species and a *presentation* of a particular animal's skin'.[64] A1298 displays this life-in-death quality, but raised up in his display case, standing still in an artificial habitat, viewed through glass, his status as 'no longer of this world' is clearly signalled. The mount was likely prepared initially to stand for his genus, to represent a taxonomic order, and to serve as spectacle. Now, he has accrued other meanings. He is exhibited to foster remembrance and to figure 'Tasmania' with its unique fauna and its particular sense of regional identity, one that is linked with, but distinct from, Australia as a whole.

The thylacine who became A1298 was purchased in 1926. A bushman, Patrick Joseph (Paddy) Hartnett, snared him in the Florentine Valley and then sold his body to the museum.[65] The mount was prepared by Alison M. Reid, whose father, Arthur R. Reid, was the curator of Hobart Zoo. Her work as a taxidermist was newsworthy as the profession was then viewed as a male preserve.[66] Reid was able to learn the art from her father who was taught it in his youth by a Scottish biologist. Arthur Reid observed that his daughter's work was superior to his own.[67] In a newspaper article published in October 1926, her work for TMAG was celebrated. As well as the thylacine, she prepared a brush opossum (*Trichosurus vulpecula fuliginosus*), a forester kangaroo (*Macropus giganteus*) and a wombat for display. Of the tiger, the article notes: 'Miss Reid's artistic talent assisted her materially in preparing the specimen of the thylacine, the skin of which has been mounted on a plaster cast in a true representation of life.'[68]

Poliquin notes that for animals, 'usually a male in the prime of life is considered the typical representative'.[69] A1298 conforms to type. In reality, of course, an ageing thylacine would not be atypical of the species. Reid has chosen to display the thylacine with his mouth shut, avoiding the need to model the interior of the maw. In mammals, as Poliquin explains, after death the fleshy parts of the mouth shrink, dry, crack and lose all colour.[70] A mount displaying the 'yawn' of the thylacine would test a taxidermist's skills considerably. Keeping the mouth shut reduces labour and acts to close down emotional cues, contributing to a sense of objectivity. Poliquin draws attention to efforts to cultivate 'neutrality of posture' in zoological specimens. This neutrality is not all-encompassing: 'animals are never just neutral, abstract markers in a classification system'.[71]

A1298, for example, was snared by a trapper. By 1926, there had been considerable debate in the press and government about snaring. It was recognised as potentially cruel.[72] Hartnett's snare of choice was a neck (or 'necker') snare.[73] These snares, often made of brass wire, are intended to catch an animal around the neck, rapidly throttling them as they struggle.[74] Writing in 1920, the journalist Albert Cherry acknowledged something of the cruelty of the trap, describing the action of the neck snare as 'simplicity itself; choking generally puts an end to torture in a little while'.[75] In 1924, an anonymous trapper from Moina wrote to the editor of *The Advocate*, a newspaper based in Pataway/Burnie on Lutruwita's north-west coast, claiming that in using a neck snare, an animal is 'easily strangled' and arguing that 'the average trapper is just as humane as the next man'.[76] Sometimes, however, neck snares fasten around a leg or the body of an animal and do not kill them. In the case of the thylacine, the Tasmanian Lands and Surveys Department issued a warning in 1888 that 'Tigers do not choke themselves with the snares – it is a very rare thing to find one dead'.[77] In 1925, a deputation met with Attorney General Albert Ogilvie, to discuss the possibility of suspending trapping. At this meeting, Henry Gould of the Society for the Prevention of Cruelty to Animals called upon Ogilvie to legislate for more humanitarian trapping methods 'so that animals should not be subjected to the torture of being left in traps for many days in snow and rain till they eventually died from their sufferings'.[78] Clive Lord, the Director of TMAG, was also present and called the instruments and techniques employed in trapping, 'the miles of carefully arranged snares, traps, and other devices', akin to modern warfare.[79] It is not known how much the thylacine snared by Hartnett suffered, whether he rapidly suffocated, or died slowly through starvation, or was found alive and then despatched. The accompanying label simply states euphemistically that the tiger was 'captured' by Hartnett.

In April 1926, the government announced a year-long close season for game, forbidding hunting and trapping to better protect the island's 'furred animals', but the policy was later abandoned.[80] Economic arguments swayed the Attorney General against the moratorium.[81] If the government had stood firm, it is possible A1298 would not be in the TMAG collection. As it was, the trapping season closed at the end of August.[82] Many trappers probably continued to trap in September and beyond. In 1927, police from Sheffield travelled to Mount Pelion West and Lake Ayr searching the region for trappers who were continuing to snare in close season. The police saw 'hundreds of snares set for fauna, which apparently had been trapped out of the mountains'.[83] The view was held that 'in a few years' time the native fauna will be as extinct in central Tasmania as the dodo' and there were calls for a complete close season for a whole year.[84] To give a sense of the destruction, in 1923 a million animals were trapped.[85] In 1926, William Crowther called the use of 'spotlighting' to kill ring-tailed opposums (*Pseudocheirus peregrinus*) a 'systematic massacre'. Some, however, blamed the paucity of kangaroos and other animals on the thylacine and the Tasmanian devil and used the disappearing game as a reason to advocate their destruction.[86] Men such as Lord, who at the same time as decrying the cruelty of trapping was also offering a bounty of £5 for a dead thylacine, were tacitly encouraging the killing of the animal.[87] I have not been able to discover in what month Hartnett snared the thylacine that he sold to Lord.[88] Given that it must have taken time for Reid to prepare the mount (which was ready by October), the tiger was probably snared in season.[89] The label for A1298 acknowledges that the thylacine was purchased but not that the museum was actively touting for dead tigers at the time.

Reid chose to show the animal with the forelegs parallel to each other, the tiger not in motion. The left hind leg is, however, posed as if he is prepared for movement or has just paused. The distinctive stiff tail has been well rendered. Poliquin notes that 'all decisions as to how an animal should be posed are necessarily selective'.[90] Choice of pose is dependent upon the taxidermist's skill, the condition of the skin and the aims behind the decision to mount the animal (for education, for example, or as spectacle). The fur on the right side of the mount is distressed, particularly around the neck and at the ribcage above the humerus. This tiger is slowly balding, showing his age. Kitty Hauser observes of photographs of stuffed animals that they sometimes draw attention to 'taxidermic realism as constructed spectacle' through showcasing seams.[91] Here the thylacine's old skin similarly makes his constructedness evident without the need for photography's 'unforgiving attention to detail'.[92]

In the adjacent display case, there is a thylacine pelt, A1287. The tiger looks like roadkill, flat, distorted, only identifiable by the distinctive dorsal bands. The accession date for the skin is unknown, as is the tiger's sex. The skin features in recent research by Stephen Sleightholme and Cameron Campbell on 'stripe pattern variation'.[93] Sleightholme and Campbell compare thylacine banding to fingerprints, suggesting that they are unique to each individual animal.[94] This important research, which traces how variations in banding manifest themselves, contributes to the ability to identify individual tigers, acknowledging their singularity.[95] A1287 has eighteen bands including bifurcated and interconnecting ones. There are other skins with the same number of bands but their form will differ to a greater or lesser extent from A1287's. The close links between animal skins and animal identities has been explored by Ann Colley. She notes that in the nineteenth century an animal's skin was frequently viewed as akin to clothing, a marker of their individuality.[96] Colley nonetheless sounds a note of caution about reducing animals to their skins, the risk of a skin-deep understanding based on surface differences rather than substantial insights.[97] She also notes how a tiger's (*Panthera tigris*) stripes were linked with savageness and butchery.[98] The thylacine as a striped 'native tiger' would probably have been tarred with these negative associations.

In the next display case, there are skeletons of a male and a female thylacine shown side by side. There are tags on the feet of the two skeletons including gender symbols, imposing binary gender, a prevalent human idea of gender, upon the thylacines: A312 is female; A315 is male. Male tigers were generally considerably larger than females (leading scientists such as Gerard Krefft to occasionally mistake female specimens for a subspecies) but there was also considerable size variety within each sex. The label to the cabinet containing A312 and A315 refers to sexual dimorphism, explaining that: 'Female thylacines were slightly smaller than male thylacines.' This was a general characteristic but not all male thylacines were bigger than all female thylacines.[99] Robert Paddle has explored how a photograph by H. Tucker of A315 alongside another thylacine in an enclosure at Beaumaris Zoo has often wrongly been used as evidence for dimorphism because the smaller tiger was incorrectly identified as female (he is actually a juvenile male).[100] The significant disparity in size is a product of age not sex.[101]

A312 came into the collection sometime in 1919–20 and her skeleton was articulated by Thomas Francis Moore of the National Museum of Victoria (NMV) in 1922.[102] A315 arrived in 1922 as part of the collection of Mary Roberts of Beaumaris Zoo. The skeleton was also articulated by Moore in 1922.[103] The tiger had been snared alive by Ted Power near Tyenna in 1911 and kept by him for a time before being sent to Nipaluna/Hobart.[104] After an impromptu exhibition of the animal at the Shamrock Hotel in Liverpool Street on the morning of Saturday 12 August, he was sold to Roberts later that day for £12 18 shillings.[105] The thylacine joined two others in the zoo's collection, a mother and her cub. Paddle identifies the male from Tyenna as the thylacine most photographed from life.[106]

A315 died at the zoo on 9 March 1915. Roberts placed his corpse in cold storage and then sought to hawk his remains to the Australian Museum and, as part of a job lot, to the NMV. James Andrew Kershaw, the Curator of Zoology at the NMV, offered to pay £10 for the assortment but the sale fell through when Roberts was unable to regain possession of four skeletons that she had loaned to the disreputable zoologist Theodore Thomson Flynn (now best known as Errol Flynn's father).[107] She prepared the bones herself and added the loose skeleton to her private collection. Paddle speculates that the tiger held significant emotional as well as scientific value for Roberts as in May 1915 she refused an offer of £10 for the skeleton. Any attachment she had for him, however, must be qualified.

A315's display label suggests that he 'became quite tame and could be walked around on a lead'.[108] Taming involves exercising power to render an animal obedient.[109] It is a disciplinary process that seeks docility as its outcome.[110] Along with the female thylacine he shared an enclosure with until her death, he was similarly described by Herbert Millingchamp Vaughan as 'quite friendly and tractable'.[111] The thylacine, enclosed within the domain of the zoo, became subject to procedures designed to modify his behaviour. This modified behaviour (interpreted as friendliness and compliance) was no doubt felt to render the tiger a more agreeable *spectacle*. It can also be interpreted as a kind of cultural genocide, destroying the tiger's 'natural' or 'wild' comportment.

Roberts exploited the fascination with tigers, their capacity to attract visitors, in full knowledge that they were in danger of disappearing. The scarcity of the thylacine was well known by 1911 when A315 (as he would later become known) was abducted and imprisoned. In an article published in the Nipaluna/Hobart-based newspaper *The Mercury* in 1910 it was noted that the tiger was now 'seldom found' but that some had been caught in recent years in the Tyenna district.[112] Roberts hoped A315 would breed with a female tiger she had in captivity but this aim was a money-saving initiative rather than a conscious effort to preserve the species from extinction. If the latter, Roberts would have immediately released A315 back into his natural habitat upon purchasing him.

Moore's decisions when articulating the skeletons of A312 and A315 are noteworthy. His choices seem informed by contemporary ideas about gender. The male has his jaws open, his teeth exposed, bared. The female has her mouth wired shut. This small difference speaks volumes. Erle Wilson's novel *Coorinna*, analysed in the next chapter, also demonstrates how ideas about human gender relations feed into visions of the thylacine. In the novel, two tigers, a male, Coorinna, and a female, Loongana, are courting: the 'fine young wolf, almost as big as a mastiff, was wooing the weedy she.'[113] The female tiger functions as a foil, enabling Coorinna's size and strength to be showcased. When she is first introduced, Loongana is described as 'a young wolf, weedy and under-sized but beautiful to eyes long deprived of the sight of his own kind'.[114] The pair of thylacine skeletons displayed at TMAG seem to embody comparable beliefs about gender difference.

There is something perverse about the positioning of A312 and A315: their skeletons face the section of wall where archival footage of thylacines in captivity is projected. The footage lasts about two minutes and plays on a loop. The pair of thylacines now continually 'view' their captive kin. Briefly, A315 'sees' himself, alive and imprisoned. He is shown with a chunk of meat, throwing it about and snapping at it, seeking to get a good purchase on the flesh. A man in boots and breeches, possibly a keeper, is visible behind a fence in the background. The man, arms behind his back, initially stands looking on but then moves rapidly to the right of the frame when he realises he is visible. The footage was shot by William Williamson on 27 September 1911.[115] Now A315's cage is replaced by a display case, but he retains the role of spectacle. In other footage, a man can be seen rattling the fencing to get a reaction from one of the captive thylacines. A315 was probably also subject to this treatment. In his account of visiting Beaumaris Zoo, the tourist Herbert Vaughan writes of some 'phalangers' (possibly eastern ring-tailed possums) that they were habitually nocturnal and '[i]t was no easy matter to compel the sleepy cross little animals to perform for my benefit; but at last we succeeded in our efforts'.[116] It is clear from Vaughan's account and from film footage that zoo staff would disturb the animals, forcing them to move around: to 'act the part' for visitors. One of the last films featuring a Tasmanian tiger, which was taken in 1935, shows Arthur Reid vigorously shaking the wire fence or 'rattling the thylacine's cage' to force him to move across the enclosure.[117] The oftentimes cruelty of film-makers and keepers towards captive thylacines is examined in more depth in Chapter 5.

As well as the film footage, there are framed photographs of captive thylacines at zoos included in the exhibition. These are hung against an expanse of exposed concrete. One of these is the photograph by H. Tucker that includes A315 and was used in the past to illustrate sexual dimorphism.[118] Another is of the 'O'May thylacine' on the day of his arrival at Beaumaris.[119] The tiger is named after the man who seized him, Billy O'May (perhaps William Percival O'May, a sawmiller and sheep farmer, who leased land in Fitzgerald near Tyenna). He is collared, eyes closed, held low to the ground by the keeper Charles 'Charlie' Daniel Newman.[120] This thylacine was bought by the zoo from O'May and displayed from 17 June 1916 to 12 October 1918, when the tiger was sold on to Taronga Zoo in Gadigal/Sydney.[121] At Taronga, the thylacine was attacked by a puma who tore off his tail. A single photograph exists of the tiger at Taronga, his tail visibly a stump. Sleightholme and Campbell speculate that the photographer, the ichthyologist Gilbert Percy Whitley, never published the image because it was too perturbing.[122]

A third photograph dated to July 1909 is by William Williamson and shows a female tiger and three pups snared at Woolnorth by Thomas Lovell and George Wainwright Junior.[123] A fourth pup and the male of the family were also snared but killed. The mother and her pups were transported by ketch from Woolnorth using the Whale's Head (a small harbour that was located near Mount Balfour) to Stanley route and then transferred to the *SS Toroa* for the voyage to Nipaluna/Hobart.[124] Like A315, the mother and her pups were treated as a spectacle during their journey to Roberts's private zoo. At Stanley, it is reported that when the tigers arrived on Saturday 3 July, they 'were the object of much curiosity among Stanley residents'.[125] Captured thylacines were clearly casually and regularly exhibited as 'freaks' or 'wonders'. Newspaper accounts make regular reference to this practice.[126] In 1844 a tiger killed by a snare on the banks of Kanamaluka (the River Tamar) was displayed for several days at the watch-house in Launceston before being preserved as a mount.[127] Cumulatively the photographs of thylacines reference the role of zoos in the genocide perpetrated against the animal, but this complicitly is not directly addressed through captioning.

Another display case, divided into seven sections, is located adjacent to where the film is projected on the opposite side of a doorway. In this case, a variety of narratives related to the thylacine are enumerated. The top left section includes two mounts, a crested porcupine (*Hystrix cristata*) and an Alpine marmot (*Marmota marmota*) that the museum received from the Royal Zoological Museum of Florence, Italy, in 1892 in exchange for several thylacines and devils. There are two thylacine mounts (numbered 1058 and 1059) in the collection of today's Museo di Storia Naturale in Florence. Alongside the marmot and the porcupine there is a photograph of one of these mounts. This display commendably draws attention to the exchange network that existed between museums at the time. Lutruwita's fauna was a valuable commodity for trade purposes in the 'specimen' economy. The tiger's scientific 'worth' fed TMAG's demand for the animal, with museums on the mainland and overseas complicit in this ghoulish economy.

The section that is below the marmot and porcupine contains three skulls, those of a dog, a thylacine and a wolf (the latter is a cast). The information panel for this section explains that the similar skulls of the three carnivores illustrate the idea of convergent evolution, the independent evolution of similar characteristics in different species. The thylacine and the wolf had 160 million years of independent evolution but have considerable cranial similarities.[128] The skull of the tiger, A1238, is of unknown gender and according to the ITSD nothing is known of the animal's provenance. Adjacent to this display is a section including a wolf trap, 'thylacine snares' and a trapper's hatchet.[129] This section is designed to be read in tandem with the one above, which contains two skulls (A1241 and A1483) with considerable damage to their craniums.[130] The label located in front of the trapper's hatchet explains that: 'To prevent thylacines preying on animals captured for the fur trade, trappers sometimes laid poison baits along their trap lines. When found, they were killed immediately by a blow to the head or shot.' A poisoned thylacine would probably not be found, but if snared and discovered alive they would doubtless have been killed in this way (until their value alive began to exceed any possible bounty payments).

The two skulls exhibit what appears to be blunt-force trauma. These thylacines may have been clubbed to death. The larger of the skulls shows massive cranial damage. The kind of fracture that is caused by a blow depends on the amount of force used and the size of the area to which it is directed.[131] Here the strength with which the skull was struck was so substantial, it caved it in. Albert Le Soeuf's and Harry Burrell's *The Wild*

Animals of Australasia includes a section on the thylacine that ends with a warning that the animal is disappearing and the observation that on killing one, some people are so superstitious of the creature they 'will even smash the wolf to pulp afterward, thus depriving science of the skeleton and skin'.[132] Something of this excessive antipathy is indexed by the skull, the fractured bones speaking of intense enmity. The other skull manifests a controlled violence, one that is functional rather than pathological. There are historical instances in which the economic value of capturing a thylacine alive proved insufficient to prevent them being killed, such as when Albert Edward Quarrell abandoned his efforts to entrap a tiger (by seizing their tail) and, on cornering the animal, shot them. Wilfred Batty also gave up trying to catch a thylacine alive and shot and kicked him to death. This shows the power of the genocidal gaze historically. The large skull seems to graphically register the effects of that gaze. Given the climate of hatred in Lutruwita towards tigers, that the blow was motivated by loathing is plausible. Here, high on a shelf in 'The Thylacine Room', animal genocide as it was perpetrated against the thylacine is quietly given its most powerful and troubling expression.

Photographs included in the exhibition provide context for the fatal injuries embodied by both skulls. In the quartet of photographs near the display case featuring trapping paraphernalia, there are two photographs of thylacine killers posing with their 'trophies'. One of these shows William George Weaver in a photographic studio posing with a tiger he killed in 1869.[133] He was evidently a keen hunter as in 1870 he was found guilty under the Native Game Act of breaking the law by killing twenty quail (probably *Synoicus ypsilophorus*) in the close season. The prosecutor Mr Graves observed that although the killing of twenty quail might strike some as inconsequential, it must be remembered that there were 'few specimens of native fauna' and there 'should be a stringent rule to preserve them'.[134] The 'Act to Provide for the Protection of Native Game and Other Birds' came into force on 17 September 1868. It only provided protection to avifauna during the breeding season (which, for quail, was 1 August to 20 April each year). The thylacine was not afforded such a legal safeguard.

The other 'trophy' image shows Albert Quarrell standing cradling a thylacine he had just shot near Fitzgerald in 1911. Quarrell, a bullock driver or 'bullocky' working at Tyenna Mill, subsequently sold the carcass of the tiger to a local amateur photographer, Charles Brown.[135] A man seated in the background behind Quarrell is often identified as the trapper David Pearce.[136] Brown, who took the photograph, may have assisted in accosting the tiger. The third photograph in the quartet is of a prone thylacine in an outdoor setting, which has been tentatively dated to the 1920s. Curled up in grass, the tiger is either dead or dying.[137] The final photograph is of

A315 standing on his hind legs, his forepaws resting against the chicken wire of his enclosure. The keeper Charlie Newman is on the other side of the fence and is probably cajoling him to 'perform' for the photographer. The image has been dated to August–December 1911 and is attributed to M. Strickland and H. Tucker.[138] Cumulatively, the four images speak to the violence thylacines were subject to, systematically hunted, either captured or killed. Dead or alive they are consistently rendered spectacle, subject to the inhuman gaze of the camera. The 'trophy' photographs are particularly sadistic, celebrating slaughter. The baiting of A315, however, is equally appalling. He is goaded the better to 'show his stripes'. I will return to the relationship between the photographs and the genocidal gaze in my Conclusion. For now, it is important to note how the vision of the thylacine as a worthless thing or as a freakish curiosity is tacitly acknowledged by the exhibition as a contributing factor in their destruction. This contrasts markedly with how the history of the tiger is represented at, for example, the Australian Museum.

Although the exhibition occasionally explicitly confronts the violence perpetrated against the thylacine, there is a reluctance to countenance the finality of the animal's destruction. An information panel on one wall is titled 'Rarity ... and extinction?' The question mark leaves the matter of the tiger's disappearance open. The text of the panel notes that although the last bounty was claimed in 1909, museums and zoos continued to fuel demand for the animal.[139] The final resting places for most of the physical remains of the animal are therefore institutions that facilitated animal genocide. In this, TMAG was more culpable than most, a veritable hub for the trade in 'specimens', a trafficking centre.[140] The text also refers to the 'functional extinction' of the thylacine, implying that it is possible a small number of the animals survive. Functional extinction, when a species no longer plays a significant role in its habitat, is reversible.[141] As mentioned in Chapter 1, the term 'zombie species' is sometimes used to refer to an animal that survives but no longer possess a 'significant ecosystem function'.[142] Calling the thylacine's extinction functional rather than total implies they may live on. The text nonetheless concludes with an affirmation of the mammal's disappearance: 'The human induced extinction of the thylacine was the start of a process that continues to have impacts on the Tasmanian environment.' The panel is studiously ambiguous, the tiger's fate left open.

The possibility of survival to the present is also supported by the top right section of the display case containing the trapping paraphernalia and skulls of slain thylacines. This section includes remote sensing camera equipment and reproductions of reward posters for the finding of living thylacines. The caption for the camera equipment makes reference to Hans Naarding's sighting of a thylacine in 1982.[143] The middle right section of the display case features the guidelines provided to park service members in 1983 to aid them in the event of their sighting a thylacine and also details searches made for the thylacine and contains plaster casts of pawprints found by an expedition to south-western Lutruwita in 1938. The trip was organised by the Tasmanian Animals and Birds Protection Board and sought '"fingerprints" of the Tasmanian tiger'.[144] The searchers, led by the naturalist Michael Stanley Reid Sharland, explored in the vicinity of Jane River and the Raglan Range. These 'fingerprints' took the form of pawprints, casts of which were made. Sharland described them as resembling 'dental plates with the claw impressions, now in reverse, projecting like teeth'.[145] The casts, if genuinely of tigers rather than another species such as wombats, are rare testimony in that they index vitality, offering traces of the life of the thylacine in the wild that are not linked to violent death.

The section also includes a hair sample and a handwritten note dated 18 March 1946 that explains that David Fleay retrieved it from near a trap. In a 1945–46 expedition lasting four months, Fleay visited the Collingwood River and the Jane River. These rivers run through much temperate rainforest and are not ideal tiger country, although the animal ranged over most of Lutruwita before European invasion. Fleay, who sighted no tigers, was probably looking in the wrong place. The same year as Fleay's expedition, thylacine tracks were found in the Vale of Rasselas.[146] Ernest Bond, who reported the pawprints, verified that they were of a tiger by consulting the 1938 plaster casts that were at TMAG. In his article for *The Mercury* describing the event, Peregrine (a pseudonym adopted by Sharland) observed that snaring in the valley and the poison baits that accompanied it would soon wipe out the tigers.[147] Fleay urged a ten-year moratorium on trapping from the West Coast Road down to Port Davey and the southwest cape.[148] The sample of hair Fleay retrieved was initially identified as being from a thylacine but subsequent tests revealed it was wombat hair. These sections relating to searches for thylacines serve to fuel the idea that the tiger somehow survives to the present day. As I will argue in Chapter 4, the refusal to accept the finality of the tiger's extinction is bound up with guilt at the loss of the species.

The bottom of the cabinet or case contains an incomplete skeleton of a tiger whose sex is unknown, which was acquired by TMAG in 1924.[149] The skeleton, which stretches the length of the cabinet, seems to have been included because the bones were retrieved from north-west Lutruwita by Theodore Thomson Flynn. The inclusion of the skeleton seems primarily designed to showcase the connection with Errol Flynn and Hollywood. The label does note that Flynn the naturalist 'was particularly interested in obtaining thylacine material for the University and the Tasmanian museum' but does not mention that the museum severed links with him when he borrowed thylacine remains and failed to return them.[150] In an article he published in 1922, Flynn drew attention to 'the great importance of securing, while there is still time, a representative collection of intra-uterine stages and pouch-young of the fast-disappearing marsupial fauna of Tasmania'.[151] He went on to observe that almost 'all the collecting has been done during the winter months, since it is during this time of the year in Tasmania that the professional trapper shows his greatest activity'. Flynn is talking about Lutruwita's fauna in general, not about the thylacine specifically (much of the article refers to the common ringtail possum [*Psuedocheirus peregrinus cooki*]), but he makes clear the Faustian bargain that scientists had made with the trappers. His concern is also shockingly not for the fauna that is disappearing but for himself. He is worried that he may not have sufficient time to carry out all the research he wishes to before his subjects vanish. His work on reproduction also requires that he source 'specimens' during the breeding season, therefore potentially hastening the demise of endangered marsupial species. The exhibition therefore obliquely references the culpability of the scientific community for animal genocide, vigorously pursuing collecting in the shadow of extinction.

The panel titled 'The Thylacine: Symbol of an Island' provides the rationale behind the organisation of the room. It traces how the animal has become symbolic of Lutruwita's identity but goes on to suggest that 'as well as being a popular symbol, the story of the thylacine is a powerful reminder of how easily a species can be lost'. The room is here framed as providing a valuable history lesson regarding what occurs when a species is not sufficiently appreciated and adequately protected. The panel also affirms the important role the museum plays in safeguarding the memory of the tiger: 'Evidence of the thylacine's existence is now found only in photographs, short excerpts of film, in bushmen's stories, and specimens in museums.' The thylacine room includes all these forms of evidence except bushmen's stories. The panel goes on to explain that '[t]his gallery tells the story of the thylacine and its interactions with society through the collections of the Tasmanian Museum and Art Gallery'. Here the thylacine is instrumentalised, used as a means, in the end, to justify the museum's function in society. TMAG's own role in the tiger's persecution is downplayed, signalled only indirectly by the exhibition.

Queen Victoria Museum and Art Gallery

The QVMAG in Launceston also devotes a display space entirely to the thylacine. Their exhibition, 'Tasmanian Tiger: Precious Little Remains', is larger than that of the TMAG and more interactive.[152] It forms an annexe to 'Tasmanian Connections', a permanent display dedicated to Lutruwita's history and natural history. There is considerable crossover in terms of approach and content between the thylacine exhibitions at QVMAG and TMAG. Both employ film footage and photographs and include 'specimens' of the animal. Both also include similar narratives in relation to the history of human–thylacine interactions post-European invasion, including discussion of the demonisation of the tiger and its persecution. As in TMAG, QVMAG places an emphasis on the thylacine's distinct morphology: a thylacine skull, for instance, is compared with a dingo skull.[153] Like TMAG, QVMAG uses the thylacine to stake a claim for the importance of the museum as a locus of remembrance and of ecological conscience. QVMAG, crucially, also acknowledges the importance of the tiger to the palawa. Additionally, the museum makes considerable effort to share oral history about the animal. Tammy Gordon, who co-curated the exhibition with David Maynard, has helped build an archive of local oral histories at QVMAG.

The entry to the alcove containing the exhibition features a composite photograph, titled *Tasmanian Marsupial Wolf*, that shows a thylacine standing atop a rocky outcrop gazing rightwards (towards the exhibition) with Warloundigerler/Cradle Mountain visible in the far distance.[154] The image was created by Herbert John King, a photographer and naturalist from Launceston.[155] King took several pictures of tigers at Beaumaris Zoo in 1933, of which the image of the tiger used in the composite is one.[156] The original photograph of the captive thylacine shows him standing beside the wire mesh fence of his cage. A boy, his left arm outstretched, is trying to attract his attention through the fence. Sunlight illuminating his face, the tiger is turned to the left, staring at someone or something. Shadows, presumably cast by part of the enclosure or by foliage, play across his back. These shadows are retained in the later composite, tacitly indexing the tiger's captivity from within his seeming freedom. King was looking down on the thylacine when he snapped him in the zoo. In King's composite, the tiger has been reversed (he now looks to his right) and repositioned so that it seems as if he is being observed from below, from slightly lower on the outcrop. The viewing position implies that we are sharing the perspective of what might be another thylacine. An image in the Norman Laird collection at the Tasmanian Archives and Heritage Office clearly shows how the picture was made.[157] King tacked a blow-up of the photograph of Warloundigerler/Cradle Mountain to a board and then cut out a reversed blow-up of the thylacine and pasted it on top. The rocky outcrop is also a cut-out from a third image. This is not King's only composite photograph. In an image titled 'A Steep Pinch', he combines two photographs to give the impression of a car climbing a tree.[158]

Christopher Orchard and James Holcombe suggest that King's 'sublime images' of Warloundigerler/Cradle Mountain were imbued with nationalist ideology and contributed to building a vision of the Australian bush as a 'paradise', an image at odds with its ecological reality.[159] This is nowhere truer than in *Tasmanian Marsupial Wolf* in which King repopulates Lutruwita's Central Highlands with tigers, 'freeing' one of the last captive thylacines for this purpose. A species subject to genocide and driven to extinction roams at liberty once again thanks to this act of compositing. The human presence in the zoo image is here excised. Sleightholme and Campbell call the original photograph the 'most endearing' of the known photographs that show human subjects 'interacting' with thylacines.[160] This is presumably because it features a child but the boy, like many adults in the zoo photographs, is trying to provoke a reaction from the tiger. This is not an adorable image but an abhorrent one, indexing the regular harassment that the native fauna incarcerated at Hobart Zoo were subject to. In the composite image, all trace of this act of persecution has vanished. The tiger is shown alone in Lutruwita's wilderness.

The logic of beginning the exhibition with a manipulated photograph, one referred to in the caption as an 'unusual image', seems to be that it foregrounds Lutruwita as a place of natural beauty. The photograph is coupled with a quotation attributed to the US President Theodore 'Teddy' Roosevelt:

> Here is your country. Cherish these natural wonders, cherish these natural resources, cherish the history and romance as sacred heritage, for your children and your children's children.
>
> Do not let selfish men or greedy interests skin your country of its beauty, its riches or its romance.

The words are sometimes linked to a speech Roosevelt gave on the rim of the Grand Canyon on 6 May 1903. He certainly expressed comparable sentiments in that speech.[161] The reasoning behind the decision to use Roosevelt's words to introduce the exhibition may be his conservationist credentials.[162] The reference to 'skinning' a country of its beauty is also redolent of the history of the thylacine, relentlessly persecuted by trappers. Roosevelt is, however, an ambiguous figure. He was a keen huntsperson and 'specimen' collector. Writing of bison as on the 'brink of extinction' because of their slaughter by humans, he sees them as subject to a 'merciless and terrible process of natural selection'.[163] He would probably have viewed the plight of the thylacine similarly.

On the opposite side of the entryway is a display case containing two thylacine skulls and some bones. The 'specimens' here and elsewhere in the exhibition appear to be chosen in part for their backstory and in part for their condition. The decision not to showcase QVM 1962.1.55, with their flesh and whiskers, their scrappy nose, is noteworthy. The flesh arguably renders the 'specimen' more valuable, more 'precious', yet the skull was probably perceived as too gory, as too reminiscent of the living animal and of their demise. The flensed bones in the exhibition, like white marble sculptures, possess a purity that residual flesh of this kind sullies or negates. In the Australian Museum, a reproduction of a wet 'specimen' seen from a distance (as occurred as part of the 200 Treasures exhibition) possesses a similar innocence. Dried flesh, when visible, acts like a stain, marking the bone as a part of a corpse. A skeleton, when cleaned, becomes devoid of traces of corporeality, transformed into something other than a body. This reality is affirmed by the way the thylacine exhibits in the 'Missing Millions' display case of the Australian Museum are labelled 'Thylacine and skeleton' as if the skeleton no longer *is* the animal but has become something qualitatively different.

The display case with the skulls and bones includes a diagram that gives details of the trapline at Aplico (discussed in Chapter 2). It is accompanied by a quotation from Robert Stevenson: 'they ruined my father ... I had erected two miles of wire netting fence against them and made automatic pit fall traps on the outside of the fence and eventually exterminated them.' An explanatory label states that at least sixty tigers were caught at Aplico between 1888 and 1906. The label also explains that the last thylacine caught at the property 'starved to death in a neglected pit fall trap'. The killing of thylacines at Aplico was premeditated, carefully planned. The netting and traps formed an extensive architecture constructed to facilitate genocide. Not all the physical remains in the display are linked to Aplico. One of the skulls was found in 1913 at Granskogen, a forested area near Warloundigerler/Cradle Mountain. This skull is noteworthy as, as the label explains, 'it is the only one in the QVMAG collection found in the wild'.[164] The location and date of discovery of the skull are written on the cranium with graphite. The other skull is from Aplico and was donated by David Cunningham, who was gifted the remains by his grandfather, Robert Stevenson. At the base of the display cabinet are assorted thylacine bones, their provenance uncertain.[165]

A text label titled 'Precious Little Remains' is located near the entrance. It explains something of the rationale of the exhibition.[166] The label begins with a short summary of the fate of the last captive thylacine (usually identified as being 'Benjamin'), drawing attention to the neglect that led to their premature death from exposure. This short biography is followed by the ensuing text:

> A handful of photos and some old film footage; some stories and recollections; and a few treasured museum specimens are all that have survived – precious little remains.
>
> This is the tragic tale of one of the world's most remarkable and misunderstood animals.

The label draws attention to the contranymic quality of the expression 'precious little remains', foregrounding both the paucity of material that exists relating to the thylacine and the immense value accorded to those things that do endure. Viewing the extinction of the thylacine as tragic promotes a narrative of sadness and suffering and encourages an affective response. The reference to 'misunderstanding' also implies that correctives will be offered to long-standing myths regarding the species.

The significance of physical remains is also signalled by their being referred to as 'treasured'. In 'Precious Little Remains', these valued vestiges work in tandem with other traces of the tiger including film footage and photographs. Among the photographs included in the exhibition is one by Margaret Tassell of the stump of a white gum (*Eucalyptus alba*) tree from the farm of Elverton located at Blessington. The caption explains that: 'At one time, up to sixteen Thylacine heads were nailed to this tree, mostly using harrow tines, some of which are still visible.' Here the genocidal gaze is obliquely referenced and, like the intricate explanation of the trapline at Aplico, something of the architecture of genocide is acknowledged. Far more sense of the amount of effort and organisation that was sometimes put into killing tigers and of the way their destruction was feted is present than at TMAG.

The Milk of Scientific Knowledge

The exhibition seeks to dispel 'myths' about the thylacine.[167] Its strong scientific component was intended to help correct 'common fallacies' and soften the significance of 'sightings that could never be substantiated'.[168] 'Precious Little Remains' therefore aims at objectivity, showing fidelity to fact rather than promulgating 'factoids' and bolstering legend.[169] Like any scientific endeavour, however, ideas and assumptions shape how knowledge is communicated. The female thylacine QVM.2002, for instance, is positioned to face off with a dingo in a composite that echoes the 'dogfight' engineered by the Smithsonian. This face-off is mirrored in the foreground where two skulls – of a thylacine and a dingo respectively – are posed with jaws agape. These are ostensibly positioned to highlight morphological differences (such as the distinct number of teeth in each animal) but having the skulls 'bare their teeth' encourages reading both animals as engaging in acts of aggression, offering a vision that is reductive, evacuating the complexity of their lives in favour of privileging one behaviour, hostility.[170]

Both species are capable of violent behaviour so the vision is not unscientific, yet nor is it 'truthful' in any simple sense. In encounters with humans, dingoes, for instance, more commonly manifest curiosity than aggression.[171] Thylacines also rarely displayed aggression towards humans unless they or others in their family were threatened. How a dingo and a tiger would react to the presence of each other is unknown. Their aggressive confrontation in 'Precious Little Remains' seems designed not to communicate facts about their behaviour but rather to reinforce the Darwinist idea of nature as a 'dog eat dog' world where only the fittest survive. In that situation, some argue that the thylacine succumbed because the dingo was better adapted to Australia's environment. Here the subtext again links the fate of the thylacine to the dingo, ignoring the role of potential human factors in their disappearance. There are no dingoes in Lutruwita so the display is motivated by a wish to address the fate of the thylacine more broadly, beyond the island's context.

The tiger has been posed in a crouching position with her head lowered and appears servile when contrasted with the dingo's raised head and erect posture.[172] These two mounts were, of course, never intended to appear together. As well as two difference species, two contrasting styles of taxidermy are confronting each other.[173] The older mount, of the thylacine, displays the technical limits of the time and is 'anatomically outmoded'.[174] The tiger seems uncomfortable in her own skin, her midriff thin compared to her rump, her neck unnaturally elongated, her fur dishevelled in places. There is a brad nail protruding from her forehead.[175] The dingo is well proportioned, their coat healthy-looking. A label explains that the thylacine mount is irreversibly faded due to light exposure and the display deliberately uses a low light to 'minimise [such] adverse effects'. The dingo seems to be more brightly illuminated, perhaps because the animal is not rare, the mount readily replaceable.

Leaving the thylacine in shadow alongside the canine bathed in light inadvertently encourages symbolic readings of the two animals. Light and dark possess powerful metaphorical associations. Michael Osborn suggests of light and dark metaphors that 'because of their strong positive and negative associations with survival and developmental motives [they] express intense value judgements and may thus be expected to elicit significant value responses from an audience'.[176] Due to their common linkage with day (light) and night (darkness), the metaphors are also deterministic, giving 'the impression that some particular series of events had to or will occur'.[177] In the context of the thylacine mount, her appearing in gloom potentially affirms the inevitability of her extinction. This message of the thylacine's inexorable disappearance is not consciously intended but nonetheless forcefully supported not solely by the lighting but by the subordinate pose of the mount and the staged clash of the thylacine with another apex predator. Although 'Precious Little Remains' was, in part, intended to dispel myths about the thylacine, the display inadvertently fuels the fiction of the 'wimp species' that invites their own destruction through inherent weakness and docility.

There is a mirror carefully positioned beneath the thylacine mount so that her pouch can be seen. It is reminiscent of a speculum. Poliquin has drawn attention to the voyeuristic dimension to all taxidermy in which animals that would likely flee and hide from humans are held fast, open to an unnatural contemplation. In the QVMAG mount, this voyeurism is amplified as a mirror brings the pouch, naturally concealed beneath the belly of the tiger, into view. The thylacine was acquired by TMAG and donated to QVMAG in 1891.[178] In 1909 Herbert Hedley Scott reset the mount to exhibit the interior of the pouch including four relatively long teats. The teats of marsupials increase in length as their young grow post-partum. The presence of the elongated teats therefore indicates that there were four pouch young or joeys, who were several weeks old (thylacines exited the pouch at about twelve weeks old), present when their mother was killed.[179] They were probably slain with her.

The pouch is clearly an important marker of morphological difference and therefore perceived as important for scientific observation.[180] The mother's body was preserved for science as, possibly, were those of her young. In her posthumous vulnerability, the mounted female thylacine was subsequently reworked by Scott to better showcase the teats as a marker of her sex. Scott's actions in opening the 'mysterious' interior of the marsupium to view work to frame this particular thylacine as a mother, as a female representative of her species. Her role as mother is rendered her truth. The use of the mirror within 'Precious Little Remains' to draw attention to Scott's handiwork also contributes to defining the thylacine (without the mirror, Scott's intervention would probably pass unremarked). The enduring fascination with gender difference and sexual reproduction that many scientists have manifested regarding the thylacine and animals in general speaks to cultural influences on scientific research. In 'Precious Little Remains', science is set up as rational and objective but this mount shows that broader cultural anxieties about sex, sexuality and the body manifest through the kind of attention accorded the thylacine.[181] In this instance, anatomy becomes the animal's destiny. Scott's exposing of the mammary glands is about conquering the unseen, about revealing and thereby coming to 'know' the animal.

Terri Kapsalis notes the close relationship between making visible and mastering in Western biomedicine.[182] A similar dynamic is in operation here in a natural history context. Exposing the interior of the pouch transforms inner enigma into a visible object available for analysis. The mirror cuts this object out from the thylacine's body, enacting a visual dissection. The excision of the pouch from the body that is performed by the mirror bears some similarity to the fetishisation of the breast in pornography. In pornography, however, breasts are reduced to sexual objects, their nurturing function usually suppressed. Here it is the pouch with its teats as a source of sustenance that is a scientific curiosity, rendered as a form of knowledge stimulation rather than erotic titillation.

The decision to expose the marsupium to general view speaks to motivations beyond science. The teats offer distressing visual evidence of the destruction of a family, something foregrounded by the accompanying label: 'the condition of the pouch indicates that the female had four pouch young (joeys) when captured.'[183] This mother's young were slaughtered, as was she. Through the way the teats are foregrounded, the lactating thylacine is made to stand in for the broader horror of the animal genocide perpetrated against her kin. The trope of the innocent child, of potential unfulfilled, is often employed to represent the horrors of genocide.[184] Here, nonhuman infants, the absent but implied joeys, perform a comparable role. Elsewhere in the exhibition, the orphaning of three thylacine pups and their ensuing fate are discussed.[185] The destruction of families is therefore a significant theme.

Owning Up

There is another mount included in the exhibition, an adult male referred to as 'Dilger's tiger'. The use of the possessive pronoun grants Dilger ownership of the thylacine and his story. The importance of this tiger is affirmed in a small information panel that emphasises the rarity of the detailed information of the animal's provenance.[186] The thylacine was mounted by William Notman, a man described in his obituary as an 'expert taxidermist'.[187] Notman was self-taught and was most noted for his avian mountings including a group of hummingbirds displayed at QVMAG.[188] Positioned towards the back of the exhibition space, the mount is posed against a verdant backdrop, rewilded. The tale of this thylacine's journey to the museum is related in an information panel located to the left of this artificial rural scene. Alfred W. Dilger, a farmer and sometimes miner, snared the tiger, killing him, close to Mathinna, where his property 'Myrtle Grange' was located.[189] The thylacine measured 5 ft. 1 in. and was described in the local press as a 'splendid specimen'.[190] As the information panel recounts, Dilger carried the dead animal nearly '16 kilometres' to the post office at Mathinna, from where he contacted Scott and arranged to sell the corpse for the low sum of 25 shillings (20 shillings for the thylacine and 5 shillings for transport costs).[191] In a mix-up, the body was initially sent to Launceston City Park Zoo (also sometimes referred to simply as City Park or as Launceston Gardens), where William McGowan (the superintendent of reserves who was responsible for the zoo) offered £3 for the dead male because of how impressive he was.[192] Realising the value of the carcass, Dilger sought to renegotiate his fee with Scott and QVMAG but to no avail.

McGowan is only mentioned in passing in the 'story' of the tiger, a tale that is more about Dilger's misfortune than that of the animal, yet the superintendent had a major role in the genocide waged against the tiger. An animal trafficker, he was always willing to acquire live and dead tigers, encouraging trade in the animals. In 1884, for instance, he placed an advertisement in the *Launceston Examiner* advising that he was willing to purchase 'live native tigers'.[193] McGowan regularly traded live animals with Melbourne Zoo and other institutions.[194] The zoological garden at City Park was also a way station for tigers transiting to overseas zoos.[195] The number of thylacines that passed through McGowan to Adelaide, Melbourne and further afield were considerable (thirty or more). He also exhibited some of the live animals he received at City Park Zoo.[196] The tigers were sometimes displayed for a short period and then traded or died while at the facility. Exhibiting native (or First) wildlife at the zoo, including thylacines, was always an aim, as a newspaper report in *The Mercury* illustrates, noting that a menagerie had been sanctioned that would include 'kangaroo and wallaby, tigers, devils, and native cats'.[197] The report also observes that the 'larrikins are likely to have a good time'.[198] Larrikins are badly behaved young men and the reference indicates how boisterous zoos were at the time.

In 1889, a pair of tigers that had been exhibited at City Park were exchanged with the Zoological Society of Adelaide for two leopards (*Panthera pardus*).[199] Later the same year, a female tiger with three pouch young was acquired by the zoo. It was reported that the mother and pups would hopefully provide 'a novelty not to be seen every day'.[200] McGowan advised the press that 'the greatest care will be necessary to bring up this interesting family of young tigers'.[201] This recognition of the need for watchfulness was borne of past failures in looking after thylacines and their young.[202] In November 1890, two thylacines, described as 'excellent specimens', were reported to be present at City Park.[203] There was clearly at least one tiger on display at the zoo in 1891.[204] In 1893 another female thylacine and three joeys (described as unfurred) were exhibited.[205] Many more tigers were incarcerated at the zoo in subsequent years. In 1924 an adult female and one pup were captured in the Western district and, after being exhibited in Melbourne, were housed at City Park.[206] Cameron Campbell dates the last thylacines to be housed at the zoo, a female and two young, to 1925. The female was sent on to Beaumaris later that year.[207]

It is probable that any captive tigers on permanent display at City Park did not survive for long as conditions at the zoo were poor. A letter published in the *Launceston Examiner* in 1895 complained of a 'beggarly array of empty cages' at City Park and attributed this to McGowan's misunderstanding of the habits, diet and housing of the mammals in his care. The writer, who used the pseudonym 'Dingo', had exotic wildlife such as 'magnificent specimens of the leopard, panther, tiger, and lion tribes' in mind, but it is probable that thylacines were equally, if not more, neglected.[208] In 1889, the columnist Tommy Trot (another pseudonym) writes of a 'lean looking native tiger' that he (the writer genders the tiger as male) had killed and eaten another thylacine because of a hunger for 'live meat'.[209] There were no witnesses to this event, only the aftermath. A thylacine consuming his own kind may point to the dire circumstances in which he was being kept.[210] In 1890 a journalist called for the Society for the Prevention of Cruelty to Animals to intervene because of poor conditions caused by chronic underfunding. The writer suggested that the thylacines (of which there were a number) at City Park were housed in better surrounds than two 'Esquimaux dogs' (*qimmit*), but many of the enclosures were 'very badly looked after'.[211] The criticism seems to have generated a response as new and more commodious cages were installed later in the year.[212] In 1896, McGowan was accused of animal cruelty for having 'tortured a bear in the City Park by keeping them in the hot sun without water and allowing a tree in their cage to impede their exercise', but the charge was dismissed.[213]

Although McGowan seems to have been a poor keeper, he clearly had some recognition that City Park was a far from ideal environment for hosting a zoological exhibition. In 1924, when a plan was mooted to move the zoo to the Punch Bowl (now Punchbowl) area south-east of Launceston, he was much in favour of the idea because at the location in City Park '[t]he cages were exposed to the wind, and the soil underneath was very wet'.[214] Concerns about the well-being of some of the zoo's birds because of their substandard accommodation were being raised as early as 1889.[215] As late as 1946, William Gellie described the small cages at the zoo as cruel.[216] Despite an awareness of the unsuitability of the site, McGowan continued to tout for tigers and other animals throughout his tenure as superintendent. City Park was a nexus for trafficking in thylacines caught in western Lutruwita, a trade that contributed to the animal's destruction. The institution's role is mentioned in 'Precious Little Remains' in a panel where City Park is identified as 'the largest known thylacine trader' and 'at least 62 thylacines' are identified as having been exhibited in or trafficked through the zoo (or both). The zoo trade in tigers is blamed for the paucity of QVMAG's collection compared to 'overseas institutions' and for the loss of 'part of our cultural heritage'. This is a partial holding to account that obscures the reality that the actions of both the zoo and the museum degraded Lutruwita's cultural heritage, contributing to animal genocide.

McGowan failed to secure the tiger killed by Dilger and he now forms the centrepiece of 'Precious Little Remains', his positioning at the far point of the alcove emphasising his importance. He is located where an exit might be expected to be found. Exits have a 'positive attraction-value' and can compete with exhibits for 'the attention of visitors'.[217] Intentionally or not, at QVMAG the thylacine mount as failed exit exploits the psychology of museum architecture. Like A1298 at TMAG, a male tiger is the 'star attraction'. From comments made by Maynard, it is clear that the sex of the animal was a motivating factor in the importance accorded him. The thylacine is 'an impressive adult male', an 'alpha male'.[218] In reality, thylacines seem to have hunted in families (a breeding pair and their young) rather than in packs of unrelated individuals so, as with wolves, the notion of alpha animals is misplaced.[219] At QVMAG, the male is nevertheless aggrandised, his perceived dominant status asserted.

The choice of mounts for exhibition was restricted (there are only five mounts in the collection and only one is identified with certitude as female), but the way those chosen are exhibited contributes to a display of male dominance. The tiger killed by Dilger is the culminant 'precious little remains' of the exhibition. Using a motion sensor, he is gradually illuminated when visitors approach him, a process of revelation that reinforces his importance. Maynard notes that the darkened case grants the tiger an aura of 'mystery'.[220] It is the museum that provides enlightenment. The light, in this sense, functions like the mirror for the other mount, feeding a narrative of discovery, of the unveiling of secrets. The female tiger is reduced to her reproductive capacity and the male to his magnitude; his size is what matters, that his remains are 'big' is what is precious.

In an information panel, Dilger is feted for his 'great service' in selling the thylacine to the museum. The snaring is described as accidental but even if the snares were intended for other animals, Dilger would have been aware of their capacity for 'collateral damage' to native wildlife. His exoneration from blame for the tiger's killing and the celebration of his actions in selling the dead body to the museum gloss over the animal's horrible end, suffocated or starved to death. Dilger's mindset is also elided. His efforts in carrying the cadaver to Mathinna attest to the financial gain he recognised in the 'happy accident' of snaring the thylacine. His primary concern was not science or cultural heritage but money. The panel also explains that the photograph that provides the backdrop to the mount is of the milieu where the tiger was killed yet as it now appears. The image works to assert the claims about knowledge of provenance given in the information label. More than this, the photograph, when combined with the mount, acts like King's composite photograph, artificially repopulating the Lutruwitan wilderness with a thylacine.

I Almost Began to Think That I Was the Monster . . .

Given the professed fragility of the two mounts on display in 'Precious Little Remains', the decision not to include the mount in the QVMAG collection known as 'Frankentiger', a taxidermy constructed of two separate animals, merits consideration.[221] As a composite it is of less scientific value than the other four mounts, it is less 'precious'. In the book *Tasmanian Tiger*, Maynard and Gordon write of 'Frankentiger' that in its making 'two specimens were effectively destroyed'.[222] 'Frankentiger' is tentatively identified as female. The nickname accorded the mount by museum staff gestures towards the tiger's 'monstrous birth' at the hands of the American taxidermist, Frank Tose, who constructed them. In *Frankenstein; Or, The Modern Prometheus*, Mary Shelley's monster is gendered as male.[223] Here, the 'wretch' is female, more the 'bride' of Frankenstein than the monster himself. The role of stitching when preparing mounts is probably a secondary motivation for the nickname. There is stitching visible on Boris Karloff's head in James Whale's cinematic adaptation of *Frankenstein* and 'Frankentiger' (like many mounts) is also clearly stitched together.[224]

Unconsciously, the reference to Mary Shelley's *Frankenstein* may also form an acknowledgement of the nightmarish potential of science, the flipside to its powers of enlightenment.[225] It is, after all, the scientist who is Frankenstein and not the 'monster' he creates. In this understanding, Tose, who yoked two tigers violently together, assumes the role of Victor Frankenstein, resurrecting the thylacine through sewing together different bodies to make the mount. In one newspaper report of Tose's 1937 visit to Australia he is described as a man who makes 'dead things "live"'.[226] Tose's actions, at the time acceptable, are now viewed as deplorable because they have compromised the 'truth' of the animal.[227] Yet, perhaps, in some ways Tose's effort is the most truthful of QVMAG's mounts, as the tiger draws attention to her constructedness, to the fantasied vision of a thylacine that she embodies. 'Frankentiger' amplifies the artificiality underpinning all taxidermy. She is far from the only mount to combine body parts from more than one individual of a given species. Skeletal mounts are often formed of multiple individuals. This is the case, for example, with the thylacine skeleton displayed at the Chau Chak Wing Museum in Gadigal/Sydney, where the skeleton (NHM498) is combined with a skull and jaw (NHM1427) from another tiger. The great auk skeleton on display in The Rooms in Newfoundland is another osteological equivalent of 'Frankentiger'. They are an amalgam of parts from many different birds that were ransacked from the boneyard that is Funk Island, a site where animal genocide was perpetrated against the auk population.[228] In *Nature's Mirror*, Mary Anne Andrei discusses how another five skeletons of the great auk were assembled by a team led by the zoologist Frederic Augustus Lucas using bones from multiple birds sourced from Funk.

Taxidermy mounts are less often constructed of remains from multiple animals, but Jules François Desirée Bailly and William Temple Hornaday added tusks to a female elephant they had mounted, transforming her into a him.[229] The Tasmanian tiger mount currently on display in the Australian Museum has a belly that has been repaired using kangaroo skin and is therefore a composite of two species, a hybrid. Their faded stripes have also been retouched with paint.[230] The Australian Museum tiger is exhibited precisely because they are compromised. They have become a sacrificial 'lamb', already damaged and therefore intentionally exploited to take on the inevitable further damage that comes with being displayed. The repair using the kangaroo hide is inconspicuous. 'Frankentiger's' melding of two bodies, by contrast, is highly visible and highlights both the workmanship underlying all taxidermy and the profession's capacity for employing ruses.

Reading the mount through the prism of *Frankenstein*, as the staff at QVMAG do, calls attention to taxidermy's dependence on death and resurrection. Ironically, Tose called Australian museums 'morgues instead of show windows for science', criticising their display practices.[231] He described the 'beauty and dignity' of contemporary natural history displays in US museums, contrasting them with 'the repellent morgues' of the past.[232] For him, Australian display practices were outmoded and unvital. He hoped to encourage a process of modernisation so that the word 'morgue' would no longer be applicable to them and 'where[by] each creature is "not dead, but sleepeth"'.[233] In the case of the thylacine, taxidermy mounts grant the species a substantial afterlife. Tigers are dead yet, thank to taxidermists, endure. The mounts act like living dead.[234] That said, Tose's own taxidermy effort is deathlier than most, visibly constructed from body parts.

The lack of importance accorded to 'Frankentiger' and the decision not to include the mount in 'Precious Little Remains' seem motivated by her unknown provenance and 'mongrel' status. From the case of the tiger killed by Dilger, it is clear that for QVMAG value is accrued through particular qualities, such as sex and size, and specific kinds of information including the locations from which tigers were derived and the chronology of their entry into the museum collection. Because there is no systematic documentation linked to 'Frankentiger's' past, she is relegated to the status of a curio. As discussed in Chapter 2, in order for the physical remains of an organism to transition to the status of 'specimen', varied accompanying data is required. The evidential value of physical remains held in museum collections is compromised if there is a lack of good documentation relating to them.[235] As some of this data is missing or corrupted in the case of 'Frankentiger' her passage to full 'specimen' status is arrested. She does not fit in in terms of becoming an 'object of knowledge'.[236] 'Frankentiger' fails to function as a metonym for the scientific idea of the thylacine.[237] This failure, however, permits her to resist being reduced to the status of 'evidence', instrumentalised as 'proof' of some aspect of her species.

Hunters and Collectors

The section of 'Precious Little Remains' that follows the mount of the tiger killed by Dilger is a wall information panel, 'Hunters and Collectors', which includes photographs and reproductions of newspaper cuttings. Maynard has described this panel as a 'narrative of destruction' that relates 'how the thylacine was perceived, persecuted and killed'.[238] I will now examine the panel in detail as it reveals much about historical attitudes towards the thylacine and about the various practices that contributed to genocide against the species.

One of the cuttings from 1887 refers to the killing of a tiger at Christmas Hills near Elizabeth Town by Edward and James Aylett and Ambrose Walsh. This killing was opportunistic; the men chanced upon the tiger, and it is therefore not readily reconciled with either hunting or collecting as intentional acts. After the murder of the thylacine, Edward Aylett did express the desire to have the animal stuffed, yet this was a retrospective decision and not motivational. The act of random killing perpetrated by the men forms part of a broader genocidal campaign against the thylacine. 'Hunting' and 'collecting' are both euphemisms that downplay the harsh reality of the casual violence that underpinned many tiger killings.

Another clipping details the poisoning of a tiger by a shepherd who worked for Henri Nicholas. The shepherd came across a partially eaten possum in a snare he had set. He laced the cadaver with strychnine and the next day found a dead tiger close by. The incident was reported to counter the belief that thylacines would not touch carrion. This event is clearly not linked to collecting, nor is it hunting in the conventional senses of the term (i.e. the killing of animals as sport or for sustenance). It is also not a direct expression of genocidal intent as the shepherd would probably not have known a tiger was responsible for eating part of the possum's carcass. The poisoning is rather an act of indiscriminate killing. The travelling correspondent for *The Mercury*, Theophilus Jones, reported in 1885 that shepherds in rural Lutruwita would often leave out 'enticing morsels', with 'a pinch of strychnine deftly introduced by way of flavouring' as a means to kill animals that prey on sheep.[239] Although earlier namechecking the 'Tasmanian tiger' as a threat to flocks, Jones specifically identifies crows, eagles and stray dogs as the shepherd's target.[240] Here, again, hunting and collecting need to be understood as euphemisms that fail to do justice to an act, this time of broad cruelty towards native and invasive species.

Other newspaper cuttings do conform to notions of hunting and collecting: there are adverts placed with the aim of procuring tigers or offering them for sale.[241] Menageries that visited Lutruwita frequently advertised their willingness to buy live thylacines and devils.[242] The panel includes one such advertisement, placed by the St. Leon's Circus and Menagerie (the circus founded by Matthew St. Leon) seeking live tigers.[243] The animals at St. Leon's were managed by 'the Great Tamer from the East', 'Professor Madden', who pioneered lion acts in Australia.[244] Like the City Park Zoo, privately run menageries were often squalid affairs. While Mr Higgins's Menagerie was based in Brisbane, for instance, there were complaints about the stench surrounding it and about the neglect of the animals.[245] The St. Leon's circus toured extensively in mainland Australia (notably in New South Wales and Victoria) and in Lutruwita, visiting rural districts as well as metropolitan areas.[246] Animals in the circus and the menagerie were called upon to entertain. The horse 'Emperor' was made to waltz, polka, walk on his hind legs, kneel and find hidden handkerchiefs.[247] There were also performing camels and monkeys (one of which would ride a horse) and a trick bear. The thylacines housed in the menagerie were feted more for their appearance than their antics, clearly viewed on the mainland as an exotic curiosity. In the 1890s, for instance, the Fitzgerald Brothers circus also touted the presence of Tasmanian tigers in its menagerie, clearly viewing them as a draw.[248] The selling power of thylacines as spectacle probably derived from myths about their ferocity and wickedness. Some reviewers clearly registered surprise at the actual appearance of thylacines, one describing 'a very kangaroo-like looking Tasmanian tiger' as an object of interest.[249]

As well as trading on their supposed maliciousness, menageries probably benefitted from the tiger's perceived 'freakiness'. An article about the thylacine published in *The Age* in 1937 suggests that 'his dull yellow skin, hung loosely over his bony body, makes him resemble more some freak animal seen stuffed in a museum' than an actual tiger.[250] Here the thylacine is cast as aberrant and monstrous. It is unclear if the journalist who penned the article had ever seen a living thylacine or if their description of the animal as resembling a taxidermy mount was ironically derived from viewing just that. The tiger is also referred to as a 'freak of history' in a review of Erle Wilson's *Coorinna*, a novel that is examined at length in the next chapter.[251] In an advertisement from 1923, a Mr Hughes announced his desire to buy 'live freaks – Tasmanian tigers preferred with young'.[252] The thylacine as a 'freak of nature' was probably something the nineteenth-century menageries exploited. In this sense, the tigers shown in these travelling spectacles were a precursor to 'Frankentiger', oddities of the natural world, albeit ones more graspable than the egg-laying duck-billed platypus, another Australian mammal often labelled a 'freak'.

Additionally, on the 'Hunters and Collectors' panel there is an advertisement placed by Edward Braddon, the Minister of Lands and Works, detailing the government bounty offered for the destruction of native tigers. Those who hunt wild animals for financial reward are sometimes referred to as bounty hunters, so the bounty scheme can be understood to constitute hunting in an expanded sense of the term. The tigers presented for bounty were to be disposed of to prevent the same animal being presented multiple times. The bounties offered by the government (£1 per adult tiger and 10 shillings per juvenile) should be viewed as blood money that facilitated genocide.

The numerous photographs that feature include ones of tigers held by zoos. There is an image of 'Benjamin', held at Hobart Zoo, the last or the penultimate thylacine to die in captivity. The tiger is sometimes referred to as having been sold to the zoo by the trapper Elias Churchill, who is said to have caught the animal accidentally while snaring. The tiger Churchill brought to the zoo is now thought by some scholars to have been an ageing female.[253] Steven Sleightholme, Tammy Gordon and Cameron Campbell have proposed that the ultimate or penultimate tiger was, in fact, one caught by the Kaine family in 1931.[254] Now Gareth Linnard and Sleightholme argue that the tiger in question was caught by Roy and Dan Delphin on 7 July 1930 and sold to the zoo in mid-October of that year.[255] Another photograph captioned 'Thylacine Asleep at Hobart Zoo, Tasmania around 1927–29' is attributed to Herbert King. In fact, the image shows two tigers as the tail of a second is visible to the right of the image. Sleightholme and Campbell offer a different date (1916) for this photograph and suggest it portrays a male thylacine that was captured by the sawmiller William 'Billy' O'May in the environs of his hometown of Tyenna in 1916 and then sold to Beaumaris Zoo. The second thylacine in the picture is identified as a female tiger, either one captured by M. Bourke (probably Michael Bourke of Somerset) or Messrs Foy and then sold to the animal trafficker James Harrison who, in turn, sold her to the zoo.

Another photograph reproduced on the panel is of one of these tigers sold to Harrison. She is shown imprisoned at a property of his. On the picture it is written: 'Tas. Hyena' and 'Sargent photo'. This indicates that the photograph was produced by Myra Bessie Sargent, who had a photography studio in Wynyard from 1912 to circa 1920.[256] The description of 'Tas. Hyena' was probably suggested by Harrison, as in the press in the period 1910–19 the term 'tiger' was much more common, followed in frequency by 'wolf'. Harrison disliked the term 'Tasmanian tiger' and argued it was incorrect.[257] Shown in an outdoor cage, sunshine casts shadows of the bars of the enclosure across the tiger's body. The photograph displayed at QVMAG is cropped, an expanse of black shadow visible to the right in the original image has been excised. The recumbent thylacine seemingly pays no attention to Sargent, her attention directed elsewhere.[258] A second photograph by Sargent of the same tiger, which the museum has not chosen, shows her with her head turned towards the photographer. The size of the tiger is much more apparent in this image and the banding is clearer, but the shadows of the cage bars are less evident. It therefore seems that the image of the resting tiger was chosen by the curators because it more powerfully indexed incarceration.

Stephen Sleightholme and Cameron Campbell have identified the thylacine in these photographs by Sargent as a female caught by Bourke in October 1915. In the newspaper account of the capture of this 'tigress', fears about the impending extinction of the species are articulated.[259] Based on anecdotes, the journalist queries this doom-laden outlook. The fact that the writer felt the need to assuage concerns about the thylacine population, however, shows there was ever-increasing recognition that the animal was endangered. Sleightholme and Campbell ground their identification of the tiger as one caught by Bourke on an October 1916 newspaper article about the nursing of war wounded in which Myra Sargent's sister Amy Ruth described showing a photograph of a thylacine in Harrison's possession to a wounded soldier. Given the timing, the photographs might equally be of a female thylacine of about eighteen months of age that was captured by members of the Foy family at a mine referred to as 'Top Cage' near the Arthur River in May 1916.[260] This tiger was described as 'nicely marked'.[261] The tiger in Sargent's photograph appears young, which lends support to this possibility.[262]

Harrison, who was based in Wynyard, secured live tigers for institutions in the British Commonwealth.[263] He advertised for 'hyenas' and other native mammals in the local press.[264] One such advert ran as follows: 'Wanted – Tigers or hyenas, alive or dead, highest price. James Harrison, Wynyard.'[265] These 'wanted' adverts and word of mouth were what brought men such as Bourke and Foy to his door. As well as purchasing tigers and other animals from farmers, prospectors and trappers, Harrison also ventured into the bush himself on collecting trips, including, in 1927, to the Arthur River on the north-west coast.[266] In March 1928, Harrison purchased a thylacine caught by two prospectors in the vicinity of the Arthur River. The tiger, who had a broken leg, was probably exhibited at the Show Day of the Table Cape Agricultural and Pastoral Society on Wednesday 21 March 1928.[267]

Harrison's love of displaying the thylacines that came into his possession is also demonstrated by his hosting players from the North Broken Hill football club of New South Wales on their visit to Wynyard in 1923. On their visit to his property, he showed them 'a very fine specimen of the Tasmanian tiger' and a 'white wallaby'.[268] The wallaby, which was newsworthy when Harrison first procured them from Trowutta in 1922, had 'half-a-dozen stripes of the thylacine (Tas. tiger) across its tail'.[269] Harrison kept the macropod because they represented a 'freak of nature'.[270] The tigers he exhibited and which were photographed also formed living curios, members of a 'doomed race'. The photographs taken by Sargent were therefore intended to record an ill-fated local oddity. She was a keen recorder of local events (parades and regattas) and the capture of a thylacine was clearly one such noteworthy occasion. As she sold postcards at her studio, it is possible she may also have viewed the thylacine as a money-spinning opportunity and vended reproductions of the image. In the photograph, the tiger is therefore commodified and burdened with representing the 'strangeness' of her species. Sargent is as much a collector as Harrison, yet it is images that she deals in.

A further photograph on the panel features a male thylacine incarcerated at London Zoo.[271] Although the caption lists the picture as undated it is included in Graham Renshaw's *More Natural History Essays* alongside a text dated August 1904.[272] Given that Renshaw lived in England, he probably visited the zoo and took the photograph (or had the photograph taken) around that time. It must be the male thylacine that arrived at the zoo on 26 March 1902, where he remained until his death in 1906.[273] The tiger was sold to the zoo by the animal trafficker William Jamrach.[274] Jamrach, who was based in Stoke Newington in England, traded animals to and from the Antipodes.[275] These photographs point towards the moral turpitude of the zoos. The rising price of tigers attested to their growing scarcity yet demand continued regardless. This renders institutions including Beaumaris Zoo (and its later incarnation, Hobart Zoo) and London Zoo complicit in animal genocide. Thylacines were purchased as faunal freaks, curios to be consumed as spectacle. Collecting these 'freaks' contributed to their destruction. The zoo photographs bear witness to this callous trade, as does the photograph of the captive thylacine at Harrison's property.

Other photographs show thylacine killers posing with their trophies. One features Albert Quarrell (misspelled Quarrel in the caption) seated, cradling a thylacine he had shot. It was taken shortly before or after the image of Quarrell standing that is reproduced at TMAG. Another shows a female tiger killed by Clement 'Clem' Isaac Penney on 1 January 1924 in the vicinity of the Arthur River. In the photograph, probably taken by John (Jackie) Robinson, the youth and an elderly man jointly hold the tiger aloft.[276] Penney came across the tiger and her mate by chance while in North Bischoff Valley travelling to look at livestock belonging to his father.[277] His encounter was subsequently reported in the *Weekly Courier*:

> While walking round he heard a muffled bark, and turning quickly found that two native tigers were advancing rapidly towards him. One was a large male and the other a rather small young female. As these animals are plentiful in this locality this bushman always carries an automatic pistol. Pulling this out he tried a shot at the leading animal, but the pistol would not go off.
>
> Throwing the useless weapon at the tigers he looked round, and fortunately saw a sturdy limb of tree at his feet. Grasping this excellent club he stood on guard. The female tiger was in the lead and was crawling forward. When about 6ft. away she sprang full at him. He met her fairly with a sweep of his heavy club, knocking her backwards on to the ground, and followed up with a killing smash on the neck. Straightening himself instantly, the bushman found the large male in the air from a mighty leap.
>
> Again the trusty club proved true, and the tiger was knocked to the ground. Gathering himself up, the animal dashed into the scrub on three legs, a foreleg being apparently broken, whimpering like a wounded dog. The bushman rushed after him but was unable to overtake the disabled beast.
>
> After the encounter the bushman saw a tiger cub in the vicinity, and this may have been the cause of the attack by the adult animals.[278]

The article is doubtless exaggerated. If read as straightforward, the report attests to the deep-seated hatred that many settler colonists felt towards tigers. Penney's first thought, even before being attacked, is to shoot the thylacines. Such a rush of blood is possible, especially as Penney's youth is emphasised earlier in the article; he is described as an 'athletic young bushman'. Yet referring to him as a bushman implies that he possesses a modicum of bushcraft, a knowledge of the wilderness and its ways. Many people would recognise that the fact that the tigers did not immediately flee signalled the presence of young in the vicinity.[279] Penney claims he came to this realisation only retrospectively. This undercuts his bush credentials and suggests his inexperience. An alternative is that he understood the situation straight away but chose to actively engage with the tigers, either in search of a trophy or because he hoped to capture one of the animals through stunning him or her. If this was the case, the subsequent report misleadingly frames his enthusiastic hostility as self-defence.

The photograph that accompanies the article is captioned 'Encounter with Tasmanian Tigers – After a Battle Royal with Two Tasmanian Tigers near Waratah, Mr. C Penney (right) eventually secured a Trophy of the Fight – a fine specimen of Tiger'.[280] It appears alongside images of Dame Nellie Melba's visit to Launceston and of boys from the Australian League of New South Wales who were then touring Lutruwita. Granting the news story an image reinforces the rarity of the tiger at the time. The killing is clearly viewed as an event. That nothing untoward was perceived in Penney's actions is affirmed by his appearing alongside Melba, the world-famous soprano and national icon.[281] Displaying the corpse of a thylacine is viewed as cause for celebration. The genocidal gaze is clearly communicated through this photograph, which was viewed as a jubilant image of an evil animal bludgeoned and subdued. Framing the encounter as a battle royal, a competitive fight in which Penney was the 'last man standing', burdens the dead tiger with representing human triumph. Through photographs such as this, 'collecting' is revealed for the murderous pursuit it sometimes historically was, with the tigers cast as antagonists.

Whether Penney was actively seeking a trophy or procured one by 'happy accident', that this is what the dead thylacine became is also evident from another photograph, one not featured on the QVMAG information panel. In this photograph taken by John (Jackie) Robinson, dated by Nic Haygarth to the mid-1920s, the Penney family is gathered in their Sunday best. Clement Penney wears a three-piece suit (probably tweed) accessorised with a pocket square and sports a shirt and tie.[282] This outfit is in marked contrast to the one-button blazer, loose-fitting muddied straight-leg trousers (perhaps corduroy, then a working-class fabric) and grandfather shirt (a band collar shirt that was a common form of apparel among labourers) of the newspaper photograph.[283] In the foreground of the later photograph, a thylacine skin lies on grass in front of two seated boys. Penney clearly kept the skin as a memento of his escapade and it had become an important part of the family's history, meriting inclusion in their photographic portrait (or, perhaps, prompting the production of the portrait).

In the photograph, one family poses before the ruins of another. As discussed in the previous chapter, trophies such as this embody the genocidal gaze. Haygarth describes the skin as 'a stripy doormat'.[284] This categorisation well captures the disdainful attitude of settler colonists towards the native carnivore. In another family photograph taken at the same sitting, Penney's mother, Theresa Mary Penney, smiling, rests her foot on the skin, assuming the pose of the conquering hunter on behalf of her son. Mrs

Penney's lowkey gesture evidences an insidious disdain. The logic that led to the genocide against the thylacine is given strong visual expression in both the photographs featuring the Penney family and the female tiger's remains, as well as by the photograph included on the 'Hunters and Collectors' panel. While euphemistic in its titling, it is the strongest museum acknowledgement I have encountered of the genocide perpetrated against the thylacine. It is not, however, the endpoint of the exhibition.

One Word More

'Tasmanian Tiger' continues with a display case entitled 'Sightings and Searches' that details various efforts to find proof of living thylacines post-1936 and material evidence for their continuing existence presented to the museum (such as scat). The display is dominated by a topographic map of Lutruwita. Artefacts are gathered at the base of the map, different examples of scat, remnants of possible thylacine kills, a photograph of an unidentified mammal (probably a Tasmanian devil), materials mistakenly identified as deriving from tigers (scat, pawprints and bones), a camera and a snare. As is the case at TMAG, there are also plaster casts of thylacine pawprints from the 1938 expedition (this time, copies). These exhibits are meant to collectively pose the question also asked by text accompanying the display: 'Are they still out there?' The text concludes that the evidence does not meet the burden of proof for the continuing survival of the thylacine in Lutruwita. The presence of this display, however, holds open the possibility, no matter how unlikely, of survival.[285] The reluctance by the museum to definitively rule out the possible existence of a relict population of tigers can be read as symptomatic of guilt regarding their destruction. In a sense then, notions of survival (like efforts to clone the thylacine) are bound up with a hope for future redemption. Believing in survival in such a context is not a sign of naivety but of compassion. When viewed in the broader context of the QVMAG exhibition, 'Sightings and Searches' is clearly not a form of denial of the violence enacted against the thylacine but rather part of a complex effort to psychologically negotiate the legacy of that violence.

The final panel of the exhibition, 'Gone but Not Forgotten', is similarly a form of reflection on the 'afterlife' of the thylacine. The panel features photographs of physical remains of tigers held in various museum collections alongside newspaper advertisements offering to buy thylacines (including, notably, one placed by TMAG) and articles celebrating their destruction. An extract from a longer frontpage article for the newspaper *The Mercury*, authored by Theophilus Jones, details a gruesome display of dead tigers created by Edward and Joseph French, farmers with land in the vicinity of Buckland in south-east Lutruwita.[286] According to Jones, the Frenchs made 'superhuman' efforts to defend their sheep holdings and had assiduously and successfully thinned out the tiger population in the local area. When visiting the Frenchs' property, Jones, accompanied by George and William Turvey, 'counted 18 heads of these destructive animals nailed to a tree'.[287] The jaws of a particularly impressive male 'were kept asunder by a peg'.[288] A female and her pups killed four days prior to the visit were also affixed to the tree. Jones notes that the four pups were too decomposed to be removed or he would have 'procured them for the Museum'.[289] This observation draws attention to the shared 'collecting' impulse of the journalist and the farmers.

The Frenchs had, in fact, created a 'countryman's museum' of the kind discussed in the preceding chapter.[290] Their collection, preserved on an improvised gibbet, was intended to showcase their successful campaign against the thylacine. Retrospectively, it provides a gruesome record of animal genocide and the feelings that accompanied it. The careful arrangement of the male tiger, with his jaws propped open in a pose seemingly intended to confirm the animal's ferocious reputation, actually presented him displaying behaviour (gaping) signalling anxiety or distress. The gape (or yawn as it is sometimes referred to) was a threat response. That the pups were included in the exhibit shows the disdain the Frenchs felt for the thylacine, capable of revelling even in the destruction of infants. Jones shares their contempt, referring to the pups as 'like four hairless rats', the simile likening the baby tigers to rodents, 'vermin'.[291] Jones's description of the rural exhibit is now itself on display, included as part of an exhibition dedicated to the thylacine. Featuring his report alongside a museum advert for tigers seems designed to draw attention to underlying connections between farmers such as the Frenchs and institutions such as TMAG and QVMAG. Both farmers and museum curators sought out tigers, looking to either kill or capture them.

The panel also includes a quotation that is attributed to Elizabeth Barrett Browning, 'Suddenly, as rare things will, it vanished', positioned above an image of a thylacine mount held at the Natural History Museum in Vienna. This is a misattribution as the lines are taken from a poem by Robert Browning, 'One Word More', which was written to Elizabeth in London in 1855.[292] Like the words attributed to Roosevelt that begin 'Tasmanian Tiger', this quotation is taken out of context. It does not refer to the thylacine but to the lost sonnets that Raphael wrote to Margarita Luti, also known as La Fornarina.[293] 'One Word More' is a meditation on loss but also a reflection on the poet's split persona, his public face as communicated through the artifice of his verse and the private man behind the poetry. In this poem, he seeks to speak as the latter, to be Browning in the raw, through the medium of the former.[294] Although unintended, this reference to Browning's effort to speak truth to himself, to move beyond the artificial and express the natural, does resonate well with aims of the exhibition 'Precious Little Remains'. Gordon and Maynard have sought to use some of the resources of natural history, the material remains of the thylacine, to trouble misrepresentations and myths that continue to have currency regarding the thylacine. Like Browning in 'One Word More', the exhibition fails to find a mode of representation that is 'fit and fair and simple and sufficient', yet the attempt to find an adequate idiom for 'speaking' the tiger is insightful.[295]

The final panel can be understood as akin to 'One Word More' in its status as an envoi. The envoi in literature provides a way for an author to send off or sign off a piece of writing such as a poem, explaining it or signalling its moral message. As an act of closure, it holds intimations of mortality.[296] The panel as a dispatch to the museum visitor operates in a similar way to literary envois. Its combination of images of thylacine remains held in various museum collections (including, for example, a photograph sourced from the ITSD of the head of a female thylacine preserved as a wet 'specimen' at the Oxford Museum of Natural History) with newspaper articles celebrating the destruction of tigers or arguing against their protection (as a letter to *The Mercury* by Robert Stevenson does) is open to being read as suggesting parallels between museum collecting practices and bounty hunting as contributing factors in the tiger's extinction.[297] The inclusion of images of the tiger's afterlife, of 'precious little remains' of the kind from which the exhibition takes its title, also seems intended to invite greater compassion towards nonhuman life in the now. In this sense, the thylacine's posthumous 'mission' is to agitate in the present for greater recognition of the need to safeguard biodiversity.

The exhibition at QVMAG comes the closest of any of the displays discussed in this chapter to acknowledging the animal genocide perpetrated against the tiger and the ways museums and zoos contributed to that genocide. Museum directors and curators exploited the genocidal gaze (without always sharing it) and are therefore no less complicit in genocide than bounty hunters even if they viewed the tiger with a different optic (as a scientific curiosity rather than a ruthless sheep killer). The majority of the museum exhibitions I examined employed the thylacine as a 'poster beast' for extinction or for the reality of Darwinian evolution. QVMAG and TMAG offer more complex representations, but both demonstrate, in their different ways, how even efforts at offering objective fact always embody beliefs shaped by social and historical context. This was particularly evident to me in relation to ideas about gender difference that manifested in their modes of presenting physical remains. The final word in all the exhibitions is never that of the thylacine. They are always put to work in the service of a cause or theory. This issue of the tiger's reduction to an instrument for expressing human ideas and values, to their symbolic exploitation, is taken up in more detail in the next chapter.

Notes

1. The Hall of Mammals is now also often referred to in museum literature as the Mammals Hall.
2. I am indebted to Darrin Lunde for sharing his insights regarding the condition of the mount. It is now not considered good enough to be openly displayed. This may in part be attributed to aesthetic reasons related to realism. I am grateful to students from the course HAR3550 *L'art du témoignage* for sharing their insights regarding the symbolic connotations of this act of veiling.
3. I am thankful to Pedro Cardoso for drawing this initiative, entitled 'Extinction Voices', to my attention.
4. Marvin, 'Perpetuating Polar Bears', 163.
5. Asma, *Stuffed Animals and Pickled Heads*, 36.
6. Ibid.
7. Maleuvre, *Museum Memories*, 1.
8. Chaumier, 'Writing and the Exhibition', 26.
9. Roosevelt gradually began to feel guilt at what he termed, in inverted commas, 'collecting'. He seemed well aware that the euphemism concealed a violent and questionable practice. Roosevelt cited in Lutts, *The Nature Fakers*, 124.
10. I am grateful to Patricia Koh for this information.
11. Construction of the building began in 1887, the same year as the founding of the museum.
12. My thanks to Patricia Koh for describing the history of display practices in relation to the collection.
13. For a discussion of small museums and how to define them, see the Introduction to Fiona Candlin's *Micromuseology*.
14. I am beholden to Patricia Koh for explaining the logic underpinning the layout of the collection. A cladistic understanding of species classification seems to inform the displays, with a move from organisms perceived as primitive to those seen to be more advanced.
15. Dodd et al., 'Introduction', 5.
16. Stark, 'The Cultural Politics of Mourning', 68.
17. Freeman, *Paper Tiger*, 175.
18. Stark uses the term 'extinction celebrity' in 'The Cultural Politics of Mourning', 68.
19. The ITSD notes that a fourth mount (3045) was exchanged with Uppsala Museum for a mouse-deer or chevrotain (an ungulate of the Family *Tragulidae*) in 1872.
20. The exhibitions have since been revamped, so the interpretations I offer here are historical.
21. The illustration of the bird is signed by Shaw, but the inspiration is an etching by the ornithologist George Edwards that was made in 1757 based on an oil painting

that included the dodo. The dodo in oil was painted from life by Roelandt Savery in 1626 using a captive bird in the possession of Hans Sloane. See Anonymous, 'George Edwards's Picture of the Dodo', 615. Edwards's etching included a guinea pig (*Cavia porcellus*) to give a sense of scale. Shaw has excised this animal and added more plant life. For a broader discussion of the history of representations of the dodo, see Hume, 'The History of the Dodo *Raphus cucullatus*'.

22. This frog is subject to a Lazarus project with efforts to clone them ongoing. See Hand, *Reviving Extinct Species*, 34.
23. Since the 1960s it had been speculated that a relict population survived on the refugium of Ball's Pyramid, a tall volcanic stack 20 kilometres from Lord Howe Island.
24. Lazarus taxa often take the form of species thought to exist only in the fossil record but then discovered living in the present day (a frequently cited example is the coelacanth, which is also displayed in the cabinet). The rodent *Laonastes aenigmamus* was thought to have been extinct for eleven million years until it was rediscovered in Laos. See Dawson et al. '*Leonastes* and the "Lazarus Effect" in Recent Mammals'. Strong arguments have been made for keeping the rediscovery of species previously thought extinct secret. See Meijaard and Nijman, 'Secrecy Considerations for Conserving Lazarus Species'.
25. Interfaces (including this one at the Australian Museum) often feature content that is intended for adults, yet receive greater use from children. Gammon notes that computer screen-based interactives in museums are especially popular with children. See Gammon, 'Visitors' Use of Computer Exhibits', 281–90.
26. A former director of the Australian Museum, Frank Howarth, notes that the impact of global heating is a contemporary concern and will be a major area of future research in natural history museums. See Howarth, 'The Future of Research in Natural History Museums', 71.
27. Wilson, *Natural History*, 61. This cultivating of conscience and reflection can be seen as motivated by the shift from collections-driven to ideas-driven curating focussed on social responsibility that accompanied the rise of the new museology. Natural history museums such as the Australian Museum, however, still of necessity adopt a collection-based model for their permanent exhibitions. For a discussion of ideas and trends associated with new museology and their uptake or lack thereof, see McCall and Gray, 'Museums and the "New Museology"'.
28. Wilson, *Natural History*, 64.
29. For a discussion of interpretative and non-interpretative labels and also different kinds of labels, see Chicone and Kissel, *Dinosaurs and Dioramas*, 83–89.
30. Ibid., 83.
31. Ibid. All these forms of text arguably 'interpret' the exhibition in some way, yet their meaning-making is more oblique. For a short discussion of the misunderstanding that captions are non-interpretative, see Chare and Frank, 'Why History and Art History?', 12.
32. Cited in Turner, *Cataloguing Culture*, 104. Turner sources the quotation from Mason's 1907–8 *Annual Report* relating to his activities at the Smithsonian.
33. Turner suggests that, for Mason, labels are 'essentially, parts of the object itself'. Turner, *Cataloguing Culture*, 104–5.
34. See Laurance et al., 'Epidemic Disease'.
35. O'Doherty, *Inside the White Cube*, 15.
36. Animal genocide would, of course, form part of the process of natural selection. It is not 'unnatural' in this sense, constituting an aspect of the changing environment that species must navigate and survive. Humans provide competition that other species

must adapt to. It is, perhaps uniquely, open to humans to 'change the rules of the game', to compete differently, less exploitatively.
37. See Grube, 'Detrimental Effects of White Valued Walls in Classrooms', 69.
38. O'Doherty, *Inside the White Cube*, 14.
39. For a discussion of the affective capacity and symbolic associations of white as a backdrop in art inspired by natural history, see Chare, 'We Murder to Attest'.
40. Farrington, 'The Rise of Natural History Museums', 207.
41. Haswell and Parker, *A Text-Book of Zoology*, 466.
42. 'Museum Staff 1829–1977', Australian Museum website, https://australianmuseum.net.au/about/history/people/museum-staff-18291977/ (retrieved 14 February 2025).
43. Sleightholme and Campbell, 'A Retrospective Assessment of 20th Century Thylacine Populations', 110.
44. Moser, 'The Devil is in the Detail', 29.
45. This female pup that the model simulates, P762, had a central role in the cloning experiment described in the preceding chapter.
46. McHugh, *Love in a Time of Slaughters*, 48.
47. A later painting on the same theme, titled 'Aborigines of Tasmania', is colder in tone. This last work was originally commonly referred to as 'The Natives'. See Plomley, 'Notes on Some of the Tasmanian Aborigines', 54.
48. For a discussion of Bock's portraits in watercolour, see Bonyhady, 'The Art of Pacification'. Bonhady notes that some of the portraits, such as that of Tongerlongter, avoid depicting injuries suffered by the palawa during the genocidal campaign of the Black War (77–78). Tongerlongter was shot in the arm and it had to be amputated. Many of the portraits are known to have faded. Plomley, 'Notes on Some of the Tasmanian Aborigines', 54.
49. Some suggest that numbats (*Myrmecobius fasciatus*) share ancestry with thylacines, although the genii diverged 3–40 million years ago. See Baker and Dickman, *Secret Lives of Carnivorous Marsupials*, 24.
50. Yates, 'New Craniodental Remains of *Thylacinus potens*', 29–31.
51. For a general discussion of the fossil record of Thylacinidae, see Wroe, 'Australian Marsupial Carnivores', 108–10.
52. For a discussion of portrayals of the thylacine in First Nations Australian rock art (including Kakadu), see Guiler and Godard, *Tasmanian Tiger*, 41–44 and Johnson, *Australia's Mammal Extinctions*, 155–56.
53. Writing not of Kakadu but of the Yaburara of Murujuga, Ken Mulvaney has suggested that the thylacine was important to ceremony there. Mulvaney describes a series of petroglyphs at Nganjarli (referred to by him as Patterson Valley) that he identifies as linked to a *thallu* (or increase) ceremony. He perceives the petroglyphs as indexing a crisis prompted by declining numbers of thylacines. In 'Extinction, Inscription and the Dreaming', Katie Glaskin uses modern ethnographic accounts of *thallu* ceremonies to question Mulvaney's interpretation of the petroglyphs (9). Mulvaney's point, however, is that the petroglyphs at Nganjarli register an unusual practice compared to other *thallu* sites (one of which is located nearby). First Nations Australians of the Burrup also acknowledge limits to their knowledge of Yaburara *thallu* sites. David Daniel, for example, explains that he does not know how a whale *thallu* site at Murujuga operated. See Daniel, *Thalu Sites of the West Pilbara*. Mulvaney also discusses thylacine petroglyphs at length in *Murujuga Marni*, considering their importance for helping to date rock art. For further discussion of the thylacine petroglyphs at Murujuga, see also Chare, 'After the Thylacine', 136–42; Freeman, *Paper Tiger*, 1–5. For a general discussion of the rock art of Murujuga, see Chare, 'Vision and Indifference'.

54. Byatt, Snæbjörnsdóttir and Wilson, 'Introduction', 14.
55. This room is a permanent exhibition. It is, however, sometimes subject to change. On my most recent visit in December 2022, a new exhibit had been added featuring the preserved remains of what Kathryn Medlock and Robert Paddle claim is the last captive thylacine.
56. In the past, a thylacine display was incorporated into a similar exhibition that included numerous examples of Lutruwita's fauna. When exhibited in such a context, the animal was made part of an ecosystem, an environment. Now, the tiger is set apart, their separateness reinforcing that they have no active role in the island's ecosystem. I am grateful to Julie Gough for sharing her insights into how the thylacine was previously exhibited at the museum.
57. This information is based on my noting the entry point used by every visitor to 'The Thylacine Gallery' over the period of an hour.
58. For a brief discussion of the way points of ingress can change the reception of works of art, see Rifkin, '"The Answers Are the Question"', 230.
59. For a discussion of museum lighting and its potential effects on display objects, see Cuttle, 'Damage to Museum Objects Due to Light Exposure'.
60. See the discussion of lighting in Moser, 'The Devil is in the Detail', 26.
61. Chicone and Kissel, *Dinosaurs and Dioramas*, 121.
62. Wilson notes a similar use of subdued lighting in the 'Salle des espèces menacées et des espèces disparues' (Hall of Endangered Species and Extinct Species) at the Muséum national d'Histoire naturelle in Paris. Wilson, *Natural History*, 86.
63. See Garside et al., 'How is Museum Lighting Selected?', 7.
64. Poliquin, *The Breathless Zoo*, 41. Emphases in the original. All emphases in the original unless otherwise stated.
65. Paddy Hartnett's preferred snaring ground was the Walls of Jerusalem in the Central Highlands. He built the Du Cane Hut in 1910 in this area. Tiger Lake is located in this locale (as is Wild Dog Creek). In the early 1920s, Hartnett was engaged in mining on the Little Wilson River. By the mid-1920s he may have returned to trapping as there is a photograph of him at Old Pelion Hut tentatively dated to 1925. See Anonymous, 'Can You Help Identify These People?', *The Western Tiers*, Tuesday 17 February 1998, 16. The Walls of Jerusalem appear prominently in *The Tale of Ruby Rose*, a film that features trappers.
66. In 1872, William Hissey, a hairdresser who moonlighted as a taxidermist, prepared a 'huge Tasmanian tiger' for the Royal Society's Museum. This museum would subsequently form the basis of TMAG's collection. The tiger was 5 ft. 6 in. long. Hissey's work as a taxidermist was described as 'artistic manipulation'. See Anonymous, 'Hobart Town,' *Weekly Examiner*, Saturday 9 November 1872, 16. A sense of the qualities that Hissey's mounts were praised for is clear from a short article on his work as a taxidermist that was published in 1868. The 'natural and life-like appearance' he gave birds and other animals was found 'very pleasing'. See Anonymous, 'Taxidermy', *Tasmanian Times*, Wednesday 20 May 1868, 2.
67. Anonymous, 'Tasmanian Animals', *The Mercury*, Friday 29 October 1926, 8.
68. Ibid.
69. Poliquin, *The Breathless Zoo*, 133.
70. Ibid., 86.
71. Ibid., 135.
72. See, for example, the anonymous article on the 'Protection of Native Animals' that mentions the 'suffering and lingering death caused by trapping and snaring'. Anonymous, 'Protection of Native Animals', *Daily Post*, Friday 21 June 1912, 4.

73. There were various types of 'necker' snare in use. See Optimist, 'Fur and Friendship', *The Mercury*, Saturday 20 December 1924, 21.
74. Snares can also be modified to capture rather than kill animals.
75. Albert Cherry, 'Sport in Bass Strait', *Daily Telegraph*, Saturday 21 August 1920, 8.
76. Anonymous [Trapper], 'Brotherhood and Fur', *Advocate*, Thursday 9 October 1924, 6.
77. Cited in Paddle, 'The Thylacine's Last Straw', 78.
78. Anonymous, 'The Game Season', *The Mercury*, Thursday 2 April 1925, 9.
79. Ibid.
80. Anonymous, 'Hunting Stopped', *The Herald*, Monday 19 April 1926, 15; Anonymous, 'Game Season', *The Mercury*, Saturday 17 April 1926, 13.
81. See William Edward Lodewyck Hamilton Crowther, 'The Open Fur Season', *The Mercury*, Friday 17 September 1926, 3. The European market for fur was depressed in 1926 so it would have been a good time to enforce a lengthy close season. See Anonymous, 'Open Fur Season', *The Mercury*, Friday 30 July 1926, 3.
82. There was apparently a two-week grace period after the close. See Tourist, 'The Hunting Season', *Daily Telegraph*, Tuesday 29 June 1926, 2.
83. Anonymous, 'Close Season Needed', *Advocate*, Saturday 6 August 1927, 2.
84. Ibid.
85. Anonymous, 'The Game Season', *The Mercury*, Thursday 2 April 1925, 9.
86. Ibid.
87. Arthur Reid was also present at the meeting and offered £20 for every tiger sent alive to the zoo. He said he had only received eight tigers in five years. As he was appointed curator in 1922, he must actually have received eight tigers in three years.
88. The catalogue entry for the Online Zoological Collections of Australian Museums (OZCAM) simply records the thylacine as entering the TMAG collection in 1926.
89. It takes several weeks to tan the hide of an animal using traditional methods, perhaps three to four for an animal the size of a thylacine. Mounting, drying and finishing a mount is relatively rapid. It is therefore conceivable that the tiger was killed in September (out of season).
90. Poliquin, *The Breathless Zoo*, 102.
91. Hauser, 'Coming Apart at the Seams', 11.
92. Ibid.
93. Sleightholme and Campbell, 'Stripe Pattern Variation'.
94. This hypothesis would only be verifiable if the skins of identical twins from a litter of thylacines were preserved, and each of the skins had distinct banding. Human identical twins share nearly the same DNA and may look similar but have different fingerprints. Nonhuman animals do sometimes birth identical twins, with a wolfhound birthing identical pups in 2016. See Segal, 'News Reports'. Colley also notes the connection between skin, fingerprinting and identity in *Wild Animal Skins in Victorian Britain*, 13.
95. The methods used by Sleightholme and Campbell to catalogue band differences are readily applicable to skins in museums and private collections but do not always translate seamlessly to efforts to differentiate thylacines by their stripes in photographs.
96. Colley, *Wild Animal Skins in Victorian Britain*, 64.
97. Ibid., 94.
98. Ibid., 95.
99. The same is true of human sexual dimorphism. For a discussion of the complexities of size and strength differences between the sexes, see Chapter 4 of Nelson's *The Stronger Women Get*.

100. Paddle, 'The Most Photographed of Thylacines', 463. William Williamson was a professional photographer based in Hobart who specialised in animal and marine photography. His studio was located at 127 Goulburn Street. In the article from 1924 by 'Snapshotter', the author states that Williamson employed a 'large old' box camera for still shots. Williamson may have used this same camera on 28 September 1911 to take still photographs (including the one thought to illustrate sexual dimorphism) of thylacines at Beaumaris. Snapshotter, 'Camera Notes', *The News*, Saturday 1 November 1924, 6.
101. A312 is known to be a female thylacine but her age at the time of her death is uncertain.
102. Moore, who was the articulator and osteologist at the National Museum in Melbourne, died on 2 September 1922. See Anonymous, 'By the Death of Thomas Francis Moore', *Nature* 110(2767) (1922): 641; Anonymous, 'About People', *The Age*, Thursday 7 September 1922, 9; Cunningham, *City of Trees*, 272. When he articulated the skeleton of the racehorse Trenton in 1907, he was praised for his skill as some of the bones he worked with had damage. See Anonymous, 'A Famous Racehorse', *Geelong Advertiser*, Wednesday 18 December 1907, 1.
103. Paddle, 'The Most Photographed of Thylacines', 465.
104. Paddle describes the tiger as Power's pet and notes that the man visited the tiger at the zoo (then located at Sandy Bay) on 11 November 1911, roughly three months after selling him. Ibid., 463.
105. A newspaper report suggests that the tiger was sent from Tyenna to the care of William Foster Trebilcock, the proprietor of the hotel. See Anonymous, 'Tasmanian Tiger', *The Mercury*, Monday 14 August 1911, 4. As Power came to Hobart to handle the sale personally, this cannot be the case. He visited Roberts on the morning of 12 August to tell her of the tiger and must have left him in Trebilcock's safekeeping during that period.
106. Paddle, 'The Most Photographed of Thylacines', 463.
107. Flynn was the University of Tasmania's first Professor of Biology and an expert on marsupial embryology. Ibid., 464. Kershaw mentions thylacines, specifically the 'Marsupial Wolf', briefly (75) in his article 'The Tasmanian Devil in Victoria'.
108. Tigers that were captured alive were often 'tamed'. A thylacine sold at auction by Roberts & Company on 31 January 1894 was described as 'tractable and a perfect specimen'. Anonymous, 'This Day, January 31', *The Mercury*, Wednesday 31 January 1894, 4.
109. See Palmer, '"Taming the Wild Profusion of Existing Things"', 355.
110. See the discussion of taming in Mackintosh, 'Crusoe's Abattoir'.
111. In 1913, Vaughan visited the Sandy Bay incarnation of Beaumaris Zoo, which he described as a small, private affair located at a villa. See Vaughan, 'The Tasmanian Tiger', 495. The zoo was relocated to the Queens Domain area of Hobart in 1923.
112. Anonymous, 'The Tasmanian Museum: The Dasyures', *The Mercury*, Thursday 3 November 1910, 8.
113. Wilson, *Coorinna*, 89.
114. Ibid., 79.
115. Anonymous, 'Obituary', *The Mercury*, Monday 9 August 1926, 6. A potential insight into Williamson's views on animal photography is provided by judges (of which Williamson was one) advising entrants to a competition of the Southern Tasmania Camera Club that 'when photographing animals', 'they should always choose a suitable background' as '[m]any fine subjects were spoilt because of unsuitable backgrounds'. Anonymous, 'Southern Tasmania Camera Club', *The Mercury*, Friday 30 June 1922, 9.

116. Vaughan, *An Australasian Wander-Year*, 129.
117. The footage was shot as part of a travel film, *Tasmania the Wonderland*, probably directed by Sidney Cook. It was rediscovered by Branden Holmes, Gareth Linnard and Mike Williams. See Alex Fox, '"New" Footage of Benjamin, the Last Tasmanian Tiger Ever Seen Alive', *Smithsonian Magazine*, 1 June 2020, http://www.smithsonianmag.com/smart-news/new-footage-benjamin-last-tasmanian-tiger-ever-seen-alive-180975000/ (retrieved 4 August 2021).
118. Tucker took a number of photographs of thylacines at the zoo. The lack of information about him suggests that he may have been an amateur or a visitor from the mainland staying in Hobart.
119. Sleightholme and Campbell, 'A Catalogue of the Thylacine Captured on Film', 158.
120. Newman worked at the zoo until at least 1935 when, at eighty-one years of age, he was injured by an escaped sun bear (*Helarctos malayanus*). Anonymous, 'Accidents: Clawed by Bear', *The Mercury*, Monday 5 August 1935, 2. He died on 13 April 1936.
121. Sleightholme and Campbell, 'Reverend George H. Judd's Thylacines', 77.
122. Sleightholme and Campbell, 'A Catalogue of the Thylacine Captured on Film', 149.
123. Lovell was the overseer at Woolnorth in 1909. Wainwright had worked there since 1901 and was himself made overseer in 1916. Wainwright's father, George Wainwright Senior, also worked for the Van Diemen's Land Company and looked after their property at Mount Cameron. He is referred to in a newspaper report about the history of the far north-west of Lutruwita simply as the 'tiger-catcher of Mount Cameron'. See Anonymous, 'The Far North-West', *Examiner*, Saturday 6 February 1909, 8. In 1903, Wainwright Snr was found guilty of the theft of 616 skins of varied species (but not thylacines) from the farmer Thomas King and sentenced to twelve months imprisonment. See Anonymous, 'Supreme Court', *Daily Telegraph*, Friday 27 February 1903, 4; Anonymous, 'The Wainwright Case', *Examiner*, Saturday 28 February 1903, 6. Wainwright Snr died in prison later that year and his widow Matilda would subsequently (in 1906) marry Lovell who became Wainwright Jnr's stepfather.
124. Anonymous, 'Tasmanian Tigers', *Daily Post* Thursday 8th July 1909, 4.
125. Anonymous, 'News of the Week', *Circular Head Chronicle*, Wednesday 4 August 1909, 2.
126. See, for example, the account of the mother and two pups caught in Parteenno/Hampshire Hills (between Burnie and Waratah) exhibited in Launceston in 1864. Anonymous, 'General News', *Launceston Examiner*, Thursday 22 December 1864, 3.
127. Anonymous, 'A Curiosity', *Launceston Advertiser*, Thursday 23 May 1844, 3.
128. For a recent analysis of the convergent evolution of the thylacine and wolf, see Newton et al., 'Ontogenetic Origins of Cranial Convergence'.
129. The snares are identified as 'thylacine snares' by the museum. In what ways they are different from snares for similar-sized animals is not clear.
130. A1241 is not included in the 2013 iteration (the fifth revision) of the ITSD, which I consulted in Lutruwita. A1483 is of unknown gender and provenance.
131. Merck, *Veterinary Forensics*, 84.
132. The reference to 'wolf' is to the 'marsupial wolf', an alternative name for the thylacine. Le Soeuf and Burrell, *The Wild Animals of Australasia*, 319.
133. This photograph is analysed again in the Conclusion.
134. Anonymous, 'Law Intelligence', *The Mercury*, Wednesday 27 April 1870, 2.
135. Cameron Campbell, 'Tales of the Old Tasmanian Bushmen', www.naturalworlds.org/thylacine/history/bushmen/bushmen_3.htm (retrieved 20 September 2023).
136. David Pearce was born in Hamilton to John Pearce and Sarah Pearce (née Jones) on 24 January 1856. He died on 23 March 1898 so he cannot be the man in the

photograph. See Scott Cook, 'David Pearce (1856–1898)', https://www.wikitree.com/wiki/Pearce-4736 (retrieved 30 July 2021). Many of the Pearce family lived in the vicinity of Fitzgerald, having settled in Hamilton and Ouse. They owned some land near Clarence River and trapped around Black Bob's and Victoria Valley. The man could be another member of the family.

137. The photograph of the curled-up thylacine forms one of a pair. The pair show the thylacine facing in different directions. It seems probable that one of these images was printed in reverse in error and they both feature the tiger in marginally different positions. This suggests that the thylacine may be dying rather than posed. If the animal had internal injuries or had been poisoned they may have curled up in this way, as mammals often die in a sleeping position.
138. Sleightholme and Campbell, 'A Catalogue of the Thylacine Captured on Film', 159.
139. In 1934, zoologists from Melbourne were seeking to secure 'specimens' to exhibit as part of the centenary celebrations of the State of Victoria. This, against the backdrop of searches to discover whether thylacines were still extant. See Anonymous, 'Native Tiger Still Exists in Far N.W.', *Advocate*, Wednesday 22 August 1934, 2.
140. The rising prices paid for thylacines by TMAG demonstrate recognition of the animal's increasing rarity.
141. De la Mare, 'On the Definition of Threats to the Survival of Species', 113.
142. Ceballos et al., 'Vertebrates on the Brink', 13597.
143. In 1996, Charlie Beasley of the Tasmanian Parks and Wildlife Service also claimed to have seen an animal resembling a thylacine. See Drollette, 'On the Tail of a Tiger', 32.
144. Anonymous, 'Tasmanian Tiger Search', *The Mercury*, Saturday 12 November 1938, 17.
145. Sharland, 'Tracking the Thylacine', 8.
146. A mention of the presence of tigers in the Vale of Rasselas also occurred in the press the year before. See Anonymous, 'Ramblings in Rasselas', *The Mercury*, Saturday 24 November 1945, 11. This article is probably by Michael Stanley Read Sharland as an accompanying section on birds is authored by Peregrine, Sharland's pseudonym.
147. Peregrine, 'Tiger Tracks in Rasselas', *The Mercury*, Saturday 16 November 1946, 6.
148. Anonymous, 'Urge Trapping Ban to Save Tasmanian Tiger', *The Mercury*, Monday 3 June 1946, 3.
149. In the fifth revision of the ITSD, this skeleton is listed as missing.
150. Flynn was honorary curator at TMAG from 1912 to 1918.
151. Flynn, 'Notes on Certain Reproductive Phenomena', 225.
152. This exhibition will henceforth simply be referred to as 'Precious Little Remains'.
153. The skulls of a Tasmanian devil, a spotted-tailed quoll (*Dasyurus maculatus*) and a white-footed dunnart (*Sminthopsis leucopus*) are also displayed nearby as the thylacine's closest living relatives (the numbat is mentioned on a label but no skull is included).
154. The palawa name for Cradle Mountain derives from Plomley, *Tasmanian Aboriginal Place Names*, 88.
155. Freeman discusses King and the composite photograph in her PhD thesis, 'Figuring Extinction' (213).
156. Three of these images are reproduced by Sleightholme and Campbell in 'A Catalogue of the Thylacine Captured on Film'. The one used for the composite is a flipped and cropped version of the image they catalogue as HBZ 23. The flipped and cropped version is reproduced on the back cover of Maynard and Gordon's *Precious Little Remains*. A copy of the photograph is part of QVMAG's collection and has

the catalogue number QVM.1983.P.1944. The original glass plate negative is held by the Tasmanian Archives and Heritage Office (NS1377/1/59). They credit the photograph, which forms part of the Norman Laird archive, to R. Davis. It seems unlikely, however, that King would use another photographer's image for his composite. Another glass plate negative in the Norman Laird archive at TAHO (NS1377/1/58) is by King and this suggests to me that the other photograph is too. The archive also includes a glass plate negative of King's composite image (NS1377/1/66, and a duplicate NS1377/1/67). I am grateful to Caroline Homer for providing me with information regarding these images and to the trustee of the Estate of Norman Laird for permitting reproductions to be made of them. Sleightholme and Campbell include a detail of NS1377/1/58 in their article (labelled HBZ 6). It is derived from an image in the QVMAG archive and is not the full image. The uncropped photograph shows not just the thylacine basking in the sun but a keeper in the background.

157. The item number of this image is NS1377/1/66. I am grateful to Caroline Homer for furnishing me with a copy of the image during the COVID-19 pandemic.
158. Cox and Ratcliff, *Tasmania Remembered*, 109. The composite is described as an example of '"trick" photography'.
159. Orchard and Holcombe, 'Photographic Reflections on Landscape Change in Regional Australia', 148. King's photographs were also put to practical use as illustrations in academic works on the geology of Tasmania. See, for instance, Austin Burton Edwards's 'Differentiation of the Dolerites of Tasmania 1', which includes several photographs by King including one of Warloundigerler/Cradle Mountain.
160. Sleightholme and Campbell, 'A Catalogue of the Thylacine Captured on Film', 155.
161. Roosevelt made reference to the capacity of humankind to mar the beauty of the Grand Canyon and advised those present that: 'What you can do is to keep it for your children and your children's children.' See Brodie, 'Report of the Governor of Arizona', 288. He also states in his 1908 address to the Conference of Governors that a conservation policy 'will preserve soil, forests, water power as a heritage for the children and the children's children of the men and women of this generation'. See Roosevelt, 'Opening Address by the President', 11. In 1903 the President of the National Livestock Association, John W. Springer, made a speech that referred to Roosevelt and spoke of respecting the land, stating that cattlemen should delight 'to guard this heritage for our children and our children's children'. See Springer, 'Annual Address of the President', 133. I am grateful to Tweed Roosevelt for his assistance in determining that Roosevelt probably never used the exact words that QVMAG attributes to him, although they certainly provide an honest reflection of his sentiments.
162. For a discussion of Roosevelt's conservation initiatives, see Izatt, 'The Precedent of a President'.
163. Roosevelt, *The Wilderness Hunter*, 248.
164. The fossil remains discussed in Chapter 2 are obviously not included in this reckoning. Wilderness is clearly understood as undeveloped land, but the label for the skull acknowledges that the Central Highlands, the area where Granskogen is located, has a 'long history of farming, trapping, snaring and poisoning'. It is therefore a 'wilderness' much travelled and marked by humans.
165. The label notes that the bones date from pre-1897 and may have been donated by 'L. Wells'. A. Lucadou Wells, a dentist, was a member of the Natural Science Association of Northern Tasmania. He travelled in the region as part of his practice and may have secured the bones from a local farm. The fifth revision of the ITSD gives the source of the remains as Moat Farm near Carrick. This farm was owned by Percy Hart in the late 1890s and by John Symmons before that. It was purchased by

E. H. Lucadou Wells (the son of Lucadou Wells) in 1931 so it seems unlikely that the bones are from there if they did enter the collection before 1897.

166. For a discussion of the function of text labels in museums, see Devenish, 'Labelling in Museum Display'.
167. Maynard, 'Tasmanian Tiger', 319.
168. Ibid., 321.
169. Ibid., 319.
170. In reality, the thylacine gape or yawn was an anxiety indicator. Dingoes often show their teeth as a sign of aggression. See Corbett, 'Social Dynamics of a Captive Dingo Pack'.
171. Rose, *Wild Dog Dreaming*, 69.
172. Maynard and Gordon speculate that this mount was acquired as part of a pair along with the male tiger QVM.2001. If this was indeed the case, the lowered head of the female would contrast with the raised head of the mount of the male, providing an example of how late nineteenth-century ideas about the role of the (human) sexes permeated taxidermy practice.
173. Taxidermy has a long history. In 1803 in his *Nouveau dictionnaire d'histoire naturelle*, Louis Dufresnes, who first used the term, identified the birth of modern taxidermy as having occurred fifty years previously (so *c.*1750). Dufresnes credited René Antoine Ferchault de Réaumur with the invention of what he called 'modern' or 'new' taxidermy. As part of his discussion, he also called attention to the aesthetic dimension to the taxidermist's art, suggesting that English efforts at mounting animals were far inferior to those of their French counterparts (512).
174. Melissa Milgrom uses this term to describe old-school taxidermy that betrays a lack of scientific understanding of a given species. See Milgrom, *Still Life.*
175. This crude addition may suggest that the skin here was puckering or too loose and needed to be joined more firmly to the 'skull' beneath.
176. Osborn, 'Archetypal Metaphor in Rhetoric', 117.
177. Ibid., 118.
178. The thylacine came into the collection prior to 1897 and may have been acquired by TMAG earlier than 1891. Alexander Morton, who had worked at TMAG (or the Royal Society of Tasmania museum, as it was initially) since 1884, served as honorary curator at QVMAG from 1891 to 1896 and the thylacine was presumably transferred from Nipaluna/Hobart to Launceston during this period. When the museum opened in 1891, it was reported to already have on display 'an excellent group of Tasmanian animals'. See Anonymous, 'The Victoria Museum and Art Gallery', *Tasmanian News*, Thursday 30 April 1891, 2. Some of these animals were 'native tigers' so it seems probable that the mount was present at QVMAG from its inception. See Anonymous, 'Launceston Art Gallery and Museum', *The Mercury*, Thursday 30 April 1891, 2. The Lutruwitan mammals including the thylacines were exhibited adjacent to a suit of Japanese armour.
179. Very young pups would have to be torn from the teats. See Newton et al., 'Letting the "Cat" out of the Bag'. See also Maynard and Gordon, *Precious Little Remains*, 61.
180. The male thylacine also possessed a pouch for protecting the scrotum.
181. I have found Terri Kapsalis's exploration of gynaecology as performance inspiring here as she ably foregrounds the cultural dimensions to scientific practice. See Kapsalis, *Public Privates.*
182. Ibid., 7.
183. The use of the term 'captured' elides the probable violent end suffered by the thylacine.
184. See Koçak, 'Gendering the Holocaust'.

185. One of the touchscreen information sections is a reproduction of typed testimony from Dick Rowe given on 25 November 1951. He details the snaring of a female thylacine and three of her pups at Christmas Hills (north-west Lutruwita) in circa 1912. The mother drowned after being snared but the pups survived and were subsequently sold to the animal trafficker James Harrison for £5 each.
186. In 'Tasmanian Tiger', Maynard refers to the tiger as 'the museum's best thylacine mount with the best provenance' (320).
187. Anonymous, 'Obituary: Mr. William Notman Centenarian Passes', *The Mercury*, Saturday 21 October 1933, 7.
188. Anonymous, '100 Tomorrow: Mr. William Notman', *Examiner*, Saturday 15 October 1932, 6.
189. Anonymous, 'About People: Late Mr. A.W. Dilger', *North-Eastern Advertiser*, Friday 30 January 1932, 2.
190. Anonymous, 'Current Topics', *Examiner*, Friday 24 May 1912, 4.
191. Maynard and Gordon, *Tasmanian Tiger*, 65. Dilger would have measured the distance in miles. The decision to use kilometres (the metric system) in the panel may be a conscious anachronism designed to make the distance (just under 10 miles) sound more impressive. Another feat of carrying was reported in 1938 when Harold Geale of George Town in north-east Lutruwita carried a kangaroo (identified as belonging to a new species) he had caught in a snare for eight miles through the bush. The journalist recounting the feat concluded that the macropod was sentenced to '"transportation for life to the Launceston City Park"', acknowledging, even if in jest, that the zoo was akin to a penal colony for nonhuman animals. See Quarterdeck, 'New Model Kangaroo', *Smith's Weekly*, Saturday 5 March 1938, 20.
192. McGowan was appointed superintendent of reserves in 1882 and remained in the role until 1937. He continued as a consultant to his son, who succeeded him, until his death at eighty in 1939. See Anonymous, 'Obituary: Mr. W. McGowan', *The Mercury*, Monday 26 June 1939, 3. William McGowan Junior held the role from 1937 until his death in 1945. See Anonymous, 'Obituary: Mr. W. McGowan', *The Mercury*, Tuesday 9 October 1945, 6.
193. 'Advertising', *Launceston Examiner*, Monday 18 February 1884, 1.
194. These trades were usually for more 'exotic' animals. In 1931, for instance, McGowan sent two Tasmanian devils to Melbourne in exchange for two monkeys. See Anonymous, 'Tasmanian Devils', *The Mercury*, Tuesday 30 June 1931, 5. In 1939, William McGowan Jnr sent a devil and a tiger cat in exchange for 'other animals'. Anonymous, 'Zoo Animals Exchanged', *The Argus*, Tuesday 27 June 1939, 2. I am grateful to Robert Paddle for sharing his insights regarding this last exchange.
195. In 1900, for example, a thylacine captured by a Mr Stephenson (possibly Stevenson) of North Esk was brought to City Park where they were to be taken care of until passage could be arranged on a steamer to Gadigal/Sydney, where the tiger would then be exhibited at the Sydney Zoological Gardens. See Anonymous, 'Launceston', *The Mercury*, Wednesday 14 March 1900, 4.
196. The Thylacine Museum website provides the most comprehensive summary I have found of thylacines exhibited at City Park. See 'City Park Zoo [Launceston] (Continued)' http://www.naturalworlds.org/thylacine/captivity/zoos/Tasmania/Tasmania_10.htm (retrieved 10 November 2021). Campbell identifies 1882 as the year the first thylacine was incarcerated at the park. Guiler refers to a thylacine exhibited at Launceston in 1879 but City Park only appears to have developed zoological exhibits in the 1880s. He also refers to thylacines featured at the Launceston Show in 1910 that probably came from the zoo. See Guiler, *Thylacine*, 66.

197. Anonymous, 'Our Launceston Letter', *The Mercury*, Friday 21 January 1881, 3. The report reads as if the tigers and other mammals have already been procured and are simply awaiting display.
198. Ibid.
199. Anonymous, 'General News', *The Colonist*, Saturday 11 May 1889, 21. The newspaper report also includes a call from McGowan for rural dwellers to send him 'healthy specimens' of Lutruwita's birds and mammals to be used for pending exchanges with other zoos.
200. Anonymous, 'General News', *The Colonist*, Saturday 24 August 1889, 21.
201. Anonymous, 'Current Topics', *Launceston Examiner*, Tuesday 20 August 1889, 2.
202. McGowan acknowledged that thylacines had been caught with young in the past but in captivity they had 'not been a success'. This indicates that thylacines and their pups had been exhibited in the past but had not lived long. See Anonymous, 'General News', *The Colonist*, Saturday 24 August 1889, 21.
203. Anonymous, 'Current Topics', *Launceston Examiner*, Saturday 8 November 1890, 4.
204. See Anonymous, 'The City Park, Launceston', *Illustrated Sydney News*, Saturday 5 December 1891, 17. The author of this article writes of the collection of wild animals including the marsupial wolf and the marsupial tiger (both of which appear to be references to thylacines).
205. The mother and her young were procured from the dairy farmer Edward Pennefather of Ringarooma. See Anonymous, 'Local & General', *The Tasmanian*, Saturday 12 August 1893, 27. Paddle discusses this quartet in *The Last Tasmanian Tiger*, 227–28.
206. The Western district as a term potentially covers a broad area and in the past, it referred to regions as far west as the Queen River and Gordon River, encompassing the Florentine Valley. See the discussion of the Western district in Johnston, *Systematic Account of the Geology of Tasmania*, 53–54. More commonly, it referred to locations nearer Launceston. The Launceston and Western Railway was sometimes referred to as the Launceston and Western District Railway and went to Deloraine with later branches to Loenarpo/Mole Creek and Limilinaturi/Devonport. The reference to 'this Western district' in the article seems to indicate that the thylacine was caught somewhere near Pataway/Burnie as opposed to the sanctuary on the west coast referred to earlier in the article. Anonymous, 'In Western Wilds', *Advocate*, Saturday 8 March 1924, 11.
207. 'City Park Zoo [Launceston] (Continued)', http://www.naturalworlds.org/thylacine/captivity/zoos/Tasmania/Tasmania_10.htm (retrieved 11 November 2021).
208. Dingo, 'The Zoo', *Launceston Examiner*, Thursday 21 February 1895, 7.
209. Tommy Trot, 'Echoes of the Streets', *The Tasmanian*, Saturday 13 July 1889, 17. As a columnist, Trot was a moraliser and proselytiser, continually commenting on the perceived ills of society. His tone seems somewhat tongue-in-cheek. The anecdote he offers of the thylacine instrumentalises the animal's behaviour in order to comment on human voracity and on the lack of gratification that comes from exploitative or rapacious conduct. Trot is briefly referenced in Wilson, 'Wandering Stars', 102n89. Tommy Trot features in a children's nursery rhyme, 'A Devoted Husband'. See Granger, *The Merrie Heart*, 144. The literary character Tommy Trot appears in a moveable book: Anonymous, *Tommy Trot and his Loves*.
210. It is not uncommon for some carnivores to devour members of their own kind, particularly when dead. The cannibal thylacine may have killed the other tiger or simply eaten their dead body. Wolves have been known to practise cannibalism in times of stress or when other wolves are injured or dead. See Rausch, 'Some Aspects of the Population Ecology of Wolves', 258–59.

211. Anonymous, 'Reading Matter', *The Tasmanian*, Saturday 24 May 1890, 28. The Eskimo dog, usually nowadays referred to as the Canadian Inuit dog, is a purebred that is at risk of extinction. From the 1950s to the 1970s, the Royal Canadian Mounted Police slaughtered Canadian Inuit dogs. These dog killings, which are referred to as *qimmiijaqtauniq* by the Inuit, may constitute an instance of animal genocide but it remains to be proven that the actions went beyond a concern for public health and were aimed at total eradication of the breed.
212. Anonymous, 'Current Topics', *Launceston Examiner*, Saturday 8 November 1890, 4.
213. Anonymous, 'Launceston', *The Mercury Supplement*, Saturday 19 December 1896, 1.
214. Anonymous, 'The Punch Bowl Scheme', *Examiner*, Friday 23 March 1924, 8.
215. Anonymous, 'General News', *The Colonist*, Saturday 11 May 1889, 21.
216. Anonymous, 'Sanctuary Suggested for Launceston', *Examiner*, Tuesday 23 July 1946, 3.
217. Melton, *Problems of Installation in Museums of Art*, 266.
218. Ibid., 327–28.
219. Our knowledge of the social dynamics of thylacines outside of captivity is scarce. See Mech, 'Alpha Status, Dominance, and Division of Labor in Wolf Packs'.
220. Maynard, 'Tasmanian Tiger', 327.
221. Frankentiger, as a composite of two animals of the same species, is qualitatively different from stuffed animals such as the 'jackalope', a cross between a jackrabbit and an antelope, a fictional beast constructed through yoking two different species violently together. The 'jackalope' was first conceived by the taxidermist Douglas Herrick. See Milgrom, *Still Life*, 12.
222. Maynard and Gordon, *Tasmanian Tiger*, 63.
223. For a brief discussion of gender and the monster in *Frankenstein*, see Chare and Willis, 'Introduction', 279–80.
224. Stitching on mounts is often most evident along the underbelly, a part of the mount that is less visible when in a display case.
225. It may equally acknowledge something of the 'freakiness' of taxidermy more generally. Bruce Schwendeman, as quoted in Milgrom's *Still Life*, has noted that taxidermy is represented in popular culture 'as a weird thing' that is 'unusual and creepy' (15).
226. John Random, 'The Two Million: Quick and the Dead', *Daily Telegraph*, Saturday 2 October 1937, 4.
227. Tose fabricated the mount during a trip to Lutruwita that included a talk at QVMAG on 'Modern Museum Methods'. The purpose of his visit to QVMAG and other museums in Australia and Aotearoa/New Zealand was 'to demonstrate new methods in preparing and displaying natural history specimens'. See Anonymous, 'Modern Museum Methods', *Examiner*, Thursday 5 August 1937, 6. Tose was highly critical of the animals displayed at QVMAG, seeing them as outdated and in need of replacement by 'specimens prepared under the latest methods'. Anonymous, 'Museum Exhibits at Launceston', *Advocate*, Monday 9 August 1937, 8. He is quoted by one journalist as stating that the exhibits 'must be mounted in a more modern fashion if they are to serve their purpose fully'. Anonymous, 'Expert's View of City Museum', *Examiner*, Monday 9 August 1937, 6.
228. For a discussion of genocide against the great auk, see Chare, 'The Beothuk, the Great Auk and the Newfoundland Wolf'.
229. Andrei, *Nature's Mirror*, 30.
230. I am grateful to Sandy Ingleby for sharing her knowledge regarding this mount.
231. Anonymous, 'Years Behind Times', *Examiner*, Thursday 12 August 1937, 3.
232. Anonymous, 'Story Told in Best Way', *Examiner*, Friday 6 August 1937, 7.

233. Random, 'The Two Million', 4.
234. Nicholas Smith has read the thylacine through the prism of the living dead, discussing how they haunt the Australian imaginary. The tiger potentially provides a displaced engagement with the atrocities perpetrated by European invaders against the palawa. Smith, 'The Return of the Living Dead', 285.
235. See Turner, *Cataloguing Culture*, 85.
236. For a discussion of how artefacts become 'objects of knowledge', see ibid., 66.
237. For an explanation of the metonymic function of specimens, see ibid., 67.
238. Maynard, 'Tasmanian Tiger', 328.
239. Theophilus Jones, 'Through Tasmania, No. 75', *The Mercury*, Saturday 7 March 1885, 2.
240. Nic Haygarth, however, notes that Jones's knowledge of rural life was not founded on extensive visits to the countryside. See Haygarth, 'Theophilus Jones', 55.
241. These include an advert placed by Percy Heazlewood Tucker offering for sale a tiger that is 'very quiet' and 'feeds well'. This advert ran in the *Launceston Examiner* on Saturday 16 January and Monday 18 January 1886. The tiger had been in Tucker's possession (snared and then chained and housed in a stable) since the middle of the preceding year. Tucker had previously caught and then sold a tiger to a buyer in Launceston. See Tasman, 'Correspondence', *Launceston Examiner*, Saturday 10 May 1884, 1.
242. See, for instance, 'Advertising', *Launceston Examiner*, Saturday 17 December 1881, 4. The menagerie that posted this advertisement toured Aotearoa/New Zealand under the name 'The Wild Beast Show'. Among various newspapers of the period, see Anonymous, 'Untitled', *New Zealand Herald*, 3 April 1880, 4.
243. The menagerie included bears and big cats (tigers and cheetahs) but also 'smaller animals'. See Anonymous, 'St. Leon's Circus and Menagerie', *Queensland Figaro and Punch*, Saturday 22 August 1885, 7. Two Tasmanian tigers were clearly part of the menagerie in March 1884. See Anonymous, 'The Gippsland Times', *Gippsland Times*, Wednesday 26 March 1884, 2–3, 3. A review of the show from April 1884 also mentions Tasmanian tigers in the menagerie. See Anonymous, 'Easter Holidays – St. Leon's Circus and Menagerie', *Bendigo Advertiser*, Monday 14 April 1884, 3. The presence of four Tasmanian tigers is promoted in an advertisement from May 1884. See 'Advertising', *Burrangong Argus*, Saturday 31 May 1884, 3. In an advertisement for the circus published in November 1884, however, this number has shrunk to three Tasmanian tigers ('procured at great expense'). The hyperbolic advert claimed these were the only living tigers in captivity. See 'Advertising', *Hawkesbury Chronicle and Farmers Advocate*, Saturday 22 November 1884, 3. A different advertisement to the one reproduced on the information panel but one also placed by the St. Leon business appears in the Tuesday 26 October 1886 *Launceston Examiner* on page 4.
244. 'Advertising', *Maitland Mercury and Hunter River General Advertiser*, Tuesday 18 November 1884, 5. Madden is identified as a loquacious Irishman in one circus review. See Anonymous, 'St. Leon's Circus', *Bendigo Advertiser*, Monday 21 April 1884, 2. A William Madden who once worked for St. Leon's is mentioned in an article about a tiger that escaped from Mr Higgins's Menagerie in Brisbane. See Anonymous, 'The Tiger Episode', Thursday 22 November 1888, 4. Madden's lion taming is mentioned in Arrighi, 'Political Animals', 611.
245. See Anonymous, 'Mr. Higgins's Menagerie', *Brisbane Courier*, Wednesday 28 November 1888, 6.
246. For a discussion of early Australian circuses, including St. Leon's, see Greaves, *The Circus Comes to Town*. Greaves foregrounds the arduousness of life in the circus, with

long journey times between venues (11). The St. Leon's circus, the first to include a menagerie, made use of the railway to travel between towns (ibid.).

247. Anonymous, 'The Advertiser', *Euroa Advertiser*, Friday 16 January 1885, 2.
248. See 'Advertising', *Morning Bulletin*, Monday 27 July 1896, 2. The presence of one thylacine in the menagerie is confirmed by a review of August 1896. See Anonymous, 'Fitzgerald's Circus', *Darling Downs Gazette*, Wednesday 19 August 1896, 5.
249. See ibid.
250. Anonymous, 'Tasmania's Tiger', *The Age*, Friday 26 November 1937, 1.
251. Hutton, '*Coorinna*: A Classic of Its Class Set in Tasmania', *The Argus*, Saturday 31 October 1953, 16.
252. 'Advertising', *The Mercury*, Friday 6 July 1923, 1.
253. Churchill had previously caught other tigers but most died before he found them or were too badly injured to survive and were killed. It is possible Churchill transported the tiger to the zoo on behalf of other snarers rather than being responsible for the capture (there is precedent for this).
254. Sleightholme et al., 'The Kaine Capture'.
255. Linnard and Sleightholme, 'An Exploration of the Evidence'.
256. See Anonymous, 'The Wynyard Shops – Miss M. Sargent's Studio', *North Western Advocate and the Emu Bay Times*, Tuesday 21 December 1915, 4.
257. In a talk he gave in 1918, Harrison averred that 'marsupial wolf' was the correct term and that the term 'hyena' usually referred to juveniles. See Anonymous, 'Burnie', *Examiner*, Thursday 22 August 1918, 3. Harrison used 'hyenas' in his early adverts for live thylacines but later switched to 'tigers' or 'tigers and hyenas'.
258. Given that a thylacine's eyes are on the sides of their head, their field of vision is wider than that of a human. The tiger is therefore probably visually aware of Sargent's presence even if she has not turned her head towards her.
259. Anonymous, 'Local and General – Tasmanian Tigress', *Daily Telegraph*, Monday 25 October 1915, 4.
260. Anonymous, 'Wynyard – Hyena Captured', *North Western Advocate and Emu Bay Times*, Saturday 13 May 1916, 2. In addition to the 1916 reference, 'Top Cage' is mentioned occasionally in the press in the 1920s in relation to mining. There is a section of the Arthur River located about 40 kilometres from Preolenna and, presumably, 'Top Cage' was somewhere not far downriver from that vicinity. A seasonal river called 'Foy Creek' is also in this area. The Foy family had a long-standing interest in prospecting and mining. Michael John Foy, for example, owned Foy's Pyrites Lode Company (as well as a sawmill) when he was resident in Deloraine in the 1890s. See Anonymous, 'Mining Meetings', *Launceston Examiner*, Saturday 15 January 1898. J. (John?) Foy was a prospector with land at Preolenna.
261. Ibid.
262. If this is the tiger captured by Foy, she was also photographed by George Judd when on display at Beaumaris Zoo. See Sleightholme and Campbell, 'A Catalogue of the Thylacine Captured on Film', 146.
263. See Anonymous, 'Funerals – Mr. James Harrison, Wynyard', *Examiner*, Friday 8 October 1943, 4. For a summary of the thylacines that Harrison possessed, see the Thylacine Museum website: 'The Thylacine in Captivity: James Harrison – Tasmanian Animal Dealer', http://www.naturalworlds.org/thylacine/captivity/Harrison/Harrison_5.htm (retrieved 21 September 2023).
264. See 'Advertising', *North Western Advocate and Emu Bay Times*, Saturday 4 May 1918, 6. Harrison places 'Hyenas' first in his list of wants, indicating their considerable value and scarcity.

265. 'Advertising', *Advocate*, Saturday 10 September 1921, 5.
266. See Anonymous, 'Wynyard', *Advocate*, Wednesday 13 April 1927, 4.
267. See Anonymous, 'Wynyard – A Tasmanian Tiger', *Advocate*, Tuesday 20 March 1928, 4.
268. Anonymous, 'Broken Hill Footballers', *Advocate*, Thursday 19 July 1923, 4.
269. The wallaby was not entirely white but had white patches. See Anonymous, 'Features of the Day: Spotted Wallaby', *Advocate*, Tuesday 1 August 1922, 2.
270. Ibid.
271. For an extended discussion of this photograph, see Freeman, *Paper Tiger*, 137–38.
272. The unnumbered page including the plate is immediately after page 116 in the book. See Renshaw, *More Natural History Essays*.
273. See Cameron Campbell's text 'The Thylacine in Captivity', http://www.naturalworlds.org/thylacine/captivity/zoos/internationally/internationally_2.htm (retrieved 13 January 2022).
274. He was one of the sons of the renowned animal trafficker Charles Jamrach. Charles Jamrach's father, Johann Gottlieb Jamrach, was also an animal trafficker. For a brief history of the Jamrach animal dealings, see Anonymous, 'The Wild Beast King', *Huon Times*, Tuesday 25 November 1919, 4.
275. See, for instance, the report of Jamrach sending bharal (*Pseudois nayaur*) and chital (*Axis axis*) to Aotearoa/New Zealand in 1908. Anonymous, 'Commercial', *Daily Post*, Tuesday 11 August 1908, 6.
276. The photograph is uncredited. The other photographs published on the same page are credited to Alfred E. Burrows who was based in Launceston. Robinson often photographed events in Waratah and would sometimes subsequently sell photographs featuring the images. See, for instance, Anonymous, 'Waratah – Jazz Dance', *Advocate*, 13 February 1924, 3.
277. Arthur Douglas Penney raised cattle and may have kept a dairy herd. An advertisement placed by Penney in *The Advocate*, a local newspaper, warns of a rise in the price of milk because of the increased cost of cattle feed. See 'Advertising', *Advocate*, Monday 5 May 1919, 4. In 1926, Penney was clearly the victim of rustlers and took out a notice in *The Advocate* warning that trespassers on his property would be prosecuted. See 'Advertising', *Advocate*, Tuesday 23 March 1926, 7.
278. A Waratah Resident, 'Fight with Native Tigers', *Weekly Courier*, 17 January 1924, 46.
279. In such a circumstance, if Penney had slowly retraced his steps any confrontation would probably have been avoided had he not been seeking it. Tigers usually (not always) actively avoided or ignored a human presence. A hunter, for example, witnessed a pair of thylacines near Latrobe in the 1890s and the animals took to the bush. See The Wild Irishman, 'Mole Creek "Wolf"' in 'Public Opinion: Letters to the Editor', *Advocate*, Monday 14 July 1952, 4.
280. Photographs page, *Weekly Courier*, 17 January 1924, 26.
281. Thérèse Radic suggests that 'Australians worshipped at the shrine of [Melba's] success'. See Radic, *Melba*, viii.
282. The suit jacket is very long in the arm and therefore probably a hand-me-down or borrowed from a neighbour. The thumb of Penney's left hand is hooked in the straight pocket of his jacket. This gesture is sometimes associated with male bravado and may show the influence of movie culture. Hooking thumbs in pockets or on a belt is associated with the figure of the gunslinger from the Old West (as well as in film, the pose is common in magazine illustrations by artists such as Frederic Remington). Mark Cooper identifies this gesture with the idea of the 'male on the make'. See Cooper, '"I Saw What You Said"', 10. Hooking thumbs in waistcoat pockets is

discussed as a signifier of bravado in the context of early cinema in Pearson, *Eloquent Gestures*, 46. The gesture also draws attention to the crotch and can be read as an assertion of masculinity. I am grateful to Alexander Fitzgerald, editor of *Irish Tatler Man*, for generously sharing his insights regarding Penney's sartorial choices.

283. Penney also wears work boots in the newspaper photograph, yet it is sometimes reproduced cropped with these excised.
284. Nic Haygarth, 'Clem Penney's Stripy Doormat', 27 November 2016, https://nichaygarth.com/index.php/2016/11/27/clem-penneys-stripy-doormat/ (retrieved 19 January 2022).
285. In his discussion of the exhibition, Maynard, however, suggests that by the time a visitor reaches this display it 'is clear that the thylacine has not survived'. Maynard, 'Tasmanian Tiger', 330.
286. I was able to trace references from the 1880s to farmers with the surname French living near Ulverstone but not to south-east Lutruwita.
287. Theophilus Jones, 'Through Tasmania', *The Mercury*, Thursday 17 July 1884, 1.
288. Ibid.
289. Ibid.
290. White, *Natural History of Selborne*, 33.
291. As Jones was born in England where ratting (the killing of rats to protect grain) was a common rural pursuit, he may have seen the slaughter of thylacines to protect sheep flocks as a comparable practice. To him, tigers were a pest akin to rats for the island's farming community. He may also have registered a connection between the killing of rats as a form of popular entertainment in England and the pleasure taken in the slaughter of thylacines. In 'Animals in Nineteenth-Century Britain', Ritvo links rat-killing with other nineteenth-century British pursuits such as cock-fighting and dog-fighting (106). For a discussion of rats as a pest to farmers in England, see Sayer, 'Rats'. In the nineteenth century, rats were also linked to criminality in the popular imagination. Jones's simile may therefore also reveal his sense of the culpability of the tigers: they were being punished by the Frenchs and other farmers for their crimes. Crofts and Hatter identify a link between rats and criminality in Charles Dickens's *Little Dorrit*. See Crofts and Hatter, '"Rats is Bogies I Tell You"', 129.
292. The original manuscript was signed and dated 22 September 1855. See Gibson, 'One More Word', 78.
293. Five sonnets have survived, which were written on the same sheets as drawings. These are now accepted as unambiguously by Raphael. Shearman, *Raphael in the Early Modern Sources*, 14. For an extended discussion of the poem in the context of Browning's views on Renaissance art, see Cheeke, 'Browning, Renaissance Painting, and the Problem of Raphael'.
294. For a thoughtful reading of the poem as Browning exploring the twin positions of artist and lover, see Gibson, 'One More Word'.
295. Browning, 'One Word More', 362.
296. Price and Bernstein, 'Whitman's Sign of Parting', 66.
297. Robert Stevenson, 'The Tasmanian Wolf', *The Mercury*, Friday 20 July 1928, 3.

Chapter 4

The Compulsion to Repeat

Erle Wilson's *Coorinna*

Tigers in the Mist

Erle Wilson's (1898–1970) novel *Coorinna*, first published in 1953, recounts the fictional story of the life of a young male thylacine, Coorinna, in the uplands of Lutruwita.[1] In 1956, *Coorinna* was listed by the journal *English* (alongside other works including the anthology *Worlds to Conquer* and a rewriting of *King Solomon's Mines*) as a prose book suitable for classroom use.[2] Although much feted upon its initial publication, *Coorinna*'s popularity has not endured (unlike other comparable animal stories such as Anna Sewell's *Black Beauty* [1877], Jack London's *The Call of the Wild* [1903] and Richard Adams's *Watership Down* [1972]).

Wilson's novel begins with an epigraph, the closing lines of the Australian writer John Sandes's poem 'The Children of the Mist', a meditation on mourning and loss for a vanished race of humans, the eponymous 'children of the mist':

> But the cloud-forms still are seeking
> For the children of the Mist.

The poem was included in a book authored by John Sandes and John Stirling (the author's name was given simply as J. S.) entitled *In Tasman's Land: Gleams and Dreams of the Great North-West*.[3] It was preceded by an illustration of the 'children of the mist', a palawa couple in a verdant landscape.[4] The man has his left arm draped over the woman's left shoulder: a gesture of possessiveness. Wilson's decision to use an extract from the poem for his epigraph implies that he viewed Sandes as a formative influence. Sandes's poem is a meditation on the extinction of the palawa. Clouds persist, continually searching for the 'children of the mist' they used to envelop. These clouds, 'white' in contrast to the 'dusky forms' of the palawa, are overdetermined and open to being read as a metaphor for Lutruwita's European invaders. In a prose section of *In Tasman's Land*, Sandes makes clear that he believed the palawa were murdered: 'The last of the Tasmanian blacks died more than a quarter of a century ago but the recollection of the sinister methods by which they were exterminated has not yet passed away.'[5] Sandes's sympathy is, however, limited as he adds that it 'is a stern and terrible doctrine ... that only the strongest shall survive'.[6] He therefore frames the extermination as somehow justified by natural selection. This vision extends to the poem, which infantilises the palawa, casting them as children, as a people who have not yet 'grown up'.

The view of Indigenous peoples as evolutionary throwbacks became prevalent in the context of Lutruwita in the late 1840s as a means to assuage local guilt at the treatment of the palawa.[7] It is a view that was fostered by Charles Darwin's observation regarding the decrease of Indigenous populations in colonised countries that: 'The varieties of man [*sic*] seem to act on each other in the same way as different species of animals – the stronger always extirpating the weaker.'[8] Darwin discusses the fate of the palawa and acknowledges that the violence they perpetrated was instigated by 'the infamous conduct of some of our countrymen'.[9] He is, however, hardly sympathetic, referring to the campaign waged against the palawa as one that has led Lutruwita to enjoy 'the great advantages of being free from a native population'.[10] John West would advance a similar view in *The History of Tasmania* where he suggested that: 'The assumption of sovereignty over a native people is justified by necessity – the law, which gives to strength the control of weakness.'[11] He noted that the distance between civilised people and 'savages' was so great that the 'savage' could never aspire to be an equal: 'his relation to that of the white can only be that of an alien, or a slave.'[12] In this choiceless choice, palawa could only accept subjugation or face fatal persecution. Shayne Breen notes that these observations imply the destruction of the palawa was a product of natural law. Breen observes that through extinction myths such as these: 'the alternative colonial narrative of a morally compromised extermination' was rejected in favour of 'a counternarrative of evolutionary unfitness.'[13] Leo Kuper has also explored how colonisation was 'linked to evolution, the conquered peoples being conceived as lower in the scale of evolution with rights and capacities by no means comparable to those of the conquerors'.[14]

Through the epigraph, *Coorinna* begins with an oblique reference to genocide and also to the notion of doomed races.[15] The connection Wilson is making between the fate of the palawa and that of the thylacine is made clear in the first chapter: 'His name was Coorinna – Sharp Fang – a name by which many fine dogs of his kind had been known to the vanished black hunters of earlier days.'[16] Coorinna (also Coorina or Corinna) was one of the Aboriginal names recorded for the thylacine in Lutruwita alongside other names including Crimererrar, Lagunta, Ka-nunnah and Loarinnah.[17] A palawa story linked to the constellations tells of how the 'corinna' (thylacine) got their stripes, which were 'drawn by the Sky Spirit Polana to honor his bravery in a fight'.[18] The translation of the name as 'sharp fang' appears to be invented, but connotes Jack London's *White Fang* (a novel about a wolfdog). The use of Coorinna as a name ensures that a memory of the palawa is preserved through the animal, just as the Latin trinomial of the Newfoundland wolf, *Canis lupus beothucus*, recalls the Beothuk.

Wilson's choice of epigraph and his decision to name the thylacine using a palawa word connect the extinction of the tiger with the perceived extinction of Lutruwita's First peoples. Nicholas Smith has noted that extinction is 'the ligature that binds Palawa and thylacine in the Australian imaginary'.[19] As part of his 1980 report on the status of the Tasmanian tiger, Steven Smith observed that 'the fate of the Aborigines had many parallels with that of the thylacine, with which they coexisted for thousands of years'.[20] As Donna Haraway has observed, the term 'endangered species' functions simultaneously 'to locate value and to evoke death and extinction in ways familiar in colonial representations of the always vanishing indigene'.[21] In the popular press, the thylacine was explicitly referred to as a 'dying race'.[22] Wilson portrays the thylacine as critically endangered, about to disappear as the palawa already have. Conjuring a vision of the 'mists of time', his chosen epigraph also figures the tiger as ancient, prehistoric, a remnant of the past persisting in the present. They are a creature out of time.[23]

The Sandes quotation is intrinsic to *Coorinna* and aids understanding of the novel, introducing both the key theme of extinction and the interrelated idea of visual disturbance or obscurity.[24] Gareth Roddy has examined the cultural meanings of mist, investigating its use as a metaphor in travelogues of the British Isles in the period 1880–1940. For Roddy, mist as it featured in guidebooks to travelling in Britain 'facilitated an imaginative escape from the modern world'.[25] In such books, 'the obscuring power of mist assisted in the escape from suburban and commercial modernity'.[26] Mist, alongside haze and twilight, comprised a prompt to the imagination, its softening of outline causing a freeing of fantasy. Through its transformative indistinctness and vagueness, it rendered temporal markers of local landscapes uncertain, fostering links with myths and legends of yore.[27] Mist, an age-old phenomenon, regular and familiar, also served to signal continuity in a period of immense social and technological change.[28] Additionally, it sometimes signalled danger.[29] In such travelogues, mist therefore possessed complex connotations and figurative potential. It provided a means to generate atmosphere and nurture emotions such as fear and wonder. Like in the travelogues, mist in *In Tasman's Land* is exploited for its potency as metaphor.

Sandes probably drew inspiration for his poem from Edwin Landseer, who produced several works titled 'Children of the Mist', including an oil on canvas work of circa 1853 that shows four stags and a hind and fawn (with two of the stags depicted as dark silhouettes enveloped by mist) and a charcoal, chalk and pastel work dated to 1866 with a stag at the centre beside two hinds and another stag and hind in the background. An 1853 crayon work uses a technique called stumping (smudging either with a rag or a stump) to shade the background and soften the forms of the deer. Alex Potts detects a particular fantasy of nature in Landseer that chimes with Darwinian ideas: 'This nature of the imagination is one in which the struggle for survival is an elemental conflict between (male) rivals is constantly being acted out in a bleak, hostile and rugged landscape'.[30] Wilson would potentially have appreciated this aspect of Landseer's oeuvres. He would also have valued the way that the notion of 'children of the mist' links to a painter who was often praised for his capacity to capture the presence and power of his nonhuman animal subjects.

Landseer's inspiration for the naming of his paintings was probably Sir Walter Scott's *A Legend of Montrose* (1819), a novel set in the Scottish Highlands in which the Children of the Mist, a band of freebooters, hide amid remote mountains and glens.[31] The novel is written in the shadow of the *Fuadaichean nan Gàidheal* (carrying away of the Gaels) or Highland Clearances, the forced eviction of tenant farmers from clan lands in the Scottish Highlands.[32] Landseer's 1866 painting is identified as a Highland setting, portraying Dunrobin in the county of Sutherland. In Scott's *A Legend of Montrose*, the mist, a feature of the Highland wilderness, is associated with freedom and isolation.[33] It is also attributed supernatural qualities, forming a protecting veil that shelters the outlaws from detection.[34] Landseer seems to primarily draw on Scott's linking of the mist with craggy, secluded terrain.

These literary and art-historical understandings of the 'Children of the Mist' feed into Sandes's poem and his vision of palawa is filtered through them. For him, Lutruwita, with its own Highlands, presented a similarly rugged and impressive landscape to the north of Scotland, although the focus in *In Tasman's Land* is on the north-western region of Lutruwita (not the island's Central Highlands). In a parallel Sandes may have recognised, comparisons have also been made between the Black Line (a military campaign designed to force palawa from their traditional lands) and the *Fuadaichean nan Gàidheal*.[35]

Sandes describes a valley 'full of mist' at Waratah and the experience of watching 'rolling mist-wreaths' creep across the landscape there.[36] Clouds wrap Partennamana/Mount Bischoff in a 'soft veil of mist'.[37] These examples of swathing terrain in mist help to conjure an image of a primordial landscape, a wild and uncivilised place, while also probably giving a faithful account of cloud cover in the area.[38] As Lutruwita has a cool temperate climate, parts of the island are frequently misty and fog regularly forms in low-lying areas such as the Huon Valley. Mist functions similarly in *Coorinna*, both describing a natural phenomenon of the uplands and acting as an unnatural figure used to lend them a primeval character: 'Tall pines loomed wraithlike in the drifting spring mists.'[39] Terrain in Lutruwita has been described by Archibald Meston as 'primeval bush'.[40] Wilson's decision to draw on Sandes's poem to introduce *Coorinna* suggests he saw potential correspondences between the persecution of the thylacine and the genocidal campaign against the palawa. In this context, the descriptions of mist offer more than a sense of realism. They are figuratively bound up with memory and the past. Early in the novel, Wilson describes a King Billy pine that had served as a familiar sight to the palawa. He writes:

> The tree had been a landmark to foraging blacks for a thousand years before Tasman was born. Through slow centuries of sun and rain and driving winds, numberless hunting parties, wandering over the misty upland of lake and mountain, had built their ephemeral bark *linés* for additional shelter close to its great bole and the surrounding earth had taken back to itself the dust of immeasurable, clean-picked bones, tossed aside from their brief camp fires.[41]

Here the tree, a point of geographical reference in the brumous terrain, stands as a temporal marker for the long history of Indigenous presence in Lutruwita, a presence it has now outlived. It was a point of continuity across generations of palawa. Now, however, they no longer visit the locale: the palawa 'had become fewer and fewer until the upland knew them no more'.[42]

Wilson's novel is one of a continuity broken, of changed times, with the fate of the King Billy pine (*Athrotaxis selaginoides*) described at the novel's beginning offering a key to understanding the broader narrative arc of the work. At the novel's opening, having fallen in a storm, the pine is 'slowly crumbling back to the earth from whence it had sprung more than a dozen centuries earlier'.[43] The King Billy is blown down in a gale. Its destruction is reminiscent of the ruin by lightning of a tree in Sandes's 'Children of the Mist': 'And the pine, by levin shattered, / Leaves an heir to take his place.' In *Coorinna*, however, the toppled evergreen has no heir. There is a finality to its loss. In this context, it is not coincidental that Wilson chooses a species of tree that is named after William Lanne, known as 'King Billy', a man who has been referred to as 'the last full-blood Tasmanian Aboriginal man to die', the male equivalent of Truganini (also known as Trucanini and Trugernanner), a woman often historically erroneously referred to as the last palawa.[44]

A thylacine has made her lair in a small cave formed by the collapsed tree. Wilson writes of the tiger emerging from this den:

> Look well at the she-wolf as she stands there in the dusk. She is one of the last of the *thylacines*, soon to disappear from the earth. A few bones and an ancestral fossil or two from Patagonia will be all that remains when she and her kind have followed into the limbo of time primitive Tasmanian man who also had once found shelter beside the ancient pine.[45]

The animal's history and fate are therefore carefully paralleled with that of the tree, which is in turn repeatedly linked with Lutruwita's (here extinct) palawa. The fate of the thylacine therefore becomes a means of repeating and seeking to work through the traumatic legacy of the genocide enacted against Lutruwita's Indigenous population. In a sense then, the animal is burdened with representing the fate of Others, a stand-in for the human. A similar approach may be at work in Julia Leigh's novel *The Hunter*, in which a hunter, known only as M, links the fate of the thylacine and the palawa through De Witt Island, which he believes was proposed as a sanctuary for both in the past.[46] The connection M makes is explored by Kylie Crane and, subsequently, Alex Philp and Fiona Probyn-Rapsey.[47]

The limbo of time referred to in the description of the tree is later figured as brume, with the 'last of the aborigines [having] passed into the mist'.[48] Mist therefore connotes disappearance, extinction, oblivion. All that seemingly shores against this obscurity are physical remains. In the case of the thylacine, bones from Lutruwita and fossils from Patagonia are the entirety of what will persist post-extinction.[49] Although privileging material remains as what is most precious and as what will provide an enduring monument to the tiger, Wilson must also be aware that his novel performs a similar memorial function. This chapter uses close reading to bring out the distinctiveness of Wilson's *Coorinna* as a memorial to the thylacine and an effort to work through Lutruwita's history of human and animal genocide. The novel tells the life story of Coorinna, a male thylacine living in the Central Highlands.[50] It moves chronologically from the tiger's birth to orphanhood, to his starting his own family, to the loss of his mate and cub and, finally, to his eventual demise, killed by a vengeful shepherd. The narrative features several fights with other native fauna and skirmishes with humans and their dogs. It is action-filled yet also includes lyrical moments, extended descriptions of the wilderness and of wildlife.

Human Genocide and Racism

Like many of Wilson's writings, the novel is aimed at a young audience. In 1939, for instance, Wilson published the children's book *The Green Frog and Other Stories*, which is a collection of short stories about animals from around the world, with a particular emphasis on the fauna of Australia and Canada.[51] Another collection, *The Grey Seals and Other Stories*, was published the following year containing stories with a more even geographical spread.[52] *Churinga Tales*, an anthology of First Nations Australian stories relating to the Dreamtime as (re)told by Wilson, was published in 1950. The stories are narrated to a young settler colonial boy by his uncle, who possesses a 'churinga' (a *tywerrenge* or tjurunga stone).[53] These stones are sacred objects owned by individuals or, more rarely, groups that are used in ceremonies and the idea of a White settler appropriating one and employing it for storytelling purposes is shocking.[54] The book, however, was written at a time when knowledge and sensitivity towards First Nations Australian culture was minimal. Wilson likely conceived of *Churinga Tales* as respectful. The anthology, illustrated by Sally Medworth, was commended by the Children's Book of the Year Award in 1951.[55]

Coorinna seems to build on 'The Story of Tonack's People', the last of the *Churinga Tales*. Tonack is one of the male names listed in Joseph Milligan's 'Vocabulary of Dialects of Aboriginal Tribes of Tasmania'.[56] The story tells of a people of the Dreamtime who move from their usual home because of a food shortage. To the far south, they find a land where penguins are plentiful and build shelters there.[57] After a time, winter arrives, and the people are cold and seek to return to the north but the sea cuts them off. They call upon their chief, Tonack, for succour. He rescues a wombat, Rowitta, who gifts him fire in gratitude: '"I have given you Ballawinna-Theeri, the Red-Fire. . . . Carry him with you and feed him well with dry wood day and night and you need never be cold again."'[58] The uncle explains at the end of the tale that Tonack's people 'lived on in their new land – which *we* now call Tasmania – keeping themselves warm in winter with the Red-Fire and living their simple, peaceful lives, unchanging, age after age, until the white men came.' The boy asks what happened after that:

> '*That* is a story our Churinga can't tell you,' his uncle replied, and the little boy thought his voice sounded very sad. 'But some day you may know.'[59]

This ending has been interpreted as evidence of a 'less cliched interpretation of race relations' than was offered by adult novels of roughly the same period, such as Allan Aldous's *Doctor with Wings*.[60] Certainly 'The Story of Tonack's People' alludes to the violence that accompanied European invasion and the human genocide in Lutruwita. *Coorinna* can be read as offering the story that is untellable in *Churinga Tales*.

Wilson's novel performs a similar role to Ernest Thompson Seton's nature writing. Brian Johnson and Manina Jones have traced how Seton's animal stories embody settler colonial subtexts regarding Indigenous peoples. Johnson suggests that when Seton is read in a 'national postcolonial context' the representations of animal victims in his stories sometimes 'evoke the indigenizing proleptic allegories of "doomed races"'.[61] Jones points out that in Seton's stories, the animals – like Indigenous peoples – are casualties of Euro-American territorial expansion. It is not insignificant that Lobo 'poses a threat to pastoral land use by Euro-American populations and must be eradicated in order for the frontier to become a safely domestic grazing space'.[62] In his preface to Tekahionwake's (Emily Pauline Johnson's) book *The Shagganappi*, Seton suggests that the collection of Native American stories tells of 'a proud race, conscious that it has been crushed by numbers, that its day is over and its heritage gone forever'.[63] Tekahionwake is 'the sad historian of her own heroic race'.[64] These kinds of sentiments find echoes in the animal stories that Seton writes, with tales such as 'Redruff, the Story of the Don Valley Partridge' providing 'a displaced version of the figure of the dying, often last, "Red Man"'.[65] Johnson has explored how the same phenomenon is at work in novels about wolves by Seton, Farley Mowat and Charles Roberts. He writes of the role of the wolf as allegory, how the animal is put to work to mitigate 'settler-invader guilt by legitimizing the displacement and dispossession of First Nations people'.[66]

Some of Wilson's earlier fiction unambiguously embodies sexist and racist attitudes that shaped his settler colonial world view. Woman as nature registers in the short story of 'Marava', a 'light-skinned Mopu girl' living with a tribe from Beni in Papua.[67] At one point, Mabu looks at Marava, whose 'gold-brown breasts rose and fell as calmly as the young wind in the swaying palms'.[68] Here Marava's body is aligned with local flora.[69] Wilson shows that she is bare-chested to emphasise the tropical locale, her breasts employed as 'colonial exotica' to signify subjugation.[70] This textual imagery is reminiscent of visual depictions of the tropics such as Paul Gauguin's paintings of Pōrīnetia/Polynesia.[71] In Gauguin's *Tahi-*

tian Women with Mango Fruits, 'woman and fruit are similarly placed as naturally available for the gratification of men's needs and desires'.[72]

'Marava' is also disturbing for the narrative template of the lighter-skinned saviour (Marava's paler skin is twice mentioned in the story), who rids the village of its greedy sorcerer Jorega through exploiting tribal superstitions to her own ends, employing the subterfuge of a fever-cleansing ritual to murder Jorega. Her lack of fear of the sorcerer stems from what she has learnt from 'god-book men' (missionaries) and the 'policimani taubada' (chief of police), namely that sorcery is 'all gammon' or pretend.[73] The source of Jorega's downfall is his appropriation of a knife; the impetus thus centres upon settler technology. As an outsider, Marava is not bound by the cultural mores of the tribe. Her character can be read as a 'White saviour' masquerading as an Indigenous Papuan.[74] In Wilson's story 'Jungle Record', also set in Papua New Guinea, a group of Papuan villagers are described as 'our primitive friends'.[75] One of the characters in the story, an anthropologist from the 'American Institute of Ethnology', Cornelius Stoner, refers to the villagers as standing and looking at him 'like a herd of scared antelope', the people cast as akin to nonhuman animals.[76]

Wilson's later writings, by contrast, sometimes offer a more thoughtful consideration of Indigenous peoples.[77] In *Minado* (first published in 1955), a novel about a wolverine in the wilds of northern Quebec, the First Nations characters, two Cree men, Lone Cloud and Little Otter, are treated sympathetically, contrasted positively with the European trapper Mort (whose name in francophone Québec would readily be understood as signalling death).[78] The men, described as '[m]embers of a conquered race, second class citizens in their own land', are shown to possess a deep understanding of the First wildlife of the region.[79] It was not until the 1970s in Canada that Indigenous knowledge of wildlife and how to oversee it began to be broadly recognised and respected.[80] Wilson credits the Cree with a profound awareness of sustainability: 'They trapped and shot from inescapable necessity, but never more than they could use and in the belief that their victims knew and understood their need.'[81] *Minado* does not offer the kind of settler-affirming narrative Johnson identifies in stories such as Roberts's 'Wolf! Wolf!', in which the destruction of a wolf pack by a backwoodsman called Purdie constructs the man as a member of an evolutionary superior species. The wolverine in *Minado* ultimately sends Mort, a seasoned trapper, packing. It is a fine example of the animal hero as *cum laude*, hyper-intelligent.[82]

In *Coorinna*, as already discussed, Lutruwita's Indigenous population is figured as extinct and described only fleetingly, with no development of individual palawa characters. The novel does not make direct reference to human genocide either. The reason for the disappearance of the palawa from the uplands is left unspoken, implicit. Wilson also describes them as primitive, yet his perspective seems to differ from that of contemporaries such as John Murphy, who referred to the palawa as being 'exterminated like noxious lower animals by the British convict settlers and colonists'.[83] For Murphy, the palawa are a crude and uncivilised people: 'nearer to the instinctive animal than recent and civilized man.'[84] Wilson's fantasying of the palawa positions them closer to the 'noble savage', a good people living an idyllic existence. His more sympathetic vision of Australia's First Nations peoples compared to peoples from Papua New Guinea may stem from their being closer to home to the Australian-based author, less readily shaped into negative Others.[85] Wilson writes of the 'primitive Tasmanian man who [like the thylacine] had once found shelter beside the ancient pine and wandered freely by deep Saint Clair – his own dreaming Leeawulena'.[86] Leeawulena is the palawa name for Lake St. Clair.[87] Here Wilson employs the settler place name and the palawa one to signal how a single locale exists as two distinct times and places. Leeawulena is a place of liberty and reverie.

In *In Tasman's Lands*, Sandes refers to naming as a form of memorial, suggesting of the river port of Teepookana that its name, meaning kingfisher, was 'was not the indigenous name of the place, but was given to it long after the last of the blacks had vanished, by some white antiquary, who decided to preserve the liquid syllables of the lost language, for picturesque reasons although the people who once used the word had disappeared'.[88] He notes that several of 'these soft, musical Tasmanian names are to be met with in this part of the country', mentioning Rinadena [Rinadeena] and Lowana.[89] Wilson's reference to Leeawulena, a name that was no longer in common usage (although well known), seems qualitatively different to Sandes's examples. The name is not employed for quaintness. It signals, rather, Lutruwita's Indigenous past. It is a past that persists in the present through the thylacine, an animal that, like the King Billy pine, links back to pre-contact times. Wilson uses the name Leeawulena more than once in *Coorinna*, which suggests that the place forms part of the thylacine's world. Coorinna does not live in the vicinity of Lake St. Clair but Leeawulena. He is a remnant of what Wilson constructs as a bygone age.

It is therefore possible to read the story of Coorinna's fate as allegorical, providing a means to explore the campaign of extermination perpetrated against the palawa. His death obliquely figures their destruction. By 1953, there had already been remarkably balanced (at least for the time) historical accounts of genocide in Lutruwita, but *Coorinna* was a work of popular fiction intended for a wide audience and allegory may therefore have seemed the best way to broach the toxic legacy of settler colonial violence without alienating publishers and readers. It is unlikely that an explicit engagement with genocide in Lutruwita would have been deemed suitable for an adolescent audience.[90] As it was, *Coorinna* received praise and went on to be translated into German (1954), Swedish (1956), French (1973) and, in abbreviated form, Spanish (1975).

If the allegorical reading is accepted then the farmer who pursues Coorinna in the narrative symbolises the settler colonists. His having named his dog after an offensive term for Black people can be interpreted as signalling the man's racism.[91] The racial slur has historically sometimes been directed at First Nations Australians.[92] The dog in such a reading can also be understood as the animal equivalent of a person deprived of their human agency, enslaved.[93] The ignoble dog, forced into servitude, functions as the opposite to the free thylacine.[94] The thylacine, adopting this reading, stands for the 'noble savage', the pre-contact palawa uncorrupted by European influence. Ultimately, however, Coorinna is hunted down and killed, potentially suggesting an inevitability to the destruction of the palawa. This allegorical dimension may have registered with an anonymous reviewer writing in the *West Australian* who mentions the 'white shepherd' and reads the novel as one in which 'white men's cunning and guns finally prevail'.[95] The reviewer alludes to the structural racism informing the novel and Wilson's tacit embrace of the logic of the 'doomed race'. Another contemporary reviewer clearly identified the presence of a Darwinian theme of the survival of the fittest, suggesting that the denouement of the novel registers a 'pitiless logic' in which 'the universal, natural law which condemns the weaker to be destroyed by the more powerful cannot be evaded'.[96]

Tigers in the Midst

Reading *Coorinna* simply or solely as a novel about human genocide, however, ignores the centrality of thylacines to the narrative. The novel is overdetermined and another theme is clearly that of animal genocide. Wilson goes to considerable lengths to flesh out his central character, *Coorinna's* titular thylacine. This effort to render the life of the tiger as faithfully as possible speaks to a wish to do justice to the marsupial rather than simply to employ them as a hollow trope. Wilson openly signals his interest in conservation in the biographical note he provides for *Coorinna*, observing that he has served as a volunteer ranger under the auspices of the New South Wales Fauna Protection Panel and describing himself as neither a naturalist, a hunter or a fisherperson but rather simply as someone whose 'interest lies in the protection of wildlife'.[97] He also declares that his wish is to 'observe and conserve rather than to try destroy life', which is far from a neutral statement in the context of a novel about the thylacine.[98] *Coorinna* therefore seems qualitatively different from his short stories. It is a work of committed literature inspired by a desire to draw attention to biodiversity loss and encourage a change in attitude to the environment.

The prose in *Coorinna* is also more worked and elaborate than in the short stories, more poetic, with repeated use made of alliteration and assonance. Wilson was, indeed, described as a 'poetic naturalist' in a contemporary review of the novel.[99] He writes at one point of a realm of air and sunlight: 'In that world, the minute florets that be-starred the moss on the roots of the pines were repeated in a multitude of forms and sizes and hues: from the tiny cream blossoms of the mountain grasstree to the singing scarlet of the waratah and the billowing white of leatherwoods.'[100] Here repeated forms in nature are embodied through a highly sibilant passage of prose, through melodious description. The euphony of this sentence is far from unique. Later, a similar effect manifests in another instance in which Wilson seeks to conjure something of the chromatics of the local flora. He viewed the colours of plant life in poetic terms, writing: 'On peak and plateau, in valley and gully and glen, blossom gave place to berries, purple, blue, crimson and white, that hung suspended in lyric beauty amongst the greenery, like captured notes of song from the throats of vanished birds.'[101] Through the rhythms of the prose, the natural world is rendered lilting, an idealised space, filled with picturesque colours and forms. There is no such lyricism present in his descriptions of the farm and farmland. Those places are postlapsarian, rendered in ugly and stilted terms.

The exquisiteness of Wilson's portrayal of the wilderness as embodied by his skilfully crafted prose has been noted. The *Sun-Herald*'s reviewer, S. H., suggested that the 'beauty of the background' helped to emphasise the ugliness of Coorinna's fate.[102] More than this, I would suggest that the beauty serves to capture what to Wilson's eyes was a vanishing Lutruwita, the cruel paradise he viewed as Coorinna's home seen as steadily shrinking, dying. Wilson's vision of a picturesque, if perilous, countryside is at odds with other descriptions of Lutruwita. Inspired by the King William Range, Michael Sharland, for instance, writes: 'There is something sinister in the vastness of the ranges, the peaks of which rise like the teeth of some gigantic cross-cut saw, scarred and broken by time. He may well marvel at the sculptural forces of Nature which have produced such a wilderness of rock, crag and valley.'[103]

Sharland's own description, which forms part of an account of an expedition to locate thylacines, also alludes to the beauty of his milieu, the mountain ranges perceived as sculpted, as artistic forms, but his choice of the simile of the saw to refer to them suggests that he equally registered his surrounds as resources.[104] Cross-cut saws were a common tool and regularly used in the timber industry. Writing under his journalistic pseudonym of 'Peregrine', Sharland made reference to the grave marker of the 'piner' (lumberjack) John Stannard that was located on the Jane River; the marker took the form of a wooden cross made from pine logs 'surmounted by two cross-cut saws'.[105] The description forms part of a broader meditation on the history and effects of the timber industry on Lutruwita, including its decimation of the Huon pine (*Lagarostrobos franklinii*). Sharland writes that the pine 'has become as rare as the gold and other precious minerals for which prospectors are for ever seeking in the wilder portions of the West'. Sharland's seeing the ranges as saw teeth therefore speaks to his fears about human exploitation of Lutruwita's wilderness.[106] Writing in 1941, he again refers to the cross-cut saw, this time in an article decrying how the best of the forests around Nipaluna/Hobart were being destroyed by timber-cutting, driving away birdlife. In that article, he recalls a black cockatoo (*Zanda funerea*) imitating the sound of a cross-cut saw: 'a strange note which might have caused anybody to imagine, as I did, that timber-workers were engaged in carving up a fallen tree.'[107] This reference reinforces the link in Sharland's mind between the saw and the lumber industry and conveys the impact of logging on the forest as soundscape. When Sharland heard the cockatoo imitating sawing, he noted a bird imitating the noise of the destruction of their own habitat. Cockatoos are known to imitate sounds audible in their surrounds, and in Lutruwita's forests after colonisation those sounds changed to include the cross-cut saw, a tool that produces a metallic humming when used skilfully.[108] Sharland represents the wilderness as already tainted visually and acoustically, made ugly by settler colonial technology.

Both Sharland and Wilson seem to recognise that destruction of habitat leads to the disappearance of species. In this context, Wilson's poetic renderings of Lutruwita's wilderness are paeans to a vanishing ecosystem. The pastoral industry contributed significantly to this destruction. Ecosystem instability in Lutruwita has been linked to transhumance grazing (the seasonal movement of livestock).[109] Overgrazing had a major impact causing erosion and pasture deterioration.[110] The burgeoning rabbit population from the1920s onwards was also detrimental. Animal genocide against the rabbits, their 'wholesale extermination', was prevented because of the value of their skins, which provided an important secondary income to shepherds.[111]

By the time *Coorinna* was published, the last captive thylacine had been dead for over fifteen years and there had already been several unsuccessful efforts to find further thylacines, including the 1938 expedition Sharland accompanied. Wilson was therefore writing about an animal he could no longer witness first-hand. The novel nevertheless hints that he had witnessed tigers in the flesh. At one point, the author seems to insert himself into the narrative. In Chapter 7, a birdwatcher makes an appearance; he is 'no naturalist but he had read a page or two of Nature's book'.[112] The man, given no name, comes across Coorinna and Loongana as they court each other: 'The watcher remained stone-still as he realised that he was watching a wilderness courtship that must rarely – might never before – have been seen by man [*sic*]; that might never be seen again.'[113] Given that Wilson denies his own credentials as a naturalist in his author biography, it seems he is encouraging his readers to connect the birdwatcher with himself and seeking to give the novel the imprimatur of truth to nature.

The thylacine as constructed by Wilson manifests many of the attributes commonly ascribed to the species, including 'staying power' rather than speed.[114] He also employs the familiar trope of the thylacine as sheep killer, writing at one point that from a lair on the edge of a high valley, Coorinna would stand 'looking out, still-eyed, across the valley to the distant sheep range'.[115] Roger Vernam's jacket illustration for the 1954 US edition of *Coorinna* portrays the thylacine from behind, showing him clambering onto a rocky ledge that overlooks plains or a valley, with mountains shrouded in mist visible in the far distance.[116] Vernam's jacket illustration is reminiscent of Herbert John ('H. J.') King's circa 1940 *Tasmanian Marsupial Wolf*, which, as discussed in the preceding chapter, superimposes a photograph of a captive thylacine that he took circa 1933 at Hobart Zoo over one of his photographs of Warloundigerler/Cradle Mountain.[117]

Like most animals in fiction, Coorinna leads a sanitised existence. He hunts and plays and sleeps but never pisses or shits.[118] The thylacine's sex life is only hinted at, with courtship described but not copulation; Coorinna's naturally rigid tail seems to function as a displaced signifier for his erect penis.[119] When the tiger first encounters Loongana, who will become his mate, he goes 'forward eagerly, his thin tail straight as a lance-wood wand'.[120] Loongana's initial response to Coorinna's entreaties is to nip his shoulder 'shrewishly'.[121] Figuring the female tiger as 'shrewish' casts her as Katherina to Coorinna's Petruchio; he must court and tame her.[122] Wilson's sexism, his personal morals and his idea of the social mores of his intended readership therefore structure the kind of narrative he provides. This includes a negative perception of animal domestication.

In *Minado*, Wilson describes Mort's hunting dogs as 'half-wild huskies' whose kind had, for generations, 'been bullied and fondled, used and abused by man; their birthright of freedom taken, until now they were dependent for happiness on man's presence'.[123] Domestication is also criticised in *Coorinna*, where Wilson writes of how 'man's herding of sheep in large numbers through the centuries' has 'changed them from individuals into brainless automata'.[124] The mob, in this reading, is an unnatural phenomenon that stifles distinctiveness. A sheep, Bunty, that has left the mob to live as a 'hermit' begins to develop her individuality, to become emancipated.[125] Wilson's views echo those of earlier nature writers such as Charles Dudley Warner, who describes a pack of hounds on the scent of the doe as 'stupids', casting the dogs as machine-like in their behaviour, keeping trail as they are supposed to but losing time because they are not sharp enough to register when the deer has doubled (turned and retraced her path, then leapt to the left or right and headed off in a different direction).[126] Doubling creates a trail of strong scent that leads nowhere, bringing the dogs to a temporary standstill. Here, domestication seems to suppress initiative. Unlike the doe, the hunting dogs are not freethinkers. Frank Dalby Davison's *Man-Shy* also contrasts the freedom of wild cattle or 'scrubbers' with the incarceration of farm cattle. At one point, the leader of the 'scrubbers' is cast as Boudica to the Romans of the farmhands who try to capture her.[127] Coorinna also represents freedom, one which Wilson, like Warner, associates with wild animals. He gifts the thylacine a personality and an agency denied the farm animals such as the shepherd's dog and the domestic sheep.

In this sense, despite readings of the novel that identify a Darwinian subtext, *Coorinna* can also be interpreted as engaging in what Robert MacDonald has referred to as 'a popular revolt against Darwinian determinism'.[128] This would position Wilson alongside earlier authors such as William J. Long, Charles Roberts and Ernest Thompson Seton. MacDonald traces how writers such as Roberts and Seton, the so-called 'nature fakers', provided an alternative to the 'awful amorality of Darwinian nature'.[129] Through foregrounding nonhuman animal learning, reasoning and morality, these authors emphasised kinship across species, the human read as on a continuum with like-minded animals such as the crow, the fox, the rabbit and the wolf (the animals are predominantly birds and mammals). Roberts and Seton did not reject Darwinism outright, they accepted the merciless kill-or-be-killed character of natural law, yet they affirmed animal wisdom and mores.[130] As will become clear, this also seems to be Wilson's position.

According nonhuman animals a personality has merits in terms of drawing attention to behavioural complexity and fostering empathy, but is also problematic. If thylacines possessed something akin to personalities (as is very probable), their unique dispositions likely exceeded human capacity for understanding. In this context, André Krebber's and Mieke Roscher's remarks on animal biographies are insightful, laying out some of the difficulties that accompany granting biographies to animals, accepting that nonhuman animals are unlikely to 'construct their lives in (auto) biographical terms'.[131] They believe that efforts to capture individuality do not necessitate efforts to reconstruct feelings or infer intentions. For them, animal biographies seek to capture the human experience of the animal and form an acceptation of the impossibility of looking and knowing and telling a life story as nonhuman animals. The value of animal biography, even if bound by necessity to a human perspective, is its capacity to resist homogenising tendencies in relation to animals, working 'with the hope of breaking the mold of identity that lumps together all animals as principally the same'.[132] The mainstay of biography is 'to *give form* to the other', with that form derived from the human animal that is the writer.

The animal in such a conception is shaped by way of a few chosen words, words that offer one possible outline at the expense of other unrealised possibilities. The first description of Coorinna provides a good example of this compact delimiting of the animal subject:

> Already many days had passed since the hour of his birth when, naked and undeveloped, blind and pink and without help, with the marvellous instinct of marsupial young, he had nosed his way through his mother's belly-fur to the warm, dark haven of her pouch. Now his own coat was soft and yellow and close and his limbs had grown until he could stand and puppyishly worry the paw that gently turned him when he would have blundered too far into the fern. He looked at the world through newly opened eyes and snarled fearlessly at every moving thing.[133]

Coorinna is the age of a pup here but the description of him as 'puppyish' also seems to introduce a canine or lupine dimension to the depiction.[134] His response to perceptual stimuli is to snarl, a vocalisation often interpreted as suggesting a propensity to aggression and reinforcing his perceived canine or lupine character.[135] Additionally, Coorinna worrying one of his mother's paws, seizing it between his teeth, also attributes wolflike qualities to him, as wolves are worriers. In Howard Pyle's story 'The Salem Wolf', for example, a wolf 'worries' a prone man 'this way and that' like 'a wicked dog worries at a sheep'.[136] When Wilson later describes Coorinna as having become 'a raving maniac' the language is reminiscent of how he portrays wolves in *Minado*.[137] His descriptions therefore serve to reinforce the image of the 'marsupial wolf', a vision of the thylacine as lupine, with all the negative connotations that identification has often held, connotations which served to facilitate their destruction.[138] In reality, although there may be cranial resemblances there are major differences in anatomy and behaviour that render comparisons with wolves misguided. Like wolves, for instance, thylacines sometimes engaged in cooperative hunting (making them unique among carnivorous marsupials).[139] Yet unlike wolves, when not hunting alone, thylacines likely only hunted in pairs, not packs.

The trope of the renegade wolf, a lone wolf using cunning to outwit human adversaries, feeds into Wilson's construction of the characters of both Coorinna and Minado. Renegade wolves are known for their 'outstanding cunning', a trait that well summarises Wilson's fictional thylacine and wolverine.[140] Stanley Young suggests that the deeds perpetrated by these renegade wolves 'were of a nature calling for prompt retribution according to the views of their human neighbours'.[141] In Wilson's fiction, Coorinna's sheep predation and Minado's thieving from Mort's traplines form the prompts for such reprisals. The shepherd who forms Coorinna's nemesis refers to him as 'One-ear', a name derived from the tiger having half an ear torn away by a rifle bullet.[142] This is the name he is given by humans. It is clear that the name 'Coorinna' is recognised only by native fauna in the narrative. Through the alternative names, Wilson thus signals the existence of nominally different worlds, that of the nonhuman and the human animal. Coorinna lives within a world of experiences and needs that is alien to that of One-ear, the shepherd's vision of a thylacine, a cunning sheep-slayer. Historically, renegade wolves have often been named after 'identifying scars or defects'.[143] Through naming Coorinna 'One-ear,' Wilson follows this tradition, rendering the thylacine akin to wolves like 'Crip', 'Old Lefty', 'Three-toes of the Apishapa' or 'Three-toes of Harding County'.[144] No doubt the Newfoundland wolf who lost his left foreleg that I discuss in Chapter 6 would have attracted a similar moniker if he had evaded hunters for longer.

Figuring wolves as renegades casts them as outlaws, ascribing them a degree of criminality. In *Wolf*, Garry Marvin details some of the atrocities perpetrated against trapped and captured wolves, treated like 'wanted criminals'. Although writing about a different species, his discussion gives a sense of how thylacines were likely handled when captured alive by farmers.[145] The animals, when kept alive, became like outlaws brought to justice, rendered spectacle. The shift in human punishment away from spectacle and towards insidious processes of control identified by Michel Foucault as manifesting in Europe in the late eighteenth century was not straightforwardly replicated in relation to the treatment of 'nonhuman animal criminals'.[146] There is a long European history (stretching back to mediaeval times) of nonhuman animals being subject to legal processes, with some condemned to death by public execution. The thylacine was never put on trial in this way, never accorded 'due process'. They were, however, still judged and condemned, with the dead tigers that were nailed to trees forming a kind of spectacle of justice echoing spectacular punishments like the public execution.[147] The thylacines that were captured were put on show, subdued and controlled. Coorinna avoids this fate. Having him survive at the end, even if condemned to a future in captivity, would undo the extinction narrative underpinning the novel. Wilson's casting Coorinna in the role of renegade indirectly positions the shepherd as a guardian of moral rectitude, bringing the rebellious 'wolf' who challenges human authority to heel.[148] The shepherd can be read as a synecdoche for the farming community more broadly and therefore *Coorinna*, in part, forms an apologia for the actions of many settler colonials towards the animal.

In his chapter on the 'Gray-wolf or Timber-wolf' in *Lives of Game Animals*, Ernest Thompson Seton includes a section on 'Famous Wolves'.[149] Seton suggests that among wolves there are 'extraordinary individuals', 'heroes', whose lives have always interested him and that he has endeavoured to record.[150] Their celebrity status is accorded because of their success in eluding capture and destruction (in this, they fit the profile of the renegade, although Seton does not refer to them as such).[151] Specific thylacines never seem to have achieved this level of individualisation in local communities. The wolves Seton describes are frequently named after the places where they roamed such as 'The White Wolf of Pine Ridge' or the 'The Winnipeg Wolf'.[152] Sometimes, in death, these 'famous wolves' have their pelt taken or are mounted. Seton is well aware that tall tales about wolves have caused 'a wholly wrong picture' of the animal to emerge and argues that the old picture of the wolf as ferocious, treacherous, rapacious, strong and cowardly 'has little resemblance to the Wolf of fact'.[153] By contrast, he contends that wolves are shy, wary and wise, but adds that there is no such thing as a typical wolf any more than there is a typical human.[154] Later, drawing on the insights of Barry Lopez, Sean Robisch similarly affirms that in studying wolves, 'isolated individual behaviour needs to be given as much credence as repeated, essentializing behavior'.[155] He contests the notion of the average animal, suggesting that '[a]veraging animals, like averaging books or people, implies a context that doesn't exist'.[156]

The 'individual variation' of wolves means that generalising about them is difficult, although Seton does state that *all* wolves are 'highly intelligent', their 'modern education' the result of 'our continual warfare against them'.[157] Through declaring 'bitter, exterminatory war on wolves', humankind has 'evolved a creature that for long has defied us'.[158] Here Seton sees what I am referring to as animal genocide as causing natural selection to come into play, with only the smartest 'varieties' of wolf surviving. Darwin's understanding of natural selection focussed more on benefits derived from morphology, physical vigour and instinct than mental acuity. He did, however, acknowledge and reflect at length on animal intelligence.[159] Darwin held out the possibility that the varied and variable conditions of an organism's life could extend beyond naturogenic environmental transformations (such as those registered by the fossil record) to encompass changes inspired by human actions, anthropogenic environmental transformations.[160] For Seton, animal genocide, or 'exterminatory war' as he characterises it, seems to constitute such a transformation, a change in the living conditions of many wolves that has benefitted the most intelligent among them.[161] His stories therefore reflect the impact of genocide but may also be motivated by a desire to undo the rationales that rendered it possible, presenting alternative, more sympathetic visions of wolves that do not conform with dominant stereotypes.

Forging Nature

The idea of the 'renegade wolf' clearly informs how Wilson shapes Coorinna's character, offering an apology for the shepherd's actions (and by extension for the farming community more broadly). The figure of the renegade animal (either historical or fictional) may, nonetheless, hold positive attributes as it involves individualisation. In *Coorinna*, the titular thylacine rises above the general to become some marsupial wolf rather than any marsupial wolf. It is possible to interpret such individualising as contributing to naturalism. As Robisch has affirmed in the context of wolves: 'Species are made of individuals.'[162] Animal individuality was something that Darwin recognised as he makes clear in *The Descent of Man* when discussing animal emotions. He notes that '[c]ourage and timidity are extremely variable qualities in the individuals of the same species, as is plainly seen in our dogs'.[163] These differences, while inherited, distinguish each dog from their kin. He also observes that dogs display a sense of humour rather than simple playfulness.[164] Additionally, Darwin credits some animals with 'imagination' and 'reason'. He states that animals 'may constantly be seen to pause, deliberate and resolve'.[165] The emotions, imaginative capacity and powers of reasoning Darwin accords certain animals individuate them (the uniqueness of each member of a species is necessary for natural selection) and also grant them a measure of individuality (a distinct character), yet popular writings about animals that followed this line of thinking were much critiqued in the early twentieth century as lacking veracity.

The nature fakers controversy was generated by nature writing such as Seton's *Wild Animals I Have Known*, in which animals were attributed individuality and written about sympathetically. William Long suggested that natural history writing required 'intense human sympathy' to render the inner lives of animals luminous. Animals that are written about without sympathy, dispassionately, 'are little better than stuffed specimens, and their actions the meaningless dance of shadows across the mouth of Plato's cave'.[166] Here then mere factual description is viewed as meaningless. It is akin to taxidermy as Long conceives it, offering up only stilted and lifeless animals. Seton's stated aim in *Wild Animals I Have Known* was to give 'the real personality of the individual'.[167] He therefore sought to escape the generalising tendencies of 'a casual and hostile human eye'.[168] Seton was open about sometimes leaving 'the strict line of historical truth' but never portrayed animals that were not 'real characters'.[169] He also acknowledged that the key literary genre underpinning his stories was that of tragedy because the 'life of a wild animal *always has a tragic end*'.[170] This implicitly cast many of the animals as heroes, flawed but admirable, often victims of circumstances beyond their control. As with many tragic dramas, Seton's animal biographies as tragedies also held a moral lesson: 'we and the beasts are kin.'[171] This vision of humans and animals existing on a continuum echoed many of Darwin's assertions in *Descent of Man*, with animals differing by degrees rather than outrightly.

Most of the animals that feature in Wilson's novels make only brief appearances, their lives established through short sketches and then rapidly extinguished. A potted biography is provided to make their ensuing death at the hands of a predator more impactful. The repeated birthing and killing that is practised in short order is unnerving. It is as if Wilson has taken Seton's observation about 'tragic endings' to animal lives too much to heart. The many fights in his novels may demonstrate the influence of Jack London.[172] Spectacles of animal violence are still common in popular cultural forms such as the animal documentary and support a 'narrow Darwinism that interprets survival of the fittest literally'.[173] Unlike most of the animal dramatis personae in Wilson's writings, Coorinna and Minado are round characters, their personalities developing across the narrative. They have minds and emotions and are capable of surprising the reader through their actions and behaviours.[174] To understand the complexity of Coorinna as a character in the nature writing tradition, it is necessary to engage briefly with the history of the genre as it developed in Turtle Island and with the nature fakers controversy. Like authors such as Seton who pioneered naturalistic animal fiction that drew on their direct observation of wild animals and presented animal characters with their own perspectives, Wilson also stakes a claim to being true to nature in *Coorinna* and to seeing the world as through the eyes of a thylacine. He invites his readers to believe he is presenting fact dressed as fiction.

Notwithstanding their efforts at naturalism, Seton's work and that of some of his contemporaries attracted opprobrium for providing unrealistic portrayals of the natural world. The naturalist John Burroughs suggested of Seton that such 'dogs, wolves, foxes, rabbits, mustangs, crows, as he has known, no other person in the world has ever known'.[175] He argued of Seton's animal biographies that 'their groundwork was fact and their finish was fiction'.[176] In the *Atlantic Monthly* essay 'Real and Sham Natural History', where he voiced these criticisms, Burroughs also discussed the nature writers William Davenport Hulbert, William Long and Charles Roberts, finding only Long and Seton romancers who were guilty of constructing sham nature. His attacks were enthusiastically supported by Theodore Roosevelt who would subsequently launch his own assault on Long, to which Long responded: 'every time Mr. Roosevelt gets near the heart of a wild thing he invariably puts a bullet through it.'[177]

Long was subsequently also taken to task for his book *Northern Trails*, which tells the story of a wolf in Newfoundland. In an article published in *Everybody's Magazine*, several naturalists attacked Long's credentials and veracity, including George Shiras, a hunter and animal photographer, who claimed that on a long trip through 'the wild, northwestern portion of Newfoundland' he saw no signs of wolves and that locals 'had likewise been unable to find any trace of these animals in recent years'.[178] Shiras makes this claim repeatedly in a stinging critique of Long's book, which he describes as 'frenzied fiction, as rare as it is raw'.[179] Wilson's knowledge of the natural world similarly sometimes occasioned criticism as his correspondence with Miles Franklin attests. She was 'almost floored' by a reference Wilson made to 'swimming wombats'.[180] Writing as Peregrine in *The Mercury*, Sharland, an ornithologist, also criticised Wilson for populating the environs of Lake St Clair with non-resident species such as the egret, kookaburra, peewit and sandpiper.[181]

Burroughs laid the blame for sham writing, in part, on the corrupting influence of market forces, financial reward encouraging sensationalist prose that abandoned fact for more marketable fancy. He appreciated nature writing that did not put in 'too much sentiment, too much literature'.[182] Writing that was excessively sentimental risked crossing into make-believe and deceiving readers with untruths. Literary style was not always viewed negatively. He praised the ornithologist Bradford Torrey's writings, for example, for their 'rare delicacy, sweetness, and charm'.[183] This he puts down to Torrey's nature, which is 'birdlike'.[184] Writing he praises therefore emulates his vision of the animals being represented, with Torrey's prose conforming to his vision of the avian. To succeed, Torrey has *become* bird, putting his humanness to one side. Being birdlike for Burroughs means being shy, gentle and alert. Shyness and gentleness would seem anthropoid attributes, demonstrating that when Burroughs discusses animals, he remains all too human, falling prey to the same failings he identifies in others.

Burroughs also praises Charles Dudley Warner for his story 'A-Hunting of the Deer', which is written, in part, from the deer's perspective.[185] He suggests that the story is 'true in the real sense of the word'.[186] Warner's tale, set in the Adirondacks, provides a naturalistic account of a doe's fate, the deer deliberately allowing hounds to catch her scent in order to protect her fawn.[187] She is pursued through woodland to a hamlet beside the

Ausable River and thence to a lake where she is finally slaughtered by a man in a boat. Does have been known to display protective behaviour of the kind described by Warner and will entice dogs to chase them, leaving their fawns in hiding.[188] The story, however, is strongly allegorical and not a simple description of nature. Warner is motivated by ethical and nationalist concerns. Strong anti-hunting sentiments manifest throughout the story and, in this context, the careful construction of the doe's personality – she is shown to be intelligent, determined and self-sacrificial – encourages sympathy for her plight and renders her death more distressing. Wilson's construction of Coorinna is also politically motivated, with ecological concerns a key motivation for his novel.

Warner's doe is also an embodiment of nationalistic ideology, her liberty reflective of the value of individual liberty more broadly.[189] Before commencing the story of the hunt, Warner compares the American deer with the Grecian (Attic) goat (perhaps the Vlahiki breed, a domesticated goat derived from Capra prisca wild goats), finding the latter species mannered in their behaviour, too acculturated and thereby too inhibited. The deer, by contrast, 'is without self-consciousness, and all his attitudes are free and unstudied'.[190] For Warner the goats have become stifled, their comportment unnatural, because of their exposure to the atmosphere of Greek culture. In this understanding Grecian (for which read European) culture is corrupting and constraining. Unexposed to this culture, the deer possesses a natural artfulness, retaining an unaffected beauty and poise.[191] Although the doe's heroic flight conforms to observed behaviour (and therefore satisfies Burroughs's desire for factuality), the dash's 'truthfulness' can also be interpreted as contrived to personify liberty. Warner uses the doe to ideological ends, celebrating new world liberty and condemning forms of hunting he views as unsporting. Burroughs does not acknowledge how these human values inflect Warner's nature writing and nor does he confront the author's mawkishness. Warner's story has been read as excessively sentimental, with Philip Terrie describing it as 'maudlin'.[192]

Of the four writers Burroughs discusses in depth, he praises Hulbert and Roberts, yet qualifies his admiration for the latter by noting that the writer's works portray 'human beings disguised as animals; they think, feel, plan, suffer, as we do; in fact, exhibit almost the entire human psychology.'[193] Wilson also grants emotion to animals. Coorinna is viewed by the shepherd as actively malevolent and constructed by Wilson as calculating in his response to the shepherd, as spited and vengeful.[194] Like Roberts, Wilson projects human qualities onto his literary creations. Coorinna is a human in a marsupial wolf's clothing. The steadfast anthropomorphism in the novel continually pulls the nonhuman animal into human frames of reference and modes of understanding. For Burroughs, animals were less complex in their behaviour than writers such as Roberts (and, by extension, Wilson) implied, although he did acknowledge that they had an emotional life and a 'rudimentary' capacity for reason.[195] He also acknowledged that animals were kin.[196] Nature writers such as Roberts were therefore guilty of according too great a sophistication to animals. Darwin operated a sliding scale of complexity and cleverness, from higher to lower animals, and nonhuman animals as conceived by Roberts, Seton, Hulbert and Long were, to varying degrees, viewed as too elevated in their attainments, too close to human, for Burroughs's taste.

The key area for Burroughs in which over-sophistication was accorded animals was in attributing them the capacity to learn. He berates Seton for conveying the idea to his reader 'that the wild creatures drill and instruct their young'.[197] He is also scathing in his criticism of Long for spreading falsehoods and deigning to downplay the role of instinct in favour of education. Burroughs castigates Long, for instance, for 'falsely' suggesting that a great blue heron (*Ardea herodias*) used morsels of frog as bait to catch fish. This criticism is highly questionable as the black-crowned night heron (*Nycticorax nycticorax*) is known to use bait such as bread to attract fish and the green-backed heron (*Butorides striatus*) has used bread and mayflies for this purpose.[198] It is not beyond the bounds of possibility that Long saw a heron employ a frog as bait instead. Burroughs, however, clearly views bait fishing as too educated a method for a bird to use. Similarly, Burroughs ridicules a 1902 story by Long about kingfishers instructing their young how to dive.[199] In 1922 another account of such a practice was given by Floyd Bralliar, who witnessed a female kingfisher 'teaching her children how to earn a living'.[200] This education consisted of the mother catching and stunning fish and then releasing them for her juveniles to try and apprehend.[201] In 1974, ornithologists were still uncertain whether Bralliar's observations constituted 'fact or fancy'.[202] More recently, kingfishers teaching their young have been observed in the Republic of South Africa.[203] Burroughs also questions Long's account of a female osprey instructing her young how to dive in the tale 'A School of Little Fishermen' from *School of the Woods*.[204] More than the tales of the kingfisher and the osprey being perceived as errors of fact, it was attributing a capacity for learning to the birds that greatly vexed Burroughs.

For Burroughs, learning is an exclusively human trait. Nonhuman animals are only possessed of instinct and 'do instinctively what their parents do and did'.[205] Here, he departs from Darwin who did countenance animal learning. Drawing on the naturalist's published and unpublished manuscripts, George John Romanes gives examples of actions of dogs where Darwin acknowledged a capacity for education. One of these involved packs of mongrel hunting dogs in the Falkland Islands attacking cattle.[206] Darwin noted that John Ross observed of them that new dogs 'introduced into a good pack take instinctively to the habits of the old'.[207] Although the term instinct is used here, Darwin interprets Ross's account to imply that the dogs engaged in social learning through imitation. Wilson accords several species a capacity for learning. In *Minado*, he describes how wolf cubs, 'like the young of many forest creatures' have to be 'taught the hunting and fighting craft of their kind: the wild joy of the chase and the combined and timed attack to bring down a powerful enemy.'[208] He also discusses Loongana training her cub.[209] For Burroughs, depicting animals as capable of learning was one way in which romancers writing about nature blurred the boundaries between fact and fiction.[210]

The essay immediately preceding 'Real and Sham Natural History' in *Atlantic Monthly* was Alfred Thayer Mahan's 'The Writing of History', an exploration of the 'spirit which should inform history writing'.[211] It is clear the editor, Bliss Perry, consciously paired Mahan's essay with Burroughs's, with some crossover existing between the essays in terms of their use of legal metaphors. Mahan states of a historian's critical faculty that 'it plays the part of judge and jury in a trial: not establishing the facts but pronouncing upon the evidence'.[212] Burroughs accuses Long of having the air of 'a witness who is trying to mislead the jury'.[213] Truth (historical or natural historical) is a central concern for both authors, yet they acknowledge that an element of artistry is always involved in presenting facts. Mahan calls history an 'artistic creation' built up out of knowledge.[214] Facts are only 'the bricks and mortar of the historian', requiring the actions of the 'mental processes of the artist to be rendered comprehensible'.[215] This artistry is not equivalent to fictionalising but there is a recognition, even if the term is not used, that facts must always be emplotted and that emplotment can take varied forms: 'differing modes of presenting the same facts give contradictory impressions of the same series of events.'[216] Burroughs also recognises the need to 'present' facts; he appreciates the literary merits of the writers he praises (and, in the case of Seton, condemns). Hulbert, for instance, employs 'invention' but never to the degree of overtaxing the reader's 'credulity'.[217] Invention here seems to mean giving narrative form to animal life without attributing made-up behaviours to each story's animal characters. These behaviours, such as the aforementioned group learning, are ones shared with humans. Burroughs is reluctant to accord substantial powers of reasoning to non-human animals; writers who do so are guilty of presenting human beings masquerading as animals. He is then concerned by the presence of anthropomorphism in nature writing.

A Human in Marsupial Wolf's Clothing

Concerns about the effects of anthropomorphism on 'impressionable readers' of animal stories were common in the early twentieth century.[218] The ways that anthropomorphism and anthropocentrism manifest in nature writing continue to cause anxiety. Krebber and Roscher acknowledge, for example, that the writing of animal biography can potentially be viewed as an 'anthropomorphic and anthropocentric construction' and therefore risks imposing a self-centredness on animals that may not match with their actual lived experiences.[219] They acknowledge that '[o]ne of the major critiques of a social and cultural reevaluation of animals that intends to recognize animals as autonomous, self-acting individuals is the anthropocentrism that is argued to be inherent in their every representation'.[220] Ralph Lutts also emphasises that animals 'are ideas as well as living, breathing creatures'.[221] These ideas, mental representations, are translated into words upon the page in animal fiction and reflect the beliefs the writer holds about a given species. This violent imposition of human values and perspectives on the animal as Other in much nature writing can nonetheless act to counter moral disengagement, something authors such as Long, Seton and, later, Wilson clearly recognised. Animal biographies, be they fiction or creative nonfiction, therefore have a politically important role to play regarding animal rights but are, equally, ethically questionable. Lutts captures something of the ambiguity of such endeavours when he refers to writers such as Seton and Roberts as merging concern for animal welfare with the traditional wildlife essay to produce works best characterised as 'Black Beauty in a wolf's clothing'.[222]

McHugh has signalled how stories such as Marshall Saunders's 1893 novel *Beautiful Joe*, which is written from the perspective of an ageing mutt, potentially helped shape 'the moral imperatives of animal advocacy'.[223] In a foreword written for an edition published in 1898, Hezekiah Butterworth of Boston, a member of the Humane Society, suggests that although about a dog the work speaks for the whole animal kingdom: 'Through it we are able to enter the animal world, and are made to see as animals see, and to feel as animals feel.'[224] Butterworth states that Saunders's 'sympathetic sight' is 'ethically the strong feature of the book'.[225] This sympathy is nonetheless qualified. It does not extend to predators. The hunting of animals (specifically bears) that 'destroy property' is seen by Saunders as qualitatively different to the hunting of animals purely for sport, which is forcefully condemned.[226] In the book's preface, Saunders emphasises that *Beautiful Joe* is 'founded on fact' and is about a 'real dog' and uses his 'real name'.[227] Although not nature writing, *Beautiful Joe* therefore demonstrates concerns about veracity that echo those voiced later by Burroughs. Saunders is clearly worried that if the book is read as fiction, its moral message about animal maltreatment will be undermined.

In his writings, Wilson seeks to establish individual identities for many species of birds and mammals, large and small. In the story 'The Shoes of the Negus-Negusti', he grants insects, some driver ants (*Dorylus*), individual (as well as collective) voices, although none are named.[228] Individuality here means a unique combination of needs and desires, of behaviours and comportments. Individuality is sometimes viewed as interchangeable with personality.[229] Personality is human-centred, emphasising the personal, the person.[230] Coorinna is presented as possessing a personality, something that foregrounds the ever-present anthropomorphic aspects of Wilson's nonhuman literary characters.

Individualising in animal fiction (or natural history writing) invites greater investment in animals from the reader and increases the possibility of empathy. This possibility is greatly amplified in Wilson's case with the characters of Coorinna and Minado, whose distinct personalities are strongly established. Because of Wilson's embrace of anthropomorphism, the thylacine and the wolverine inevitably remain bound to human ways of seeing and feeling. This is a form of subjugation of the animals, the alien singularity of each species disavowed and replaced by familiar human behaviours and emotions. In this sense, Wilson's *humanising* of Coorinna is destructive. The price of individuality (and the consciousness raising it enables) is the erasure of the thylacine's alterity.

The conditions that facilitate animal genocide often involve a violent generalising, a lumping together of a group of animals as, for example, abject or evil or doltish. This is frequently accompanied by a hatred of them for the negative characteristics they have been attributed with. Writing in the context of human genocide, Thomas Szanto refers to this process as 'social typification' and 'collectivization' in which 'stereotyped negative properties' are attributed to a group. He does not believe this process of coming to hate can apply to nonhuman animals.[231] The animosity directed towards the thylacines as a species suggests differently. Individualising of the kind practised by Wilson potentially works to dull the effects of this kind of blanket negative stereotyping of a given group. The shepherd, however, despite seeing Coorinna as exceptional does not deviate from his murderous intent.

In the Bible, the shepherd metaphor has traditionally been used to signal a transition from doom to salvation, with the Lord as shepherd in the Old Testament capable of showing 'remorse for his wrath'.[232] Coorinna moves from being a general incarnation of the thylacine to a particular individual for the shepherd, yet this does not blunt the latter's animosity, which never transforms into caring. Despite having accorded 'One-ear' a name, thereby contributing to a sense of his individuality, the sheepherder still shows no hesitation in killing the tiger. The narrative is one in which the genocidal gaze, predicated on a hatred of tigers as a group, is supplanted by personal animus. By the end of the novel, Coorinna is subject to individual-directed hatred rather than simply hatred as a thylacine. He no longer 'represents' tigers as a species but stands for himself, by himself. The shepherd's hatred shifts from being for 'wolves' in general (for the thylacine as a species in the abstract) to being for 'One-ear' (a specific, guileful tiger). The reader is invited to participate vicariously in the shepherd's malevolent admiration of the predator, as he moves from hatred *tout court* to a more complex esteeming hatred. As the shepherd is well aware of the rarity of tigers his decision to kill 'One-ear' remains genocidal, yet his behaviour is unusual as he has allowed him to transcend the general typecasting of tigers. There is also a moment in the narrative where the power of individuality to, at least briefly, disrupt bloodlust is shown. It is an instant in which Coorinna outwits the shepherd and his dogs. The man is so impressed that 'it is doubtful whether he would have used his rifle had he held it at that moment'.[233] He refers to 'One-ear' as a 'trimmer', an Australian slang term of approval indicating distinction. This instance, while it ultimately fails to undo the shepherd's deadly purpose, demonstrates the potential capacity of individualising to cloud the genocidal gaze. It is when seen outside the group, as a singular individual, that Coorinna is granted a moment of virtual reprieve.

Wilson goes further than the shepherd. He gives us Coorinna not solely as a clever, cunning and elusive sheep-slayer but as a defiant and loving, wise and determined being. He casts the tiger as a tragic figure, one undone by his love of Loonganna and their cub and, at the last, by his boldness. Gordon Sayre has explored the use of the tragic form by settler authors to represent Native American resistors as heroes. Sayre reads these efforts as part of a complex response to settler colonial culpability regarding the mass murder of Native Americans that sought to 'assuage or dismiss responsibility'.[234] As he notes, perpetrators of destruction of Indigenous peoples seek to frame their actions as part of a natural process. A similar operation is at work in *Coorinna*. The eponymous hero shows goodness and achieves a measure of nobility; he is granted a kind of 'humanity', a quality that makes the reader pity him when he is defeated.

As Wilson is writing a tragic novel, he must set his hero apart from other thylacines. In tragedy as it is traditionally conceived, the hero is required to appear superior (if imperfect) to those around them.[235] Coorinna is presented as a cut above the rest of his kind. His father, for instance, is shot with ease by the shepherd but Coorinna proves to be far cannier and elusive as an opponent. The cleverness and courage Wilson attributes to him works to counter the long-standing stereotype of the tiger as meek and a dullard. An 1867 press report, for instance, refers to the thylacine as 'by nature a coward'.[236] The press report of Wilfred Batty's 1930 killing of a tiger also notes that the animals are 'great cowards'.[237] In 1909, an article about a tiger held at London Zoo referred to the 'Tasmanian wolf' as 'one of the most stupid [of] animals'.[238] A 1934 essay on the thylacine and the Tasmanian devil written for children suggests that the former possesses a 'dull stupidity'.[239] Through Coorinna, Wilson contests characterisations of this kind that construct the thylacine as doltish, wimpish and worthless.

Notwithstanding his positive portrayal (or because of it) of Coorinna, however, there is still an air of inevitability about the tiger's vanquishment. There is never a sense that Coorinna will prevail, whereas in *Minado*, by contrast, the wolverine's potential triumph against Mort the trapper is prefigured in his driving away of Lone Cloud and Little Otter. As Wilson's *Coorinna* explores extinction through the literary form of tragedy, the tiger's downfall is preordained, necessary to summon the sympathy of the reader. The story is designed to encourage a re-examination of settler colonial attitudes towards native fauna in Lutruwita and more broadly. For the violence of settler persecution of native fauna to register as forcefully as possible, the hero's death must occur. It is, nonetheless, a death that takes place out of sight, implied rather than described. Coorinna does not actually die on the page. The fatally wounded thylacine hauls himself into nearby bauera for shelter.[240] The plant derives its name from the Bauer brothers, Ferdinand and Franz, who were botanical artists. Franz accompanied Joseph Banks to Australia and was with him when he first recorded a species of the plant in Botany Bay.[241] One of its colloquial names is dog rose because it reminded settler colonists of *Rosa canina*, a climbing rose found in Europe. Helge Vedel suggests that the canine reference is meant to be disparaging, indicating worthlessness 'in comparison with cultivated roses'.[242] Pliny offers an alternative explanation, believing that the rose could cure the bite of mad dogs.[243] Whatever the origins of the name, Coorinna is made by Wilson to expire among the Canidae.

Endings and Endlings

The last thylacine to die in Lutruwita, often referred to as an endling, is usually identified as 'Benjamin', a male tiger held captive at Hobart Zoo.[244] Recently, it has been suggested that the tiger that expired during the night on either 6 or 7 September 1936 was an elderly female.[245] Whether the last captive thylacine was 'Benjamin' or another tiger, at their death they were not the last of their kind. Early expeditions post-1936 found evidence (mainly tracks) of thylacines in the wild. Sharland reported that the 1938 expedition he was a part of located clear thylacine foot impressions at Thirkell's Creek, Royle Creek (as it was named by the expedition) and elsewhere in the vicinity of Jane River.[246] At the time, it was readily accepted that the expedition 'had proved that the district examined was inhabited by [thylacines]'.[247] Sharland indicated the presence of 'definite evidence' of the species' survival.[248] Sharland also remarked that the tiger was 'making a bold stand against encroaching settlement and other factors which have driven it from most districts in the island'.[249] Attributing the thylacine boldness, granting them bravery or daring, provides a different picture of the animal to the, at the time, still commonly held view of them as craven and almost inviting destruction.

David Fleay's account of a November 1945 expedition describes hearing what he believes was a tiger, the cry suggesting 'the brief, sharp creak of a door'.[250] On a trail towards Lighting Plain the expedition found 'an unmistakeable and fresh series of tracks made by a Thylacine'.[251] Efforts were made to capture the tiger yet to no avail. Later searches also found tracks or heard vocalisations but these accounts are less compelling. Hans Naarding's chance sighting of a thylacine at Togari on the north-west coast in March 1982 was viewed as highly reliable, with the Parks and Wildlife Services director Peter Murrell calling it 'irrefutable and conclusive'.[252] Survival into the 1950s therefore seems highly probable and into the 1980s, quite possible. Yet 1936 is still the accepted temporal marker for disappearance. The refusal to shift the extinction date to post-1936 may be for psychological as much as evidentiary reasons. There is a desire to 'know' the date of disappearance even if that knowing is knowingly flawed. The night of 6 to 7 September 1936, while not pinpoint accurate, is preferable to 'sometime in the 1950s' or 'at some time in the 1980s' or 'at some point in the early 1990s'.[253]

In fictional works such as *Coorinna* and *The Hunter*, there is a desire to relocate the death of one of the last (if not the last) of the thylacines from a zoo to the wilderness. *Coorinna* echoes many historical accounts of farmers hunting down 'marauding' tigers. The stockman Wilfred Batty, who shot a thylacine on 13 May 1930, provides one of the most well-known instances.[254] Batty's act is usually credited as the last killing of a thylacine at liberty. The tiger in question is sometimes referred to as being 'in the wild' but this is misleading as wilderness is a term usually applied to areas where there is no permanent human settlement. Mawbanna, where Batty lived, is a small farming community. The area surrounding Mawbanna is pastureland. *Coorinna* relocates the shooting of one of the last thylacines to a more remote location, although whether it constitutes 'true wilderness' is questionable.

The denouement to the novel involves the shepherd outwitting Coorinna using traps concealed around the corpses of Loonganna and the cub. Coorinna is badly injured by a gin trap but limps away, leaving a trail of blood. The shepherd's dog picks up his scent and gives chase. The shepherd is elated when he senses he is close to killing Coorinna, 'swept by the exultation of the human hunter with destruction in his hands'.[255] A comparable joy is present in one of the photographs of Batty posing alongside the tiger he has shot. In an article that appeared in the *Canberra Times*, he is referred to as 'the smiling hunter'.[256] The reference to the shepherd's excitement in *Coorinna* gestures towards a long-recognised link between hunting and sex in which hunting is understood to sublimate sexual energy.[257] Interpreted in this way, the shepherd's euphoria represents a plateauing of his sexual excitement prior to the ecstasy of the kill. Wilson, however, withholds this final stage of the hunting-as-sex cycle from the reader. Feelings of ecstasy have also been linked with acts of atrocity. Dan Stone, for instance, writes of 'ecstatic communities', of groups of perpetrators who find shared joy in transgressing, in going beyond the bounds of 'ordinary rules of behaviour', through acts such as genocide and mass murder.[258] The shepherd and Batty do not engage in transgression as the killing of thylacines, the extermination of vermin, was commonplace and readily accepted. Through their acts they do, however, affirm their belonging to the community of thylacine haters and may be seen to find ecstasy in simply conforming, performing their animosity for the species.

The shepherd's response to firing the fatal rifle shot that despatches Coorinna is not described. The climactic moment is only rendered through the body of the tiger, his strength ebbing as he slips away. Here Coorinna's phallic masculinity falters, he can no longer stand erect and he has to heave himself into a shielding patch of scrambling shrubs. At the last, he is unmanned. In a battle of competing masculinities, tiger and human, it is the shepherd that prevails. Hunting, as the shepherd does, is frequently associated with the construction of a masculine identity and with demonstrations of human dominance over the natural world.[259] The man's success contrasts with *Minado*, in which it is the wolverine who outwits his human foe. It is possible that the victory of the wolverine over the trapper makes good a failing Wilson identified in *Coorinna*, where the 'natural' order of human dominance is reasserted. After *Minado*, he would abandon animal fiction and natural history. His final book, *Adams of the Bounty*, was about the mutiny on *HMS Bounty* in the South Pacific in 1789.

The end of *Coorinna* mirrors its beginning; Coorinna, injured, disappears into shrubs as mist descends. Mist here, like at the novel's beginning, incarnates 'primitive' pre-contact Lutruwita but, like the brume in John Carpenter's *The Fog*, it can also be interpreted as signifying repression and a refusal to accept responsibility for past misdeeds.[260] Slightly earlier, the tiger is described as almost 'the incarnate spirit of that enduring antique land', 'torn and battered and bleeding'.[261] In Wilson's eyes, Lutruwita is wounded, bearing the physical traces of settler violence. Embodying the traumatic impact of European invasion, Coorinna is made to vanish, uniting with the 'children of the mist'. Sandes's poem was published in a book funded by the Emu Bay Railway Company. Mutawaynatji/Emu Bay is located on Lutruwita's north-west coast. Laylawina, named the Emu River by Henry Hellyer, reaches the sea at this location. Hellyer, a surveyor for the Van Diemen's Land Company (VDLC), chose the name for the river because of emu tracks he saw in its vicinity.[262] The tracks had guided him away from very rough terrain to 'better travelling'.[263] While near the Laylawina/Emu River, Hellyer and his party had to seek refuge after a 'thick, impenetrable mist' came upon them.[264] He first saw the river on 13 February 1827, so the tracks were those of the Tasmanian emu (*Dromaius novaehollandaie diemenensis*) rather than of emus introduced from mainland Australia.[265]

Two Indigenous names for the emu are recorded by George Augustus Robinson in his journals as Gonanner (the South East nation) and Pun. nune.ner (the Paredarerme nation).[266] Robinson's immediate reaction on encountering emus in the wild was to set one of his dogs on them and also to send men in pursuit. Along with loss of habitat, the introduction of hunting dogs to Lutruwita is thought to have contributed to the extinction of the bird.[267] As the decline in the emu's numbers was regularly noted by settlers, their continued killing for food or as a perceived pest (consciously contributing to their ongoing decimation) can be interpreted as becoming genocidal from the 1820s onwards. By the late 1850s the bird was probably extinct in the wild. A captive bird is reported as having died in 1873.[268] By 1910, when Sandes's poem about human extinction was published by the Emu Bay Railway, the name of the railway therefore referred to a bird that was extinct, extirpated.

During his journey of exploration, Hellyer's camp was often passed by a 'heavy animal', a creature that Ida Lee speculates was the 'Tasmanian wolf'.[269] While staying at Parteenno/Hampshire Hills near the Laylawina/Emu River in 1830, Robinson also alluded to the presence of a 'hyaena'.[270] On his way to Parteenno/Hampshire Hills, he recorded seeing 'several carcases of sheep which had been killed by hyaenas or wild dogs'.[271] Robinson was in the area to investigate reports of the massacre of palawa from the Northern nation by employees of the VDLC. The VDLC had extensive landholdings in the region including 50,000 acres along the coast at Mutawaynatji/Emu Bay, 100,000 acres at Pilri/Cape Grim and 10,000 acres at Parteenno/Hampshire Hills.[272] Robinson wrote of the VDLC employees that 'they would shoot [palawa] without mercy to make a merit of it'.[273]

There were three incidents in a three-month period (from December 1827 to February 1828) in which palawa were murdered.[274] The Pilri/Cape Grim massacre of 10 February 1828 in which approximately thirty palawa died was the most deadly of these incidents. It was perpetrated by the VDLC. Knowledge of the massacre derives in significant part from information Robertson was able to glean on 16 June 1830 from one of the perpetrators, the shepherd Charles Chamberlain. Chamberlain's account frames the massacre as an act of *lex talionis*, as revenge for the killing of sheep: 'We threw them down the rocks where they had thrown the sheep.'[275] He suggests that thirty palawa were killed just as thirty sheep were killed.[276] Chamberlain's implicit equating of the life of a palawa man (he claims the victims were all, or nearly all, men) with that of a sheep demonstrates that he viewed the former as no more than 'beasts'. Ian McFarlane suggests that the massacre, coupled with other contemporaneous murders, probably caused the clan's extinction.[277] The shepherd's account of the massacre was given 'with such perfect indifference that the blood chilled'.[278] Robinson chastises Chamberlain, yet his own virulent racism is made plain by the journal entry, as he affirms that all palawa look the same.

The Emu Bay and Mount Bischoff Railway Company was founded by the VDLC in 1887.[279] The railway replaced a tramway that was built on the cheap to profit from the discovery of tin deposits at Partennamana/Mount Bischoff.[280] Sandes therefore published his poem in a book sponsored by a company with direct ties to an organisation whose employees were historically guilty of genocide against the palawa. The company was also responsible for large-scale habitat destruction and land degradation through farming. Wilson therefore relocated an excerpt from Sandes's poem to *Coorinna*, lifting it from a book published by a firm affiliated to the company that first introduced bounties for thylacines. For a time, both private and state-funded bounties were available for the killing of thylacines.

Wilson's own book reflects elements of the genocidal logic that made the bounty possible. Settler colonists viewed the thylacine as a malign presence in need of eradication because of the threat they posed to sheep. The bounty was framed as a rational response to a pest species threatening the pastoral industry. In reality, the thylacine was a scapegoat and there was no rational justification for the campaign of extermination. Wilson's *Coorinna* is therefore disappointing as it maintains the genocidal logic of the tiger as a sheep killer and a threat to human interests.[281] The shepherd, however, acts at a time when bounties are a thing of the past. His drive to kill is therefore wholly pathological, uninfluenced by financial incentives. He does not receive structural support from the state or private enterprise for his campaign against the local tiger population. He acts as an individual. Historically, the persecution of thylacines was carried out by individuals or very small groups but, crucially, they represented a broad group who harboured animosity for the animal and could count on financial, political, moral and emotional support. *Coorinna* is set in a period when intergroup (human–thylacine) conflict seems a thing of the past. The thylacine's rarity means that the main rationale for their extermination (supposed significant stock losses) is no longer extant. In these circumstances, political, moral and emotional support is also potentially diminished.

During the 1930s, there was an increase in newspaper articles discussing the protection of the thylacine and the possibility of their extinction.[282] Dennis Colbron Pearse wrote a letter to the editor of the newspaper *The Mercury* calling for greater efforts to safeguard the thylacine and worrying about the 'slur which is bound to fall upon us by its extinction'.[283] In this climate, killing a tiger could no longer be openly celebrated; it would generally be viewed with condemnation.[284] Susanna van Essen suggests that *Coorinna* is set in this time period.[285] There is therefore an atavism in the shepherd's early contempt for tigers, a contempt manifested through his scornful treatment of the corpse of Coorinna's father. This part of the novel describes a reactivation of traditional negative emotions towards the thylacine as a species (a group), feelings of anger and hatred. Such feelings would no longer be generally shared; the shepherd therefore represents a minority, embodying the emotional embers of enmity. Through the sheepman's private loathing, Wilson examines the emotional mechanics of genocide and the genocidal gaze. In the novel's denouement, there is no hesitation once the shepherd has Coorinna in the crosshairs of his telescopic sight: the wolf must die.

The palawa survived European efforts to eradicate them but the tiger is now extinct. Wilson, however, cannot quite close the book on his creation. He leaves open the possibility of the thylacine's persistence because he does not confirm his death and because Coorinna's status as a lastling, as endling, is never made explicit.[286] Here, Wilson joins a widespread refusal to sign the thylacine's death certificate. As discussed in the last chapter, even the Tasmanian Museum and Art Gallery hedges its bets regarding extinction. The continuing sightings of the thylacine in Lutruwita are symptomatic of an unwillingness to confront the finality of disappearance. They index a desire for extinction not to be a reality, perhaps bound up with guilt at the human role in the animal's destruction.[287] In this sense, the cryptozoologist who keeps the flame of the thylacine alive manifests humanity. Theirs is a denialism that may be viewed positively, as motivated by compassion. Yet, like Wilson, who arrests the thylacine's destruction, those who argue for survival also seem unable to fully confront the destructive violence endured by the tiger and acknowledge human culpability for their vanishing.

Conclusion: Beyond Words

Although Wilson seeks to garner sympathy for Coorinna's plight, his efforts form a kind of violence in themselves. Coorinna is subject to a 'nonvoluntary martyrdom', to borrow Karyn Ball's phrase, which she uses to describe how the figure of the 'animal' is endlessly forced to rehearse human concerns.[288] The tiger becomes a mere agent for voicing Wilson's apprehensions and grief. He uses *Coorinna* to process his own settler guilt, articulating a recognition of colonial violence and its bloody legacy. Even if the human triumph over the nonhuman animal that closes the novel is overlooked and Wilson's individualisation of Coorinna is foregrounded, the failure to seek to transcend anthropomorphism has severe repercussions. Wilson words the thylacine in human terms although he ostensibly writes from Coorinna's point of view. His *idea* of the animal is all too human. This, in itself, enacts a kind of violence towards the thylacine.

Wilson's predicament is that he must address the thylacine through the medium of human language and the tiger therefore acts, in part, as a mirror for him, constantly reflecting his own desires and values. Writing about nonhuman animals necessarily involves overwriting them, an overwriting that forms an insidious kind of violence, a murder by way of naming. In the *First Philosophy of Spirit*, as part of a meditation on the nature of memory, Georg Wilhelm Friedrich Hegel mentions Adam's naming of the animals as something that nullified them as beings.[289] Glossing this, Maurice Blanchot explains that this nullification occurred because the animals were transformed from simply being to being as ideas as well as living entities.[290] It is a nullifying that, as Peter Schwenger notes, building on Blanchot, involves an inventive aspect.[291] Naming murders to create, forging meaning, finding significance, at the same time as it annuls the things it refers to, the sensed being. Blanchot admits that no literal death is involved in this process, but it does lead to detachment, to a destructive disconnect. The word can circulate in the absence of that which it names, a situation that Blanchot links to death. He suggests that the ideal negation that occurs through naming in language, what he calls language's capacity for 'deferred murder [*assassinat différé*]', depends for its existence upon the potential for literal negation that is physical death.[292] The name's capacity for detachment is readily demonstrated by 'thylacine'; the word persists despite the animal's disappearance. The word 'thylacine' refers to a no longer living thing, a thing that lives on in and as that word. Words therefore refer to things and, as Schwenger observes, are things, objects. Writing about animals (and other things) is thus to participate in their ongoing objectification.

Faced with this situation, is it possible to write of an animal, of the thylacine, without ideating them? Difficulties pertaining to the very act of writing may, on some level, have registered with Wilson. Like a blanket of mist, his descriptions and personifications stifle and obscure the thylacine. It is possible to read the ending of *Coorinna* as, in some way, an acknowledgement of this issue:

> [Coorinna] sank to the earth, then, with failing strength, dragged himself stubbornly into the deep green tangle of the bauera as the mist of the mountains rolled down muffling sight and sound.[293]

As the dying thylacine crawls into the shrub, he simultaneously crawls out of language. The rolling mist of the description is replaced by the flecked wood pulp of the page that refuses to reveal the tiger's last breaths. The blank part of the page, impenetrable, acts like the bauera that Coorinna hauls himself into, rendering him unseeable. Bauera is composed of thin twining branches forming a natural barrier that is difficult for humans to walk through. The cellulose fibres of the pulp similarly block insight. Most novels, of course, end with a stop followed by nothing. *Coorinna*, however, precedes this conventional ending with a description of disappearing from view, therefore drawing attention to the process of becoming absent and, by extension, to the blank page, the ever-present undergrowth of the word, the pulp, now affirmed as figurative mist. The first edition, in fact, has a blank space between the final paragraph and the words 'The End', words that are absent in later editions. There is therefore a gap, a space of invisibility and silence, before the close.

The pulp that undergirds this space, whatever its actual geographical origin, has significance in the context of a story set in post-invasion Lutruwita.[294] Fine paper production in Australia began on an industrial scale in Pataway/Burnie in Lutruwita when Australian Pulp and Paper Mills (APPM) opened a mill there in 1936.[295] The lumber was supplied by forests in Wulaninka/Surrey Hills, from land previously owned by the VDLC but purchased from them by Gerald Mussen. The VDLC had failed in their efforts to farm livestock profitably there because of the cold climate.[296] When the VDLC took control of Wulaninka/Surrey Hills, it was a mix of grassland and eucalypt, which had been carefully constructed and maintained by the palawa using fire management.[297] This habitat was well suited to the thylacine.[298] After the expulsion of the palawa and the VDLC seizure of the lands and in the absence of a burning regime, much land was deconstructed, becoming rainforest (habitat less suited to tigers). Once the demand for timber increased to fuel the paper mill, the habitat changed again with large expanses dedicated to silviculture. As noted earlier, Sharland and other naturalists were worried about the destruction of Lutruwita's forests because of the demand for timber.[299] Here trees were not destroyed but replaced, a practice that is also destructive. In Wulaninka/

Surrey Hills, local eucalyptus trees did not regenerate quickly enough to meet demand, so a species derived from Victoria on the mainland was planted to aid these silviculture ventures.[300] Eucalyptus was used to make paper pulp.

Although the pulp gestures to the natural world as it is exploited as a resource, to timber felled and processed, the ground of the page is also open to being read as a 'wild' space, one not occupied by Wilson's prose. It is unsettling in its minimal immensity, a blank hole into which Coorinna/*Coorinna* disappears. Wilson, of course, probably had no thought of such things when he composed the novel's ending. He would also likely reject comparisons between himself and the shepherd who has Coorinna in his sights. That said, parallels between writing and shepherding are not unusual. In the context of glosses on Proverbs 10:19 ('In the multitude of words there wanteth not sin: but he that refraineth his lips *is* wise'), words have been compared to sheep: 'The more you have, the more likely there will be some that go astray.'[301] In this understanding, it is sage to be tight-lipped. Wilson's waxing lyrical about Lutruwita's flora and fauna does the opposite. He is therefore usually a bad shepherd, allowing his flock of vocables to run loose. The sheep he tends is Coorinna, a wild animal who is named, mastered and domesticated through his writing. Yet, at the last, Wilson loses sight of his creation, his figuration of a tiger. Coorinna retreats into a space beyond words.

This space can be understood as affectively charged, resistant to representation and knowability. Milena Marinkova has noted how affect-driven art discourages 'personal identification or subjective ownership by the reader/viewer' but does encourage 'openness to otherness'.[302] Art that privileges affective experiences, as Susan McHugh has emphasised, can speak to cross-species connection.[303] McHugh encourages a reading of animals in stories that acknowledges their capacity to function as more than mere metaphor, with animals transcending the figurative role of 'humans-in-animal-suits'.[304] Drawing on the ideas of Gilles Deleuze and Félix Guattari, she traces the emergence of 'nonhuman-agency forms' in animal narratives.[305] Stories for her form 'spawning grounds for forms of species and social agency'.[306] The study of animal stories also provides 'new, contingent foundations for ethical relatings' and the cultivation of positive passions in cross-species entanglements.[307] Her approach attends to aesthetic dimensions in animal stories, looking beyond content to consider 'the complicated operations of affect' as a means to formulate 'an ethics premised on feelings honored as concrete, intense and shared'.[308]

Coorinna, however, is not a novel that systematically seeks to promote affect (although, of course, affect is continually present). It is therefore also useful in this context to continue thinking about the blank space from within a semiotic framework, seeing it not solely or simply as an outflux of affect but also a withholding of signification. The space signifies as absence, signalling a refusal to say. By contrast, the novel *The Hunter*, in so many ways remarkable as a form of speaking *to* the thylacine, differs in the bloody substantiality of the *unnamed* tiger's demise. There is a visceral finality to that death. We grasp the loss of the thylacine unambiguously. In *Coorinna*, there is no sentence stating Coorinna's death, no death sentence, at the novel's close. The tiger is left to slip away, avoiding the fatal trap of the word. In this eluding of Wilson's word-grasp, *Coorinna* therefore offers a moment of potential where the thylacine escapes the death grip of language, persisting as an absent presence upon the page. The possibilities of a creative practice that presents the tiger through uses of absence will be examined in more depth in the next chapter.

Notes

1. As well as being published as a novel, *Coorinna* was also serialised in the *Junior Express* newspaper. Tim Jetson suggests that Erle Wilson provides the best rendering of Lutruwita's high country in prose that there is. Jetson, 'The High Country', 21.
2. Anonymous, 'For the Classroom', 36.
3. The main part of the book, which is a lyrical account of a railway journey through Lutruwita, is by Sandes but the unpaginated four-page description of the Emu Bay Railway Company towards the end is likely by Stirling or an acolyte.
4. Sandes and Stirling, *In Tasman's Land*, 9. The man is erroneously shown holding a boomerang. Historically, the palawa did not manufacture boomerangs. See Baird, *Voices of Aboriginal Tasmania*, 10.
5. Sandes and Stirling, *In Tasman's Land*, 10.
6. Ibid.
7. See Breen, 'Extermination, Extinction, Genocide', 76–77.
8. Darwin, *Journal of Researches*, 435.
9. Ibid., 447.
10. Ibid.
11. West, *The History of Tasmania*, 92.
12. Ibid., 96.
13. Breen, 'Extermination, Extinction, Genocide', 88.
14. Kuper, *Genocide*, 89.
15. For an extended discussion of the notion of 'doomed races', see Brantlinger, *Dark Vanishings*.
16. Wilson, *Coorinna*, 12. The use of First Nations Australian names in the book can be read as a kind of 'salvage linguistics'. Salvage linguistics, usually associated with early twentieth-century Turtle Island anthropological practices, involves the 'collection' of a language as a means to retain traditional knowledge. For an overview of the salvage paradigm in anthropology, see Wajsberg, 'Notes on the Failures of Salvage Linguistics'.

17. See Bailey, *Lure of the Thylacine*, 126. In palawa kani, the name for the thylacine is kaparunina. Corinna (deliberately chosen to echo Coorinna) was the civilian name given to the character Thylacine in the DC comic series *Suicide Squad: Bad Blood.* The character Thylacine is from the Ngarluma, a First Nations Australian people from the Pilbara in Western Australia. The Yaburara (also Jaburara), whose language is similar to the Ngarluma, created many petroglyphs of thylacines in Murujuga in the Pilbara. Shari Shebbens has explained that the civilian name chosen for Thylacine is intended 'to acknowledge the thylacine's place in Tasmanian Aboriginal history and culture'. See Teresa Tan, 'New DC Comic Book Superhero Thylacine is an Indigenous Australian from the Pilbara', *ABC News*, 12 April 2020, https://mobile.abc.net.au/news/2020-04-13/aboriginal-superhero-thylacine-new-dc-comic-book-hero/12136148 (retrieved 3 February 2022). Using a palawa word as a name for a person of the Ngarluma is problematic as it suggests that one Australian First Nations people is substitutable with another. The name Wilson gives Coorinna's mate, Loongana (which he translates as 'Fleetfoot'), is also a palawa word meaning 'swift'. The twin-screw steamship the *Loongana* operated between Melbourne (Victoria) and Launceston (Lutruwita) in the early twentieth century. See Cox and Ratcliff, *Tasmania Remembered*, 138.
18. The constellation is called 'Wurrawana Corinna'. The constellation's European equivalent would be a region near Gemini. Penprase, *The Power of Stars*, 73. As Dianne Johnson explains, there is no one overarching First Nations Australian astronomy because of differences in time, space, language and culture across Australia's vast land mass. The star knowledge in Lutruwita was likely different from that of mainland Australia. See Johnson, *Night Skies of Aboriginal Australia*, 6–7.
19. Smith, 'The Return of the Living Dead', 283.
20. Smith, *The Tasmanian Tiger*, 21. Both the palawa and thylacines were subject to bounty schemes, although for the palawa the bounties were, at times, for specific individuals. See Lawson, *The Last Man*, 74.
21. Haraway, *When Species Meet*, 18.
22. See Martin Hambleton, 'For the Boys and Girls: The Tasmania Wolf and Devil, Strange Survivors of a Dying Race', *Sunday Mail*, Sunday 7 January 1934, 8.
23. The tiger is like the palawa as Terry Goldie discusses them in *Fear and Temptation*, a people who are rendered 'less and less real' by the colonial imaginary, becoming an 'indigenous past without a present' (158).
24. Dieter Berger notes that epigraphs are sometimes viewed as 'ornamental' and sometimes as 'intrinsic' to a novel's meaning. See Berger, '"Damn the Mottoe"', 377.
25. Roddy, 'Mist in "the West"', 45.
26. Ibid.
27. Ibid., 46, 48.
28. Ibid., 50.
29. Ibid., 46–47.
30. Potts, 'Natural Order and the Call of the Wild', 20.
31. Scott's novel inspired a two-canto poem by Annette Turner entitled 'The Children of the Mist', which was published as part of a collection of Turner's verse in 1827.
32. The novel includes an oblique reference to this forced depopulation. See Ferris, 'The Historical Novel and the Problem of Beginning', 77.
33. Scott, 'A Legend of Montrose', 525. The identification of mist with isolation in relation to the Scottish Highlands is an enduring one. Shots of a mist-enveloped landscape, for instance, are used in the detective show *Taggart* as the titular character travels from Glasgow into the Highlands to investigate a murder. This choice of shot is clearly intended to signal remoteness. See *Taggart: Love Knot* (dir. Peter Barber-Fleming,

UK, 1990). The film *Restless Natives* (dir. Michael Hoffman, UK, 1985) about a pair of highwaymen also occasionally depicts the Highlands as shrouded in mist.
34. Scott, 'A Legend of Montrose', 527.
35. See Ryan, 'The Black Line', 7. Ryan also notes that the historian Charles Esdaile compared the processes involved in the Black Line to a Scottish Highlands shooting party (9).
36. Sandes and Stirling, *In Tasman's Land*, 21.
37. Ibid., 20. The palawa name for Mount Bischoff is sourced from Plomley, *Tasmanian Aboriginal Place Names*.
38. Much like the way fog operates to connote the primordial in a film such as *King Kong* (dir. John Guillermin, USA, 1976), in which a fog bank hides a land of megafauna that time forgot. Fog, which meteorologically describes a phenomenon that more greatly reduces visibility, is often figured as malevolent. See, for example, John Carpenter's *The Fog* (USA, 1980) in which fog portends acts of violent revenge by the ghosts of a massacred leper colony.
39. Wilson, *Coorinna*, 65.
40. Meston, *The Van Diemen's Land Company*, 46.
41. Wilson, *Coorinna*, 4. As well as temporary structures of the kind described by Wilson, some palawa (such as clans in the north-west of the island) built permanent structures. See McFarlane, 'Aboriginal Society', 25–27.
42. Wilson, *Coorinna*, 4.
43. Ibid., 3–4.
44. Murray, 'The Childhood of William Lanne', 514. Lanne's palawa name is forgotten as he was very young (seven years old) when he was uprooted from his home in Burghley and removed to Flinders Island. Burghley is located in Wulaninka/Surrey Hills.
45. Wilson, *Coorinna*, 7.
46. De Witt Island was certainly proposed as a sanctuary for thylacines, although the habitat is not ideal for them. The idea does not seem to have advanced greatly. See Anonymous, 'A Vanishing Race, the Marsupial Wolf: Sanctuary Proposed', *Examiner*, Wednesday 5 October 1932, 7. See also Guiler, *Tasmanian Tiger*, 30.
47. Crane, 'Tracking the Tassie Tiger', 113–14; Philp, 'Hunted, Now Haunting', 86–87; Probyn-Rapsey, 'Animals and Extinction', 105.
48. Wilson, *Coorinna*, 8.
49. In fact, the borhyaenids of Abya Yala, long thought to be the closest ancestors of the thylacine, are more distant relatives than dasyurids such as the Tasmanian devil and the Australian tiger cat (also tiger quoll). See Thomas et al., 'DNA Phylogeny of the Extinct Marsupial Wolf', 465. Writing in the 1950s, however, Wilson was not to know this. There is also much fossil material from Australia.
50. Wilson refers to the uplands but presumably means the Central Highlands.
51. The flyleaf states that a number of the stories had been broadcast, suggesting they were initially written for the radio.
52. Wilson, *The Grey Seals and Other Stories*. The only copy of this book I have been able to locate is held by the National Library of Australia and I have been unable to consult it. I am grateful to Dominic Williams for his efforts in seeking to trace alternative copies.
53. Emile Durkheim's lengthy discussion of the churinga (as he calls it) in *Les formes élémentaires de la vie réligieuse* may have been one of Wilson's sources on the topic. Durkheim's book was translated into English in 1915 as *The Elementary Forms of Religious Life*. Durkheim discusses how the unique status of the churinga is bound up with its being inscribed with a totemic sign (*The Elementary Forms of Religious Life*,

172). He recognises their importance as sacred objects and identifies their connection to ancestors (363). For a discussion of the churinga and totemism in Durkheim, see Mitchell, *What Do Pictures Want?*, 123.

54. Focussing on the writings of Walter Baldwin Spencer, Susan Lowish discusses settler colonial understandings of the *tywerrenge* in Chapter 6 of *Rethinking Australia's Art History*. For further important discussions of the *tywerrenge*, see Batty, 'The *Tywerrenge* as an Artefact of Rule' and Gibson, '"You're My Kwertengerl"'.
55. This was the first book Medworth illustrated and she was just seventeen when the book was published.
56. Milligan, 'On the Dialects and Language of the Aboriginal Tribes of Tasmania', 271.
57. This land is Lutruwita. The little penguin (*Eudyptula minor*) is a native species of the island. For a major survey of the penguin population in Lutruwita, see Stevenson and Woehler, 'Population Decreases in Little Penguins'.
58. Wilson, *Churinga Tales*, 94.
59. Ibid.
60. Dewar, 'In Search of the Never-Never', 164.
61. Johnson, 'National Species', 339. Nyman makes a similar point, suggesting that Seton's narratives 'are parallels of the Social Darwinist narratives of the allegedly inevitable extinction of non-Western races'. Nyman, *Postcolonial Animal Tale*, 87.
62. Jones, 'Wildlifewriting?', 138. Nyman offers a compelling reading of the ethnic and racial tensions that are played out through 'Lobo', with the marauding wolf, constructed seemingly as Mexican, undone by his desire for whiteness as embodied in his mate Blanca. In Nyman's reading, 'Lobo' is implicitly a condemnation of miscegenation. Nyman, *Postcolonial Animal Tale*, 80–83.
63. Seton, 'Introduction', 9. Johnson identified as Mohawk (her chosen term) or Kanien'kehá:ka (also Kanyen'hehà:ka).
64. Ibid.
65. Jones, 'Wildlifewriting?', 142.
66. Johnson, 'National Species', 337.
67. Bonaputa-Mopu, also known as Miani, is a language of Papua New Guinea (henceforth PNG). It is spoken in the Madang Province, where the River Ramu (also mentioned in the story) is located. Beni is a place in Morobe Province in PNG. Madang Province and Morobe Province border each other.
68. Wilson, 'Marava', 7. Palms hold considerable economic importance to local communities and the simile, if taken to be Mabu's, may reinforce his sense of Marava as a symbol of his wealth.
69. The representation of 'woman as nature' also occurs elsewhere in Wilson's corpus. He compares a woman's reddish brown hair with jarrah wood, for instance, in the short story 'Combined Operations', which is set in rural Australia.
70. For a discussion of the 'exotic breast', see Gripsrud et al., 'Psychosocial and Symbolic Dimensions of the Breast', 219.
71. The French artist Raul Pommet in Wilson's short story 'Haeremai!', who resides in Papeete in Tahiti, is possibly inspired by Gauguin. In this story, the French man sees through the scheming of a Chinese man when the Indigenous population are unable to.
72. For a discussion of Paul Gauguin's paintings of Polynesian women see Parker and Pollock, *Old Mistresses*, 119.
73. Here Wilson seems to be trying to use terms from Tok Pisin, a creole language of PNG. Alys Brown uses similar language in the story 'The Reformer', published in the 1929 Christmas edition of *The Bulletin*. This may have been one source that Wilson used when composing the story. Wilson provides adverse depictions of PNG's Indig-

enous peoples in at least two other short stories, 'The Squealing Devil of Kitori' and 'The Pay Back'.

74. The linkage of Marava with European settler colonists is implied by her husband Mabu identifying his dearest possessions as his 'light-skinned wife' and a trade-knife bought from 'white men' ('Marava', 7).
75. Wilson, 'Jungle Record', 360.
76. Ibid. The American Institute of Ethnology is a fictional entity seemingly modelled on the Bureau of American Ethnology and the American Institute of Physical Anthropology. No antelope species is native to Australasia.
77. His racism towards the Indigenous peoples of Papua New Guinea, however, remained deep-seated. See, for instance, a 1953 letter to the editor of the *Sydney Morning Herald* about pidgin English. Wilson, 'Replacing "Pidgin"', *Sydney Morning Herald*, Thursday 30 July 1953, 2.
78. Mort is a common name and is sometimes used as a diminutive for Mordecai, Mortimer or Morton. In French, *mort* means death.
79. Wilson, *Minado*, 106.
80. See Foran, *The Subjugation of Canadian Wildlife*, 39.
81. Wilson, *Minado*, 50–51.
82. Lutts, *The Nature Fakers*, 97.
83. Murphy, *The Origins and History of Religions*, 25.
84. Ibid., 17.
85. At the time Wilson was writing, PNG was, however, administered by Australia.
86. Wilson, *Coorinna*, 7.
87. Plomley, *Tasmanian Aboriginal Place Names*, 22.
88. Sandes and Stirling, *In Tasman's Land*, 50–51.
89. Sandes might also have referenced the former mining town of Corinna. This township is mentioned for its goldfields in the 1886 edition of *Walch's Tasmanian Almanac* (307). The town is listed in the almanac as the last one possessing an electric telegraph on the route from Waratah and Strahan. Gustav Thureau first inspected the settlement in 1881. Thureau, 'Report on the Corinna Goldfield', unpaginated.
90. The book is identified as for adults by Wilde, Hooton and Andrews but seems to fall somewhere between children's and adult literature. See Wilde et al., 'Wilson, Erle'.
91. The racist epithet was a commonly used name for black dogs into the 1950s and therefore also contributed to the novel's realism. The word was in common usage in Lutruwita in the 1940s and early 1950s and appears in newspapers of the time (employed, for example, by the farmer and notorious thylacine killer Robert Stevenson in press correspondence). Stevenson uses the word twice in letters sent to the *Examiner* in the 1940s. See Robert Stevenson, 'In Other Days', Letters to the Editor, *Examiner*, Saturday 30 August 1941, 6; Robert Stevenson, 'Mowbray Swamp', Correspondence, *Examiner*, Wednesday 10 January 1945, 4. Stevenson's use of the term cannot be viewed as exceptional. The era's casual racism is encapsulated by the Reuter's fashion correspondent Barbara Stevenson, who describes the colour of a worsted suit using the term. See Stevenson, 'The Women's Forum', *Advocate*, Wednesday 24 August 1949, 12. For a specific reference to the use of the epithet as a name for a dog, see Anonymous, 'Black – but not…', *The Mercury*, Tuesday 4 December 1951, 5. The name given this news story alludes to a case of mistaken identity regarding a black dog but may also register growing recognition of the term's identification as racist (and of a refusal, by some, to accept the fact).
92. See Mellor, 'Contemporary Racism in Australia', 476.

93. The link between animal domestication and slavery is alluded to in Marshall Saunders's *Beautiful Joe*, where the dog's name is compared to human slave names in the United States (1).
94. If Wilson had acknowledged the continuing existence of the palawa he would probably have regarded them as 'fallen', unfree, corrupted by European settlers.
95. Anonymous, 'Marsupial Wolf', *West Australian*, Saturday 20 February 1954, 21.
96. Anonymous, 'Books and Writers: A New Animal Classic?', *Sun Herald*, Sunday 1 November 1953, 63.
97. Wilson, *Coorinna*, unpaginated [183].
98. Ibid.
99. Howarth, 'Short Views', 230.
100. Wilson, *Coorinna*, 97.
101. Ibid., 133.
102. S. H., 'Books and Writers: A New Animal Classic', *Sun-Herald*, Sunday 1 November 1953, 63.
103. Sharland, 'In Search of the Thylacine', 24.
104. Referring to mountain ranges as akin to saws is not uncommon. In *Echo in the Wilderness*, a novel that features a thylacine, Hesba Brinsmead describes the Frankland Range in Lutruwita as 'a distant, fantastic outline of serrated peaks, stabbing against the sky in demented-looking, saw-toothed ranks' (5). Here the mountains are fantasised as an unhinged attacker, the wilderness figured as a place of madness and violence. The thylacine is associated with the mountains and, by extension, these qualities (27). At another point, a thylacine is described as of the same colour as a hilltop, Lonely Hill, cementing the animal's relation with the feature and reinforcing their isolation (107).
105. A photograph of Stannard's grave was taken during the expedition to find the thylacine. Peregrine (Michael Sharland), 'Huon Pine', *The Mercury*, Saturday 5 August 1939, 5. Sharland made further reference to the grave a decade later. See Peregrine (Michael Sharland), 'Birds in the North-West', *The Mercury*, Saturday 26 February 1949, 11. The grave is now in disrepair. See Melissa Mobbs, 'Tasmanian YouTube Stars Rob Parsons, Levi Triffitt, Hugh Calvert Uncover Lost Grave of John Stannard', *Examiner*, 20 June 2021, https://www.examiner.com.au/story/7300130/prospectors-discover-piece-of-forgotten-history-on-banks-of-remote-jane-river/ (retrieved 16 March 2022).
106. The cross-cut saw was a favoured comparison used by Sharland for mountain ranges. In 1936, he writes of the 'ranges' near the Tyenna Valley (presumably the Sentinel Range) that they 'traverse the country in a series of peaks and escarpments that from a distance resemble the serrated edges of a crosscut saw'. The newspaper article makes reference to the possibility of exploiting mineral resources and timber in the remote region. Sharland, 'Mountains and Button-Grass Plains', *Sydney Mail*, Wednesday 4 March 1936, 41. In 1922, Sharland had also used the simile to describe the Flinders Ranges in South Australia. Sharland, 'Lake Eyre Expedition', *The Mercury*, Saturday 9 September 1922, 11. Concerns about the overexploitation of timber in the Wirrabarra Forest of the southern Flinders Ranges had been an issue since the 1880s, so in this context it is possible that the figure of speech was similarly motivated by conservation concerns intersecting with descriptive aptness. For an example of such concerns, see A. S., 'Forestry in South Australia', 436.
107. Peregrine (Michael Sharland), 'The Black Cockatoo', *The Mercury*, Saturday 24 May 1941, 4.

108. The cross-cut saw was already nearing obsolescence in the 1940s, soon to be replaced by the raucous and strident chainsaw. References to cross-cut saws in Lutruwita's press peak in the 1920s to 1940s. They are very rare from the 1960s onwards. Advertisements for chainsaws feature in newspapers from the 1950s onwards. See, for instance, the advert for the Danarm petrol-driven chainsaw in *The Mercury*, Wednesday 26 August 1953, 19. As well as its capacity to 'hum', the cross-cut saw has been described as a 'saw that sings'. See Alex Weinberg and Lisa Romano, 'Forest Service Tool Fells Trees, Slices through Massive Logs – and Sings', US Department of Agriculture website, https://www.fs.usda.gov/features/forest-service-tool-fells-trees-slices-through-massive-logs-%E2%80%93-and-sings (retrieved 16 March 2022).
109. Shepherd et al., 'Management of the Central Plateau of Tasmania', 1.
110. See Scott, 'Transhumance in Tasmania', 170.
111. Ibid., 171.
112. Wilson, *Coorinna*, 87.
113. Ibid., 89.
114. Ibid., 79.
115. Ibid., 99.
116. The image is similar in composition to engravings of the thylacine that Freeman discusses in *Paper Tiger*, which depict them as 'a focused, predatory animal that is looking at something in the landscape below' (96). Vernam was an American artist who was, at one time, involved in the Works Progress Administration (WPA) federal Art Project initiated by the administration of Franklin D. Roosevelt. He seemingly had little knowledge of Lutruwita's landscapes or its flora and fauna and the tiger he has painted looks down upon a red rock valley with what appear to be maples or pines visible in the middle distance. The setting is more Sierra Nevada than Lutruwita's Highlands.
117. Lake Dove is visible to the left, close to the tiger's forepaws. The composite photograph, which is held in the collection of the Queen Victoria Museum and Art Gallery, Launceston, is variously dated by them to *c*.1935 and 1940.
118. The nearest to a depiction of excretion in Wilson's writings is an owl coughing up pellets.
119. Mary Allen remarks upon the lack of sex lives in animal stories of Turtle Island. Wilson seems to have followed in this tradition in his Australian animal imaginings. For Allen the dearth of sex accorded to Turtle Island's animals may reflect puritan values but is more likely influenced by a desire to preserve a sense of freedom: 'there is no greater threat to liberty than a mate.' Allen, *Animals in American Literature*, 15. Jopi Nyman, however, has shown that in Jack London's writings a sexually charged gay subtext is sometimes in operation. See Nyman, *Postcolonial Animal Tale*, 61–63.
120. Wilson, *Coorinna*, 79.
121. Ibid., 80.
122. William Shakespeare's *Taming of the Shrew*, in which the bad tempered 'shrew' Katherina is tamed by Petruchio, is a comedy, but the gender politics presented in the play nonetheless resonate with how Wilson constructs the relationship between Coorinna and Loonganna. For a useful discussion of gender relations in the play see Boose, 'Scolding Brides and Bridling Scolds'.
123. Wilson, *Minado*, 142.
124. Wilson, *Coorinna*, 120.
125. Bunty also provides Wilson with the opportunity to represent how free-range domestic animals can potentially turn feral, a common occurrence in both Australia and Canada where livestock such as pigs have escaped and live as wild animals. Wild pigs

pose a considerable problem to farmers in both countries. For a discussion of Canada's wild pig problem, see Aschim and Brook, 'Evaluating Cost-Effective Methods'.

126. Warner, *A-Hunting of the Deer and Other Essays*, 16.
127. Davison, *Man-Shy*, 110.
128. MacDonald, 'The Revolt Against Instinct', 225.
129. Ibid., 226.
130. For another discussion of how Darwinism informs the writings of Roberts and Seton, see Dunlap, 'The Realistic Animal Story'. In Dunlap's reading, the two authors are more conformist to Darwin's ideas than MacDonald believes.
131. Krebber and Roscher, 'Introduction', 2.
132. Ibid. The wish to define animals as possessing individuality, however, retains a sense of the identitarian, which some (such as Carla Freccero) regard as problematic for reasons outlined by me in Chapter 1.
133. Wilson, *Coorinna*, 12.
134. Most Canidae young are referred to as pups, although juvenile wolves are sometimes erroneously referred to as cubs and fox young are at times called cubs or kits. The common decision to call thylacine juveniles 'pups' can be interpreted as a legacy of the 'marsupial wolf' designation. Hyenas have cubs as do tigers.
135. The word 'snarl', which describes growling and the baring of teeth, is Germanic in origin (*schnarren*) and probably imitative, an example of the effort to bring nonhuman animal vocalisation into human language.
136. Pyle, 'The Salem Wolf', 11.
137. Wilson, *Coorinna*, 85. See Wilson, *Minado*, 177.
138. For a discussion of negative views of the wolf, see Marvin, *Wolf*, 7–8.
139. Baker and Dickman, *Secret Lives of Carnivorous Marsupials*, 177–78.
140. Young and Goldman, *The Wolves of North America: Part 1*, 275.
141. Ibid.
142. Wilson, *Coorinna*, 62.
143. Young and Goldman, *The Wolves of North America: Part 1*, 275. See also Marvin, *Wolf*, 109–11.
144. The wolf called 'Crip' had a clubbed forefoot. The name 'Old Lefty' was given to a wolf who twisted off much of his left foot escaping from a trap and 'adopted a peculiar gait' thereafter. See Young and Goldman, *The Wolves of North America: Part 1*, 284, 277.
145. Marvin, *Wolf*, 108.
146. Foucault, *Discipline and Punish*.
147. Writing of the European practice of nailing birds of prey to the doors of houses, Jen Girgen suggests that the motivation for this public spectacle of punishment was probably to act as a deterrent to other raptors. See Girgen, 'The Historical and Contemporary Persecution of Animals', 118.
148. Like the hunter-trapper-poacher M in *The Hunter*, the shepherd acts alone. Hunting is often a social affair but the shepherd and M are 'lone wolves'.
149. It is noteworthy that there is no comparable section in, for example, his chapter on the woodland caribou, no 'Famous Caribou'. The carnivore, particularly the apex predator, is often conceived as more 'heroic'.
150. Seton, *Lives of Game Animals: Volume 1*, 313.
151. Renegade is a term that Seton reserves for dogs that join wolf packs.
152. Young, who cites Seton's examples, also adds examples such as the 'Aguila Wolf', the 'Pryor Creek Wolf' and the 'Unaweep Wolf'. See Young and Goldman, *The Wolves of North America: Part 1*, 275–85.

153. Seton, *Lives of Game Animals: Volume 1*, 319.
154. Ibid.
155. Robisch, *Wolves and the Wolf Myth in American Literature*, 47.
156. Ibid., 90.
157. Seton, *Lives of Game Animals: Volume 1*, 319.
158. Ibid.
159. Ibid. Darwin saw instinct as distinct from reason. In *The Descent of Man*, Darwin describes animal intelligence as it manifests through emotional sophistication. The 'higher animals' exhibit many 'complex emotions' (102). In the context of the thylacine, arguments for the species' survival to the present involve claims that the tiger has learnt from interacting with humans, becoming more shy and retiring.
160. For an overview of human-driven evolution, see Hendry et al., 'Human Influences on Evolution'.
161. Darwin writes of 'beneficial variations' in *The Variation of Animals and Plants under Domestication*, 6.
162. Robisch, *Wolves and the Wolf Myth in American Literature*, 44.
163. Darwin, *The Descent of Man: Volume 1*, 99.
164. Ibid., 102.
165. In *The Descent of Man*, Darwin notes that reasoned responses may become instinctual over time (97). He also accepts that differentiating reason from instinct is often difficult (107).
166. Long, 'The Modern School of Nature Study and Its Critics', 693. This is not the only taxidermy metaphor employed by Long in the essay. He also observes of Gilbert White's nature writing that it is 'invariably dry as a stuffed owl' (698).
167. Seton, *Wild Animals I Have Known*, 9.
168. Ibid., 10.
169. Ibid., 9.
170. Ibid., 12. In her essay 'Animal Victims', Margaret Atwood reads stories such as Seton's as operating not in a tragic but an elegiac mode to which pathos is central (217).
171. Seton, *Wild Animals I Have Known*, 12.
172. Allen suggests that the killing of one wild animal by another forms the most exhilarating violence in London's works. Allen, *Animals in American Literature*, 90.
173. DeKoven, 'Guest Column', 364.
174. See Forster's discussion of round characters in *Aspects of the Novel*, 73–78.
175. Burroughs, 'Real and Sham Natural History', 301.
176. Ibid., 303.
177. Cited in Lutts, *The Nature Fakers*, 113.
178. Shiras in Clark, 'Real Naturalists on Nature Faking', 426.
179. Ibid.
180. Roe, *Her Brilliant Career*, 532. Wombats have been known to swim across rivers or in flooded areas. See, for example, Charlotte Gore, 'Wombat Nicknamed Ian Thorpe', ABC News, Tuesday 21 November 2023. https://www.abc.net.au/news/2023-11-21/act-wombat-rescued-yerrabi-pond-woman-in-kayak-ian-thorpe/103133288 (retrieved 21 March 2025).
181. Peregrine (Michael Sharland), '"Tiger" Theme in Novel', *The Mercury*, Saturday 12 December 1953, 19.
182. Burroughs, 'Real and Sham Natural History', 299.
183. Ibid., 298.
184. Ibid.

185. The story appears in *In the Wilderness*. Written in 1875, it was first published in 1878 as 'The Adirondacks Verified: A-Hunting of the Deer'.
186. Burroughs, 'Real and Sham Natural History', 300.
187. The story was initially published as part of a series by Warner about the Adirondacks and should be read as contributing to a general portrayal of the place and of people who live and visit there.
188. Gavitt, 'Disturbance Effect of Free-Running Dogs on Deer Reproduction', 44.
189. Additionally, David Herman has convincingly identified a Christian subtext in the story. See Chapter 2 of Herman's *Narratology beyond the Human*.
190. Warner, *A-Hunting of the Deer and Other Essays*, 4.
191. Warner here implicitly affirms the myth of the United States as virgin territory, ignoring the existence of pre-colonial Native American culture.
192. Terrie also calls Warner 'tearful'. See Terrie, 'Urban Man Confronts the Wilderness', 20fn23.
193. Burroughs, 'Real and Sham Natural History', 299.
194. The construction of Coorinna as an avenger for his slain kin echoes how palawa attacks on settlers were interpreted by men such as Robinson. He ascribed such attacks to acts of retaliation for past wrongs, laying the blame for intergroup violence on a few bad settler apples rather than acknowledging the broader provocation constituted by colonisation, with its accompanying acts of land and resource theft.
195. Burroughs, 'Real and Sham Natural History', 299–300.
196. Ibid., 300.
197. Ibid., 303.
198. See Preston et al., 'Green-Backed Heron Baits Fish with Insects'; Riehl, 'Black-Crowned Night Heron Fishes with Bait'. The great blue heron has been seen catching fish attracted to floating bread but not manipulating the bread. See Zickefoose and Davis, 'Great Blue Heron (*Ardea herodias*) Uses Bread as Bait for Fish'. The heron has also been observed vibrating water with their bill to attract fish, simulating the presence of bait. See Combs, 'Great Blue Heron Forages for Fish by Bill-Vibrating'. Whether a heron would sacrifice an amphibian as bait remains questionable, but that Long may have seen a heron bait fishing is credible.
199. Burroughs read Long's account of kingfishers teaching their young in the story 'The Kingfisher's Kindergarten' published in *Country Life in America*. The story was reprinted as 'Kingfisher's Kindergarten' in Long's *A Little Brother to the Bear* (177–90). Kingfishers also feature in *Secrets of the Woods* but they do not engage in educating their chicks.
200. Bralliar, *Knowing Birds through Stories*, 163.
201. Long first witnessed kingfisher adults using dead minnows in their diving lessons. Days later, he saw live minnows being employed for this purpose.
202. See Kilham, 'Biology of Young Belted Kingfishers', 246.
203. Krog, 'The Learning Curve'.
204. Burroughs uses the term 'fishhawk' instead of osprey ('Real and Sham Natural History', 308). Ospreys have been observed engaging in behaviour of the kind portrayed by Long. See Meinertzhagen's 1953 (published 1954) study, 'The Education of Young Ospreys'. A more recent study by Megan Howard and Will Hoppitt that focussed on ospreys in the nest, 'Ospreys Do Not Teach Offspring How to Kill Prey at the Nest', failed to witness teaching but acknowledged that in ospreys 'diving is probably the component of hunting behaviour that would benefit most from parental teaching' (3).
205. Burroughs, 'Real and Sham Natural History', 305.

206. The other featured a dog assailing a badger. Romanes himself extensively discusses animals learning through imitation. For the reference to Darwin, see Romanes, *Mental Evolution in Animals*, 221.
207. Ross, *Voyage of Discovery*, 246.
208. Wilson, *Minado*, 136.
209. Wilson, *Coorinna*, 138.
210. For a useful discussion of this controversy, see MacLulich, 'The Animal Story'.
211. Mahan, 'The Writing of History', 290.
212. Ibid.
213. Burroughs, 'Real and Sham Natural History', 309.
214. Mahan, 'The Writing of History', 290.
215. Ibid., 291.
216. Ibid. Hayden White explains emplotment as the '*kind of story*' (his emphasis) used to narrate a history. See White, *Metahistory*, 7.
217. Burroughs, 'Real and Sham Natural History', 300.
218. McHugh, *Love in a Time of Slaughters*, 212.
219. Krebber and Roscher, 'Introduction', 8.
220. Ibid.
221. Lutts, 'The Wild Animal Story', 2.
222. Lutts, *The Nature Fakers*, 32.
223. McHugh, *Love in a Time of Slaughters*, 212.
224. Butterworth, 'Introduction', 7.
225. Ibid.
226. The character John Wood, uncle to Joe's loving owner Laura, is used by Saunders to discuss the different kinds of hunting that exist and to denounce fox hunting and hare coursing (Saunders refers to 'rabbit', but this seems unlikely) (*Beautiful Joe*, 186). He also disparages the use of steel traps, calling them cruel and 'an abomination' (187–88).
227. Ibid., vii.
228. Wilson, 'The Shoes of the Negus-Negusti', 106–7.
229. See Hogg, 'Personality, Individuality, and Social Identity', 178.
230. My preference for the term singularity in earlier chapters when referring to the unique characteristics of specific thylacines is motivated, in part, by a desire to resist the kind of anthropomorphising that accompanies discussion of individuality or personality in relation to nonhuman animals.
231. Szanto specifically names cats, which, as domestic pets, seem qualitatively different to many groups of animals. See Szanto, 'In Hate We Trust', 461.
232. Van Hecke, 'Metaphor in the Hebrew Bible', 15.
233. Wilson, *Coorinna*, 153.
234. Sayre, *Indian Chief as Tragic Hero*, 4.
235. See Moses, 'The Tragic Novel', 7.
236. The same article incongruously describes a tiger putting up a tremendous fight against a human aggressor. See Anonymous, 'Encounter with a Tasmanian Tiger', *Illustrated Sydney News*, Tuesday 16 April 1867, 5.
237. Anonymous, 'A Destructive Animal: Hyena Killed at Mawbanna', *Advocate*, Wednesday 14 May 1930, 6.
238. Anonymous, 'Too Stupid to Tame: Tasmanian Wolf in London', *Great Southern Leader*, Friday 27 August 1909, 6.
239. Martin Hambleton, 'For the Boys and Girls: The Tasmania Wolf and Devil, Strange Survivors of a Dying Race', *Sunday Mail*, Sunday 7 January 1934, 8.

240. This kind of 'death without dying' ending features in films such as *Thelma and Louise* (dir. Ridley Scott, USA, 1991), in which Thelma and Louise drive off the edge of the Grand Canyon but the film cuts before they crash to the ground, and in 'Blake', the final episode of the television series *Blakes 7*, when there is a cut to the end credits while the defiant anti-hero Kerr Avon is surrounded by enemy troops, leading to an acoustically ambiguous last stand played out across the credits that leaves Avon's fate unresolved. Coorinna is more akin to the last example, with a smidgen of doubt as to whether he dies able to sustain the true believers who can speculate: 'perhaps the bullet missed his vital organs.'
241. This species would subsequently be named *Bauera rubioides*. It is the same species that is native to Lutruwita. None of the palawa names for bauera have survived. I am grateful to Daisy Allan for sharing her insights relating to this.
242. Vedel and Lange, *Trees and Bushes in Wood and Hedgerow*, 170.
243. See Haas, 'Neurological Stamp', 470.
244. Considerable debate has been expended on 'Benjamin's' gender. Stephen Sleightholme gives a good summary of the arguments and provides a compelling case for 'Benjamin' being male in 'Confirmation of the Gender of the Last Captive Thylacine'. Recently, Ross Benjamin and Eline Lorenzen identified 'Benjamin' as the 'last thylacine' in their article 'Thylacine Tales' (7).
245. See Paddle and Medlock, 'The Discovery of the Remains of the Last Tasmanian Tiger'.
246. Casts were made of some of these prints. Sharland also reported seeing the tracks of wild dogs. See Sharland, 'In Search of the Thylacine'. Despite Sharland's claim, it is possible the prints are those of a wombat and not a tiger.
247. Anonymous, 'Tasmanian Tiger: Habitation of Jane River District Proved by Expedition', *The Mercury*, Wednesday 23 November 1938, 6.
248. Sharland, 'Tasmania's Rare "Tiger"', 88.
249. Ibid., 84. Sharland nonetheless still refers to the 'primitive characters' of the species (88).
250. Fleay, 'On the Trail of the Marsupial Wolf', 159.
251. Ibid., 174. The tracks were found in the vicinity of Poverty Plain.
252. Quoted in Simon Bevilacqua, 'Talking Point: Look a Tiger in the Eye', *The Mercury*, 7 September 2016, https://www.themercury.com.au/news/opinion/talking-point-look-a-tiger-in-the-eye/news-story/1d16bf6d4b9b39904ac2cc327cde5c60 (retrieved 2 March 2022).
253. Sightings have, of course, continued, but after Naarding's none have been as persuasive.
254. The killing was reported as having occurred 'yesterday' in *The Advocate* on 14 May 1930. In 1980 in a magazine article that included excerpts from an interview with Batty the date was given as 6 May. See Brass, 'The $55,000 Search to Find a Tasmanian Tiger'.
255. Wilson, *Coorinna*, 177.
256. Anonymous, 'Tasmania's Problem Is: First Catch Your Tiger', *Canberra Times*, Thursday 5 December 1968, 15.
257. Kalof, 'Animal Blood Sport', 441; Kheel, 'The Killing Game', 38.
258. Stone, 'Genocide as Transgression', 58.
259. See, for example, the discussion of the significance of the tiger in Colonial India in Sramek, '"Face Him Like a Briton"'.
260. On repression in *The Fog*, see Leeder, 'Skeletons Sail an Etheric Ocean', 72.
261. Wilson, *Coorinna*, 180.

262. Meston, *The Van Diemen's Land Company*, 39–40. The palawa name for Emu Bay, Mutawaynatji, bears no resemblance to known words for emus in clan languages, so Hellyer's name is not a translation of an existing Indigenous place name.
263. Lee, 'Early Days in North Tasmania', 165.
264. Ibid.
265. Dudley Le Souëf discusses records of emus being sent from Victoria to Lutruwita in the 1850s. See Le Souëf, 'Extinct Tasmanian Emu', 230, 231. He also suggests that there may have been earlier imports (231).
266. Robinson, *Friendly Mission*, 323. Robinson's transcription of palawa words is often questionable. For a discussion of Gonanner in this context, see Crowley and Dixon, 'Tasmanian', 405. Drawing on an 1826 manuscript by Thomas Scott, Milligan gives an alternative rendering of the Paredarerme (Oyster Bay) nation word for emu as 'Padanawoonta'. Milligan, *Vocabulary*, 60.
267. See, for instance, the remarks of William Vincent Legge as recorded in Anonymous, 'Australian Ornithologist's Union', 158.
268. Le Souëf, 'Extinct Tasmanian Emu', 230.
269. Lee, 'Early Days in North Tasmania', 173.
270. Alexander McKay receives 7/- (seven shillings [the equivalent of roughly £25 today] was about a day's wages for a skilled tradesperson) for a 'bitch hyaena' from George Robson, an employee of the VDLC who lived in the Parteenno/Hampshire Hills. See the entry for 25 August 1830 in Robinson, *Friendly Mission*, 241.
271. Entry for 24 August 1830 in Robinson, *Friendly Mission*, 240. Wild dogs in the region were identified by James Backhouse in *A Narrative of a Visit to the Australian Colonies* as 'very destructive to the sheep' (122). Backhouse also refers to the 'Hyena' or 'Tiger' as a sheep killer and recounts a tale of a man who was confronted by a thylacine on the Emu Bay Road, 'probably having had its young ones in the bush' (122).
272. In addition to the name Pilri for Cape Grim (proposed by the Tasmanian Aboriginal Centre), the name Kennaook is sometimes used. Haygarth, 'Mining the Van Diemen's Land Company Holdings', 94.
273. Robinson, *Friendly Mission*, 236.
274. See McFarlane, 'Cape Grim'.
275. The sheep were killed in response to an incident a few days prior in which VDLC shepherds had tried to rape palawa women after luring them into a hut. See McFarlane, 'NJB Plomley's Contribution to North-West Tasmanian Regional History', 132. The cliff (the 'rocks') where the murders occurred is called Taynayuwa in palawa kani.
276. Chamberlain seems to belatedly be trying to construct some kind of logic behind the atrocity as a means to justify it. According to Ian McFarlane over one hundred sheep were killed. See McFarlane, 'Cape Grim Massacre', *Companion to Tasmanian History*, Centre for Tasmanian Historical Studies, 2006, https://www.utas.edu.au/library/companion_to_tasmanian_history/C/Cape%20Grim%20Massacre.htm (retrieved 10 March 2022). For a semi-fictionalised account of Robinson's meeting with Chamberlain, see Mudrooroo, *Doctor Wooreddy's Prescription*, 97–99.
277. McFarlane, 'NJB Plomley's Contribution to North-West Tasmanian Regional History', 139.
278. Robinson, *Friendly Mission*, 207.
279. James describes the railway company as a virtual subsidiary of the Land Company, sharing the same offices, secretary and directors. See James, 'The Emu Bay and Mount Bischoff Railway', 97.

280. The tramway opened in February 1878, offering a thrice-weekly service between Mutawaynatji/Emu Bay and Partennamana/Mount Bischoff. See Walch, *Walch's Tasmanian Almanac* (1879), 213.
281. The genocidal logic that 'rationalised' attacks on the palawa was similar in this regard. The various First Nations peoples of Lutruwita were construed as a threat to the colony. There was no place for the nations in the settler colonial project so they were murdered or, later, exiled from their lands and pressured to assimilate, subject to cultural genocide long after physical genocide had ceased.
282. This insight is based on searches of the Trove database of Australian newspapers using combined search terms such as 'tiger', 'protect', 'Tasmania' and 'extinction'.
283. Dennis Colbron Pearse, 'Letters to the Editor: The Tasmanian Tiger', *The Mercury*, Monday 4 January 1937, 6.
284. The same went for capturing a tiger, hence Elias Churchill's purported quiet sale of a female thylacine to Beaumaris Zoo in 1936. See Paddle and Medlock, 'The Discovery of the Remains of the Last Tasmanian Tiger'.
285. Van Essen, 'Denying Extinction?', 14.
286. Wilson may also be unwilling to write of Coorinna's death agonies, viewing such a description as too voyeuristic and therefore as unethical, but he readily gives other (often graphic) accounts of animal death, so this explanation seems unsatisfactory.
287. While hiking in a remote area of Lutruwita, I came across tracks that looked like those of a thylacine (they were probably from a wombat). It was a reminder of the power of desire to inflect vision.
288. Ball, 'Primal Revenge', 544.
289. Hegel, *System of Ethical Life*, 221–22.
290. Blanchot, *La part du feu*, 312.
291. Schwenger, 'Words and the Murder of the Thing', 100.
292. Blanchot, *La part du feu*, 313.
293. Wilson, *Coorinna*, 181.
294. The first edition of the novel was published by Andre Deutsch and printed in England then imported to Australia, so the paper did not derive from Lutruwita.
295. Trade publishing began earlier, in the 1890s, and many titles were printed in Australia (such as the Australian editions of *Man-Shy*) on locally made paper. Pataway is the palawa kani name for Burnie. Pataway is in the lands of the Pirinilaplu people. The territory was historically associated with the Plairhekehillerplue clan.
296. Onfray et al., 'Surrey Hills, Northwest Tasmania', 14.
297. Fletcher et al., 'The Loss of an Indigenous Constructed Landscape', 140.
298. In 1937, the stockman-trapper Luke Etchell claimed to regularly snare thylacines in the area during the hunting season. See Haygarth, 'The Myth of the Dedicated Thylacine Hunter', 36. In his obituary, Etchell was said to be 'most proficient at snaring' and to have 'waged eager war on [kangaroos and wallabies] every time the game season was declared open.' Anonymous, 'Mr. Luke Etchell', *Advocate*, Thursday 1 July 1948, 2.
299. The impact of logging on the environment in Lutruwita is discussed in more detail in Chapter 5.
300. Onfray et al., 'Surrey Hills, Northwest Tasmania', 7–9.
301. Proverbs 10:19, in *Holy Bible*, 663. See Buchanan, *Parent/Teacher Handbook*, 58; Koptak, *Proverbs*.
302. Marinkova, *Michael Ondaatje*, 14–15.
303. McHugh, *Animal Stories*, 219.
304. Ibid., 212.

305. Ibid., 14.
306. Ibid., 19.
307. Ibid., 20.
308. Ibid., 218.

Chapter 5

Coloured by Suffering

Capturing the Thylacine on Film

Of Wolves and Norsemen

The film *Severed Ways: The Norse Discovery of America* is loosely based on the Vinland sagas, a series of thirteenth-century accounts detailing Viking trips from Greenland to Canada in the early eleventh century.[1] *Severed Ways* follows the fate of two Vikings, Orn and Volnard, stranded in the wilderness somewhere in Turtle Island. The setting of the film is vague although it is often suggested that the locale is Newfoundland in Canada. Some of the footage used was shot at the reconstructed Viking staging post at L'Anse aux Meadows in the north of the island.[2] Orn and Volnard had been reconnoitring the forested land when their comrades became engaged in a skirmish with some *skrælings* – the Norse term for Turtle Island's First peoples – and fled.[3] The two, left for dead by their companions, decide that their best option is to head north to a Norse settlement that they refer to as Vinland, presumably L'Anse aux Meadows.[4]

To briefly summarise the narrative, early in their journey the men encounter two monks, escaped Viking slaves from Iceland, who have built a small timber church. Orn kills one of the monks and Volnard catches the other but then lets him go. Orn then torches the church.[5] The tensions between Old Norse religion and Christianity are a major theme in the film and a feature of the Vinland sagas. The two Vikings continue their journey. Volnard meets the surviving monk again and soon separates from Orn and begins to learn about Christianity. Orn, meanwhile, is drugged and captured by a First Nations woman, who transports him back to her camp.[6] He later escapes and finds Volnard and the monk, killing the latter. The two Vikings fight. Volnard then heads off alone and is fatally arrowed by a First Nations man. Orn rushes to the scene, killing the archer. He then builds a floating funeral pyre for his dead comrade. The lone Viking continues north, eventually dying of exhaustion and exposure in a snow-covered forested landscape.

Severed Ways is noteworthy for a scene early in the film that takes place at a lean-to the two Vikings have constructed in the forest. Orn stands guard by firelight while Volnard sleeps. He is disturbed by strange noises, clearly animal, coming from nearby and, taking a burning brand, investigates. A pack of wolves is fighting, perhaps over scraps of foods. Orn throws the brand at them, causing them to disperse. As he settles once again, a wolf howl can be heard. This brief scene may be the only cinematic portrayal of Newfoundland wolves. Orn does not appear particularly disconcerted by their presence, probably because at that time wolves were present, if not common, in much of the Viking world, including Iceland and Greenland.[7] In *Severed Ways*, it is Indigenous people rather than wolves that Orn and Volnard are fearful of. At one point in their journey northwards, Orn spots a *skræling* by a stream. His reaction is to try and sneak up and attack them, but they disappear and Orn takes flight, panicked. The First Nations person who eludes Orn, the woman who will subsequently abduct him, is accorded a circular, restricted field of vision by the camera, a point of view that indexes her alterity. She is shown not to see in the same way as the Norsemen.

The disturbance in the field of vision that the First Nations woman is made to embody also draws attention to *Severed Ways* as a film, enacting a kind of distancing between the spectator and the screen image through drawing attention to its representational status. Reflecting on history on film, Natalie Zemon Davis associates this kind of approach with art cinema.[8] History films in art cinema can potentially generate a more reflexive response from their audience through discouraging full immersion in a given story. Another way that *Severed Ways* rejects a seamless vision of the past is through the employment of an anachronistic score including compositions by black metal and death metal bands. Davis suggests that good history films should find aesthetic ways to signal the inventiveness at work behind the story being presented.[9] There is a need to satisfactorily frame its 'truth status'.[10] Director Tony Stone explained that this choice of music was motivated by a perceived cultural continuity. Heavy metal is described as embodying a 'warrior spirit' and manifesting a harshness and paganism in keeping with the Vikings.[11] For Stone, the metal music is also historiographical, it 'recalls history', but the history it offers is affective, providing the feel of Viking experiences in Turtle Island rather than offering a narrative of events. The influence of Norse culture is marked in heavy metal subgenres including black, death and Viking metal.[12] Through the figure of the 'heroic warrior', Viking metal has been linked with expressions of hypermasculinity.[13] The music in *Severed Ways* therefore works to buttress a vision of Norse masculinity as embodying exaggerated aggression and bravery.[14] It combines with the film's portrayal of the Vikings living off the land to produce an image of Norse masculinity as akin to frontier masculinity, a masculinity involving skill at bushcraft and hunting and also a ready capacity for violence.

Drawing on Jock Phillips's research, Raewyn Connell contrasts 'the brawling single frontiersman' with the 'settled married pioneer farmer'.[15] Historically, the Norse who explored Turtle Island came from farming communities, but in *Severed Ways* it is the Viking as warrior-marauder that is emphasised. The monks are the ones portrayed as operating a smallholding, keeping poultry. Orn and Volnard are, however, incapable of adequately performing their frontier masculinity. Volnard goes 'soft', growing increasingly interested by Christianity, and Orn clearly feels emasculated. He dreams that his wife leaves him for another man because of his cowardice. His capture and subsequent violation by the First Nations woman – who straddles and rides him while he is tethered and supine – reinforces his unmanning.[16] Both Vikings therefore do not conform to the acoustic vision of masculinity offered by the musical score.

The celebration of hypermasculinity in some metal music is channelled through references to animality, particularly lupine behaviour. Aspasia Stephanou has thoughtfully explored this phenomenon, focussing on the use of the wolf and the werewolf as themes in black metal.[17] She foregrounds the importance of Norse mythology and particularly of the monstrous wolf Fenrir as important thematic sources for metal musicians.[18] Stephanou notes that the founder of Burzum (one of the bands that features in *Severed Ways*), Kristian Vikernes, changed his name to Varg, which means wolf in Norwegian. Ashley Walsh suggests that the wolf as it is figured in metal music emerges from two distinct sources, Norse mythology and Christian literature. In Christian literature, the animal is associated with sin and vice.[19] As part of his account of the raid on Lindisfarne in 793 CE, the monk Symeon of Durham described the Viking raiders as like 'stinging hornets' and 'fierce wolves'.[20] The analogy is meant to affirm the inhumanity of the Norse: they behave like 'animals'.

The brief scene in which Orn scares off the wolves seems designed not to foreground his 'animality' but rather his 'primitiveness'. Orn's lupine encounter has been read as a homage to *Quest for Fire* (dir. Jean-Jacques Annaud, Canada, 1981), where a similar event is portrayed.[21] In *Quest for*

Fire, Amoukar, who is a member of a group called the Ulam (seemingly Neanderthals), confronts a pack of wolves with a burning brand, causing them to flee.[22] If the cinematic parallel is accepted, then Orn's repeating of the gesture of Amoukar links him with 'primitive' humans.[23] The wolves figure the wilderness as primeval but through the reference to *Quest for Fire*, Orn is also constructed as primitive, a throwback to a bygone age. The wolves in both these films are used as ways of figuring the dangers of the 'natural world'. Their growls and howls become a ready acoustic shorthand for the perils of the backwoods. Growling signals aggressive intention and serves to foreground a wolf's ferocity.[24] Long before *Severed Ways*, the wolf's association with viciousness was exploited for marketing purposes in cinema. The short 1903 film *North American Grey Wolves*, for instance, which featured caged wolves at feeding time, was promoted by its director, Siegmund Lubin, on the basis that: 'Every one knows how ferocious wolves are from what has been written about them, but many have never seen the beast, and to this element our picture appeals.'[25] The film here is conceived as illustrating and affirming pre-existing perceptions of wolves. A film distributed by Pathé entitled *The Wolf Hunt* was screened in the United States in December 1909 and similarly exploited the fear of wolves for commercial ends.[26] In this fiction film, the storyline involved a wolf abducting an infant and then being pursued by cowboys. The wolf is shot and beheaded and the head is displayed on a tree stump. A review of the film in the *New York Dramatic Mirror* describes the scene in which the wolf carries off the newborn baby as one that would have been 'too horrible for picture presentation' except that '[w]e know it is not real'.[27] It is clear that the baby shown being left unattended earlier has been substituted for 'a bundle of white rags' such that 'the poor wolf, which is later shot, has [the audience's] heartfelt sympathies'.[28] These last comments, while droll, indicate that at times audience members may not obey the dictates of the narrative, identifying with the nonhuman animal rather than the human protagonists. Both the documentary and the fiction film cash in on the public fascination with wolves and their purported brutality. The early animal actors are made to figure ferocity, always burdened with representing human ideas about their behaviour. This typecasting continues in much later films such as *Severed Ways* and, more recently, *The Grey* (dir. Joe Carnahan, USA, 2011).

In metal music, as with cinema, wolves and werewolves are figures of violence.[29] Stephanou and D. Ferrett note, however, that as symbols of anarchic excess they are always reliant upon a paternal figure for their transgressive power: 'The thirst for annihilation, the violent transgression of limits, the disruption and the illicit crossing of boundaries are only manifested and secured with the presence of the phallus.'[30] As Ferrett observes, this means that fantasies of transgression are founded in a 'topography of grand patriarchs', of figures such as Christ, Satan and Odin.[31] This same logic operates in *Severed Ways*, in which Orn's heathen rejection of Christianity requires the presence of the monks and of Volnard, all of them serving as the baseline against which to measure his riotousness. In *Severed Ways*, the responses of the marooned Vikings to their fate are starkly contrasted: Orn is the wolf and Volnard the lamb. The former seems to regress to base instincts, embracing aggression and carnality, while the latter seeks to elevate himself, turning to spiritual solace. Despite its brevity, the appearance of the wolves, their walk-on part, is therefore a crucial moment in the film, signalling Orn's kinship with them and aiding his character development. They are employed as a plot device, instruments used to make Orn more rounded.

In the analyses that follow, the marsupial wolf, the thylacine, rather than the grey wolf will be my focus. There is, however, often crossover between how wolves and Tasmanian tigers have been represented in cinema, something unsurprising given how European lupine iconography influenced portrayals of thylacines in art.[32] The clearest example of such crossover is the low-budget horror film *Howling III*, the third film in *The Howling* franchise, a franchise centred on werewolves. *Howling III* features a werewolf colony descended from thylacines. The horror film *Dying Breed*, again featuring thylacines, also draws on lupine-inspired imagery for effect. It tells the story of two couples who travel to remote Lutruwita in search of the tiger and stumble upon a group of cannibals descended from an escaped convict.[33] The cannibals have deformed, canine-like teeth. Although not werewolves, their feasting on human flesh and their teeth make them redolent of lycanthropes. The fiction film I analyse in detail here, a movie adaptation of Julia Leigh's novel *The Hunter*, dispenses with lupine iconography. The thylacine in the film *The Hunter* is still sometimes burdened with symbolism in ways comparable to the wolves in *Severed Ways*, but the film also gestures towards an alternative way of envisioning the tiger, one where they register as telling absence rather than symbolic presence.

Public 'Benjamin'

Howling III opens with a sepia-tinted version of archival footage shot by the naturalist David Fleay of a captive thylacine at Hobart Zoo (the successor to Beaumaris Zoo). Fleay's footage also features in *Dying Breed* and in *The Hunter*.[34] In each of these examples, the monochrome archival footage is used to foreground the animal as a 'thing' of the past, as historical rather than actual. This makes the discovered survival of the thylacine (or of a hybrid) in each motion picture more impactful. Fleay's film also shows a living tiger, granting the extinct carnivore a filmic afterlife. The tiger lives on as a kind of afterimage through the moving picture.[35]

The footage was shot by Fleay on 19 December 1933 using a small ICA 35 mm cine camera, presumably the ICA Kinamo.[36] He probably operated with a positivist understanding of camera technology as granting an unfiltered, mechanical copy of reality.[37] This kind of ocular empiricism accords the camera eye significant evidentiary value.[38] Through the lens, the 'facts' of the animal would be preserved for posterity, the footage providing a neutral examination of the thylacine that could subsequently form the basis for natural historical interpretations. In this sense, although short and without commentary the effort can be classed as of the genre of the nature or wildlife film, aiming to reveal something of the life of another species.[39] The camera, however, does not afford simple empirical observations of the thylacine. The positioning and motion indexes Fleay's beliefs about tigers and how 'Benjamin' (or 'it' as the naturalist perceived him) should be seen.[40] As Scott MacDonald observes of nature films in general, 'the moment a nature filmmaker begins to construct a particular film, there is no escaping point of view: filmmakers must choose what to show us and determine a filmic structure that exhibits a particular set of conclusions, whether they are those of an individual scientist, a group of scientists or science-interested-laypeople'.[41] Through close attention to the camerawork, we get a sense of what coloured Fleay's vision, of his ideas and values.

Fleay's footage is often approached uncritically as unmediated visual evidence. The lack of voiceover or of intertitles makes the film appear raw, embodying the tiger in his simple 'isness'. The 'truth claims' of the thylacine footage are, however, suspect. Robert Paddle has drawn attention to the atypical behaviour that tigers manifested in captivity compared to in the wild, discussing 'abnormal behaviour in inappropriate contexts and environments'.[42] The stress and distress (the trauma) and the boredom of incarceration are indexed by the comportment of filmed tigers. In addition to this, some of the thylacines in the archival films are shown being harassed, being provoked into action by their keepers. In other films of 'Benjamin', he is openly shown being coaxed, cajoled and bullied into 'performing'.[43] Through technical necessity, the primarily nocturnal animal is also filmed in daylight. 'Benjamin' is forced into a day job.[44]

Sometimes the film that is screened today (which lasts roughly seventy-five seconds, depending on the projection speed) has been subject to clear editing.[45] It is not the uninterrupted raw footage that Fleay shot. Using the colourised version of the film held by the National Film and Sound Archive of Australia (NFSA) as a benchmark, clear differences are visible in the ordering of shots in the version of the footage that currently appears on the website of the Thylacine Museum (TM).[46] The section of film that bookends the TM version, for example, occurs roughly thirty seconds into the NFSA version. This section is a close-up of the tiger standing and gaping; in the background a pile of dry leaves and twigs is visible behind a mesh fence. The decision to include the famed gape at the beginning and end of the TM footage renders it more attention-grabbing. The NFSA film concludes with 'Benjamin' turning his back to the camera and padding off. It begins with the tiger shown very close to the camera, head on and in profile, footage that appears eleven seconds into the TM version. The NFSA film then cuts to a shot of the tiger (about ten seconds in) in the same part of the cage, this time on his haunches, mouth agape. This shot occurs eighteen seconds into the TM version. The differences between the NFSA and the TM films demonstrate that the latter is a montage rather than raw footage. In this sense, the film as it is reproduced on the TM website has a *history*.[47] The editing that has occurred (which is not acknowledged on the museum website) creates relationships across the sequence that do not appear in the original. The motivation seems to be to make the film more engrossing and visually stimulating.

The NFSA footage, by contrast, appears to consist of separate shots produced chronologically. Even this apparently unedited film, however, 'interprets' the thylacine, reflecting how Fleay understood 'Benjamin'. As Stephen Sleightholme and Cameron Campbell remark, Fleay 'shoots' Benjamin from various angles rather than keeping the camera static.[48] The other known films employ a fixed viewpoint. Moving around while filming transforms the process from simply recording a chosen framing of a scene to repeatedly actively composing shots. As Thomas Eberle observes in the context of still photography, acts of composition increase the aesthetic dimension of a given work.[49] The ways Fleay moves through the space of the enclosure are guided by his aesthetic judgement as he seeks 'good' shots of 'Benjamin'. This aesthetic judgement works in tandem with his scientific motives, his desire to document. Eberle notes that aesthetic criteria are culturally derived.[50] They are also historically contingent.

When filming 'Benjamin', Fleay concealed himself under a curtain, a makeshift, portable hide.[51] The first shot of Fleay's film seems to record the tiger examining this hide, sniffing around it and looking at the camera. 'Benjamin' would have recognised the smell of humans but not the form he was confronted with. Fleay filmed three short initial segments of footage of the tiger standing and then sitting beside the hide. He is seemingly exploiting the tiger's proximity to get good footage of his head, focussed on recording morphology. As mentioned previously, the tiger gapes at this point, distressed by the alien presence in his enclosure. From this agonistic response, the hide was clearly perceived as a threat.[52] The tiger was striving to discourage the alien presence from remaining in situ yet, unfortunately, this probably had the opposite effect as the ability to open the jaws so wide has regularly been identified as one of the thylacine's remarkable characteristics. The 'yawn' is iconic and Fleay was no doubt delighted to film it.

Ignoring this sign of anxiety, the naturalist kept at his task. In the next section of film, Fleay has shifted position and captures the tiger standing beside a mesh fence. He seems to be trying to fit 'Benjamin's' whole body in the frame (the tail is only partly visible), again motivated by a need to record anatomical form. Once he has filmed 'Benjamin's' body in profile for a few seconds, he stops. The next shot is of the tiger lying down, basking, alert but at ease. There is a cut and then 'Benjamin' is filmed in the same position but apparently startled, raising his head and then standing up. It is possible that someone outside the enclosure intervened here as 'Benjamin' was not 'doing anything' for the camera. Yet basking *is* doing something. It is a behaviour that can have several motivations including to ensure thermal comfort. Besides basking, many marsupials (including the numbat and, possibly, the Tasmanian devil) also enter states of torpor (a state of reduced physiological activity).[53] Basking, as a comfort and an energy-saving mechanism, is an important behaviour, yet for Fleay or one of his companions (or both) it is clearly not of suitably filmic interest.[54]

Film-worthiness guides the kind of animal comportment that is deemed apt for recording. This worthiness is likely bound up both with scientific goals and with Fleay's aesthetic sensibility. He discussed the efforts of news camerapersons to secure a 'good shot' of captive animals and the pleasure that accompanies this aim. In the Hobart Zoo film, he also seems to be seeking pleasing images. A sense of what he conceived of as a 'good shot' can be gleaned from his nature writings. He is, for example, thrilled by a carpet python (*Morelia spilota*) found near the Murray River who 'put on a display of bluff and savagery' for the movie camera.[55] Fleay also mentions filming the 'comedy' of a male brush turkey 'taking fright at the appearance of one of his *own offspring*'.[56] Here animals are conceived in theatrical terms, they 'act' for the camera, feigning ferocity or putting on slapstick. 'Benjamin's' basking is not dramatic enough and he is therefore prodded into playing a more gripping and engaging role.[57]

Although we do not see what startled 'Benjamin', other films of him give a sense of what probably happened. In footage taken in 1934 by Eric Bierre, at one point the keeper Arthur Reid repeatedly rakes the wire mesh of the enclosure with a paling to goad a reaction from 'Benjamin'. A similar form of provocation was probably employed when Fleay was filming. Fleay's willingness to disturb a resting animal is indicated by his account of the carpet python, which he described as 'basking' prior to being forced into a show of bravado. In this sense, there is a degree of 'direction' in the 1933 film. 'Benjamin' is not left to his own devices. He is prodded into action and the film thus indexes deliberate acts of aggravation. Animal cruelty is often conceived as physical violence but, in this instance, 'Benjamin' is subject to psychological abuse.[58] Historically, zoos and menageries regularly endorsed brutality of this kind. A menagerie located at the Quadrant on Brisbane Street in Launceston in the early 1870s, for example, possessed a thylacine, 'the king of the menagerie', who would hiss and growl 'when stirred up, and requested to walk round and show his muscle'.[59] Fleay and Reid seemed to similarly stir 'Benjamin' so he would exhibit his physique. Many of the creatures displayed at Hobart also endured regular jabbing by members of the public. A 1925 news report mentions visitors 'poking walking-sticks, umbrellas, etc., at the animals'.[60] In raising concerns about animal neglect at the zoo, 'Bell Bird' alluded to Marcus Clarke's story *For the Term of His Natural Life*, implying that the nonhuman animals, like Clarke's hero Rufus Dawes, were being wrongfully held captive.[61]

To view Fleay's film is to see a victim of animal genocide ('Benjamin' was not purchased to form part of a breeding programme but purely to function as a spectacle for people engaging in recreation, this at a time when it was well known his species was critically endangered) and to bear witness to animal cruelty. The ready reproduction of the film in various contexts demonstrates that the presence of this callousness is not fully recognised or is deemed unimportant. In addition to the stress and boredom of his incarceration, when being filmed 'Benjamin' was subject to emotional distress. After being disturbed, 'Benjamin' walks leftwards and the film cuts. When shooting resumes, he has returned to the same part of the enclosure where he now stands, gaping. Fleay's continual pursuit clearly troubled the thylacine. Here, the wish to record, to adequately capture 'Benjamin' on film, overrides respect for his welfare. Pacing backwards and forwards, he seeks to escape the camera-presence that assails him but always finds his path blocked. At the forty-seven-second mark in the colourised version of the film, just before the film cuts, 'Benjamin' can be seen checking himself midstride, turning back, unmistakeably responding to the over-proximity of the camera. Cumulatively, Fleay's film therefore records a merciless pursuit of the 'good shot' that subjects 'Benjamin' to prolonged anguish.[62] There is further footage of him basking and being stirred to his feet. The camera looms over him at this point (roughly sixty-five seconds in) and continues to do so until the end of the film. It cuts for the final time as he is once more walking away.

Much of Fleay's footage may be intended to record comportment and it does certainly document 'Benjamin's' movements, but the behaviour it records is directly instigated by Fleay's actions. The film is therefore a historical record of human–thylacine interaction, of human cruelty and nonhuman animal upset. Additionally, the regular pacing that 'Benjamin' performs is a behavioural trait that is common among captive carnivores. It is an example of stereotypy, a persistently repeated act by an animal that seems to serve no purpose. Pacing probably has multiple causes. In the film, it appears to be a direct response to 'Benjamin's' aversion to the invasive presence of the camera. Other explanations for pacing as a locomotory stereotypy are that it registers boredom, repeated frustrated escape attempts or unfulfilled efforts at foraging.[63] Fleay's film therefore portrays comportment induced by captivity, with pacing potentially registering 'Benjamin's' poor psychological welfare. Zoos today monitor stereotypy as a 'mental health yardstick'.[64] The footage potentially attests not just to 'Benjamin's' immediate perturbation at the film-maker's intrusion but also to underlying well-being issues. Fleay acknowledged that the thylacine was '[s]hamefully treated as a zoo inhabitant in the past'.[65]

While making the film, Fleay was famously bitten on the rump. He flippantly attributed this behaviour to the tiger being hungry, yet it is another clear sign that 'Benjamin' felt threatened. In his writings, Fleay shows awareness that taking pictures of animals could be harrowing for nonhuman animal subjects. A 1935 newspaper article about possums described a female as naturally timid and recounts her rescuing her infant from a group of camerapersons and bearing 'him out of the picture'.[66] From an account of a later experience with a kangaroo called 'Peter', it is obvious that Fleay wilfully ignored evidence of anxiety when photographing or filming animals. The kangaroo charged Fleay, knocking him over, and then smashed his camera.[67] Fleay's presence clearly similarly unsettled 'Benjamin'. Reid had to 'ward him off continually' using a paling while the naturalist took photographs and, perhaps, while he filmed.[68] The still and moving images were therefore conceived in a climate of violence. While they do not directly record the thylacine being thwacked with a wooden stake or, at the very least, threatened with it, 'Benjamin's' discomfort is still highly evident.

Off-Colour

The enduring fascination with archival films of thylacines is made clear by the centrality they are granted in museum displays (as discussed in Chapter 3) and by recent efforts to track down additional footage. This fascination is partly explained by the seeming proximity each film grants to the vanished species. They proffer an imagined intimacy. Griselda Pollock notes how archival materials in general fuel fantasies of retrieving the past.[69] Archival film, with its visual resemblance to reality, its iconicity in the Peircean sense, fosters a complex response regarding the past, simultaneously holding forth and withholding what once was.[70] Peter Kilroy has drawn attention to how the indexicality of documentary film gives the impression of touching the referent.[71] Because of this, in footage such as Fleay's, the sense of being close to a thylacine is palpable. The absence of colour, however, also acts to distance the spectator from the sights on screen, emphasising pastness. The grain of the film, the white threads and specks that punctuate its projection, also register its age, acting like wrinkles that index the passage of time. These signs of age that are caused by dirt and by imperfections in the emulsion are often excised during processes of colourisation.

In this context, the decision to painstakingly colourise a digitised copy of the Fleay footage, a project led by Samuel François-Steininger, is noteworthy. The remastered version was released in 2021 and is readily available on the internet. Laura Mulvey has examined how digital technology has changed our relationship to film. Digital spectatorship is more interactive and permits an easier 'reading in detail'. Through technologies such as the DVD player and streaming, spectators gain greater 'possession' of films. This technologically enhanced possessiveness is linked to sadism (in the psychoanalytic sense of the term), to a desire for mastery and a will to power.[72] Mulvey describes how hitherto unnoticed elements in film can emerge and become invested with significance.[73] Key to this exposure of the optical unconscious of the film is the capacity to easily pause and rewind footage. Mulvey notes the link between pausing and death, writing of the 'stop of death' and of how the way film can blur the boundaries between the living and the not living resonates with unconscious anxieties, prompting fear and fascination.[74] The capacity to still the image, to freeze-frame 'Benjamin' as he pads and frets within his cage, renders the animate inanimate.

Colouring film images has occurred since early in the history of cinema, yet digital colourisation projects are distinctive and form another technological innovation that potentially transforms film experience. Seeing 'Benjamin' in colour considerably alters the film's potential reception. Digital colourisation produces a qualitatively different image to manually added colour. In early film, tinting was characterised by imperfections and imprecision and was limited and muted. Later computer-based technologies of colourisation such as the digitally colourised Fleay footage are more controlled and varied.[75] Such colourisation does not involve adding material to the film stock in any simple sense but rather altering pixels. Various digital techniques were used to create the 'Benjamin' that appears in the colourised film. These included compositing (replacing pixels from the original with pixels sourced from a different image), colour grading (employed to adjust colour ranges), the use of AI algorithms to reduce noise (removing graininess, often referred to as denoising) and to enhance the appearance of movement in the film (adding new frames to make Benjamin's motions more seamless), rotoscoping (another means of perfecting realistic motion in film, this time through manual editing), 2D animation (digital drawing) and lighting (which can rectify changes in luminosity across numerous frames caused by the film's ageing or by inconstant ambient light).[76]

The colourisation project therefore involved both machine and human interventions. Algorithms can fill in missing data with the additional data based on machine-learned conjecture. Humans then check the work created through artificial intelligence and, if necessary, make changes to it. François-Steininger and his team spent over two hundred hours colourising the eighty seconds of footage shot by Fleay.[77] The time consumed in getting the colourisation 'right' is a manifestation of the weight of responsibility François-Steininger felt towards the subject matter. He has described the project as 'paying tribute to the last representative of a species'. This paying tribute is bound up with issues of accuracy. Discussing two earlier projects, *America in Color* and *Australia in Colour*, François-Steininger talks of 'restoring' monochrome films to 'the closest original state of the scene'.[78] For him, colourisation aesthetically enhances a film and makes it more historically accurate.[79] In this understanding, black and white footage, while authentic, corrupts the historical record because it leaches an event of its local colour. As Elizabeth Edwards has noted, it is a commonplace to view black and white historical images as embodying a 'deficit'.[80] Monochrome does not acknowledge chromatic 'fact' and is therefore lacking. It portrays a 'desaturated world'.[81]

Through careful research into the colours of the past gleaned by way of oral accounts, writings and visual materials, a colourisation project such as the one dedicated to the film of 'Benjamin' can produce a 'fuller' historical record, the past now made available in true to life 'technicolour'. Restoration as a practice is often described using medical metaphors; it is about healing wounds and restoring health.[82] This implies that whatever is being restored is ailing or injured. 'Benjamin', in such an understanding, is off-colour. François-Steininger's role is to give him a *restorative*, nursing him back to health. Colourisation is conceived as 'reviving' the past in the present. The *Australia in Colour* project brought historical events 'to life for the first time in stunning colour'.[83] The colourised Fleay footage similarly animates 'Benjamin' through a process of revivification that not only involves enacting chromatic changes but also 'cleaning' the tiger. Perceived impurities in the original film stock have been excised.

For François-Steininger, this 'detritus' taints the image of the past. Viewed from a different perspective, however, the 'visual muck' is a key element of the film's history and speaks to its age. The spots and slivers of light that 'interrupt' the footage are indices of its authenticity. Blotches (often caused by dust or warping to the emulsion) and scratches (created by erosion of the film during projection) are part of a film's life story.[84] They register dimensions of the history of its production and reception. To excise the dirt is to create a de-historicised visual fiction. Changing the 'grain' of the film is to engage in modification of its body (with grain here understood in an expanded sense encompassing not just the film grain but also the dirt and scratches it has accrued). Liz Watkins writes of colourisation as giving greater delineation to what is perceived to be important. This occurs through a process that sacrifices visual texture (such as blotches and scratches) and sensuousness for a smoother visuality.[85] In the 2021 film, the slicker, less noisy footage grants greater visual mastery over 'Benjamin', working to render him more 'knowable'. This knowledge comes at the expense of the affective power of the original footage.

The film stock Fleay employed had a nitrate base.[86] Gelatine is usually present in such film. It is used as an adhesive and is a component of the emulsion, which is formed of silver halide crystals embedded in a gelatine coating.[87] Gelatine, a protein, can be sourced from animals or plants, but plant-based gelatine lacks elastic properties and would therefore not be ideal for use in film stock. Animal gelatine derives from bone, hide and connective tissue. The gelatine is produced through a 'conditioning process' of biochemical denaturation and hydrolysis effected on the collagen present in animal remains.[88] When used in photography, gelatine is usually sourced from cattle bones and pig skin.[89] The emulsion of Fleay's film therefore probably includes bovine or porcine physical remains.[90] The film embodies animals as products, destroyed to serve human ends, to aid image-making.[91] This means that animal death undergirds 'Benjamin's' likeness. He persists across time as a combination of free silver atoms and residue from collagenous animal tissue, as an image grounded in killing and loss.[92] The blotches in the original footage are caused by a lack of uniformity in the texture of the gelatine. The colourisation process works to hide this 'animal history' embedded in the film stock. It removes imperfections that serve as unintended memorials to animal death, indexing mass loss. Kit Heintzman has argued that histories such as this, histories in which individual animals are absent but animals are preserved in a fragmented and reconstituted form, also merit our attention.[93] Although these animals are reduced to 'stuff', to a product, they once were singular.

When narrative films have been colourised, considerable controversy has ensued. Colour versions of classic black and white films have been deemed inauthentic and disrespectful.[94] By contrast, when archival footage has been colourised there has not been a comparable level of criticism. In part, this seems to be because such footage is not viewed as embodying artistry or to be narrative-driven. Changes to archival films are therefore not seen to alter or diminish a creative vision. The colourisation of archival footage is also not motivated by commercial concerns (the colourisation of classic black and white films was initially viewed as an ethically dubious effort at increasing profits). The colourisation of Fleay's film, however, also raises contentious issues. Although he saw 'Benjamin' in colour, Fleay, for example, probably 'thought' his film in monochrome, viewing the enclosure in shades of grey, considering luminosity rather than reflecting on hues.[95] He anticipated the film looking a particular way (but never its being in colour) and handled his camera accordingly. The black and white footage indexes this 'handling'.[96]

Eberle has written of the 'photographic gaze' that takes over when engaging in still photography, a gaze that is multifarious in nature but commonly involves the wish to take 'good pictures'.[97] The gaze is informed by a knowledge of technical possibilities. The film-making gaze is similar in nature to the photographic gaze. Fleay filmed in the knowledge of the technical possibilities of his cine camera, reflecting on issues such as light and shadow. He used ambient light (in contrast to the stylised lighting of a studio production) and filmed on a day when the forecast was for fine weather (with a sea breeze).[98] He also had specific motives for filming that guided his gaze. Fleay's intentions (in as much as they can be reconstructed) form part of the traces of the past that are incarnated in the film. It is therefore simplistic to view colourisation as making the 'past' more present, as the process simultaneously erases facets of that past, notably the influence of a historically specific cine camera technology and, bound up with that, a 1930s Australian naturalist's mindset, on the form and content of the footage. In this context, the benefits of colourisation are not black and white.

Fleay's efforts to keep 'Benjamin' at the centre of the frame clearly reflect what he believed to be crucial. Capturing the thylacine was at the heart of his endeavour. Colouring the footage makes 'Benjamin' stand out against the muted tones of the enclosure, reinforcing his importance. Colour influences what a viewer looks at when viewing a film.[99] In this sense, the process resonates with Fleay's aims, with his gaze, although it changes how viewers perceive the film, introducing colour contrasts where none existed. The absence of colour contrast gives the greyscale film a greater sense of flatness, making 'Benjamin' blend with his enclosure. Colourisation foregrounds him, suppressing the structure that imprisons him. Studies of monochrome and colour photographs have previously noted this phenomenon. Colour images with animals as their subject foster greater recognition of details relating to the animals but colouration has a far less significant impact on the perception of background.[100] Through increasing figure–ground differentiation, colourisation transforms how the film is read, giving 'Benjamin' even greater emphasis. At a phenomenological level, this shifts how viewers perceive him. The proximity that film and photography encourage between viewer and subject is augmented in the colourised footage. The 2021 version closes the temporal distance between viewer and viewed: 'The time gap between the past and the present fades away.'[101]

In this sense, the colourised film de-historicises 'Benjamin', making him too *present*. The removal of a historical marker (as it was embodied in the greyscale) changes the probable reception of the footage. The implications of this are manifold. The colourised film, for instance, probably registers differently emotionally for viewers. Relatively low luminosity in film encourages feelings of anxiety. This is why it is a common staple of horror films, including *Dying Breed*, which includes night scenes. Poor light in narrative cinema is a mood cue.[102] Better lighting, which grants visual mastery, can foster a sense of security. Fleay's black and white footage possesses fair luminance but the colourised version is much brighter. Although the film is not narrative cinema and the lighting is not employed to support a story that is being told, luminosity potentially still inflects how the footage is viewed. The more muted black and white film, when watched in the knowledge that thylacines are extinct (or, in 1933, on the verge of extinction), inspires a downbeat response. Making light of Benjamin by excising obscurity and shadow (qualities figured as negative in the Western imaginary) encourages a more positive reception.[103] The 2021 footage, with its 'miraculous' colourising of 'Benjamin's' pelage, seems to revive the tiger in the present and has a potential for uplift not present in the original. The greater detail afforded by the increase in luminosity also augments the sense of visual mastery. 'Benjamin', in this way of seeing, is more fully 'owned' by the viewer. The black and white footage as a mode of representation can be conceived as ethically preferable as it 'refuses' to give 'Benjamin' entirely over to the spectator. This is one of many ethical issues raised by the interpretations and transformations accompanying 'Benjamin's' colourisation.[104]

The smoother motion (at least compared to other online digital versions of the footage) of 'Benjamin' in the colourised film also potentially shifts reactions to him. In cinema, jerkiness is associated with the beginnings of film. Jonathan Burt has discussed 'jerky' camerawork in relation to early efforts to film animals in movement, with such joggling drawing attention to the camera technology.[105] The Fleay footage doesn't manifest this kind of jolting but there is still something choppy about 'Benjamin's' movements in the 1933 film. Thylacine locomotion is a subject of ongoing investigation and a motional stiffness has been attributed to the animal.[106] Some speculation on locomotion derives from studying the extant films of the animal. Technological issues such as spatial jitter (vibration or wobble caused by film movement in a projector) and projection speed mean any data extrapolated about motion must be treated with an abundance of caution.[107] Additionally, small enclosures of the kind 'Benjamin' was imprisoned in do not permit the cursorial locomotion (endurance running at speed) that thylacines probably employed to chase prey (unless they were ambush predators).[108] 'Benjamin' could thus never get into his 'stride' for the camera. His surrounds were too cramped. Whatever the truth of thylacine locomotion (either suited to extended chase or to pounce-pursuit), the motion of 'Benjamin' as he appears in Fleay's film is as much the product of the limits of the technological apparatus as of his own morphology. The slight visual stuttering of the earlier version invites recognition that the film is not recent and that it offers a mediated encounter with the marsupial. The more fluid motion in the 2021 film works to conceal the age of the film and reduces the possibility of critical reflection on the seen. Viewers therefore come to 'know' 'Benjamin' differently in the colourised footage. Neither form of knowledge should be regarded as superior but their qualitative differences need to be recognised and examined.

The softer focus of the initial footage is important. Writing of anthropological practices of representation, James Peacock describes 'soft focus' as associated with openness and holism. Objects that are seen in soft focus are not as 'sharply differentiated from background and foreground but instead [melt] into [their] surroundings as though seen in a mist'.[109] Soft focus embodies 'a holistic breadth of vision that includes the world as well as the perceiver'.[110] Peacock contrasts 'soft focus' with 'harsh light', with the latter associated with a 'no nonsense realism'.[111] Peacock does not directly define realism but suggests that it is 'a way of seeing favoured by natural science' and concerns 'the basic and elemental'.[112] A realist, in this understanding, seeks to see things 'in the raw'. The sharper focus and the cleaner image of the colourised film are efforts to amplify its sense of realism. They do enable a greater reading in detail of the film. The puckered wire at the back of the cage where 'Benjamin' (or one of his predecessors) has tried to push through to the adjoining enclosure is made far more evident. The scuffs on the concrete floor are also more differentiated. Does this heightened clarity make the past more present in colour? The 'purer', crisper footage is conceived as giving a more historically accurate portrayal of 'Benjamin'. The reality is more complex.

Watkins has examined the fictive aspects of colourisation processes. Colour in these processes should not be equated with historical veracity but with reasoned plausibility. Colours from the past are presumed rather than known. In black and white footage, for example, hues are less perceptible than the intensity of colours: colours such as dark red and dark blue will both appear a very dark grey. As Hobart Zoo closed in 1937 and only a few of the concrete and iron structures remain standing, the colour given to the tiger enclosure and its surrounds is, of necessity, a work of conjecture. The light greys that form the enclosure in the original footage reflect the absence of bright colours to be found there. The dry silver timber and drab olive foliage in the 2021 version embody common sense decisions about probable colouration. The ground visible outside the enclosure looks arid, although Hobart experienced considerable rainfall in the run-up to 19 December, so the earth may not actually have appeared quite so parched.[113] Other than his dark bands, 'Benjamin' is also light grey in the original film. François-Steininger and his team, however, had some sense of the hues of his pelage based on archival records. Thylacines were described as having fur of a 'pale greyish-brown colour, with a tawny tint', with the back 'transversally banded with blackish brown stripes'.[114]

'Benjamin's' colouring is based on that of skins held in different museum collections, on colour sketches and paintings and on written descriptions. In a sense, he is an amalgam, wearing the colours of several of his kin. The physical remains that François-Steininger studied were probably mainly those of former zoo animals sourced from Lutruwita.[115] Some of these tigers were undoubtedly captured when their rarity was already common knowledge. They were victims of animal genocide. Through his coat of many colours, 'Benjamin' comes to embody a kind of unintended chromatic monument to other victims of the genocide perpetrated against his species. He is now tinged with the colours of others, with foreign bodies, while his own remains are lost. They are believed to have been thrown out with the zoo's daily rubbish when he died in May 1936 or forwarded to a museum on the mainland.[116] Here, 'Benjamin' comes to stand for the general, he signals 'thylacine', his colour seen as substitutable rather than singular. There is therefore a violent chromatic levelling at work in the colourisation process, one that suppresses difference, constructing artificially generated 'general' tones out of multiple tigers, their singular appearances and existences mixed together, sacrificed to 'type'.

François-Steininger's team also worked from sketches and paintings of the tiger. As Carol Freeman explores at length in *Paper Tiger*, these representations do not so much reflect the thylacine as work to 'shape' how they are perceived and understood. Artworks portraying tigers are never neutral and embody various values and beliefs. As Freeman examines, depictions of the thylacine were frequently inflected by negative sentiments regarding the marsupial. The influence of specific drawings and paintings cannot be identified in the colourised 'Benjamin' but it can be assumed that some of the materials François-Steininger used embodied the 'destructive ideas' about the animal that contributed to their extinction.[117] 'Benjamin's' colouring therefore draws on imagery infused with fear and loathing. The tiger is tinged with negative emotions.

Despite the tremendous effort expended in creating 'Benjamin's' pelage, a slow viewing of the digital film (of the kind envisaged by Mulvey) shows creasing in the fur: crimping is particularly visible around the toes and wrists. At certain moments in certain places, the fur also seems matted or to reflect a metallic finish. François-Steininger observed of the project that the very high resolution of the digital copy emphasised detail, making the dense fur a challenge to animate.[118] Occasionally the labours to make the fur lifelike seem to have failed. The illusion of seeing a tiger filmed in colour falters. Details reveal something not quite right about the 'painted' tiger and therefore index instances where technology and human endeavours reached their limits as they strove for chromatic and textural fidelity. The body of pixels that passes for 'Benjamin' therefore comes up wanting. This digital unconscious to the image, repressed in real-time viewing, emphasises the tiger's constructed nature.

A more obvious failure and extreme sense of disjunct between digital body and actual body is the CGI tiger that appears in *The Hunter* towards the film's close. This thylacine was initially seen as cringeworthy by Freeman, who notes her stilted movements.[119] The anonymous tiger in *The Hunter* was manufactured by the now defunct Gadigal/Sydney-based visual effects studio Fuel VFX based on archival footage including the Fleay film. She too is haunted by her dead kin. They contribute to her tenuous motions. 'Benjamin', unsurprisingly, is a more solid and fluid incarnation. He is not cut and pasted onto a backdrop but appears in his 'home environment'. Nonetheless, something unlifelike also manifests in François-Steininger's pixel tiger. Although the colourisation project's aim was to show the past as it really was, the digitally manufactured thylacine doesn't quite inhabit his body. Up close, he is abstract, no more incarnate than the emulsive body repeatedly pacing the concrete in Fleay's analogue film. Neither the 1933 nor the 2021 film can give us (back) 'Benjamin' but something of the motivation behind François-Steininger's efforts seems indebted to notions of resurrection. Colourisation has often been framed as 'resurrecting' old films and when this revivification occurs in footage of an extinct animal it is doubly apposite as a descriptor.[120] The idea of resurrection is also helpful when considering *The Hunter*.

On the Trail of *The Hunter*

Julia Leigh's novel *The Hunter* centres on M, a mysterious mercenary who will track down rare fauna for a fee. M is engaged by a bioweapons company to kill the last thylacine, a female, and secure biological material from her. He travels to Lutruwita where, using the alias Martin David and posing as an academic researching Tasmanian devils (*Sarcophilus harrisii*), he lodges with Lucy Armstrong and her two children, Sass and Bike. Lucy's husband, Jarrah, disappeared the year previously while conducting research in rough country. M settles into a routine, scouring the wilderness for signs of the tiger and intermittently returning to the Armstrong household to replenish his supplies. Gradually he grows close to the family, becoming Lucy's lover and acting as a surrogate father to the children. After many weeks, M is called away at short notice for another job. When he returns, the Armstrong family is no longer there as an accident has occurred that left Sass badly burned and caused Lucy to have a breakdown and become institutionalised. Bike has therefore been placed in foster care. Beset by grief and loss, M resumes his search for the tiger. He is ultimately successful in finding her and fulfils his mission.

At one point in the novel, M fantasises about becoming the thylacine, suggesting that he sees continuity between himself and her.[121] When he ultimately confronts her, however, as she lies dying from his rifle shot:

> He circles tight, until he can look into her eyes. They are a deep yellowy-brown and when he does look into them he realises that she . . . does not seem to see him – her eyes are blank and vacant and say nothing.[122]

M appears to desire a flicker of recognition, a sign of commonality, but he is confronted by irreducible difference, by his not existing to the tiger in terms he can understand.[123] Blankness here should not be taken as an absence of consciousness but as opacity. M is describing seeing radical otherness.[124] The tiger sees through M and rather than look at *him*, refuses him the affirmation he craves.[125] This does not mean he becomes indifferent to her. Even after he finishes her off, she remains 'more than an animal to him'.[126] He continues to invest human significance in her. The novel's complex exploration of fantasies of animality and of how ideas of the animal are used to human ends informs the film adaptation.

Randy Malamud suggests that dimensions of the power dynamics of the gaze as explored by Mulvey can be transposed onto the human gaze as it is directed towards nonhuman animals. They are to be looked at in a similar way to how spectators adopt a masculine standpoint as they contemplate women in mainstream cinema: 'the image of the animal [is] passive raw material for the active gaze of the human.'[127] Malamud is concerned by the implications of the objectification of nonhuman animals in cinema as it is reductive, dividing them into either friend or foe, useful or expendable. His concerns resonate in the context of wolf films such as *The Grey*, in which a wolf pack is portrayed as nothing more than a group of relentless, ruthless killers. The closing scene is one in which the wolf pack leader contrives a combat *mano a mano* or *zampa a zampa* (paw to paw) with his human adversary. The human fashions artificial claws from a knife and broken bottles, manufacturing a prosthetic paw that renders him 'wolflike'. Wolves use their claws for gripping rather than ripping so this moment of paralleling, of becoming 'like', is only seeming. In the film adaptation of *The Hunter* similarities are established between the human protagonist and the thylacine, both of whom are constructed as a 'lone wolf'.

The central human character is a mercenary, here hired by an international biotech company called Red Leaf. Like in the novel, he poses as Martin David and gradually becomes attached to Lucy and her children.[128] In the film, Bike seems to have been rendered mute by the loss of his father. A local farmer, Jack Mindy, does his best to care for the family. Unbeknown to Martin, Jack is in contact with Red Leaf and keeping an eye on the local situation. Martin commences searching for the tiger, setting traplines. Several weeks into his mission, he discovers Jarrah's skeleton in the bush, the skull displaying a bullet wound. Among the dead man's possessions, there is stationery with a Red Leaf letterhead.

Martin's growing proximity to Lucy causes Jack to become jealous and he contacts Red Leaf claiming that the hunter has become a liability. In another deviation from the novel, Red Leaf despatch a rival hunter. While Martin, who has located what he believes to be a thylacine den, is in the bush, the unnamed rival breaks into the Armstrong household while the family are sleeping, searching for clues as to the den's whereabouts. Chancing across Martin's coordinates, he then commits arson. The rival hunter locates and ambushes Martin, insisting at gunpoint that he reveal the location of the den. Martin offers to guide the rival there and lures him into a leg-hold trap before shooting and killing him. Upon returning to the Armstrong homestead, he finds it burnt out and empty. Confronting Jack, he discovers that Lucy and Sass died in a fire. Bike, who survived, has been taken into care. Martin travels into the bush for a final time and waits in the den for the thylacine. When she finally returns to the refuge, he kills her and cremates her body, ensuring that there is no surviving genetic material for people to profit from. He then scatters her ashes and informs Red Leaf that the animal is gone forever. In the film's final scene, he reunites with Bike who mouths 'Martin' while racing towards him.

The film's redemptive ending mutes the moral complexity of Leigh's novel; in the latter M fulfils his mission, taking the physical samples he requires, and then coolly burns the thylacine's remains before heading homeward to brew a coffee.[129] The film adaptation does, however, share the novel's themes of mourning and melancholia. The mystery of Jarrah's fate leaves Lucy and her children unable to move on. Martin is also left grief-stricken by the death of Lucy and Sass. More broadly, *The Hunter* can be read as expressing melancholia linked to the extinction of the thylacine. The restaging of the tiger's disappearance enacted by the film, a restaging that artificially revives the animal, gestures towards a wish to explore and, perhaps, work through that loss. In this context, the act of cremation is telling. Cremation was common among palawa clans.[130] The film *Manganninie* (dir. John Honey, Australia, 1980) ends with the palawa woman Manganninie being cremated. In *The Hunter*, the ritual seems designed to mourn the loss of the Armstrong family, the tiger and, potentially, the palawa.[131]

Alongside the topic of extinction, environmental concerns more broadly are an important theme. The Green movement features repeatedly and tensions are shown to exist between local industry and environmentalists. In this context, Steven Allen suggests, 'the ecological and economic values of the wilderness' are explored and found to be irreconcilable.[132] At the local pub, a 'Save Our Native Forests' sticker on the paper towel dispenser has been defaced so it reads 'Save Our Jobs'. These jobs are linked to the commercial lumber industry. Martin's search takes place in the Upper Florentine Valley, where there are protesters opposing logging.[133] The valley supports old growth forests and has historically been a real flashpoint for environmental activists.[134] The immensity of Lutruwita's highlands is first communicated to Martin through a giant fallen tree that he encounters while ascending to a plateau with Jack Mindy. The locals believe that Martin is an environmental activist working against their interests and write 'Go home greenie scum' in shit on the windshield of his 4x4.[135] Jarrah is also constructed as an activist but his correspondence with Red Leaf is never explained, rendering his motives and character questionable.

Martin employs both active (tracking with a rifle) and passive hunting methods (snares and traps) in pursuit of his 'target'. As he is employed by Red Leaf, he is more poacher than hunter, intentionally seeking to illegally kill an animal for his paymasters.[136] A wildlife offender, he illicitly pursues the tiger as an exploitable resource. Wildlife crime, occasionally referred to as 'camouflage-collar crime', involves infractions committed against the environment, notably against animals and plants.[137] Martin, however, does not conform to common conceptions of the wildlife offender as aggressive, thrill-seeking and thirsting for adventure. He is capable of controlled violence and clearly enjoys working in the outdoors, yet his activities do not appear designed to buttress his masculinity. Angus Nurse has suggested that much wildlife crime is bound up with the exercising of masculine power and involves men targeting large nonhuman animal males as a means to showcase their manliness.[138] As Nurse notes, such activities often resort to speciesism, with criminals objectifying animals in order to 'neutralise the cruelty inherent in their actions'.[139]

Martin does not practise objectification, on the contrary he asserts the thylacine's subjectivity. By referring to the tiger as 'she' he shows that he views her as an elusive personality rather than a mere object. His choice of vocabulary (more evident in the novel but present in the film) also resists the distanced and reductive language employed by corporations and legal entities in the context of animal abductions for research, where they are referred to as 'resources' and 'specimens' that are 'captured' or 'extracted'.[140] In a biomedical or biotechnological context, nonhuman animals are often referred to as 'objects' or 'goods'.[141] This echoes the historical objectivising language that contributed to the conditions that made genocide possible, comments such as the trapper H. Pearce referring to thylacines as 'them bloody useless things'.[142] As Jennifer Maher and Tanya Wyatt observe, even the language used by critics of trading in wildlife risks reproducing 'the instrumental nature of people's relationship with wildlife'.[143] Terms such as 'collecting' and 'harvesting' commodify wildlife, framing them as consumer goods and depriving them of their individuality.[144] Despite his refusal of this kind of vocabulary, Martin still pursues the thylacine on behalf of Red Leaf, seeking to source animal parts for their weapons research. As the species is critically endangered (and thought, by almost all, to be extinct), they will clearly pay handsomely for the thylacine, which they conceive as an extremely rare commodity.

The Image Hunter

This aspect of *The Hunter*, the commodification of the last thylacine, echoes the historical extinction driver of museum collecting. As discussed in earlier chapters, once the tiger was in significant decline the bounty ceased and it was museums and zoos that sustained trade in the animal, this despite the species' clear scarcity. Without this consumer demand, some of the later thylacine captures would not have occurred. As the museums and zoos knew the animal was critically endangered, they are guilty of genocide. Their financial aid abetted efforts to either catch (in the case of zoos) or kill (in the case of museums) tigers. The thylacine is not the only species to have suffered museum-driven genocide; Dolly Jørgensen has explored the role of museums in what she refers to as the 'erasure' of the Caribbean monk seal (*Monachus tropicalis*).[145] This 'erasure' was animal genocide, with collecting expeditions to the breeding ground on the coral reefs known as Los Triangulos (The Triangles) in the Bay of Campeche in the Gulf of Mexico perpetrated in full knowledge of the species' declining numbers.

In the context of the thylacine, it is possible that in December 1911, 33-year-old Albert Quarrell of Tyenna, for example, would not have tried to catch the thylacine that spooked his bullocks near the town of Fitzgerald if live animals did not command a high price.[146] Unable to trap the tiger, his decision to kill them may have been an act of vengeance after they bit him or a decision to settle for less money given that even a carcass had value.[147] He sold the dead animal to a local photographer, Charles Brown, for £5.[148] It was Brown that took two photographs of Quarrell showing off his kill. In one of the photographs Quarrell is standing, holding the thylacine's corpse, displaying his prize. The tiger's mouth is slightly open, a death grimace. Behind their hind legs, a seated figure is visible, their face in shadow. In the second photograph, in which Quarrell has moved to his left and sat down, this person's face can be seen. They are usually named as the trapper David Pearce. It is probable that the photograph of Quarrell standing was taken first. Then, burdened by the dead weight of the tiger, he rested on something, perhaps a tree stump. Both photographs are posed with the tiger exhibited in profile so that their stripes are clearly visible. In the first, Quarrell is shot against the backdrop of a nearby tree. He stands in front of the trunk, giving a pleasing symmetry to the image. In the second photograph, the hand of a third figure is visible holding the tail of the tiger, the better to show the length of the animal.[149] Quarrell grips the muzzle of the thylacine, supporting the head for display (and closing the tiger's mouth). He is joined by his work dog (perhaps a kelpie) for this picture. Quarrell has his left arm around the domestic animal who, like their master, looks at the cameraman. The dog's inclusion suggests, to me, that they played some role in the events immediately preceding the tiger's death. The two photographs are celebratory, feting Quarrell for his killing of the marsupial 'pest'. His genocidal gaze fixes the viewer.

For Quarrell, as well as an 'annoyance', the thylacine presented economic opportunity, a welcome bonus income. His initial aim to take the tiger alive suggests that he foresaw selling them to a middleman or directly to a zoo. They were a rare commodity. The trees that form the backdrop to the photograph were also commodities. Quarrell's bullocks, 'his suffering and stubborn team', were used to haul timber, probably employing a practice called 'snigging' (dragging a log using a chain).[150] The townships of Fitzgerald and Tyenna benefitted from logging and the harvesting of trees such as eucalyptus and myrtle. In 1911, there was only one timber mill in

Tyenna, owned by C. Browning.[151] Substantial sawmills were established at Fitzgerald (owned by Gourlay's) and Tyenna (owned by the Huon Timber Company) in the following decade. Quarrell therefore aided in habitat destruction as well as killing rare fauna.

In bounty times, Tyenna was probably where payments for all the thylacines killed in south-west Lutruwita were made.[152] When Quarrell killed the tiger, the government bounty had ceased but Beaumaris and then Hobart Zoo purchased thylacines sourced from Tyenna in the period 1912 to 1925.[153] The rural settlement therefore became an important staging post for the trade in live tigers. Brown, the man who purchased the corpse, clearly saw enduring value in the physical remains. He may have intended to take a photograph of the dead animal and then sell copies of it (although no such photographs seem to have survived). He may also have sold on the remains to a dealer or collector for a profit. His two photographs of the dead tiger embody a vision of the thylacine as *expendable*. In this they incarnate the genocidal gaze as Baer conceives it.[154] The photographic apparatus is complicit in genocide here because it produces imagery that buttresses human domination and thylacine denigration.

In *The Hunter*, Martin studies photographs of this kind alongside films of the thylacine. Once he has accepted his mission, he reviews a montage of archival footage on a laptop in his hotel room. The first excerpt he watches is from a film probably made in 1933 that shows 'Benjamin' bounding and pawing the ground in front of the link fence of his enclosure.[155] In the full film, Reid can clearly be seen on the other side of the fence taunting 'Benjamin' with the carcass of what may be a dead chicken. In *The Hunter*, this vision of cruelty is kept out of sight. Although 'Benjamin' is obviously in captivity, living in a human-made environment, the presence of people around him is suppressed. He is being made to stand for his species and acknowledging Reid's presence would be a distraction. The violent actions employed to coax 'Benjamin' into life are concealed. Brief footage from about a minute into the Fleay film is then shown, featuring 'Benjamin' on his haunches. Then footage from 1930 of a captive thylacine at London Zoo plays on the laptop. The film was shot by Frank Percy Smith and features a female tiger eating a chunk of meat. She was one of two thylacines exported to London in 1925 (as part of an animal exchange undertaken by the trafficker George Chapman), living as a captive until 9 August 1931.[156] More footage excerpted from the Fleay film (about forty seconds in) then appears. It is cropped so that, at times, 'Benjamin' fills most of the screen (in the unmodified footage much more of the wire fence behind him is visible). In a series of shots, he paces his cage and gapes. The footage then fades out, with 'Benjamin' gaping replaced by a shot of the upland where Martin searches for the tiger.

Viewers are therefore presented with a montage but most will not be aware that what they are seeing are excerpts from different films. The fragmentary nature of the footage may be intended to suggest Martin viewing and reviewing the surviving films. The montage therefore offers a generic introduction to the thylacine, designed to 'introduce' them to the screen audience. It focuses on particular actions and behaviours to construct a specific vision of the tiger as an energetic carnivore with fearsome jaws. The decision to end with the gape seems intended to reinforce the thylacine's dangerousness, their worthiness as the hunter's quarry. This fearsome creature is who awaits Martin in the wilderness.

Later in the film, Martin again views parts of the Fleay footage and some still images. The first of the stills may be an inverted frame from one of the mid-1933 films shot by an unknown person, as it shows a thylacine being harassed by Reid through a fence. This image is followed by the photograph of Quarrell seated flaunting the tiger he has shot. The next one is a cropped version of a 1924 picture by John (Jackie) Robinson of 16-year-old Clem Penny (see my discussion in Chapter 3) and another man showing off the corpse of a female thylacine that the youth had beaten to death.[157] An 1869 studio photograph follows, showing William George Weaver seated with his rifle, the tiger he has killed suspended from the ceiling in front of him. The next photograph in the series is of a dead or dying tiger curled up in a patch of grass. Cameron Campbell and Stephen Sleightholme date this image, one of a pair, to the 1920s.[158] In *The Hunter* this snapshot is obscured by a battery low warning icon, as are the succeeding images, one from 1930 of Wilfred Batty with the tiger he shot propped dead against a fence, more stills from Hobart, a photograph of a tiger in a pen at the animal trafficker James Harrison's property in Wynyard, a picture of a newspaper cutting about Beaumaris, a January 1910 image by William Williamson of a family group (a mother and her three pups) at Beaumaris (which was sold as a postcard at the time), a 1911 picture of an adult male tiger at Beaumaris, another 1911 picture of the male, this time alongside a juvenile male, an uncropped version of the Penny image and, finally, a still from the Fleay film of 'Benjamin' gaping. The screen then goes blank as the battery dies. Martin's desperate tapping of the cursor to 'flick' through the images as his battery life ebbs away can be read as figuring extinction, forming a digital restaging of the tiger's disappearance.

The photographs in this scene also construct a particular vision of Martin's quarry. Here, the depiction of the tiger that is offered is more varied, showing the violence the species suffered at human hands, enduring captivity and death. The family group also works to 'soften' the image of the thylacine, although this softening is tempered by the final image of 'Benjamin's' jaws. The second, later engagement with archival imagery is therefore more likely to generate a sympathetic audience response. The photographs nonetheless continue to embody the genocidal gaze. Any audience is therefore faced with a question Pollock has raised in a different context: 'what kind of backward gazing can avoid killing a second time?'[159] Writing in the context of the Shoah, Pollock is concerned with how, ethically and politically, we, as 'belated viewers of a historical documentation of atrocity [are] now to position ourselves before it, given its undecidable but treacherous conditions of production'.[160] She cautions that whatever our own political or personal affiliations and identifications, by viewing such imagery we risk subjecting the vulnerable anew to a gaze that cannot be secured from the sadism or voyeurism that was inherent in its creation.

With this issue in mind, as well as repeating the species' vanishment, the screen going blank can be read as figuring the refusal of a particular kind of visual practice that does violence to the thylacine by framing them in human terms. This refusal, which I will explore further in the final section of this chapter, is only temporary. An egregious representation of the tiger occurs as part of the film's denouement. In a scene of heavy symbolism, Martin's prey is rendered an instrument to assuage human guilt over extinction. Martin has concealed himself in one of the thylacine's dens and, eventually, she returns to it. She notes his presence but rather than run away merely ambles off. He follows and, as he is preparing to shoot her, she turns to look at him. After a time, rather than fighting or fleeing, she lowers her head, a gesture that seems an invitation to kill her.[161] This seeming summoning of death implies suicidal intent on behalf of the thylacine, a kind of 'suicide by hunter' akin to 'suicide by cop' (but of a more muted kind where she is doing nothing that invites the use of lethal force). There is little evidence for suicide in nonhuman animals so this must be read as one among multiple instances of anthropomorphism in relation to the film's representation of the thylacine.[162] As Martin simply assists her dying, the film provides a vision of her killing as, if not justified, potentially self-induced. Others, including Sally Borrell, have read the tiger's death as a mercy killing (a kind of 'justifiable genocide'), with Martin putting her out of her misery (a wretchedness linked to her solitude and to her being perpetually hounded).[163]

The tiger's death differs markedly from the novel, in which the tiger scents M and attacks him.[164] His initial shots do not kill her outright and, after pausing, he finishes her off. In the film, a single shot is sufficient. After murdering her (this is murder, unlawful, because she is a protected species and her killing is premeditated), Martin cries. Freeman reads these tears as representing anguish 'for the last member of all extinct species'.[165] Borrell rightly emphasises how Martin also projects his grief over Lucy's and Sass's deaths onto the thylacine.[166] His tears are as much for the Armstrong family as for the tiger. The killing is also heavy with Christic overtones.[167] By inviting her death, the thylacine seems to be assuming humankind's sins, expiating them. When Martin carries her lifeless corpse, the scene is reminiscent of iconography associated with the Deposition of Christ. The thylacine is therefore reduced to a redemptive symbol, used to help the film audience work through the horror of the extinction of her species. She is anthropomorphised and instrumentalised, made into a mouthpiece for human concerns.

Finding Fact in Fiction

The need to engage strategies that seek to avoid instrumentalising animals is a key concern in animal studies. Tania Aguila-Way suggests that challenging the instrumentalisation of animals can potentially provide 'a more ethical acknowledgement of animal others'.[168] Rosemarie McGoldrick associates wildlife film and photography with observation, with what she refers to as 'scoping': an invasive lens-based investigation that contributes to human fictions of animal lives. Films such as wildlife documentaries reduce animals to dramatis personae, 'dewilding' them. *The Hunter* is already a work of fiction, but the thylacine's death as it is represented towards the end of the film is clearly also an expression of human fantasy. In this way, like nature documentaries, *The Hunter* distances us from nonhuman animality and shows scoping to be a compromised form of visibility. In her roles as an artist and curator, McGoldrick prohibits working from 'visual documents' of live animals such as photographs because of their 'compromised visibility'; they are always already outdated and represent another's vision of the animal.[169] She also prohibits the use of traditional genres of art, citing portraiture as an example. Such genres cast animals in human terms, 'domesticating' them.

McGoldrick calls for new modes of representing nonhuman animals that resist reducing them to symbols of human concerns and obscuring their 'animality'. Her inspiration for seeking new ways of portraying animals was a sudden chance encounter with a wild animal, a European badger (*Meles meles*). This unplanned meeting shocked McGoldrick out of her familiar frames of reference and manners of reception. She felt 'danger', or danger was the term she retrospectively applied to describe this chance coming together. McGoldrick characterised her feeling as 'a "wild" response'.[170] Wildness here can be understood as describing an uninhibited, unmediated 'natural' reaction to an event, although such an idea is one that risks equating 'animality' with body and instinct at the expense of reason.[171] It is this kind of raw encounter, a 'shared animality', that Mc-

Goldrick believes needs to be carried into art.[172] In her unpreparedness, her shock upon meeting the badger, she temporarily lost her 'humanity'. 'Humanity' then in the varied senses of the term (humankind, human nature, humaneness) gets in the way of experiencing animality and human–nonhuman animal commonality. Art must see beyond human concerns to link with animality.[173]

For McGoldrick, the alternative to the violent effects of 'scoping' is 'visualisation'. Her understanding of the term is influenced by Russian formalism and by the thinking of Robert Kosara. For Kosara, artistic visualisation is about communicating a concern about something rather than simply rendering data intelligible (which is the goal of much scientific visualisation).[174] It therefore potentially involves ethical and political dimensions. For Kosara, artistic visualisation continues to be data-reliant (data is the raw material that is worked upon as part of the creative process) and data authenticates whatever is cause for concern. McGoldrick does not emphasise the need for information or facts but rather encourages their paucity. For her, visualisation is, above all, a making strange of the hitherto familiar, a rewilding of the everyday wild as it is presented in many nature documentaries and much animal art. This making strange involves embracing 'the invisible, the implied, the off-stage or the unidentifiable'.[175] For McGoldrick, '[c]ontemporary conditions favour visualizing the animal via absence or the unidentifiable'.[176] Visualisation involves rejecting lens regimes as mediating media of representation. Lens-based regimes subjugate nonhuman animals to human understandings and needs.

While sympathetic to McGoldrick's argument, Steve Baker queries any total injunction against pictorial representation.[177] Baker shifts the discussion away from technologies of looking and the logics they embody or support and towards different ways of seeing – viewing nature attentively, appreciating beauty – as promising modes of engagement. There may be a dimension of this kind of heightened regard in *The Hunter*, with its many shots of the countryside, often framed as harsh, at times shown to be immense and majestic. These vistas give a sense of place and of the passage of time. The panoramic and birds-eye views also convey the enormity of Martin's task of locating the thylacine in the vast and rugged surroundings. Additionally, they gesture to human insignificance, with the poacher appearing small in the huge space. Yet Martin is able to live in and off this at times unforgiving land, so it also functions to foreground his powers of endurance. The film therefore exploits landscapes to express varied meanings and encourage assorted emotions. The spectator's repeated encounters with rough country, however, also potentially enable an excess to emerge, an *unintended* terrain.[178] The wilderness is thus able to grow wild.

There are occasional shots in *The Hunter* of animals doing their thing, not playing a part. Although they are instrumentalised, used to showcase the local 'fauna', they nonetheless do so in their own world, being themselves.[179] They are not observed in a quest for insight nor are their lives dramatised. They are filmed as backdrop, an approach that affords them a measure of liberty denied the thylacine. Through how they are filmed, they seem qualitatively different to the animals as ciphers for human concerns in the representations discussed by McGoldrick. Even the animal 'extras' in *The Hunter* are, however, framed and subject to film editing, made to mean in terms that are not their own. They are already humanised in a phenomenological understanding of the term, as they have been singled out as meaningful to humans and made part of the human environment that is cinema.[180]

The nonhuman animals living in the 'margins' of *The Hunter* remain marked by culture, subject to representation.[181] One of the issues with McGoldrick's approach is that she retains 'representation' as the foundation of artistic practices of image-making. Using such an approach seems to render accessing the 'real' (the stuff that undergirds representation) impossible. Representation is always transformative and predicated on distance and difference from that which it references, re-presences. It is a way of imaging ideas about things. McGoldrick acknowledges that visualisation is about 'communicating complex *ideas* quickly'.[182] Visualisation as a kind of representation therefore involves the same problem of ideation discussed in the last chapter. To visualise animals, like observing them, is to transform them into ideas, even into the idea of 'animality'. Yet although caught in this quandary, McGoldrick's emphasis on images of absence does gesture to a way beyond this impasse.

Beyond Disappearance

The first two thirds of *The Hunter* offer a narrative that is, in a sense, in suspended animation, with Martin repeatedly setting and checking traps, the tiger at this stage figuring as an 'absent presence', as a suggestion rather than a fleshly entity. The traps do not catch the creature, becoming markers of failure, signalling the tiger's stubborn elusiveness. Through the traps, *The Hunter* points to the possible presence of the thylacine while simultaneously withholding them, implying their existence without lending them substance. This withholding can, of course, be interpreted as a means to build suspense, a technique akin to that adopted in classic films such as *Jaws* (dir. Steven Spielberg, USA, 1975), where showing less of the predator protagonist increases tension and expectation. The relatively lowkey appearance of the thylacine towards the film's close, however, belies such a reading. In the end, there is no combat between Martin and his quarry comparable to the battle between the crew of the *Orca* and the great white shark (*Carcharodon carcharias*). The tiger's final appearance (she is spotted briefly and fleetingly by Martin earlier in the film) is not sudden and scary and she meets her death meekly. The final third of the film, in its conventionality, acts instead as a counterpoint to what precedes it, operating in a different, straightforwardly Symbolic (in the Lacanian sense of the term) psychic register.

The Symbolic forms one of Lacan's three registers of psychic reality along with the Real and the Imaginary. The Real refers to the time of maternal plenitude when an infant lacks for nothing and has no notion of difference from the mother figure (a figure whose role can be fulfilled by a man, woman, trans or non-binary person) who encompasses them. The Imaginary is linked to processes of identification and fantasying. It involves the well-known mirror phase or stage (postulated by Lacan to occur when an infant is between six and eighteen months old), when a child identifies and introjects the image of themselves that they see reflected in a mirror (real or figurative). This reflection has stable contours and is at odds with how they experience their own body, which registers as a chaos of sensations, of variable ingestions and excretions. The ideal self-image in the mirror, once internalised, becomes the psychic support for the emerging ego and introduces ideas of difference to the nascent self. The infant begins to separate from the Other, the mother figure, with this violent cut of difference concealed through fantasies of identification: I am *distinct* from You yet I am also *like* You.[183] Pollock crucially notes that this identification is linked to anthropomorphism, with the infant finding their likeness in the reflected human form.[184] The infant therefore also comes to recognise their difference from other forms at this stage, moving from sensations of amorphousness to incorporating a vision of humanness. Prior to the mirror phase, the infant is therefore not *formally* different from other organisms, including nonhuman animals.

For Lacan, the Imaginary is the part of the human psyche that is most closely allied with nonhuman animals.[185] Like many thinkers of his time he believed that what differentiates humans from other animals is their possession of language. Language, for Lacan, is associated with the Symbolic Order.[186] In contrast to the Imaginary, which is dualist and centred upon the infant and the (M)Other, the Symbolic is triadic and includes a symbolic father figure. Lacan refers to the *Nom-du-père*, the Name-of-the-Father, an expression that, when heard in French, connotes both the name (*nom*) of the father and the no (*non*) that is the paternal prohibition against incest. Naming occurs in the Symbolic, a procedure that Lacan indicates constitutes a becoming human. For him, nonhuman animals exist in nameless nature, cut off from language.[187] Accession to language, in turn, dissociates humans from their animal natures. Lacan therefore continues in the tradition of Hegel (discussed in the preceding chapter) in which naming nullifies animality. Once the human is a subject in language, they are positioned at an unbridgeable remove from the nonhuman animal.

Bracha L. Ettinger conceives psychoanalysis differently, refusing to accept as totalising the alienating effects that, for Lacan, accompany entry into the Symbolic Order. She identifies another dimension to subjectivity, the Matrixial, that is characterised by linking rather than the cut of difference. Judith Butler notes in 'Disturbance and Dispersal in the Visual Field' that Ettinger 'concerns herself with the ethical and political question of what it means to make a link across a border when the dominant ethos maintains that no link can be made'.[188] In the context of nonhuman animals, this link is one that challenges the linguistic separation that safeguards human distinctiveness and privilege. The Matrixial acknowledges 'other subjects as unknown but not hostile or intrusive' and as connected to the self.[189] It therefore holds vital potential for negotiating relations with foreignness, admitting an Animal Other yet without appropriating them to the familiar, without making then mean in human terms.

In the specific context of the thylacine, the communication of trauma across species is also a central issue. I want to suggest here that *The Hunter* opens an aesthetic space where encounter with something of the thylacine's suffering becomes possible. The film creates moments that are conducive to what Ettinger calls wit(h)ness-Thing, a perceptual process that encourages encounters with the Thing (the Real) that are registered affectively by way of aesthetic experience. Lacan regarded the Thing, maternal plenitude as a lost object desired by the subject, as a kind of traumatic kernel haunting the subject but remaining unknowable from within the Symbolic. For Ettinger, by contrast, something of the Thing can be conveyed through aesthetic encounter by way of the registration of intensities, affects, rather than the recognition of signs. This felt communication of affect is called co-poiesis.

In *The Hunter*, encounter occurs through repeated gestures that begin to insist in the viewing experience. Such gestures include the building and setting of traps, a recurring motif that functions to lull the audience into a state that Ettinger refers to as 'floating attentiveness' and Pollock as 'receptive reverie'.[190] Receptive reverie is, to some extent, analogous to an analyst listening to the analysand in a clinical setting in psychoanalysis in that it describes an openness to 'unspoken undertones' as they register through speech and gestures, rhythms and caesuras.[191] Receptive reverie is also nurtured by the many scenes of Martin walking across rainforest and moorlands. Through time – signalled by the common cinematic shorthand of clouds scudding across the sky, salmon pink, mauve, deep blue, grey, varied colours and moods at different times of day – this walking becomes wandering, drifting, as the hunter finds no trace of his prey. It is a wandering occasionally interrupted by obvious purpose, but often Martin seems to be simply traversing terrain.[192] He is circling an absence that neither he nor the viewer can master or possess. This absence is not a lack, not a deficit, but rather an affective force, an intensity, that functions outside enunciative processes.

John Paul Ricco's exploration of a 'disappeared aesthetics' is enlightening here. Writing in the context of what he refers to as the time of AIDS, Ricco notes of loss that it becomes lost when subject to symbolisation and narrativisation, when exposed to recuperative logics of representation.[193] In a sense, once a traumatic death such as that of an AIDS victim or, in this instance, a victim of animal genocide is represented, lent substance, something of the trauma they endured dissipates. Pollock has written compellingly of how processes of signification including becoming representation provide a kind of balm, distancing us from 'the overwhelming, undigested thingness of trauma'.[194] In this sense, representation disappears trauma. Archival footage such as the Fleay film and digital resurrections of the kind used in *The Hunter* nullify pain and loss and thereby numb the aching immensity of extinction. The absence signalled by Martin's gestures and wanderings, by contrast, refuses to cohere into sense. It operates, rather, as an unsettling intensity. As intensity without sign, shadow without formal source, this affective force does not disappear loss but rather impresses it on the beholder.[195]

When Martin is out in the field (rather than studying archive film) there is no image of the tiger to view, describe, experience or consume. Rather the thylacine is registered by way of an allusive, elusive approach, one that resists 'semiotic closure'.[196] Through the repeated unsuccessful setting of traps, the fruitless hiking, through Martin's patient failure, something begins to insist. Jacqueline Rose convincingly argues for repetition 'as the constant pressure of something hidden but not forgotten'.[197] Here, repetition acts to signal an 'other' to that Other that we normally recognise as the thylacine. I am referring here to a radical alterity, to something that is other to our languages of entrapment, beyond the distancing and destructive effects of the sign.

Cumulatively, the physical actions of setting the traps gesture towards the film's efforts to communicate something outside the orbit of the Symbolic, a thing resistant to the lure of definition, refusing the confine of outline, rejecting the appeasing sense and substance of the sign. By way of this process of allusion, this indexing of potential presence, the film permits moments of positing the thylacine that refrain from possessiveness. Often, lens-based technologies such as cine cameras 'capture' images of things. In France, news camerapersons were historically referred to as *chasseurs d'images* (image hunters). The word *chasse* (hunt) implies not just finding images but also apprehending or seizing them. Image hunting suggests a particular kind of relation to the images, one that involves a use of force, an assertive appropriation. In this vein, Matthew Brower argues that photographing wildlife should be viewed as a kind of 'hunting'.[198]

The violence of visual technologies, their invasiveness, is a clear theme in *The Hunter*. In one instance, Martin falls prey to an infrared trail camera. The device captures him investigating a wombat carcass strung between some branches. He extracts the camera's memory card and later views it on his laptop, finding himself keeping company with creatures such as Tasmanian devils and quolls, caught alongside them. He too has involuntarily become archived, his presence catalogued by a modern image-recording technology. The event can be understood to give Martin insight into the dynamics of hunter and prey that are played out through image capture; he comes to know first-hand what Susan Sontag calls the 'soft murder' of the camera.[199] Sontag is referring to the violent, violating objectification that accompanies being photographed. She describes cameras as sublimations of firearms; as such they are akin to the tactical rifle used by Martin.

In *The Hunter*, the camera occasionally incarnates a predatory look. There are shots of the poacher being watched by his rival, who registers as a blur at the bottom of the frame as he observes Martin in a point of view shot. Mindy also views Lucy rapaciously at one point: he either desires her or desires Martin and covets her relationship with him.[200] Such scenes, however, present a camera consciousness aligned with the human, even if it is an inhuman humanity.[201] Film more broadly as a mechanical form of representation might be conceived of as embodying an inhuman gaze that is ill-suited to fostering ethical encounter. Certainly, the Fleay footage discussed earlier should be regarded negatively as bound up with efforts at mastery and as embodying a violent logic of difference, both enabling factors in the nonhuman animal genocide of the thylacine. Some films have sought to develop ways of seeing that are not human-centred. In *Severed Ways*, as well as seeking to differentiate First Nations and Viking ways of seeing, the director, Tony Stone, has suggested that the point of view established by some of the shots is also intended to be that of the encompassing wilderness rather than of specific individuals: 'As the film goes on and gets into nature it slows down and is shot from the point of view of the trees and the woods.'[202] Here, then, the trees, the forest, are accorded a perspective and observe the people, particularly Orn, who travel through them. In

The Hunter there are also instances where the point of view is ambiguous and may be that of nature itself appraising the poacher. Whether these perspectives truly embody a non-anthropomorphic gaze is questionable, yet they do gesture to the need to see beyond human perspectives regarding nonhuman and human phenomena.

The meditation on image capture and its violence that is interior to the narrative of *The Hunter* runs parallel to a challenge to this will to visual mastery, the aforementioned gestural economy that points towards the possibility of aesthetic encounter. In the initial two thirds of *The Hunter*, this film does not objectify the thylacine as subject, does not scourge the tiger with meaning. It refuses to impose narrative fictions of the kind discussed by McGoldrick or to represent and, thereby, nullify the thylacine. The refusal to force signification upon the tiger (which is also a feature of the novel) and the decision instead to nurture aesthetic encounter provide a means to grasp something of the 'after-affects' of animal genocide, the trauma of the thylacine's extinction.[203]

Notes

1. For a recent discussion of the sagas that sensitively examines their depiction of Indigenous peoples, see Aylward, *Beothuk*, 16–20.
2. In his essay 'Who's Savage Now?!', Kevin Harty, for example, identifies the setting as Newfoundland (117). Other footage was shot in Vermont and Maine. The Vikings may have visited the coast of Maine. Indigenous trade also seems to have brought Viking objects to what is now the state of Maine sometime after 1067, as a penny struck during the reign of Olaf the Peaceful (Olaf Kyrre) (1067–93) was discovered during excavations at a settlement at Naskeag Point. For a recent discussion of the coin, see Gullbeck, 'The Norse Penny Reconsidered'.
3. Skræling was also used to refer to the Indigenous people (perhaps the Dorset) of Greenland. Eirik the Red's *Saga* tells of the Vikings battling with the Indigenous population at Hóp, a place sometimes identified as in Newfoundland.
4. L'Anse aux Meadows was probably the Viking settlement of Straumfjord referred to in the Vinland sagas. Vinland was a region likely to the south of Straumfjord. The Vikings did have another temporary settlement that they used in the summer months called Hóp. Some have suggested that Hóp was in what is present-day New England or Nova Scotia. See, for example, Larsson, 'The Vinland Sagas and Nova Scotia'. Birgitta Linderoth Wallace argues that Hóp was in eastern New Brunswick. See Wallace, 'An Archaeologist's Interpretation of the *Vinland Sagas*', 230.
5. This action echoes the arson attacks on churches perpetrated by Norwegian right-wing extremists with connections to the black metal scene in the early 1990s.
6. She is listed as an Abenaki woman in the film credits. The Abenaki live in the regions now known as the Maritimes and Québec but are not usually associated with Newfoundland.

7. The Greenland wolf (*Canis lupus orion*) is likely the same subspecies as the arctic wolves (*Canis lupus arctos*) that inhabit Ellesmere Island in Nunavut, Canada. For a discussion of the history of the wolf in Greenland, see Dawes et al., 'The Wolf (*Canis lupus*) in Greenland'.
8. Davis, *Slaves on Screen*, 133–34.
9. Davis, *A Passion for History*, 23.
10. Davis, '"The People of the Past Come First"', 169.
11. Reichert, 'Headbanger'.
12. For a general discussion of Viking metal, see Trafford, 'Viking Metal'; for a discussion of Viking metal as (polar) history, see Sellheim, 'Black and Viking Metal', 512–14.
13. Von Helden, 'Barbarians and Literature', 257.
14. In popular culture, Viking masculinity is associated with emotional restraint. The Viking cartoon character Hagar the Horrible, for instance, claims not to have 'shed a tear in twenty-five years'. See Hagar the Horrible, *Washington Post*, 17 June 2018, SC3.
15. Connell, *Masculinities*, 194.
16. The sex scene risks perpetuating the popular stereotype of First Nations women as sexually wanton. See Weaver, 'The Colonial Context of Violence', 1559.
17. Stephanou, 'Playing Wolves and Red Riding Hoods in Black Metal'. Amanda Digioia provides a broader discussion of the wolf in heavy metal in her essay 'A Cry in the Dark', 296–300.
18. For a discussion of Fenrir in the prose version of the Norse mythological work the *Edda*, see Kousoulis et al., 'Violent Death and Trauma in Norse Mythology'.
19. Walsh, '"A Great Heathen Fist from the North"', 59.
20. Cited in Wylie, *History of the Scottish Nation*, 347. See also Walsh, '"A Great Heathen Fist from the North"', 62.
21. See 'Severed Ways', *Wikipedia*, https://en.wikipedia.org/wiki/Severed_Ways (retrieved 9 July 2020).
22. For a recent discussion of Neanderthals as they feature in *Quest for Fire* and also the film's overall problematic portrayal of archaic humans, see Marshall, 'Imagined Communities of Prehistory'.
23. Something of the complex reality of Neanderthal culture is recognised in *Quest for Fire* but they are still contrasted harshly in the film with a more culturally sophisticated group of hominids referred to as the Ivaka (who seem intended as *Homo sapiens*).
24. For a discussion of growling and its significance, see Harrington and Asa, 'Wolf Communication', 67, 69.
25. Cited in *North American Grey Wolves*, 1903, *American Film Institute Catalog*. https://search.proquest.com/docview/1746369429?accountid=12543 (retrieved 9 July 2020). Derek Bousé notes that the film was marketed on the 'ferocity' of the wolves. See Bousé, *Wildlife Films*, 199.
26. This film seems to be different to *La chasse aux loups en Russie*, made by the Pathé brothers, which was first screened at the Omnia Pathé in Paris in July of 1910 and includes footage of men on horseback and of a pack of dogs chasing and attacking a wolf in snowy terrain. It is also not to be confused with the 1908 nonfiction film of the same title *The Wolf Hunt*, which was produced by Oklahoma Natural Mutoscene Company. The latter film is briefly referenced as an example of the 'cinema of attraction' in Nanna Verhoeff's *The West in Early Cinema*, 53. It features the wolf catcher Jack Abernathy (renowned for catching wolves with his bare hands) demonstrating his prowess. See Wooley, *Shot in Oklahoma*, 26–27. The logo of the Oklahoma Natural Mutoscene Company was of a human hand seizing the lower jaw of an open-mouthed wolf.

27. Anonymous, 'The Wolf Hunt', 17.
28. Ibid.
29. Stephanou, 'Playing Wolves and Red Riding Hoods in Black Metal', 159–60.
30. Ibid., 161.
31. Ferrett, *Dark Sound*, 151.
32. See Chapter 4 of Freeman, *Paper Tiger*.
33. There have been suggestions that the thylacine also engaged in cannibalism. See, for example, Courtney, 'Cannibal Dogs'. Such speculation has been used by those who believe in the tiger's continuing survival to justify the lack of discovery of recently deceased thylacine remains. See, for instance, Harris, 'Hold that Tiger!', 31.
34. Narraway and Stark discuss the use of archival footage in *Dying Breed* and *The Hunter* in 'Re-Animating the Thylacine', 17.
35. For a discussion of film as after-image, see Bienvenue and Chare, 'Introduction', 2–4.
36. The film capacity of the Kinamo varied between 50 and 80 feet depending on the model.
37. Ray, 'Social Theory', 140.
38. For a discussion of ocular empiricism, see Childress, 'Visualizing the Other', 10–11.
39. See the definition of the nature film in MacDonald, 'Up Close and Political', 4.
40. Fleay refers to him as 'it' in his newspaper articles 'Hobart Zoological Collection' and 'Bush Notes'. Fleay, 'Hobart Zoological Collection: Unique Marsupial Wolf', *The Australasian*, Saturday 20 January 1934, 43; 'Bush Notes: The Vanishing "Tasmanian Tiger"', *The Australasian*, Saturday 30 March 1940, 36.
41. MacDonald, 'Up Close and Political', 5.
42. Paddle, *The Last Tasmanian Tiger*, 9.
43. As Sleightholme and Campbell note, in one film Arthur Reid presses a lamb against the mesh fence to coax movement from 'Benjamin'. Sleightholme and Campbell, 'A Catalogue of Motion Picture Films of the Thylacine', 781.
44. This is also the case in recently rediscovered footage of 'Benjamin' from 1931. For a discussion of this footage, see Linnard and Sleightholme, 'The Rediscovery of the 1931 Stewart Film of the Last Captive Thylacine'.
45. I have been unable to inspect the Screen Australia copy of the film first-hand to see if the version that commonly circulates betrays evidence of editing (glued sections).
46. The black and white version of the footage on the Australian Screen website loops back at the end to the beginning of the film (repeating about ten seconds of footage). 'Tasmanian Tiger Footage', Australian Screen website, https://aso.gov.au/titles/historical/tasmanian-tiger-footage/clip1/ (retrieved 1 June 2022). The best version to watch to get a sense of the original length and ordering of the shots is therefore the colourised version available on the NFSA website: 'Tasmanian Tiger in Colour', NFSA website, https://www.nfsa.gov.au/latest/colourised-footage-last-tasmanian-tiger (retrieved 29 February 2024). The NFSA film begins with 'Benjamin' investigating the camera, which supports viewing this as raw footage unedited by Fleay. I am very grateful to André Gaudreault for his help and advice regarding the analysis of the different versions.
47. Pierre Sorlin is at pains to emphasise of historical films that they all have their own histories. The same can be said of archival footage. Sorlin, *The Film in History*, 22.
48. Sleightholme and Campbell, 'A Catalogue of Motion Picture Films of the Thylacine', 783.
49. Eberle, 'The Art of Making Photos', 316.
50. Ibid., 317.
51. His 'hide' may have resembled the umbrella tent that Paul Bartsch employed for the still photography of birds. A photograph of this 'hide-and-seek tent' was reproduced

in *Popular Mechanics Magazine* in 1930. See Anonymous, 'Hide-and-Seek Tent', 272. The US magazine had an office in Melbourne and was widely distributed in Australia, although there is no evidence that Fleay was a subscriber.

52. Unlike a yawn, which is a sign of boredom or fatigue, the gape as agonistic display is motivated by fear. The two words have a shared etymology but distinct modern meanings.
53. See Nicol and Maskrey, 'Thermoregulation, Respiration and Sleep'; Cooper and Withers, 'Patterns of Body Temperature Variation'. Baker and Dickman suggest that devils do not enter torpor. See Baker and Dickman, *Secret Lives of Carnivorous Marsupials*. Torpor can be spontaneous or induced through lack of food or water. See Geiser, 'Hibernation and Daily Torpor in Marsupials', 6.
54. Torpor similarly involves reduced expenditure of energy. It can also protect against heat stress.
55. Fleay, 'Catching Snakes: Thrills of a Dangerous Sport', *The Australasian*, Saturday 3 October 1936, 51.
56. Fleay, 'Turkey with a College Education', *The Argus*, 28 February 1936, 10.
57. 'Benjamin' was, of course, already 'in character', typecast from the outset, 'performing' his species.
58. There has been a recent effort to rehabilitate Hobart Zoo and suggest that 'Benjamin' was not neglected. See Linnard and Sleightholme, 'An Exploration of the Evidence', 326–30. The psychological torture 'Benjamin' was subject to may not constitute 'neglect' (the failure to provide basic necessities), but is animal cruelty.
59. Anonymous, 'The Tasmanian Menagerie', *Cornwall Chronicle*, Wednesday 14 December 1870, 2.
60. Anonymous, 'Beaumaris Zoo', *Examiner*, Tuesday 16 June 1925, 4.
61. 'Bell Bird' [pseudonym], 'Some Unconsidered Creatures', *The Mercury*, Thursday 27 December 1923, 9.
62. The brevity of the footage (just over a minute) may show sympathy from Fleay towards the thylacine. There would be at least four minutes worth of film in the camera (if not considerably more). 'Benjamin's' ordeal could therefore have lasted far longer (although the naturalist may have shot more footage of the tiger that day that was subsequently discarded and lost).
63. Clubb and Vickery, 'Locomotory Stereotypies in Carnivores', 61–62. Katy Shepherd's video artwork *Loopus* (2014) eloquently captures the ennui of 'Benjamin's' captivity through portraying him in watercolour in an animation repetitively pacing against a white backdrop. See 'Loss and Extinction', https://katyshepherd.co.uk/films/loss-and-extinction/# (retrieved 20 June 2022).
64. MacDonald, 'The Effects of Crowd Size and Keeper Presence', 3.
65. Fleay, 'On the Trail of the Marsupial Wolf', 129.
66. Fleay, 'Infant Possum with a Future', *The Argus*, Thursday 3 October 1935, 10.
67. Anonymous, 'Camera Shy!', *Herald*, Wednesday 2 January 1935, 1. See also Fleay, 'Bush Notes: Fallow Deer in the Ranges of Victoria', *The Australasian*, 29 June 1940, 36.
68. Fleay, 'Hobart Zoological Collection: Unique Marsupial Wolf', *The Australasian*, Saturday 20 January 1934, 43.
69. Pollock, *Art in the Time-Space of Memory and Migration*, 19.
70. Icons in the Peircean sense are forms of representation that share resemblance with what they signify (such as a realistic portrait painting of a person). Film and photographs are often, in part, iconic. For a useful summary of Peircean semiotics, see Silverman, *The Subject of Semiotics*, 14–25.
71. Kilroy, 'Reading the Writing-Machine', 130.

72. Mulvey, *Death 24x A Second*, 171.
73. Ibid., 28, 144.
74. Ibid., 32.
75. The colourisation project for the Fleay footage used ultra-high-resolution digital scans. Digital colourisation processes have changed through time and have their own history.
76. 'Tasmanian Tiger in Colour', NFSA website, https://www.nfsa.gov.au/latest/colourised-footage-last-tasmanian-tiger (retrieved 1 June 2022).
77. Colourising is not just time-consuming but also expensive, and it looks like efforts were concentrated on rendering 'Benjamin' as impressive as possible at the expense of his surroundings. The notably skilful rendering of his fur can be seen to embody monetary concerns and aesthetic and historiographic decisions.
78. 'Australia in Colour', NFSA website, https://www.nfsa.gov.au/latest/australia-colour-landmark-sbs-series-nfsa-content (retrieved 8 June 2022).
79. Ibid.
80. Edwards, 'Book Review: *The Colour of Time*', 331.
81. See Marina Amaral's comments on her colourisation project *The Colour of Time*: 'The Colour of Time: A New History of the World, 1850–1960', 9 August 2018, https://marinamaral.com/the-colour-of-time-a-new-history-of-the-world-1850-1960-2/ (retrieved 1 December 2024).
82. Chare, 'Material Witness', 85.
83. 'Tasmanian Tiger in Colour', NFSA website, https://www.nfsa.gov.au/latest/colourised-footage-last-tasmanian-tiger (retrieved 1 June 2022).
84. For a discussion of the nature of blotches and scratches, see Silva and Corte-Real, 'Removal of Blotches and Scratches'.
85. Watkins, 'The Politics of Nostalgia', 134.
86. 'Tasmanian Tiger in Colour', NFSA website, https://www.nfsa.gov.au/latest/colourised-footage-last-tasmanian-tiger (retrieved 29 February 2024).
87. See Calhoun, 'The Preservation of Motion-Picture Film', 518; Cherchi Usai, *Silent Cinema*, 1–2; Sclocchi et al., 'Fungal Biosorption', 367. When nitrate film is developed the halide crystals are chemically reduced to metallic silver with the developed silver image remaining fixed in the gelatine. See Schreiber and Gareis, *Gelatine Handbook*, 262–63.
88. See ibid., 48.
89. Calixto et al., 'Gelatin as a Photosensitive Material', 2.
90. If the film stock was manufactured in Australia it is also possible that the gelatine was of ovine derivation. Australia Kodak Limited, for instance, produced film stock. See Leggio, 'A History of Australia's Kodak Manufacturing Plant'. For a discussion of the materials used to make gelatine in Australia, see Schreiber and Gareis, *Gelatine Handbook*, 26. Gelatine with the greatest uniformity is derived from piglets; see Kite, 'Collagen Products', 193.
91. For a discussion of animal death in pre-photographic image-making practices, see Bienvenue, 'From the General to the Particular', 346.
92. Early photographic (but not film) processes sometimes did not employ gelatine. Palladium prints, for instance, did not involve animal death. This mode of making photographs is still sometimes used today, including, notably, by the Chiapas-based Mexican artist José Angel Rodríguez. Rodríguez develops his palladium prints, which often portray members of Chiapas's Indigenous communities including the Tzotzil, on amate paper (a paper handmade using bark, employing techniques perfected in pre-contact times). See Corkovic, 'La Cultura Indígena', 891–92.
93. Heintzman, 'Material Cultural Studies', 359–60.

94. See Grainge, 'Reclaiming Heritage'.
95. Edwards notes that monochrome shaped how past photographers (and, it could be added, cinematographers) saw the world. See Edwards, 'Book Review', 331.
96. In art, 'handling' refers to the characteristic ways that an individual artist manipulates their chosen medium (often referring to traditional media such as paint [gouache, oil, watercolour] and sculpture [marble, wax, wood]). I am suggesting that as a cameraperson Fleay also 'handled' the camera in specific ways that index his personality.
97. Eberle, 'The Art of Making Photos', 314.
98. Anonymous, 'News in Brief', *The Mercury*, Monday 18 December 1933, 6. In Fleay's film, shadows cast by tree branches on the floor of the cage can be seen moving in a breeze.
99. Armstrong and Cutting, 'On the Physical Variables Filmmakers Use to Engage Viewers', 2.
100. Reid et al., 'The Effect of Ability, Colour and Form', 15.
101. Žeger et al., 'Greyscale Image Colorization Methods', 113326.
102. See Armstrong and Cutting, 'On the Physical Variables Filmmakers Use to Engage Viewers', 2.
103. Zavagno, 'The Perception of Light from a Phenomenological Perspective', 349.
104. For an important discussion of ethical issues in relation to the colourisation of Holocaust photographs, see Watkins and Williams, 'The *Faces of Auschwitz*'. Martin Jolly has discussed the unethical use of frames from the François-Steininger version of the Fleay footage. The frames are reproduced in newspapers with no acknowledgement that they have been colourised. See Jolly, 'Pygmalion in the Archive', 53.
105. Burt, *Animals in Film*, 51.
106. Dixon, 'Thylacinidae', 3.
107. For a discussion of spatial jitter, see Buisson et al., 'Motion Compensated Film Restoration'.
108. The thylacine is described as a cursorial predator in Newton et al., 'Letting the "Cat" out of the Bag', 17. Others argue against the thylacine's capacity for cursorial motion and suggest that the tiger resists categorisation as either a cursorial or pounce-pursuit predator, being more generalised than specialised in their behaviour including their locomotion. See Figueirido and Janis, 'The Predatory Behaviour of the Thylacine', 938–39; Janis and Figueirido, 'Forelimb Anatomy', 1336–37.
109. Peacock, *The Anthropological Lens*, 144.
110. Ibid., 145.
111. Ibid.
112. Ibid.
113. Anonymous, 'Hobart Weather', *The Mercury*, Monday 18 December 1933, 5.
114. Westell, *The Book of the Animal Kingdom*, 207.
115. I contacted François-Steiniger about the specifics of the skins that were studied but the requisite information was not forthcoming.
116. See Paddle and Medlock, 'The Discovery of the Remains of the Last Tasmanian Tiger'; Linnard and Sleightholme, 'An Exploration of the Evidence', 330–32.
117. Freeman, *Paper Tiger*, 22.
118. 'Tasmanian Tiger in Colour', NFSA website, https://www.nfsa.gov.au/latest/colourised-footage-last-tasmanian-tiger (retrieved 29 February 2024).
119. Freeman, 'The Last Image', 196n7.
120. In 'The Colorization of Black-and-White Motion Pictures', Michael Landau, for instance, describes colourisation as an attempt at resurrecting and revitalising old black and white films (61).

121. Leigh, *The Hunter*, 91.
122. Ibid., 164.
123. In *Wild Dog Dreaming*, Deborah Bird Rose writes of gazing into the eyes of a wild animal as looking into 'mystery' because 'we know we have almost no idea how the animal sees us' (69).
124. A similar encounter with otherness occurs in Bill Viola's 1986 film *I Do Not Know What It is I Am Like*. See Burt, *Animals in Film*, 50–51.
125. Elizabeth Bishop's 1946 poem 'The Fish' recounts the narrator hooking a 'tremendous fish' and looking into his eyes, which 'shifted a little, but not to return my stare'. This non-exchange of the gaze echoes that of M and the thylacine. Bishop's narrator lets the fish go.
126. Leigh, *The Hunter*, 164.
127. Malamud, *An Introduction to Animals and Visual Culture*, 74.
128. M's employer in the novel is simply referred to as the Company. The generic name has the same sinister, overarching connotations as the Company in *Alien*. For a discussion of the Company in *Alien*, see Chare, *After Francis Bacon*, 149.
129. Tony Hughes-d'Aeth notes in 'Australian Ecology, Deep Ecology and Julia Leigh's *The Hunter*' that many critics were disappointed by M's lack of a 'moral epiphany' in the novel (22). Like Hughes-d'Aeth, I believe that far from a failing, M's lack of a Damascene conversion to nonhuman animal rights and ecological care is part of what grants the novel its power.
130. See Hiatt, 'Cremation in Aboriginal Australia', 104–8.
131. Martin collects the tiger's remains in a thermos flask that belonged to Jarrah, a gift from his family, and scatters them from a clifftop. He therefore uses a practice (cremation) associated with the palawa and stores the thylacine's ashes in an artefact associated with the Armstrong family.
132. Allen, 'Tasmania', 53.
133. The tiger is seemingly located in the vicinity of Chaplin Creek, relatively near the Gordon River in the Franklin-Gordon Wild Rivers National Park.
134. See Krien, *Into the Woods*.
135. This echoes a sledgehammer attack on a car in the Upper Florentine that is discussed by Krien in *Into the Woods* (13–14, 227), although in that instance two people were in the vehicle.
136. Hill, 'A System Thinking Perspective', 190.
137. The term 'camouflage-collar crime' appears, for instance, in Crow et al., 'Camouflage-Collar Crime'.
138. Nurse, 'Collecting Wildlife', 204.
139. Ibid., 206.
140. Goyes, 'Denying the Harms of Animal Abductions', 173.
141. Ibid., 172.
142. Guiler, *Thylacine*, 115.
143. Maher and Wyatt, 'International Trade in Animals and Animal Parts', 223.
144. Ibid.
145. The genocide was also perpetrated by commercial hunting enterprises that killed the seals for their blubber. See Jørgensen, 'Erasing the Extinct'.
146. Quarrell was living in Tyenna in 1908. He seems to have later moved to nearby Fitzgerald.
147. Batty similarly tried to capture the tiger he encountered but the latter was too quick for him. When the tiger seemed likely to escape, he shot him. As the thylacine did not die immediately, Batty at first hoped that he would recover and he could then confine him and profit from selling him alive.

148. Brown was presumably an amateur or part-time photographer given the remote location.
149. Although thylacines had stiff tails, it seems probable that this person was also performing a similar role in the first photograph. In that image, the tip of the tail is out of frame.
150. Wright, 'Bullocky', 25. Quarrell's profession was that of 'bullocky', the driver of a bullock team. His son followed him into the profession. In 1934, Albert Junior, who was working for Tyenna Mill, was fined for temporarily blinding one of the bullock team after striking them with a stick. Anonymous, 'Bullock Flogged', *The Mercury*, Tuesday 29 May 1934, 7. Albert Senior was also once in trouble with the law, fined for allowing horses he owned to stray.
151. See Anonymous, 'Tyenna Railway Extension', *Daily Post*, Wednesday 18 October 1911, 11. Browning's mill was established in 1904.
152. Guiler, *Thylacine*, 71–72.
153. Ibid., 61.
154. Baer, 'A Reconsideration of Sexual Violence', 653.
155. For the dating of this film, see Sleightholme and Campbell, 'The Earliest Motion Picture Footage of the Last Captive Thylacine?'
156. Her companion died in transit. See Sleightholme and Campbell, 'A Catalogue of Motion Picture Films of the Thylacine', 783–84.
157. For more information on this photograph, see Sleightholme and Campbell, 'A Catalogue of the Thylacine Captured on Film', 151.
158. Ibid.
159. Pollock, *Art in the Time-Space of Memory and Migration*, 110.
160. Ibid.
161. In 'The Last Image', Freeman suggests three possible readings for the gesture: disinterest, fatalism or submission (196).
162. See Preti, 'Animal Suicide'.
163. Borrell, 'Mercy Killing', 53.
164. Freeman reads this attack as unrealistic but there are accounts of active aggression by thylacines (admittedly sometimes when their young were nearby) when confronted by humans.
165. Freeman, 'The Last Image', 198.
166. Borrell, 'Mercy Killing', 53.
167. Chare, 'After the Thylacine', 152.
168. Aguila-Way, 'Beyond the Logic of Solidarity as Sameness', 6.
169. McGoldrick, 'Unscoping Animals', 265. McGoldrick does not discuss visual documents of *dead* animals or of extinct animals imaged alive (animals that now live on only through archival materials), but her insights also have relevance in these contexts.
170. Ibid., 261.
171. McGoldrick presumed the badger shared her instinct-driven 'fight or flight' response (ibid.) and does not countenance that the badger might have deliberated. It is also possible that rather than having an automatic physiological reaction, the badger's response was habitual. For a discussion of the possibility that nonhuman animal behaviour presumed to be instinctual may be habitual, see Lemoine, *L'habitude et l'instinct*, 81.
172. McGoldrick, 'Unscoping Animals', 261.
173. Alternatively, nonhuman animals must be accorded their humanity. This is something McGoldrick does not consider. She therefore seems bound to the dualist logic that Tim Ingold highlights in 'Humanity and Animality' (21). Ingold notes that many cultures accord nonhuman animals personhood (24).

174. Kosara, 'Visualization Criticism', 634.
175. McGoldrick, 'Unscoping Animals', 261.
176. Ibid., 263.
177. Baker, 'Fish Market, Lagos'.
178. In her article 'Crocodile Dundee', Jeanette Hoorn discusses a similar phenomenon in *Crocodile Dundee* (dir. Peter Faiman, Australia, 1986), in which visual imagery of outback Australia works in excess of the narrative to 'fortify a sense a place' (122). This 'placefulness' is supported through camerawork that tarries on the countryside. In *The Hunter* it is more through repetition that the locale affirms itself, exceeding narrative and symbolic concerns.
179. I am grateful to Florence Sabourin-Bordeleau for sharing her insights in this regard.
180. See the discussion of Edward Husserl's understanding of the humanisation of animals in Ferencz-Flatz, 'Humanizing the Animal'.
181. In 'Humanizing the Animal', Ferencz-Flatz lists cinema as one of the ways that animals become the object of humanisation (223).
182. My emphasis. McGoldrick, 'Unscoping Animals', 263.
183. Pollock, 'The Visual', 183.
184. Ibid.
185. For a discussion of the nonhuman animal's relation to the Imaginary in Lacan, see Buse, 'The Dog and the Parakeet'.
186. Language, however, is not the exclusive preserve of the Symbolic Order and also relates to the Imaginary and the Real. Julia Kristeva has explored these relations in particularly nuanced and productive ways. See Chare, *Auschwitz and Afterimages*.
187. Lacan accords nonhuman animals codes (to which they are captive) rather than languages. See Buse, 'The Dog and the Parakeet', 137.
188. Butler, 'Disturbance and Dispersal in the Visual Field', 151.
189. Pollock, 'Introduction', 18.
190. Ettinger, 'Matrixial Trans-Subjectivity', 221; Pollock, 'A Matrixial Installation', 195.
191. Pollock, 'A Matrixial Installation', 195.
192. Explanations can obviously be offered for his behaviour (for example, he is using a grid search pattern) but I am interested in the psychic effects of his repetitive actions on the spectator here rather than in narrative causes.
193. See Ricco, *The Logic of the Lure*.
194. Pollock, *After-affects/After-Images*, 3.
195. Pollock describes trauma as a shadow without a form. See ibid.
196. Pollock, *Art in the Time-Space of Memory and Migration*, 121.
197. Rose, *Sexuality in the Field of Vision*, 228. Rose is here referring to how repetition can cause unconscious material that is usually obscured to manifest to the self.
198. For a discussion of the history of equating wildlife photography with hunting in Turtle Island, see Brower, 'Trophy Shots'.
199. Sontag, *On Photography*, 15.
200. This last reading is encouraged by Mindy wiping his cheek after Lucy kisses him, seemingly in disgust, which could be interpreted as evidence of a homoerotic subtext.
201. Deleuze, *Cinema 1*, 21.
202. Reichert, 'Headbanger'.
203. In *After-affects/After-images*, Pollock uses the term 'after-affects' to describe the affective traces of trauma that can potentially emerge through aesthetic encounter.

Chapter 6

The Newfoundland Wolf

> The Wolf is now exceedingly rare in Newfoundland and I have never seen a specimen.
>
> —Outram Bangs, 'The Land Mammals of Newfoundland'

A Beothuk Wolf

In 1937 Glover Morrill Allen and Thomas Barbour used skull comparisons to identify the Newfoundland wolf as a distinct subspecies of grey wolf or *Canis lupus*.[1] Allen and Barbour worked from thirty wolf skulls held in the collection of the Museum of Comparative Zoology at Harvard University, including four sourced from Newfoundland. They found that the upper carnassials (teeth designed for shearing meat) of the Newfoundland wolf and also their anterior pillar (part of the roof of the mouth) and premaxilla (small cranial bones in the upper jaw) differed from those of other wolves of Turtle Island and speculated that the creature, with its 'local peculiarities', was 'a southern relict of the Arctic white wolf'.[2] The four skulls from Newfoundland that they studied had been received by the naturalist Joel Asaph Allen from the trapper James M. Nelson in June 1865. They also examined a wolf skin (lacking feet and a tail), described as whitish in colour, that Elwood Worcester gifted to the museum in 1932.[3] He had travelled in Newfoundland many times but believed, from memory, that he had caught the wolf near Grand Lake in the 1890s.[4] Until donating it, Worcester had used the skin decoratively as a rug.

Allen and Barbour proposed that in 'allusion to the now extinct aborigines of Newfoundland, the Beothuks [*sic*]', the newly identified subspecies be named *Canis lupus beothucus* (Beothuk dog wolf). They chose this trinomial to classify the new subspecies in the knowledge that the wolf was extinct, clearly drawing parallels between their fate and that of the Beothuk, a First Nations people of Newfoundland who were commonly believed to have died out.[5] The choice of Latin grammatical form (the Latinising of 'Beothuk', a Beothuk word roughly translated as 'the people') conforms to standard scientific practice for the naming of species, a practice rooted in the systems of classification devised and perfected by European scientists such as Gaspard Bauhin, John Wilkins and, most notably, Linnaeus.[6] In Linnaean classifications, the second name in a binomial such as *Canis lupus* has no meaning if separated from the first (the genus, here *Canis*). Similarly, the third name (the subspecies, here *beothucus*) has no meaning if separated from the first and second names. In mammalogy, thanks to Allen and Barbour, the Beothuk have become bound to the grey wolf for their significance; the unambiguously extinct wolf nominally continues to serve as a figure for their vanishing.

Exactly how significant the grey wolf was to Beothuk culture is difficult to fathom. There may be an account of wolves living at a Beothuk camp. A man known only as E. S. who was a member of the party that murdered the Beothuk leader Nonosabasut (also Nonosbawsut) and captured his partner Demasduit (also Demasduwit) gave an account of the Beothuk camp that mentioned finding 'a bitch and her whelps, about two months old' there.[7] These animals are not named and could be dogs, but there are no other records indicating that the Beothuk kept hounds. The term bitch has also historically been used to refer to she-wolves.[8] In a major article on the Newfoundland wolf, the naturalist Leslie Tuck quotes from an account given in 1612 by the then governor of Newfoundland, John Guy, of a meeting in Trinity Bay with the Beothuk in which he describes his party being approached by a man in a canoe 'with a flag in his hand, of a wolfe skin'.[9] Tuck does not provide further context for this encounter, but as he relates it immediately after a discussion of the likelihood that the Beothuk traded furs including wolf skins with early settler colonists, it is possible to surmise that the naturalist interpreted the Beothuk's gesture as signalling that he wanted to trade skins.[10] Guy's narrative is, however, relatively lengthy and the wolf skin (which he refers to as white in colour) features repeatedly in it. The Beothuk man, for example, did not simply carry the skin, he shook it and made 'a loud noise' as he came near. Guy's party interpreted this as a call for a parley and put out a white flag. Later, the action of jiggling was repeated: 'the sauage [savage] made a loud speech, and shaked the skin which was answered ... in like manner, and as the sauage drew neare, he threw downe the white skin on the ground the like was done by Master Whittington.'[11] After a series of exchanges of food and tools between the groups, the skin is eventually given to the aforementioned Whittington and the Beothuk take the white flag.[12] The wolf pelt in this account seemingly possesses symbolic import. It is used as a means of opening a channel of communication. Unlike the beaver, sable and bird skins that Guy's group will subsequently barter for, it is invested with communicative power. It is not just any skin.

What happened to this white wolf skin, one of the first recorded artefacts exchanged with colonists by the Beothuk, is not known. Over three hundred years later, Allen and Barbour use a different white wolf skin to aid their research into the possible distinctiveness of the Newfoundland wolf, one that was also given as a gift (this time by a leisure hunter to a natural history museum). In outlining the characteristics of *Canis lupus beothucus*, they refer directly to this hide, describing it as 'entirely pure white with a tinge of ivory yellow on the head and limbs' and as having 'dense and woolly' underfur.[13] The skin contributes to their knowledge of the new taxon they have identified – a taxon to which they give the name *beothucus*. Taxonomy as a system of ordering has historically been entwined with colonialism and possession. Harriet Ritvo notes that 'the competitive colonization of soldiers and diplomats had its analogue in the nomenclatural activities of zoologists'.[14] Naming in this context forms part of a process of dominion. Allen and Barbour draw attention to imperial conquest by way of their reference to the Beothuk.

Naming the wolf after the Beothuk, finding a name for them, neglects the reality that they had already been named *moisamadrook* (this is, at least, how the spoken Beothuk word was transliterated).[15] As European influence in Newfoundland increased and the Beothuk people diminished, the nominal *moisamadrook* gradually ceased to be used or heard, replaced by 'wolf', an English word derived from Old Frisian.[16] Before the arrival of Europeans with their 'wolves' and 'loups', hundreds perceived *moisamadrook* trotting Newfoundland's barrens and walking its forests.[17] The skin shaken towards Guy and his party was a 'wolfe' to them but not to the man who held it. The Newfoundland 'wolf' is a post-contact invention. The rejection of Indigenous names for species was openly called for by colonial historians of Turtle Island.[18] Perceptions of animals like wolves by Indigenous peoples also differed greatly from those of colonists. In the context of exterminating wolves in Turtle Island, Andrea Smalley reports that '[l]awmakers found that they had to offer significant inducements for wolf hunting since native peoples found the idea of eliminating an entire species strange' and 'killed wolves only when individual animals became a nuisance to their own hunting activities'.[19] The genocidal gaze of settler colonials towards wolves was not shared by First Nations peoples such as the Beothuk. They saw wolves differently; they saw not wolves but *moisamadrook*, *mahihkan* (Cree), *paqtism* (Mi'kmaq) or another animal.[20]

On 14 May 1937, the date they officially made their identification of *Canis lupus beothucus*, Allen and Barbour also invented an animal. Drawing on speech act theory, Yves Cambefort has noted how naming in zoology is never merely a means of describing a species (or, by extension, subspecies) but rather forms an 'assertive declaration', an act that generates that species: 'By the formula "new species", it is not only a new name but also a new taxon – i.e. a new concept – which is being created.'[21] Allen and Barbour did not simply put a name to a skull. They did not stumble across a new subspecies; rather, by way of their words, they actively conjured it into being. Cambefort discusses how questionable it is to think that scientists adjust their words to fit the objects of their perception (as if the object shapes the words rather than vice versa). He also signals the limits of zoological naming as a speech act. The declaration of the existence of *Canis lupus beothucus* does little, for example, to convey what is distinct about that species. The words require a material substrate, a physical supplement, to make sense. This supplement is the holotype: a specific 'specimen' used when a species (or subspecies) is formally described and named. As Cambefort explains, an interested person reading about a new species 'must see an exemplar, a [holotype] in order to "realize" both the original description and the meaning of a (new) name'.[22]

In the case of the Newfoundland wolf, the holotype is Skull Number 351 from the Museum of Comparative Zoology. This skull was initially entered into the *Catalogue of Bones of Mammals in the Museum of Comparative Zoology* simply as *Canis lupus* (as a grey wolf). Later the word *occidentalis* was added to the name, the skull now designated more specifically as that of *Canis lupus occidentalis*, otherwise known as the north-western wolf, a Turtle Island subspecies.[23] Presumably later still (although the chronology can only be speculated at), another name was added in pencil: '*Canis lupus beothucus*'. At some point in the remarks was inked: 'Type *Canis lupus beothucus*' and above it in smaller script 'G.M. Allen + T. Barbour'. The skull has therefore possessed at least three names in addition to its numeric identifier, demonstrating the veracity of Cambefort's assertion that nomenclatural acts are open to contestation and cancellation.[24]

The skull these names refer to, the bone tissue behind the words, should not simply be regarded as a passive surface upon which these various names are figuratively inscribed and erased.[25] The cranium and mandible of the wolf possess a kind of agency. Like the white skin shaken by the Beothuk man used to initiate an encounter, they can be understood to cause events to happen. The skull named *Canis lupus beothucus* sustains its classification rather than merely receiving it. It is a classification that may be brought into question if further study with new technologies fails to substantiate claims that the wolf really is a distinct subspecies.[26] The taxonomic validity of the Newfoundland wolf was already identified as borderline by Rudolph Anderson in 1943.[27] The way subspecies are identified is usually suspect, with much depending on the criteria that are being used to define them, and even if morphologically the Newfoundland wolf was found to be highly similar to wolves in Labrador and northern Quebec there is always the possibility that compartmental differences existed that are now lost to natural history.[28] As discussed in Chapter 2, species identification also necessarily involves the reduction of animals to types, to groups of the same that differ from other groups, repressing singularity or difference within a given group.[29] To counteract this levelling, the 'one of them' logic of Western natural history that has inadvertently aided genocidal practices against nonhuman animals through facilitating typecasting, I want to devote much of the remainder of this chapter to the physical remains of one specific victim of animal genocide, a wolf shot by William Whiteway whose remains are held in the collection of the Newfoundland and Labrador Provincial Museum at The Rooms in St. John's in Newfoundland.[30] The chapter also includes the most detailed overview to date of the bounty scheme that contributed to the extinction of the Newfoundland wolf.

Speaking in Bones

In the late nineteenth or early twentieth century, Whiteway shot and killed a wolf, possibly female, on the Gaff Topsail Plateau, a plain featuring several monadnocks (known colloquially as topsails) that is located north-east of Hind's Lake in Newfoundland's interior.[31] The wolf's skin was subsequently given to a tanner, James Ewing, to be treated.[32] Whiteway then used the tanned skin, which was backed with felt, as a rug. John Maunder suggests it was hung on a wall.[33] In a 1923 letter to Ernest Thompson Seton, Gower Rabbits recalled that '[m]ore than 20 years ago, 8 wolves were seen in one bunch near Gaff Topsails'.[34] A wolf was reported killed in the Gaff Topsails area in 1911. They may well be the wolf that Whiteway shot. If this was the case – as Tuck observes – they were probably the last Indigenous wolf killed in Newfoundland, rendering them 'an important (but sad) historical relic'.[35] By the 1930s, the Newfoundland wolf was considered extinct.[36] All that now physically endures of the subspecies from recent times are the remains used by Allen and Barbour for their identification, comprising two complete skeletons, an additional two skulls and a skin (of which two complete skeletons, one skull and the skin are held by the Museum of Comparative Zoology and one skull by the Smithsonian National Museum of Natural History), and the skull and skin of the wolf that Whiteway shot.[37]

Here, I want to provide a history of the genocidal violence perpetrated against wolves in Newfoundland and, alongside this, trace the post-mortem history of the skin and skull of the wolf Whiteway killed. To do the latter, I will engage in what is referred to as object biography. This is an approach to the study of material artefacts that, as Jody Joy explains, tracks the histories they accrue over time and works 'to reveal relationships between people and objects'.[38] The origins of object biography can be traced back to works such as Annie Carey's *Autobiographies of a Lump of Coal &c.* Carey provides 'autobiographies' of a lump of coal, a grain of salt, a drop of water, a bit of iron and a piece of flint. Each of these is designed to bring young readers 'to that *truth of things* which we have at present reached'.[39] Her aim is therefore to share leading-edge scientific knowledge yet in such a way that the things she writes of ostensibly speak for themselves. A lump of coal, for example, addresses a group of children, advising them that it desires to tell 'my own history'.[40] As an act of personification, Carey's style of address can be read as amplifying (albeit unconsciously) the way that all scholars deal with objects, with things, giving voice to them in some way. The act can, however, also be interpreted as an unconscious recognition of the agency of things.[41]

Often object biography as an approach to the study of material culture is linked with Igor Kopytoff's 1986 essay 'The Cultural Biography of Things', which focusses predominantly on commodities as things. Kopytoff believes that it is possible to ask similar questions of the biography of a thing as of the biography of a person.[42] These include tracing 'life changes', for example, changes in the use of a thing. He suggests that '[b]iographies of things can make salient what might otherwise remain obscure'.[43] Biographies can, for instance, bring to light how objects are culturally redefined and used across periods of time. It is this dimension of Kopytoff's biography of things that I draw on here, combined with Carey's insights into the agency that material artefacts possess. This agency, as I will demonstrate, manifests in relation to the life stories of objects. The biography of things provided here is, in this sense, part biography, part autobiography. Things tell aspects of their own story; they have life narratives embedded within them that words help to *realise*. The physical remains of the wolf shot by Whiteway, for example, still bear traces of the fatal gunshot that despatched them. A considerable part of the left side of the cranium is missing, the bone having shattered with the impact of the bullet. The skull bodies forth this knowledge of the wolf's demise. The damage is visible, tangible. My words may draw attention to the injury, yet they cannot be said, in any simple sense, to call it into existence.

In Chapter 2, I explored the stories that the physical remains of thylacines – including their bones – can tell us about animal genocide. Zoë Crossland has described how, in forensic anthropology, bones are popularly said to 'speak for themselves'.[44] For Crossland, this belief in vocal bones suggests 'the idea that the body provides unmediated access to past events and reflects an empiricist faith in the ability of the facts of the body to speak for themselves'.[45] Gérard Wajcman has explored this phenomenon in relation to *CSI: Crime Scene Investigation*, tracing how popular television drama has come to reflect changes in actual crime investigation in which the confession has been replaced by the gathering of material evidence as the primary objective of criminal investigation.[46] For Wajcman, however, it is not that confession has disappeared. It is rather that the role of confession has now been transferred from the suspect to material evidence. In *CSI*, he states, 'only things confess', only things 'speak and tell the truth'.[47] In this sense, the drama helps to vehicle the kind of naïve conviction (identified by Crossland) in the facts of the body speaking for themselves. Crossland justly critiques this belief.

Any episode of *CSI* or, more recently and pertinently, *Bones* presents multiple instances of scientific narratives (what Wajcman calls 'confessions') giving meaning to what would otherwise simply be bone fragments, blood, tissue. The human remains in *Bones* have complex stories coaxed out of them. These stories require scientific technologies (and scientists who can extrapolate and interpret the data these technologies produce) in order to be told. In *Bones*, Temperance Brennan (Emily Deschanel) and her team provide what could be called abbreviated object biographies for the physical remains they investigate. These narratives are prompted by qualities held by the remains.[48] In this sense, contra Crossland, rather than depriving objects of all agency in the stories they tell, I prefer to view the stories that emerge in actual forensic practice and here (in the context of studying nonhuman animal remains) as collaborative endeavours. Bodies, as Crossland and Rosemary Joyce note, convey traces of the past, yet the traces are not 'self-evident'.[49] They require articulation. In the powerful concluding paragraphs of their essay 'Anthropological Perspectives on Disturbing Bodies', Crossland and Joyce work hard to express the complex relationships that exist between physical remains, narrative and the past. They foreground the 'dynamic relationships' that enable human remains to register as signs, relationships that involve engaging with material traces such as bones, attending to their stories, translating them into words as part of a memorial process.[50]

Trophy Wolf

In a letter to Leslie Tuck, Herbert Maxse Knowles Whiteway explained that his brother shot the wolf, part of a group of 'four or five', 'as it was running away'.[51] This account is confirmed by the skull, which tells of violent death through the jagged bone of the left side of the rostrum, the rough, diamond-like hole just in front of the orbit.[52] Entry wounds to the head often cause circular holes with smooth, rounded borders.[53] The wound sustained to the maxilla of the wolf, however, is fan-shaped with a bevelled edge.[54] This suggests that the bullet entered the skull at an oblique, shallow angle. It came from behind. Whiteway was, perhaps, also on higher ground. The wolf may have been running away having sensed the presence of the hunter. A narrative, hesitant but not unreasoned, emerges through close analysis of the cranium, the suffering bone, in tandem with the written account. Knowing the circumstances of the killing of the wolf corroborates aspects of the story the skull itself expresses. Two narratives, one written in broken bone and the other in the words of the hunter's brother, combine to communicate the end of the creature's life. The object has the capacity to generate a kind of posthumous autobiography, partial but insightful. The matter of the skull therefore *matters* for the wolf's story. Like the Newfoundland wolf holotype in Massachusetts, this skull shapes the words that refer to it as much as it is shaped by them.

Many of the circumstances of Whiteway's encounter with the wolf are clearly still sketchy. It seems likely he was out hunting although less probable that his chosen prey was a wolf.[55] The writings of James Patrick Howley are enlightening in this context. Howley was a keen huntsperson and in his 'Reminiscence of Forty-Two Years of Exploration and Survey in and about Newfoundland, 1869–1916' he gives many accounts of his hunting exploits. These pursuits are usually aimed at deer: at caribou and moose. For Howley, wolves are irritating because they drive off game. He recounts of a hunting trip to the upper part of the Northern Feeder, a stream forming a tributary of Robinsons River in western Newfoundland:

> We fully expected to run across Caribou every minute, the country was ideal for them and we saw much footing, often quite fresh, but not a deer could we see. It seemed strange where they could be as there was very little cover for them anywhere. At last the mystery was solved when we came across a place where the footing of wolves and the torn up surface showed us that these ravenous brutes had been here, that a struggle had taken place and that the poor deer had been chased out of the country hereabout. While cruising around by myself I saw a fox sunning himself on a hill-side and succeeded in getting near enough to shoot him. He was a poor specimen of a red fox not of much value.[56]

Here Howley shoots a fox simply because he was there to be shot. His intended quarry was caribou. Doubtless if wolves had still been in the vicinity, he also would have endeavoured to shoot them.[57] The wolf shot by Worcester, the skin of which was gifted to Harvard, 'happened to come over the side of a hill where I was sitting with my rifle waiting for caribou'.[58] Wolves were clearly often spur of the moment kills, with hunters taking advantage of chance, rather than engineered encounters.[59] In places such as Newfoundland where a bounty was on offer, wolves could provide a useful source of additional income.[60]

The kind of language Howley uses to describe hunting deer is markedly different from that he employs here to talk about wolves. Wolves are referred to as 'ravenous brutes', as voraciously hungry and savage. Caribou, by contrast, are superior animals. They furnish 'fine heads' for hunters.[61] In 1904, Howley recalls shooting 'a fine doe', likely fatally injuring her; she then eluded him. Although most mammals were fair game, hunters such as Howley (and probably Whiteway) operated with a hierarchy. In addition to his clearly negative perception of wolves, Howley also refers to lynx as 'ugly customers'.[62] Animals such as bears, foxes, lynx and wolves often likely formed collateral prey in the quest for bigger game.

Despite the existence of such a hierarchy, Whiteway nonetheless clearly felt killing the wolf was an achievement that merited commemoration. Joseph Hatton and William Harvey describe wolves and bears as appealing to the 'more adventurous sportsmen', although 'both are scarce and difficult to find'.[63] Perhaps in recognition of this rarity, if not nobility, Whiteway commissioned a tanner to prepare the skin as a rug. He wanted a trophy by which to remember this particular kill. Worcester behaved similarly in 1896. The relative rarity of wolves (coupled with the wealth of the hunters, which rendered bounty payments unneeded) may have been the motivation for both men to keep the skins as mementos. Another might have been the perceived quality of the pelts. When Howley writes of the red fox he has shot not being of much value, he draws attention to how not all animals of the same species are viewed in the same way. There is an aesthetico-economic dimension to Howley's observation. Factors such as size, colouration and (linked with these) monetary value, for example, may influence a hunter's decision to preserve a particular kill. Whiteway could therefore have viewed the wolf he shot as a superior 'specimen', as healthy and of a good size, as worthy of admiration and worth holding onto. In the novel *Trap-Lines North*, an embellished account of trapping in northern Ontario, Jim Vanderbeck shoots a very large wolf that has been causing anxiety around the town of Nakina. Vanderbeck contemplates turning the wolf into a rug because of his immense size and notoriety.[64]

Whiteway may also have retained the wolf skin because of a heightened animosity towards wolves. As discussed in preceding chapters, the display of nuisance animals by landowners is not uncommon. Johann Knobel writes of the wedge-tailed eagle that it 'was common practice in rural Australia to hang eagle carcasses in rows on roadside fences, after the birds had been shot, trapped or poisoned'.[65] There was also a history of similar acts in Newfoundland. In the winter of 1838, a wolf thought to be a child killer was slaughtered in the vicinity of St. John's and put on display: 'It was kept for some time at a house, and a charge of sixpence made from every person who went to view it.'[66] The wolf shot by Whiteway was not linked with any specific wrongdoing but it is probable that he nevertheless viewed them negatively as his father, William Vallance Whiteway, had been an enthusiastic supporter of the bounty scheme initiated to encourage the killing of wolves. The father's animus towards wolves may well have been shared by his son. In such circumstances, the wolf skin would be a trophy, a record of victory achieved over 'vermin'.

An Inventory of Genocide

At the time Whiteway shot the wolf, the bounty that had intermittently been offered since 1839 in Newfoundland for the killing of wolves was still in effect. Based on legislative records, it seems that no request for payment was submitted by him. Tuck lists the last bounty payment claimed (for $12, which was, by then, the agreed bounty for a single animal) as having occurred in 1896.[67] He may be mistaken, as bounty claims are difficult to track as they were sometimes listed as sundries in government accounts, yet the fact that Whiteway kept the skin supports the idea that he did not make a claim. It was standard practice for Justices of the Peace, those who were empowered to issue certificates authorising bounty payments, to retain or destroy the skin of the wolf for which the claim was made. There were moves to change this requirement in 1879, when a Select Committee appointed to consider encouraging sheep farming in Newfoundland made a list of recommendations, one of which was that 'in order to ensure the more efficient protection of sheep and to encourage the killing of wolves, it is desirable that the person killing a wolf should be permitted to retain the skin, as well as to receive the bounty, as the skin has a considerable mercantile value, and its surrender materially detracts from the amount of the bounty'.[68] This recommendation would seem difficult to implement without raising the possibility of multiple claims being submitted by different persons for the same skin and was seemingly ignored. Potential abuses of the bounty system were recognised.[69]

The bounty was first instituted by the Newfoundland Legislate Assembly through an Act passed in September 1839.[70] During its passage, the wording of the bill was amended by Her Majesty's Council, with one suggested change to the preamble involving substituting 'depredations' for 'inroads' in the section that read 'much injury has arisen from the Inroads of Wolves in this Island'.[71] The council also advised adding 'to the Inhabitants' after 'has arisen'. These changes were implemented. The preamble to the bill as passed read: 'Whereas much injury has arisen to the Inhabitants from the Depredations of Wolves in this island; and it is expedient to encourage the destruction of the said Animals :- '. These seemingly minor changes are highly telling. The addition of 'inhabitants' clearly pits the wolves against Newfoundland's human population. If the harm they have caused is to livestock, the injury is nonetheless to the human owners of that stock. 'Inroads' with its sense of hostile incursion is also replaced by 'depredations', with its connotations of plundering or pillaging. The latter term seems designed to give a sense of the wolves engaging in wrongful appropriation, to frame them as committing the crime of theft. The motivation behind the bill is made clearer by these slight yet significant alterations; the wolf was perceived to be a threat to people's livelihoods.[72]

The Governor, Henry Prescott, who had to give his assent to the bill, made a speech on 14 September 1839 (transcribed by the speaker) in which he stated that the legislation's 'speedy operation has been rendered necessary by the increased daring of these destructive animals'.[73] Writing in 1846, Philip Tocque recalls that the animals were numerous around St. John's 'three or four years ago, and were prowling about so near the dwellings, as to endanger the lives of the inhabitants'.[74] He details how the disappearance of a young child of five or six years of age on his way to school was attributed to a wolf.[75] Tocque links this sense of danger in the island's major town with the introduction of '"The Wolf-killing Act"'.[76] In *Newfoundland in 1842*, Richard Bonnycastle also makes reference to the harmfulness of the wolves, writing in a footnote that the wolf 'grows very large, is frequently traced near the capital, and does much injury to cattle; a price is put by the legislature on its head.'[77] Tocque likewise describes the wolf as proving 'very destructive to the cattle, destroying a number of cows and sheep'.[78] Prior to European invasion, Newfoundland's wolves preyed predominantly on caribou. As they are able to switch easily between wild and domestic prey, they probably did kill sheep and cattle, although distinguishing between a wolf bite and the bite of a wild dog is next to impossible for a farmer. The pastoralist Michael Downey, for example, complained to James Hayward 'of having lost a number of lambs this spring to wolves'.[79] Some lambs (approximately 15 per cent) die shortly after birth and wolves will consume carrion in the absence of fresh meat so it is quite possible a quantity of Downey's losses were not caused by wolves killing his livestock.[80] He would be unable to discern if wolf (or dog) bites to dead lambs were post-mortem or not.

Although Col Bailey suggests that the scheme did not take effect until 1841, two wolf bounties were claimed in the period from 14 September 1839 to 3 January 1840, by William Tavenor (or Treanor) of Trinity (16 November) and Martin (or Benjamin) Day of Salmon Cove (24 November) respectively.[81] Eight certificates were presented to the Colonial Secretary, James Crowdy, for wolves killed in the period 3 January 1840 to 2 January 1841.[82] Denis Quigley (of Harry Cove) claimed for three wolves and single claims were made by Michael Fennelly (of St. Shott's), William Sevior (of Trinity Bay), Thomas McDonald (of Gaultois), Samuel Dobson (of Hant's Harbour) and Joseph Roswell (of New Bay).[83] The accounts for the period are, however, incomplete, with the certificates presented for wolves killed carefully recorded but only three payments clearly detailed: 18 April (£5), 20 June (£5) and 26 October (£6 for Coroners and Wolves).[84] The other five payments were likely recorded as sundries. The 26 October payment for coroners and wolves shows how different expenses were often combined in a single account entry. Through attending to the island's financial accounts, however, it is possible to engage in a calling to account, to reconstruct a relatively detailed, if imperfect, history of the bounty scheme and those who participated in it. I have chosen to include the names of all those for whom bounty claims are recorded, as behind each of these human names are (sometimes multiple) individual wolf lives. This is a catalogue of singular lupine deaths. The probability that the bounty scheme would cause total extermination was high (based on historical precedents in Europe and the United States) so this must also be read as a roll call of the culpable, of genocidaires. Many, if not all, of these men (no women made claims) would have known from European precedent that bounty schemes caused extinction.

There is a payment of £20 in the Treasurer's Cash Account for the quarter ending 30 June 1840 that was made on 9 May to Patrick Burke. It is listed as 'Wolves per Patrick Burke'.[85] Leslie Tuck believed this to be a bounty payment but it is, in fact, a sum provided to Burke, a soldier, 'in consequence of [his] being incapacitated to perform any duty by his endeavouring to destroy wolves'.[86] The payment was suggested on 8 April 1840 by the legislative assembly.[87] On 13 April 1840, a letter from G. B. Frazer (Burke's commanding officer in the Royal Artillery), dated 10 April and written in support of Burke receiving financial aid, was read out to the Assembly:

> Sir, – I have the honor to request that you will be pleased to submit for the favourable consideration of his Excellency the Governor the case of gunner and driver, Patrick Burke, of the Royal Artillery, who is about to be invalided in consequence of a severe fracture of the left arm, which he

> accidentally met with when employed in October last, at the suggestion of his Excellency, with other volunteers of the Company under Lieut. Geary, in pursuit of wolves, then infesting the immediate neighbourhood of St. John's.
>
> I beg to state that this accident was not occasioned by intemperance or neglect of any kind. That his former service, (six years) will be lost to him, and that it is more than probable he will be discharged without any reward whatever. Gunner Burke has moreover a wife and family dependent upon him for support, and any aid that can be afforded will be bestowed upon a deserving soldier.
>
> I have, &c.
>
> (Signed) G. B. Frazer,
>
> Major, Commanding Royal Artillery

The letter is revealing because it clearly shows that before the institution of the bounty scheme a military campaign had already been undertaken at the instigation of the Governor to eradicate wolves in the Avalon Peninsula. Wolves in the letter are described as 'infesting' St. John's, as attacking or harassing the city. Burke was, in this instance, granted £20 and would subsequently petition for further assistance in 1845.[88]

For the period 1 January 1841 to 31 December 1842, twenty-five bounty payments were paid.[89] Michael Shelly of Barr'd Islands Fogo claimed for three wolves and the other claims were for single wolf kills.[90] The single claims were made by Maurice Lewis (of Port aux Basques), James Jeans (La Poile), Francis Poulett (La Poile), Michael Carroll (King's Cove), Andrew Paul (Burgeo), Gabriel Billard (Red Island), Martin Walsh (Goose Cove, White Bay), Thomas Joe (White Bay), John Johnston (Greenspond), Henry Rowsell (Hall's Bay, Fogo), Benjamin Finnamore (Bard Island, Fogo), William Hadden (Gander Bay), James Gudger (Muddy Hole), James Dunn (St. John's), Maurice Brien (Broad Cove, Gasters), James Cull (Bard Island, Fogo), John Bowles (Bard Island, Fogo), Benjamin Finnamore (Barr'd Islands, Fogo), Henry Rowsell (Hall's Bay, Fogo), William Blackmore (Red Island, Fortune Bay) and John Doran (Outer Cove).[91] The geographical spread of these claims is broad. Few claims, however, come from the west or north of Newfoundland. This probably reflects population densities at the time rather than where wolves inhabited. The relative paucity of claims from the Avalon Peninsula suggests that wolves were already scarce in that heavily settled region.

In *Wandering Thoughts, Or Solitary Hours*, Tocque gives an account of a trio of men claiming £5 from the Justice of the Peace at Brigus for a wolf kill in this period: 'On the 18th of March, 1842, three men, natives of the neighbourhood of Colliers in Conception Bay, brought to R.J. Pinsent, Esq. J.P. at Brigus, the skin (and head attached) of a male wolf, for the purpose of obtaining £5 under the Wolf-Killing Act.'[92] The wolf had previously escaped from a trap (losing one of his forelegs in the process). He was blamed for killing several cows. The claimant for this wolf must be Maurice Brian of Broad Cove as his claim is the only one certified by Pinsent in the period. Broad Cove is about eight hours' walk from Brigus. Tocque reports that Pinsent preserved the skin and had it stuffed and that it was 'for some time kept in the court-house at Brigus, as a fine specimen of the Newfoundland wolf'.[93] He also reports that 'on inspecting the skin of the wolf, after his death, fifty-six shot holes were discovered in it'.[94] Maunder rightly draws attention to what this fusillade reveals about attitudes to the wolf: 'The ferocity with which it was despatched is a good indication of the hatred and fear the settlers had of wolves.'[95] Given the considerable damage the pelt must have sustained, it seems unlikely that the skin was appreciated for its physical fineness, but rather for its symbolism of human triumph over a 'dangerous beast'. Bonnycastle remarks of this same wolf that 'his skin would have been valuable if it had not been pierced with many swan shot'.[96]

The Brigus wolf was probably stuffed and preserved because his remains told quite a tale by way of their damage. Like that of the wolf killed by Whiteway, the skin carried aspects of this wolf's life story, of his death. Bonnycastle and Tocque both note that the wolf was silvery in colour. Bonnycastle additionally notes that this 'pure silver grey' colouring was 'very rare'.[97] He lists the wolf's dimensions, which are copied by Tocque in his own account. The length from the nose to start of the tail is given as 5 ft. and including the tail as 6 ft. 6 in. The height of the fore-shoulder was 2 ft. 9 in. and at the haunch, 2 ft. 8 in. Additionally, Tocque notes that the wolf's remains were subsequently given to the Museum of the Agricultural Society of St. John's. This museum was, for a time, curated by E. L. Moore, a corresponding member of the Zoological Society of London.[98] This skin, like others that are recorded, has since vanished.[99]

From 1 January 1843 to 1 January 1844, four bounties were claimed by Charles Sheeves (White Island), John Fowlow (Trinity), James Carter (Little Cod Roy) and John Nachaile (Bay Despair).[100] A bill to continue the bounty scheme was presented by John Valentine Nugent on 22 March 1843. The Act to continue the Act (and thence the scheme) for another two years was passed on 22 May 1843. There is, however, a gap in disbursements from 1844 until 1847, when payments totalling £50 were made.[101] Single wolf kills were claimed by Peter Yeado (Fogo), Thomas Jeffers (Hooping Harbour), Joseph Jedore (Bay Despair), George Styles (White Bear Bay) and Michael Salome (White Bear Bay). Henry Roswell (Fogo) and John Jure (Exploits Bay) claimed for two wolves each.[102] For 1848, single payments were made on 5 June to George Castle (for killing a wolf in Fogo) and on 2 September to John Johnston (White Bay) and

John Strickland (Seal Cove). On 12 November, Joseph Paul was paid £10 for killing two wolves at River Exploits.[103] In 1849, payments are recorded to four different people, Maurice Lewis (£5), Bernard John (£10), John Stevenson (£5) and Henry Cull (£5).[104] The Act must then have been allowed to lapse.

In 1858, a bill to revive the Act to kill wolves was successfully sponsored by John Casey.[105] A new Act entitled 'An Act to revive and continue an Act entitled "An Act to encourage the Killing of Wolves in this Colony"' was passed on 10 May 1858.[106] This Act of revival was good for ten years. Richard Howley, the acting financial secretary, details the expenditure on wolf killing in 1859 as totalling £55. Nine individuals received payments, with Andrew Paul and Noel Paul receiving £10 each for killing two wolves apiece.[107] Their payments were both made on 30 December. The other payments were made in April (Joseph Bernard, Edward Poulette), August (William Billard), October (William Blackmore, Newell Mitchell) and November (Peter Joseph Cope, John Mitchell).[108] For 1860, Tuck lists a total of £50 of bounty payments divided among nine individuals, with John Stevens receiving £10 on 15 August for the killing of two wolves.[109] The other payments were to John Mitchell (7 March), Daniel Ryan (20 March), Noel Paul (26 March), Martin Billard (23 April), Frank Joe (7 May), Patrick Fitzgerald (19 June), John Mitchell (28 June) and John Steacy (14 July).

The payments for 1861, however, show a marked increase. The total expenditure is £140, with six people claiming for two or more wolf kills and twenty-eight kills in total. Payments were made to Michael Dunn (30 March), Edward Leonard (30 March), John Stevens for two kills (25 April), Simon Cull for two kills (8 May), Noel Paul (8 May), John Mitchell for four kills (3 June), Patrick Fitzgerald (29 June), Samuel Valis (13 July), Martin Villiers (22 July), Newel Mitchell (22 July), Lewis John for two kills (7 August), John Brazil (7 August), John Fennymore (27 September), William Mead (30 September), Poleto Salmon (30 September), Andrew Joe (19 November), Noel Paul for three kills (5 December), Alfred Beaton for two kills (14 December) and John Stevens (14 December).[110] Noel Paul claimed twice in this period (in May and December) and also made claims in 1859 and 1860, demonstrating that some people were potentially going to considerable effort to kill wolves for bounty. Paul, a Mi'kmaq trapper, was a regular visitor to the interior of the island.[111] The territory of the wolves he killed probably did not extend to areas with extensive livestock holdings. A bounty motivated by the need to protect livestock therefore encouraged the killing of wolves far from farmland.

In 1862, ten payments were made totalling £60.[112] On 23 September, George House received £10 for two wolf kills and the same happened on 8 November for Andrew Paul. The rest of the payments to Denis Ryan (6

May), Nicholas Jeddore (27 May), Patrick Henfiant (30 June), Joseph Paul (3 July), Nicholas Jeddore (16 July), William Henry Crew (23 September), William Domany (1 October) and Joseph Paul (8 November) were each for the killing of a single wolf. On 4 March 1863, the Attorney General introduced a bill to suspend the bounty. The bill's passage through the legislature was not as smooth as that of previous wolf-related bills. Although passed without amendment by committee on 11 March 1863, when the motion was put to the House of Assembly that the committee's report be received, William Vallance Whiteway, seconded by Thomas Knight, moved in amendment that the report not be received.[113] A vote was taken and the amendment was defeated by fourteen votes to ten.[114] The report was then accepted and the bill moved forward to its third reading. The bill was subsequently passed by the house and then on 19 March by the Legislative Council. On 25 March, the Governor gave royal assent to the bill. William Whiteway, who had tried to scupper the bill's progress, had been elected to the House of Assembly in 1859. He was the father of the William Whiteway who killed the wolf whose remains are now exhibited in The Rooms, the 'Whiteway wolf'.

Payments were still made under the wolf killing Act in 1863 to a total of £45. Peter Jeddore received £10 for killing two wolves (1 May) and payments for single wolf kills were made to Noel Joe (1 May), Joseph White (23 May), Francis Honore (4 July), Samuel Gauding (4 July), Thomas McDonald (4 July), James R. Knight (31 July) and Thomas Downing (30 September).[115] The scheme was seemingly not revived again until 1872 at which time the reward offered was listed as $12 (dollars had become the island's currency in 1865). The initial payments made in 1873 to Jas Burden (29 May), William Collis (29 May), Frederick Shears (13 June) and Eli Jewel (13 June) were for just $10. A fifth payment of 13 dollars and 85 cents was made to the merchant Philip Brown of Sound Island 'for killing a Wolf'.[116] The money he was granted was part of a sum advanced to the Governor by the Colonial Treasury and the payment therefore does not seem to be part of the bounty scheme.

In 1874 payments to Edward Foley (21 May) and John Henderson (23 October) were also just for $10. Payments for $12 only seem to have commenced in 1877. The names of those who received bounties from 1877 onwards are not given in the government records, only the amounts for each year: 1877 ($12.00 = 1 wolf), 1878 ($24.00 = 2 wolves), 1880 ($72.00 = 6 wolves), 1882 ($132.00 = 11 wolves), 1883 ($36.00 = 3 wolves), 1885 ($48.00 = 4 wolves), 1886 ($12.00 = 1 wolf), 1887 ($12.00 = 1 wolf), 1888 ($24.00 = 2 wolves), 1890 ($12.00 = 1 wolf), 1894 ($108 = 9 wolves). A solitary payment of $12 is recorded for the financial year ending 30 June 1896.[117] This is the last payment identified by Tuck and by myself.

In the 1920 *Journal of the House of Assembly of Newfoundland*, the Wolf Act was still budgeted for by the Department of Colonial Secretary, although combined with a variety of other sundries (including cab hire, cartage and telephones) estimated together at $800 for the periods 1919–20 and 1920–21.[118] The killing of wolves then clearly ceased to be a budgetary concern because no claims were being made. The bounty, however, persisted. Tuck notes in a 1954 newspaper article about the early natural history of Newfoundland that the Act was still in force in that year.[119]

Poisonous Politics

The wolves claimed for bounty are not an accurate gauge of the total number of wolves killed in Newfoundland in the period post-European invasion and prior to their extinction. The wolf killed by Whiteway, for instance, was seemingly not claimed for. There may have been other instances where hunters killed wolves and either abandoned the carcass or retained the animal's pelt for themselves. In his *Reminiscences*, James Howley recalls encountering a wolf on 4 October 1888 in central Newfoundland, stating that 'this was the first, and indeed the only wolf I had ever seen alive, though we had frequently come across very fresh footing of the brutes. I should very much like to have got a shot at him and save the skin for the Museum.'[120] Here Howley sees the wolf not as ready money but as a potential scientific 'specimen'. He observes of the wolf: 'There was no mistaking the long, slinky, foxey-grayish brute, which seemed to walk along with that peculiar hang dog gait, characteristic of his kind.'[121]

There would also have been occasions when hunters injured a wolf but the animal escaped, dying of their wounds later. Furthermore, livestock owners may have employed strychnine to poison wolves.[122] From 1860, its use was 'wide and promiscuous' in Canada.[123] The poison was certainly in use on the island of Newfoundland when it was a colony.[124] The 1 June 1885 issue of the *Patriot and Terra-Nova Herald* includes an article summarising legislative proceedings in the House of Assembly to amend the Game Preservation Act. The article condemns the use of strychnine to kill foxes as 'not sportsmanlike' and also as detrimental to the island economy because of the value of fox fur.[125] The 28 March 1885 issue of the *Harbour Grace Standard* also reports on this legislation, noting that strychnine does not act instantaneously: 'A fox may run for a long distance after taking this poison before succumbing to its effects.'[126] Strychnine is, of course, indiscriminate and if it was in use to kill foxes then wolves would also have fallen victim to the poison. Because its effects are not instantaneous, animals poisoned by strychnine frequently die undiscovered so ascertaining the number killed is impossible.[127]

In the 1885 debate about the bill to amend the Game Preservation Act, Robert John Parsons Junior, a Liberal member of the Coalition of Conservative and Liberal Parties led by William Vallance Whiteway, argued passionately against the bill:

> The people require something that will give them employment instead of an Act to prevent the poisoning of wolves, otters, beavers, muskrats and weasels, and I would advise the hon member to try to devise some measure for the purpose of keeping the wolf from the poor man's door. Why should people who make a living by destroying these animals be deprived of their living simply to suit a hobby of the hon and learned doctor?[128]

Although the motivation behind the bill was clearly in major part a desire to protect the island's foxes, Parsons's rhetoric deliberately reframes the debate as one about the use of poison to eradicate wolves, the first species he mentions. His use of the expression 'keep the wolf from the door' to refer to poverty is a sly rhetorical manoeuvre as it simultaneously conjures images of wolves close to human habitation. The economic and political considerations behind the killing of wolves are made plain here. David Mech has drawn attention to the 'vote-producing gimmick' of wolf bounties, in which threats about the animal's destructivity are combined with the enticement of potential financial rewards for hunters and trappers.[129] Mech also foregrounds the headline-winning appeal of touting anti-wolf legislation. It is this appeal that Parsons appears to be capitalising on, setting himself up as the defender of the poor and protector against wolf attacks.

The Keeper of the Skin

The skin of the wolf shot by Whiteway was not exchanged for bounty but rather became a private souvenir, displayed on the wall in a house. Rachel Poliquin writes of hunting trophies that they have been 'created specifically to focus and retain memory'.[130] In such artefacts, 'meaning and materiality are forever merged'.[131] The meaning of a hunting trophy, Poliquin goes on to explain, is one of triumph. It forms a 'celebratory display of deliberate death'.[132] Trophies are often 'the products of passionate, hot-blooded, very personal killing'.[133] As already discussed, the wolf shot by Whiteway may not have been sought out by him, may not have been what he set out to stalk that day. Their killing, however, would have had personal resonance, relevance, given his father's historical link with the animal.[134] Whiteway Senior's negative view of the animal, his hatred, was probably bequeathed to his son. The father was also a keen hunter and probably took his son with him on hunting trips.[135] In this context, the dead wolf may even have provided a means for the son to remember his father and honour him.

The meanings attached to the wolf's remains are context-specific and open to change and contestation.[136] The scientific value of the wall-hanging was recognised by Tuck. He initiated correspondence with Whiteway's brother to see the skin and then to offer to purchase it. Tuck had a long-standing fascination with the Newfoundland wolf and wrote about the subspecies several times. In September 1947, for instance, he published an article in *The Newfoundlander* that listed many references to the wolf in accounts of Newfoundland.[137] He also discusses the wolf at relative length in an article for the *Evening Telegram* published on 17 February 1954, making reference to the wolf shot by Whiteway, the skin of which 'is in the possession of the Newfoundland Natural History Society'.[138] Tuck's final publication on the wolf, an essay for *Newfoundland Quarterly* that was still in draft form when he died, is also his most comprehensive.[139] This essay probably formed the paper he advised the zoologist Alexander Banfield in 1958 that he wanted to write.[140] Tuck also gave a radio broadcast on the wolf in April 1955.

Whiteway's brother, Herbert Whiteway, sold the skin to Tuck, who was acting on behalf of the Newfoundland Natural History Society, on 13 March 1952.[141] Tuck, who was a member of the Society, thenceforth became what he referred to as 'Keeper of the Skin'. The physical remains therefore changed hands, shifting from hunter's trophy to zoological relic. At the time, Tuck was working as Newfoundland's Wildlife Officer. He held this post from 1949 to 1976. A keen writer on the avifauna of Newfoundland, he published a bestseller, *The Murres*, about the habits and distribution of the seabirds, a kind of auk. Mention is made in *The Murres* of the disappearance of the flightless great auk, which 'became extinct within historical times'.[142] Great auks had inhabited Funk Island in Newfoundland and formed a food source for the Beothuk until the bird colony was wiped out by European settler colonists. Tuck attributes their disappearance to 'over-exploitation'.[143] He observes that the 'tragic story of the extinction of the great auk in Newfoundland, largely due to a commercial venture when its oil and feathers were utilized, is fairly well known'.[144] He therefore also wrote about a second animal genocide perpetrated in Newfoundland.[145]

As a naturalist, Tuck was particularly sensitive to species endangerment and loss and to the need for conservation. In his 1955 radio broadcast regarding the wolf, he laments their passing:

> I often regret the fact that there are no wolves at the present time on our Island. Such a predator, fulfilling its normal function would ensure sturdy and disease-free races of caribou and moose.[146]

Here, Tuck's regret is framed in terms of the wolf's benefit to the ecosystem. He is drawing attention to how their existence was closely interrelated with their primary prey. Through killing individual caribou and moose, wolves strengthened the populations of those animals. This is because most of the ungulate prey that wolves consume are 'young, weakened, debilitated, or older animals'.[147] As carnivorous large mammals, wolves form what Otto Frankel and Michael Soulé call an umbrella, an element in an ecosystem that provides security for 'the more dense species'.[148] Animals such as the Newfoundland wolf potentially have what Soulé referred to elsewhere as long coat-tails: many other species are dependent on them.[149]

In another talk, this time to the St. John's Rotary Club, Tuck writes of ecology, of the need to recognise the 'balance of nature' and 'the relation of each organism to its environment'.[150] This position is echoed in a paper about conservation entitled simply 'Proposed Talk', in which Tuck observes: 'the welfare of everything is somehow bound up with everything else.'[151] During the Rotary Club talk, he also mentioned the wolf, 'the only land mammal [in Newfoundland] which has become extinct in historical times'.[152] He speculates on how they came to be native to the isle, rejecting the idea that they crossed the Strait of Belle Isle when it was frozen and arguing that the 'peculiarities of the wolf' must have 'originated in stock isolated on this island for a very long time, perhaps from the original stock remaining when Newfoundland became separated from the Mainland'.[153] Here he states a case for the singularity of the wolf, proud of their uniqueness, their distinct identity. They form a part of Newfoundland's natural heritage. This gives a sense of how Tuck perceived the skin of the wolf killed by Whiteway, as a vital part of the region's environmental history.

The talk to the Rotary Club also refers to other extinct animals associated with the island: the great auk, the Eskimo curlew and the Labrador duck.[154] The address is, in significant part, a reflection on species loss and on the broader ecological impact that accompanies it. It closes with a clarion call for greater environmental responsibility:

> We have reached a point in colonisation in Newfoundland in which we have to be visionary. No longer can we use up our wildlife, or carelessly destroy our forests with fire, with the delusion that these resources are inexhaustible. Furthermore, we need to preserve for our children a part of these resources which is their birthright. We have also reached a point in culture, in which we are finding aesthetic enjoyment from many things in nature. Therefore, we should make a test of the highest culture, the capability to enjoy the earth and its creatures, without the desire to wantonly destroy them.
>
> In conclusion, I would like to say that the Newfoundland Natural History Society, recently formed, is building its policy along those lines, and I am sure you join me in wishing it every success.

The talk was given in December 1950 and the Newfoundland Natural History Society (now Nature Newfoundland and Labrador) had been founded earlier that year.[155] Tuck identifies a key aim of the society as environmental preservation. In March 1951 in a talk to the McDonald Fellowship Club, he foregrounds the importance of conservation, stating that the natural resources of Turtle Island are not inexhaustible and 'we have seen too many forms of wildlife become extinct or reach a dangerously low population'.[156] This section immediately follows a discussion of the Newfoundland wolf. The wolf's extinction and the urgent need for conservation were therefore clearly linked in Tuck's mind. In this context, the purchase of the wolf skin in 1952 – early in the life of the society – must relate to that aim. The animal's physical remains served as a tangible reminder of the passing of a once key element of Newfoundland's ecology. In Tuck's keeping, the wolf skin therefore shifted from being a trophy to a symbol of tragedy. The remains represented a potential rallying point for greater ecological responsibility and a statement of intent by the Society of their seriousness regarding Newfoundland's natural heritage.[157] The wolf shot by Whiteway changed from memorialising a politics of anti-lupine hatred to becoming part of a process of soliciting a politics of care.

Tuck was attuned to how perceptions influence human sympathy or animosity towards other animals. In an undated paper on conservation, he noted: 'It is not unnatural for us to prefer small sin[g]ing birds to the cruel birds of prey, but if we destroy the latter indiscriminately we may be confounded by grave imbalances in the natural community which react most unfavorably on the very creatures we wish to protect.'[158] The Newfoundland wolf, Tuck no doubt recognised, had been perceived in similarly unsympathetic terms to the 'cruel' raptors. The skin of the wolf shot by Whiteway was tangible proof of such negative attitudes. Now, in the aftermath of the wolf's extinction on the island, the skin also had considerable scientific value as the physical remains of a vanished subspecies. For this reason, in 1958 the wolf skin was loaned for safekeeping to the National Museum of Natural Sciences (now the Canadian Museum of Nature) in Ottawa, Ontario. For a period, all known physical remains of the wolf were therefore housed outside Newfoundland. The Society, however, had loaned the skin to Ottawa on the understanding that if Newfoundland ever had a suitable museum the wolf would be returned for display there.

Shortly after loaning the remains, Tuck received a letter from Alexander Banfield, then chief zoologist at the National Museum, which shared information on the wolf gleaned from close study. Like Allen and Barbour, Banfield identified morphological differences between the wolves of Newfoundland and those of other regions. He noted that the 'canines, premolars and molars seem to be rather light in construction compared to wolves from Quebec and Ontario'.[159] The wolf 'was smaller than average size for Canadian wolves'.[160] Banfield also went to some length to describe the colour of the pelt, remarking that although it was generally grey, 'on each side is an unusual narrow band of reddish-brown tipped hairs'.[161] He speculated that based on the thick pelage, the wolf was killed in winter. The lack of wear manifested by the teeth suggested that it was a young animal. He indicates that the 'specimen may have been a female'.[162] The killing of female wolves of breeding age is more harmful to wolf populations than that of males. Given that Herbert Whiteway refers to the wolf forming a member of a pack of four of five, she may have been the breeding female, with the other wolves representing her mate and their offspring of less than three years old.[163] Through close study, Banfield therefore revived part of the life story of the wolf.

Under the Rug

In 1977, the Newfoundland Museum began to think seriously about acquiring the skin of the wolf shot by Whiteway for their collection. In January of that year, the cataloguer for archaeology and ethnology at the museum, M. J. Thistle, viewed the skin in Ottawa and discussed the feasibility of mounting it. Three months later, Thistle wrote to the museum's director, Martin Bowe, suggesting that the skin be obtained as an exhibit. On 27 October 1977 at a meeting of the Natural History Society, a motion was passed: 'That we request the National Museum returns the Newfoundland wolfskin, and that it be put on permanent loan to the Provincial Museum, with appropriate arrangements made.'[164] John Maunder, the Curator of Natural History at the Newfoundland Museum, contacted the Society (of which he himself was a member) shortly afterwards formally requesting the skin for display at the newly renovated Natural History section of the museum.[165] The President of the Society, Charlie Horwood, wrote to the National Museum of Natural Science asking that custody now be transferred to the museum in Newfoundland.[166] Prior to its transfer, however, the Newfoundland Museum asked for the National Museum's taxidermist, Terry Morgan, to 'refurbish the skin and mount it as a taxidermied specimen'.[167] This decision would radically alter the appearance of the wolf.

In a letter sent by Louis Lemieux, the Director of the National Museum, to Martin Bowe, Lemieux encloses '"before" pictures' of the wolf and opines: 'Let's hope that Terry's talents can do something for the poor thing.'[168] The expression of sympathy for the animal is unsurprising given the wolf's appearance in the photographs.[169] The skin, spread-eagled, is attached to drab olive felt with scalloped edges. Patches of felt are visible through the hide, between the neck and back. The wolf's remains are photographed against a blue, seemingly linoleum floor.[170] Although referred to by Maunder as a wall rug, the artefact has the appearance of something designed to be viewed from above.[171] The wolf's head was posed with jaws agape, a relatively complicated pose that would have little visual effect if the skin was attached to a wall.[172] The lower incisors are all missing and the fake tongue is broken, exposing a tan-coloured material (perhaps papier mâché coated over clay), damage that, combined, indicates the skin may have been kept at ground level, trampled underfoot.[173] The dark amber glass eyes with their heavily dilated pupils bulge unnaturally from the wolf's head; too big for the eyelids to contain, they lend the face an ugly weirdness. The upper incisors are caked in dark matter, either dried tissue or, the skin perhaps having receded, glue or paint. The taxidermist Albert Farnham wrote of mounted rug heads that they are 'usually intended to register rage or anger' and that appears to be the case here.[174] The wolf Whiteway killed was given a fixed growl, made to symbolise aggression.

By 1980, however, far from inciting fear, awe or triumphalism, the wolf skin, worn, dirty, distressed, prompted only pity in Lemieux. To use the remains as an exhibit in this state, retaining the form of a hunting trophy, would nonetheless have provided historical insight into local attitudes towards the animal in the early twentieth century. The rug embodied the genocidal gaze, offering a vision of the wolf as vicious brute. The Newfoundland Museum though had other ideas. Maunder sent a letter to Morgan with very precise instructions for how to mount the wolf:

> The specimen should be posed in a normal German Shepherd Dog-like alert position: perhaps with one front paw raised and the head turned about 45 degrees to the side. The specimen will be viewed from one side as far as I know. If one side is better than the other, please regard this as the display side and turn the head accordingly. It might be more interesting if the front paw was 3 or 4 inches higher than the hind two; as if the fore-quarters were supported by a rock to give the animal a better view. The tail might be best curled over the back. The impression I would like is that of a healthy alert animal as opposed to something just standing there looking bored.[175]

Maunder's primary concern here is to ensure the wolf's vitality. Although the skin derived from an animal that was shot dead, the wolf must look as alive as possible for future gallery-goers. Some idea of movement, of vigour, must be injected into the mount.

The curator's call for the wolf to be posed as a dog is counterintuitive. Dogs are descended from wolves but a German shepherd is a domesticated animal. There is nonetheless a tendency to link the two canines because of similarities in physical appearance. Jane Packard, for example, suggests that 'newborn wolf pups look much like German shepherd pups'.[176] Experiments into behaviour using rope tasks conducted with German shepherd dogs and wolves by Nancy Hiestand nevertheless found marked behavioural differences between the two. She refers repeatedly to 'species differences' revealed by the research in the related areas of cognition and comportment.[177] Maunder likely uses the analogy in the belief that Morgan will be more familiar with dogs, but privileging similarities risks sidelining the differences that grant singularity. In the taxidermy mount, fidelity to the actual physical comportment of wolves in the wild, to the specificities of how they show alertness, is not a primary concern. A general sense of liveliness is more important than realism.

Given the condition of the rug when he received it, Morgan's achievements are remarkable. Martin Bowe would write to Lemieux expressing the Newfoundland Museum's satisfaction with the mount, describing the taxidermist as 'a master craftsman'.[178] He stated that the 'result of Terry's labors is quite amazing, and beyond our expectations'.[179] Morgan, however, did not follow all of Maunder's suggestions. This may have been because of practical limitations caused by the condition of the skin, the technical difficulties of some of the requests, the taxidermist's own aesthetic sensibilities or a combination of these. The wolf now appears with their mouth closed, their head raised slightly and turned to the right. Morgan has designed the mount for the right flank of the animal to be seen and they are currently displayed in such a way as to encourage that. Viewed from the left side, gaps are visible where there was insufficient skin remaining to cover the forelegs (although the plaster bones are carefully coloured to blend with the pelt, minimising how obvious this is). The taxidermist ignored the wish to have the front paws raised. The vitality Maunder demanded is conjured through the turn of the head (as if something has just caught the wolf's attention) and the positioning of the hind legs, the right slightly in advance of the left as if the wolf has paused momentarily. The use of varnish is also careful, a dab onto the lacrimal caruncle of each eye to give a glisten, a hint of life-affirming moisture; the nostrils also have dabs in the interior, the suggestion of mucosity, of oozing vitality. The eyes are now smaller and recessed, more 'natural-looking'.

On the left side of the muzzle, stitching has been left visible – a white suture peeking out from beneath the fur – as if Morgan wanted to display some sign of his craft, of his artistry. The thread suggests the labour of the taxidermist lying behind this practised vision of a creature in life. Morgan has, nonetheless, gone to considerable lengths to hide any readily visible traces of the animal's former fate. Damage to the wolf's skin caused by the shot to the head was seemingly already not highly visible when it was a rug. Farnham remarked that '[b]ullet holes of large caliber destroy considerable skin and in order to close them it is best to cut them to a triangular shape and draw together by sewing up from the corners of the triangle'.[180] The entry wound was likely sewn up like this when Whiteway had the rug made. Morgan also worked carefully to conceal any visible sign of the wolf's violent death. Evidence of the hole would index hate and its genocidal effects. Additionally, the gap between the fur of the neck and the back was repaired. Maunder realised that 'some inserts [would] have to be added here and there to patch holes and things'.[181] This seems to have been the case. Morgan has restored the wolf rather than simply conserved their remains, an important distinction.

Restoration means returning something to a preceding state rather than simply preserving it, keeping it as it is.[182] Clearly Morgan could not bring the wolf back to life, but Maunder asked him to make them as lively as possible, to replace the flat, inert dead wolf of the rug with a mounted wolf that still appeared to be among the living. To achieve this revivification, the taxidermist radically altered the wolf, the object that was the rug. The mount gives a much clearer visual impression of the wolf as they appeared in life, what might be seen as a 'truer' likeness, although, of necessity, it remains a stilled life. The wolf is shown in the act of attending to something but they can never spring to action. They have no continuity of behaviour to call their own. Transforming the physical appearance of the remains – giving the wolf a body again, putting them back on their feet, reanimating them – also changed their symbolism. The mount as they are displayed in the museum is not intended as a hunting trophy but as a natural historical document. It records aspects of what one young Newfoundland wolf in winter looked like. Its documentary value as evidence of the attitude of a particular man towards the wolves of Newfoundland in the early twentieth century, its embodiment of environmental history, has dissipated, if not disappeared completely. The human responsibility for the death of this wolf, one of the last of its kind, and the celebration of that death has been obscured.

The wolf is currently displayed in a nook, up some stairs at the rear of the main exhibition space. The nearest section of the main exhibition is titled 'Boreal Basics' and is a section about the resilience of the island's boreal forests. In the nook, the other artefacts on display seem unrelated to each other, important items from the museum's collection that don't quite fit elsewhere. There are black bear (*Ursus americanus hamiltoni*) skulls used to illustrate the role of 'voucher specimens' in natural history. There are bald eagles (*Haliaeetus leucocephalus*) employed to show the morphological variations that can exist within a given species. There is a double-crested cormorant (*Phalacrocorax auratus*) nest accompanied by the information that the birds have been known to nest near both saltwater and freshwater.

The label for the nest advises that: 'Nothing is ever as simple as it seems.' Then there is the Newfoundland wolf.

The wolf is marginalised, reduced to an oddment.[183] They are seen to possesses value but not as crucial to the narrative of Newfoundland's history. A great auk skeleton, by contrast, is displayed in the main exhibition space, integrated into an exploration of the island's environmental history that includes sections on the Beothuk and other peoples. On a positive note, isolating the wolf from the Beothuk in spite of the Latin trinomial that binds the First Nations people and the subspecies avoids buttressing the narrative of their shared extinction. Extinction narratives relating to the Beothuk are, as Fiona Polack and others have explored, highly problematic and open to contestation.[184] Yet sidelining the wolf works to conceal the animal genocide perpetrated against the subspecies.

The mount and skull in The Rooms belong to a direct victim of a genocidal campaign yet a museum visitor will usually have no conception of this. The label accompanying the exhibit, its wording printed against a forest rendered in ghostly silhouette, makes no mention of the bounty scheme. Hunting is given as a cause of extinction alongside 'prey availability' and 'natural population fluctuations'. Given that the wolf was clearly suffering 'unnatural' population decline in the nineteenth century thanks to the efforts of bounty hunters and as collateral damage in poisoning campaigns, mentioning natural fluctuations seems ill-judged. In the aftermath of animal genocide against the Newfoundland wolf, there is a studied reluctance to confront the reality of the active role of settler colonists in the extermination of the subspecies. This part of the biography of the wolf slaughtered by Whiteway goes unmentioned, but the bullet-shattered skull speaks for itself.

Notes

1. The issues of which groups of wolves can justifiably be referred to as distinct subspecies is a fraught one. See Paquet and Carbyn, 'Gray Wolf', 482.
2. Allen and Barbour, 'The Newfoundland Wolf', 230.
3. The colour of the wolves in Newfoundland was often remarked upon but, as Paquet and Carbyn note, the colouration of wolves is 'highly variable and usually of little use in ascertaining the geographic origin of specimens'. Paquet and Carbyn, 'Gray Wolf', 484. In his 1829 book *Fauna Boreali-Americana*, John Richardson writes of the 'races of wolves' of Turtle Island that he eschewed naming any of them after their colouring as colouring only denotes 'casual varieties of a species' (62).
4. Based on Worcester's recollection, Leslie Tuck suggests 1896 as the year the wolf was shot. See Tuck, 'The Newfoundland Wolf' [1947], 6.

5. For a useful summary of debates regarding the perceived extinction of the Beothuk, see Polack, 'Introduction'.
6. Linnaeus's 1758 edition of *Systema Naturae* is, as Jack Burke explains in 'A Brief History of the Taxonomy of Mammals', 'taken as the zero point for zoological nomenclature' (78).
7. Cited in Howley, *The Beothucks or Red Indians*, 100.
8. In the fifteenth century, Edward of Norwich wrote that 'the vixen of the fox bears as long as the bitch of the wolf bears her whelps'. Norwich, *The Master of Game*, 64. In the eighteenth century, Samuel Johnson defined 'bitch' as a 'Female of the canine kind (wolf, fox, &c.)' in his dictionary. Johnson, *A Dictionary of the English Language*, 143.
9. Tuck, 'The Newfoundland Wolf' [1979], 23.
10. Tuck died before he had finished writing the article and may therefore have envisaged reworking this section.
11. Cited in Howley, *The Beothucks or Red Indians*, 16. Punctuation as in the original.
12. Ibid., 17.
13. Allen and Barbour, 'The Newfoundland Wolf', 230.
14. See Ritvo, 'Zoological Nomenclature and the Empire of Victorian Science', 342.
15. See Gatschet, 'The Beothuk Indians', 413.
16. There is a long-standing Mi'kmaq community in Newfoundland. The Mi'kmaq word for wolf is *paqtesm* or *paqt'sm*. The Innu word for wolf is *maikan*. The Innu were regular seasonal visitors to Newfoundland for hundreds of years. They may also have overwintered on the island. See Aylward, 'The Beothuk Story', 202.
17. I include the word 'loup' as French colonists also settled in Newfoundland and after the Seven Year War (1754–63) they were granted the right by the British to catch and process fish on part of the shoreline of Newfoundland. See Aldrich, *Greater France*, 12–13. The 'barrens' is a term used in Newfoundland to refer to rocky areas with low vegetation. As a name that conjures images of emptiness and desolation, 'barrens' can be seen as an example of terminology that breeds what Halsey calls 'contempt for the earth'. See Halsey, *Deleuze and Environmental Damage*, 231. As Burzynski et al. note in *Exploring the Limestone Barrens of Newfoundland and Labrador*, the barrens are far from wastelands and comprise an important ecosystem that 'supports a very high percentage of the Island's rare plant species' (1). As well as locomotion described as trotting and walking, wolves are also commonly ascribed a capacity for loping and galloping.
18. See Smalley, *Wild by Nature*, 35.
19. Ibid., 103.
20. For a recent discussion of the limits of a Western ontological conception of the wolf, see Mcphie, 'Nazi Werewolves from Outer Space'.
21. Cambefort, 'Zoological Nomenclature and Speech Act Theory', 163.
22. Ibid., 165.
23. The ink used to write *occidentalis* is darker and the handwriting appears different (notably the 'a' and 's'). This subspecies was first identified in 1829 by Richardson, who published a description of their characteristics in *Fauna Boreali-Americana*, 60–65.
24. Cambefort, 'Zoological Nomenclature and Speech Act Theory', 165. For a recent discussion of the taxonomic complexities relating to wolves (which includes a reference to the Newfoundland wolf), see Davis, 'The Wolf Pack'.
25. Or, in the case of the identification number, literally inscribed.
26. These technologies frequently focus on form or sequencing as the 'truth' of (sub)species identity rather than on behaviour and are therefore limited.
27. Anderson, 'Summary of the Large Wolves of Canada,' 391.

28. Morphology is usually privileged over behaviour when identifying subspecies, but behaviour offers another way of grouping together animals as similar or different. Behaviour can be linked to the idea of animal cultures and to 'animal ethnicity' while morphology is more 'race'-oriented. Karen Wonders uses the term 'animal ethnicity' in the context of dioramas to describe how nationalist ideas can be mapped onto the natural world through museum exhibition practices (Wonders, 'Habit Dioramas'). That is not how I understand the term here. For me, 'animal ethnicity' refers to a group of animals that disclose a common culture manifested through shared behaviours that distinguish them from other groups of animals of the same species. Cultural differences and cultural transmission have been noted, for example, in several species of whale. See, for instance, Whitehead, *Sperm Whales*.
29. In a human context, the reduction to types also forms part of racial abstraction. See Savoi and Chare, 'Beyond First Impressions'.
30. Throughout this chapter I expressly refrain from referring to the wolf as Whiteway's, although they are commonly referred to with his surname used as a possessive. My aim here is to free the wolf from Whiteway to some degree. The wolf is probably female but as this has not been established with certainty, I use the gender-neutral third-person pronoun 'they' rather than 'she' to refer to them.
31. In his 1982 article 'The Newfoundland Wolf', Maunder suggested that the wolf was shot in 1911 but has since offered 1894 as the year. The year of 1894 is provided in Maunder's revised version of the 'The Newfoundland Wolf', The Rooms website, https://web.archive.org/web/20240421051959/https://www.therooms.ca/the-newfoundland-wolf-0 (retrieved 27 March 2025). For a discussion of the geology of the Topsails see Jennifer Organ's 'Quaternary Investigations of the Topsails'.
32. Recent analysis of the skin found that it carries traces of arsenic and mercury, which may derive from Ewing's tanning process. I am grateful to Nathalie Djan-Chékar for this information.
33. Maunder, 'The Newfoundland Wolf', 41.
34. The letter is dated 14 November 1923. Rabbits is therefore talking about a sighting some time around the turn of the century. Seton, *Lives of Game Animals: Volume 1*, 263. In a 1951 letter to Leslie Tuck, Whiteway's brother Herbert suggested that the wolf was killed 'over 40 years ago'. Handwritten letter from H. M. K. Whiteway to Leslie Tuck dated 22 December 1951, Newfoundland Wolf Archive, The Rooms.
35. See Tuck, 'The Newfoundland Wolf' [1979], 22. A male wolf from Labrador that must have crossed to Newfoundland on sea ice was shot by Joe Fleming in 2012. Although initially identified as a coyote, genetic testing confirmed the canine was a wolf.
36. The first reference to the wolf being extinct occurred in the 17 January 1912 issue of the *Evening Telegram* (now *The Telegram*). There are, however, recorded sightings up to the winter of 1922–23.
37. Due to an exchange of materials with the Museum of Comparative Zoology, one of these skulls is now held by the Smithsonian National Museum of Natural History (USNM 264482). The skull in the Smithsonian collection was examined by Edward A. Goldman when he prepared his entry on the Newfoundland wolf for his section on the classification of wolves in Young and Goldman, *The Wolves of North America: Part II*, 435–36. There are also ancient remains including a maxilla discovered at a Maritime Archaic burial site. See Tuck, 'The Newfoundland Wolf' [1979], 37.
38. Joy, 'Reinvigorating Object Biography', 540.
39. Carey, *Autobiographies of a Lump of Coal &c*, 6. Italics are in the original.
40. Ibid., 10.

41. Helpful conceptualisations of the agency of things are provided by Alfred Gell in *Art and Agency* and Karen Barad in 'Posthumanist Performativity'. Barad's conception of agency differs markedly from Gell's in that for her, agency is not something possessed by pre-given entities but rather a quality that emerges from matter's intra-acting. Gell thinks of agency as a property of things: 'Social agency can be exercised relative to "things' and social agency can be exercised by "things"' (*Art and Agency*, 17–18).
42. Kopytoff, 'The Cultural Biography of Things', 66.
43. Ibid., 67.
44. Crossland, 'Of Clues and Signs', 75.
45. Ibid. As part of a discussion of the reconstruction of the history of a trophy skull (a skull sourced as a keepsake during military combat), Willey and Leach note that we rarely 'know for sure that the events reconstructed from the bones' are 'what actually happened'. Willey and Leach, 'The Skull on the Lawn', 187.
46. Wajcman, *Les Experts*, 43.
47. Ibid., 45.
48. In 'Forensic Anthropology in the 1990s', Reichs (149), for instance, discusses how in forensic analyses of bones skeletal lesions can indicate ante-mortem health problems and stress and strain indicators in bones can potentially provide insight into the leisure and professional activities of a given individual.
49. Crossland and Joyce, 'Anthropological Perspectives on Disturbing Bodies', 26.
50. Ibid.
51. Handwritten letter from H. M. K. Whiteway to Leslie Tuck dated 22 December 1951, Newfoundland Wolf Archive, The Rooms.
52. I had already deduced that the gunshot likely came from behind when examining the skull.
53. For a discussion of gunshot trauma in relation to bone, see Smith et al., 'Look Until You See', 139–44.
54. The wound caused by this kind of gunshot trauma is often referred to as a 'keyhole defect' because of its shape. See Berryman and Symes, 'Recognizing Gunshot and Blunt Cranial Trauma', 348–49. See also Dixon, 'Keyhole Lesions in Gunshot Wounds'.
55. Writing in 1842, Bonnycastle suggests that hares and wolves were the only mammals that were hunted in winter. Bonnycastle, *Newfoundland in 1842: Volume 2*, 148.
56. Howley, 'Reminiscences of Forty-Two Years of Exploration and Survey in and about Newfoundland, 1869–1916; 1899', 7. Howley Family Papers COLL-262, Series 1 Personal Papers 1769–1958, 3.0 Correspondence and Writings, 1854–1917, 3.03.033.
57. Howley's willingness to gun down almost any (nonhuman) mammal that moved is evinced by his shooting of a porpoise on 17 September 1902 with an explosive bullet that killed 'the brute instantly'. Howley, 'Reminiscences of Forty-Two Years of Exploration and Survey in and about Newfoundland, 1869–1916; 1902', 44. Referring to a cetacean as a 'brute' seems unusual now but porpoises were described as 'murderous' in a near-contemporary essay by the naturalist Nicolas Pike. See Pike, 'The Struggle for Life in the Deep', 633.
58. Letter to Thomas Barbour dated 22 October 1932. Cited by Tuck in 'The Newfoundland Wolf' [1947], 5.
59. In 'Wolf Hunters', the contemporary Alaskan hunter Christopher Batin observes that '[m]ost wolves are taken incidentally to hunting other species' (146).
60. In the novel *Trap-Lines North*, wolf bounties are a welcome supplement to income. Meader, *Trap-Lines North*, 282–84. As the actions of Whiteway and Worcester demonstrate, the novelty of possessing a trophy sometimes outweighed financial incentives.

61. Howley, 'Reminiscences of Forty-Two Years of Exploration and Survey in and about Newfoundland, 1869–1916; 1903', 4.
62. Howley, 'Reminiscences of Forty-Two Years of Exploration and Survey in and about Newfoundland, 1869–1916; 1899', 5.
63. Hatton and Harvey, *Newfoundland*, 249.
64. The skin is actually purchased by the Royal Canadian Mounted Police to be stuffed and mounted. Meader, *Trap-Lines North*, 269, 283–84.
65. Knobel, 'The Conservation Status of the Wedge-Tailed Eagle', 295.
66. Tocque, *Wandering Thoughts*, 112. The wolf as a potential child killer is also a theme in *Trap-Lines North*, where a wolf's tracks are described as following a little boy's (248). Lindsay Vanderbeck refers to tales of wolves stealing children from cabins in Québec (248).
67. In an appendix to his 1979 article 'The Newfoundland Wolf', Tuck carried out pioneering work regarding the extent of the bounty payments. Researching specifics is sometimes difficult as while some individual payments (name, date, amount) were recorded in the 1840s, others seem to have been logged under sundries. When the name, year and amount of the payment is known, other information (such as the location of the claim) often cannot be ascertained. Tuck was working in a time before records were digitised, which makes his achievements all the more remarkable.
68. *Journal of the House of Assemble of Newfoundland* (St. John's: F. W. Bowden, 1879), 171.
69. Robert Barlow McCrea mentions the possibility that wolf skins might be imported and submitted for bounty. See McCrea, *Lost Amid the Fogs*, 281.
70. Some date the start of the bounty to 1839 because the Act was passed then and payments began in 1839. The Queen, however, only approved the Act on 5 March 1840. The 1843 journal containing the entry for the first reading of the Wolf Killing Bill, a bill pursuant to the renewal of the Act, confirms this, describing the Act as having been passed in 'the third year of the Reign of her Present Majesty' (i.e. Queen Victoria, who ascended to the throne in 1837).
71. *Journal and Proceedings of the House of Assembly of Newfoundland* (St. John's: n.p., 1839), 129.
72. Wolf livestock depredation is, in reality, often overestimated. See, for example, the discussion of Minnesota in Bath, 'Identification and Documentation of Public Attitudes', 2-18. In their review of depredation over several biogeographical areas, 'A Review of Wolf Depredation on Livestock and Implications for the Yellowstone Area', John Mack, Wayne Brewster and Steven Fritts found that predation losses are greater for sheep than cattle. In the case of cattle, depredation was highly variable and on average affected only a small number of available livestock. The review concluded that in 'areas where wolves and livestock live together, wolf depredation occurs at relatively low levels but individual producers may experience significant losses' (5-40).
73. *Journal and Proceedings of the House of Assembly of Newfoundland* (St. John's: E. D. Shea, 1839), 205.
74. Tocque, *Wandering Thoughts*, 112.
75. Ibid.
76. Ibid.
77. Bonnycastle, *Newfoundland in 1842: Volume 1*, 224–25.
78. Tocque, *Wandering Thoughts*, 111–12.
79. Hayward, 'Report of James S. Hayward', A-631.
80. Lamb mortality can be as high as 25 per cent in some circumstances. See Scales et al., 'Lamb Mortality'.

81. Col Bailey, *Lure of the Thylacine*, 135. The two payments from 1839 are not noted by Tuck in his list of bounty payments in 'The Newfoundland Wolf' [1979]. There are discrepancies between the spellings of the names as they are listed on p. A-26 and p. A-35 of the appendix of the *Journal and Proceedings of the House of Assembly of Newfoundland* (St. John's: E. D. Shea, 1840). All references to pages in the appendices to the journals of the House of Assembly are prefixed with an A.
82. *Journal of the House of Assembly of Newfoundland* (St. John's: E. D. Shea, 1841), A-89.
83. Tuck does not name these bounty recipients and lists seven not eight claims for this period.
84. *Journal of the House of Assembly of Newfoundland* (St. John's: E. D. Shea, 1841), A-9; A-11; A-14.
85. Ibid., A-10.
86. *Journal of the House of Assembly of Newfoundland* (St. John's: E. D. Shea, 1840), 140.
87. Ibid. Burke's surname is erroneously spelt Burk in the records.
88. *Journal of the House of Assembly of Newfoundland* (St. John's: A. Shea, 1845), 89.
89. Two years of expenditure are detailed in the 1843 *Journal of the House of Assembly of Newfoundland: First Session* (although catalogued with this title at Memorial University, the journal is actually titled *Journal of the General Assembly of Newfoundland*) and, except for isolated instances, it is not possible to ascertain when over the two-year period the bounties were claimed.
90. *Journal of the House of Assembly of Newfoundland* (St. John's: W. R. Shea, 1843), A-501.
91. Michael Carroll is listed as Michael Connell in Tuck's list of bounties because that is the name used in the treasury accounts for 22 June. Carroll is the name given in the Return of the number of certificates presented for bounty. I have opted to follow the latter. Tuck only identifies five bounty payments in this period and offers names for three (Connell, Walsh, Dunn).
92. Tocque, *Wandering Thoughts*, 112.
93. Ibid., 114.
94. Ibid.
95. Maunder, 'The Newfoundland Wolf', 40.
96. Bonnycastle, *Newfoundland in 1842: Volume 2*, 259.
97. Ibid.
98. John E. Maunder, 'The Newfoundland Museum: Origin and Development', The Rooms website, https://web.archive.org/web/20240414234014/https://www.therooms.ca/the-newfoundland-museum-origin-and-development (retrieved 27 March 2025).
99. William Eppes Cormack gifted a Newfoundland wolf skin to John MacGregor. It measured six feet from 'the snout to the point of the tail'. Macgregor, *British America*, 95.
100. *Journal of the House of Assembly of Newfoundland* (St. John's: A. Shea, 1844), A-232.
101. *Journal of the House of Assembly of Newfoundland* (St. John's: E. D. Shea, 1848), A-482.
102. Tuck, 'The Newfoundland Wolf' [1979], 25.
103. Ibid. See also the *Journal of the House of Assembly of Newfoundland* (St. John's: A. Shea, 1848), A-105.
104. *Journal of the House of Assembly of Newfoundland* (St. John's: E. D. Shea, 1850), A-26. Tuck lists Lewis's payment as £155 as per the accounts. Tuck, 'The Newfoundland Wolf' [1979], 25. This would, however, amount to a claim for thirty-one wolves. The total payment of bounties for the year is given as £25, which means £155 is a typographic error and Lewis's payment should have been entered as £5 (one wolf).

105. *Journal of the House of Assembly of Newfoundland* (St. John's: E. D. Shea, 1858), 54. After being considered by committee on 7 April, the bill was passed without amendment at its third reading on 8 April 1858.
106. Ibid., 107.
107. The location of each claim was no longer detailed, so understanding the geographical spread of the killings is not possible.
108. *Journal of the House of Assembly of Newfoundland: First Session* (St. John's: n.p., 1860), A-63.
109. *Journal of the House of Assembly of Newfoundland: Second Session* (St. John's: E. D. Shea, 1860), A-54.
110. Noel Paul appears to have claimed for four (one; three) wolves in total in this period. See *Journal of the House of Assembly of Newfoundland: Third Session* (St John's: James Seaton, 1862), A-68.
111. Noel Paul is a missionary name and not Paul's birthname. See Bryan Marsh, 'The Search for Old Tom Joe', Anglo Newfoundland Development Company website, 3 April 2020, https://anglonewfoundlanddevelopmentcompany.ca/2020/04/03/the-search-for-old-tom-joe/ (retrieved 20 July 2023).
112. *Journal of the House of Assembly of Newfoundland: Third Session* (St John's: J. Seaton, 1863), A-80.
113. *Journal of the House of Assembly of Newfoundland: Third Session* (St John's: J. Seaton, 1863), 79. The lawmaker with 'stentorian lungs' (281) described by McCrea in *Lost Amid the Fogs* as seeking to filibuster the legislation to suspend the bounty may well be Whiteway.
114. *Journal of the House of Assembly of Newfoundland: Third Session* (St John's: J. Seaton, 1863), 80.
115. *Journal of the House of Assembly of Newfoundland: Fourth Session* (St John's: J. Seaton, 1864), A-79.
116. *Statutes of Newfoundland, 1873*, Capitulus XV., 2. This payment is not noted by Tuck.
117. *Journal of the House of Assembly of Newfoundland: Fourth Session* (St. John's: Evening Telegram Office, 1896), A34–35.
118. *Journal of the House of Assembly of Newfoundland* (St. John's: Daily Star Office, 1921), A-14.
119. Tuck, 'Early Natural History of Newfoundland,' *Evening Telegram*, 17 February 1954, 2, 14.
120. Howley, *Reminiscences of James P. Howley*, 272.
121. Ibid., 271.
122. The 1 June 1885 issue of the *Patriot and Terra-Nova Herald* includes an article in which the use of strychnine to kill foxes is condemned.
123. Young and Goldman, *The Wolves of North America: Part 1*, 324.
124. Newfoundland became a British Dominion in 1907 and confederated with Canada in 1949.
125. On Tuesday 23 March 1886, the House received petitions presented by a Mr Bond on behalf of the inhabitants of Harbor Briton, St. Jacques and Gaultois 'praying for the restraint of the use of Poison'. *Journal of the House of Assembly* (St. John's: Dowden & Sons, 1886), 71–72.
126. Anonymous, 'House of Assembly: Synopsis of Proceedings', *Harbor Grace Standard*, 28 March 1885, 4.
127. In *Lives of Game Animals: Volume 1*, Seton notes that when poison is used 'the victims [are] seldom found' (264). He suggests that in places where bounties are offered,

the number of animals for which claims are made should be doubled to account for deaths by poisoning.

128. Cited in the 28 March 1885 *Harbour Grace Standard.* The doctor Parsons is referring to is presumably the pharmacist John Joseph Dearin who had spoken just prior to him.
129. Mech, *The Wolf*, 333.
130. Poliquin, *The Breathless Zoo*, 150.
131. Ibid.
132. Ibid., 151.
133. Ibid., 152.
134. Whiteway's act might best be described as an assassination give the then rarity and thus importance of the wolf.
135. A 'drogue' (the name sometimes used for clumps of trees in the barrens of Newfoundland) located near the Terra Nova River and used as a geographical feature by hunters was named after William Vallance Whiteway. Hesketh Hesketh-Prichard reported seeing a bear close to the drogue on a hunting expedition in 1904. Hesketh-Prichard, *Hunting Camps in Wood and Wilderness*, 148. Hesketh-Prichard writes that in the interior of Newfoundland can be found, 'few and far between, the dying remnant of the wolf-packs, for whose slaughter a heavy Government bounty is still . . . offered, though very rarely claimed' (85).
136. The taxidermy 'specimen' acts as 'a site of contested narratives', with these contestations often concealed. See McHugh, 'Taxidermy's Literary Biographies', 143.
137. Tuck, 'The Newfoundland Wolf' [1947].
138. Tuck, 'Early Natural History of Newfoundland', 2. Leslie Tuck Papers COLL-152.
139. This essay was prepared for publication using Tuck's drafts by William Montevecchi.
140. Tuck stated that he 'had some grand material on the Newfoundland Wolf and hope[d] to write a paper on it eventually'. Letter from Leslie Tuck to A. W. F. Banfield, 16 May 1958, Whiteway Wolf Archive, The Rooms.
141. Letter from Leslie Tuck to Léo Edward Francis English, 6 June 1955, Whiteway Wolf Archive, The Rooms.
142. Tuck, *The Murres*, 19.
143. Ibid., 24.
144. Ibid., 57.
145. For an assessment of when the mass killing of auks became genocidal, see Chare, 'The Beothuk, the Great Auk and the Newfoundland Wolf'.
146. A transcript of the radio broadcast is held within the Leslie Tuck Papers COLL-152.
147. Paquet and Carbyn, 'Gray Wolf', 490.
148. Frankel and Soulé, *Conservation and Evolution*, 121.
149. Soulé, 'What is Conservation Biology?', 729.
150. Tuck, 'Newfoundland Fauna', unpublished talk given to the St. John's Rotary Club on 14 December 1950, Leslie Tuck Papers COLL-152. 4.03.003, Archives and Special Collections, Queen Elizabeth II Library, Memorial University.
151. Tuck, 'Proposed Talk [Conservation]', 3, unpublished and undated draft, Leslie Tuck Papers COLL-152. 4.03.017.
152. Tuck, 'Newfoundland Fauna'.
153. Ibid.
154. The Eskimo curlew is officially critically endangered although it may be extinct. It had clearly not been seen in Newfoundland for many years when Tuck gave his talk and has not been reported there since. A sighting was reported in Nova Scotia, Canada by Randy Hoffman in 2006.

155. Montevecchi, 'In Memoriam'.
156. Tuck, 'Talk to the McDonald Fellowship Club, 8th March 1951', unpublished paper, Leslie Tuck Papers COLL-152. 4.03.005.
157. As part of the terms for loaning the skin to the National Museum of Natural Sciences, Tuck insisted that references to it in publications (scientific or otherwise) be made with the concurrence of the Society as a means 'to establish the identity of the Society'. Letter from Leslie Tuck to Austin Cameron dated 8 April 1952, Newfoundland Wolf Archives, The Rooms.
158. Tuck, 'Proposed Talk', 3.
159. Letter from Alexander Banfield to Leslie Tuck dated 20 January 1960, Newfoundland Wolf Archives, The Rooms.
160. Ibid.
161. Ibid.
162. Ibid.
163. Handwritten letter from H. M. K. Whiteway to Leslie Tuck dated 22 December 1951, Newfoundland Wolf Archive, The Rooms.
164. Minutes of the Meeting of the Natural History Society of Newfoundland on 27 October 1977, Newfoundland Wolf Archives, The Rooms.
165. Letter from John E. Maunder to Charles Horwood dated 31 October 1977, Newfoundland Wolf Archives, The Rooms.
166. Letter from Charles Horwood to Louis Lemieux dated 2 January 1978, Newfoundland Wolf Archives, The Rooms.
167. Letter from John E. Maunder to Terry Morgan dated 16 March 1978, Newfoundland Wolf Archives, The Rooms.
168. Letter from Louis Lemieux to Martin Bowe dated 10 January 1980, Newfoundland Wolf Archives, The Rooms.
169. I am grateful to John Maunder for sharing these photographs with me.
170. The colour contrast between the *bleu de France*-coloured linoleum flooring and the grey and tawny skin is jarring. Photographed against a different backdrop, the skin's colouring might appear less ugly.
171. Ernest Seton, however, states that although wolf fur is 'rich, full, and beautiful', it 'is not very durable as a rug'. In this context, given the relatively good condition of the pelt, Maunder may be correct. Seton, *Lives of Game Animals: Volume 1*, 254.
172. Wolf skins are, however, regularly hung on walls by hunters. Christopher Batin recounts having a 'wolf rug hanging in [his] library den'. Batin, 'Wolf Hunters', 148.
173. Nevertheless, in his *Home Taxidermy for Pleasure and Profit*, Albert Farnham includes a picture of a timber wolf rug in which the head is positioned perpendicular from the skin. If the head of the wolf shot by Whiteway was initially placed similarly, the skin could have functioned well as a wall hanging. Farnham, *Home Taxidermy for Pleasure and Profit*, 213.
174. Ibid., 116.
175. Letter from John E. Maunder to Terry Morgan dated 16 March 1978, Newfoundland Wolf Archives, The Rooms.
176. Packard, 'Wolf Behavior', 47.
177. Hiestand, 'A Comparison of Problem-Solving and Spatial Orientation'.
178. Letter from Martin Bowe to Louis Lemieux dated 5 January 1981, Newfoundland Wolf Archives, The Rooms.
179. Ibid.
180. Farnham, *Home Taxidermy for Pleasure and Profit*, 114.

181. Letter from John E. Maunder to Terry Morgan dated 16 March 1978, Newfoundland Wolf Archives, The Rooms.
182. Muñoz Viñas, *Contemporary Theory of Conservation*, 16–18.
183. This marginalisation was noted by the artist Marie-Soleil Provençal, who described the mount as exhibited 'in a back room, as if it is meant to be forgotten'. In 2022, in response to this sidelining, Provençal created a temporary mural, a transient memorial to the Newfoundland wolf, entitled 'Shadows of an Epoch: They Were Wolves'. Mandy Cook, '"Share and Learn": Street Wall Mural a Response to Climate Emergency, Colonization', *Gazette*, 15 September 2022, https://gazette.mun.ca/campus-and-community/share-and-learn/ (retrieved 29 February 2024). A photograph of this mural is used as the cover image for this book.
184. Harries suggests that the Beothuk are extinct as a culture, leaving open the possibility of genetic continuance. See Harries, 'A Clearing with a View to the Lake', 100.

Afterword

Afterimages of the Genocidal Gaze

Wilfred 'Wilf' Batty, a farmer and labourer, is often described as having killed the last 'wild' thylacine. Batty shot the male tiger on 13 May 1930 on his father's property in Mawbanna in north-west Lutruwita.[1] His initial intention was to catch the thylacine alive so he could sell him. When he realised the marsupial was too nimble, Batty put a bullet in him. Injured, the tiger tried to bite Batty, who kicked him twice in the jaw. The wounded animal would take approximately half an hour to die.[2] There are at least eight photographs of the dying or recently deceased thylacine.[3] These were taken no less than two and probably four different photographers. Four of the images show the dying or dead tiger on a woodchip pile. Four show him dead propped against some fencing. In two of the images featuring the propped tiger, Batty poses kneeling beside his 'kill', holding one of his farm dogs. He is carefully displaying his rifle. These two 'snuff' photographs taken by Patrick 'Pat' O'Halloran are trophy images, memorialising Batty's victory over a varmint. The farmer is smiling in both pictures. Marianne Cooney describes him as 'overjoyed'.[4] The adjective well summarises his toothy grin. Batty would have known the thylacine was incredibly rare yet he shows no remorse for his actions. Rather, he is satisfied with himself for a job well done. His beaming countenance is a symptom of the mentality that made animal genocide against the thylacine possible. The tiger, to Batty, was readily expendable. He only thought twice about killing him because the financial rewards were greater if he could take him alive.

O'Halloran's portraits of the 'glorious' hunter are horrible. Their horror, however, pales in comparison with at least two other photographs. These ghastly souvenirs record the life of the tiger slipping away. In one snapshot he is still capable of standing, in another he is prone, barely able to raise his head.[5] One of these photographs was taken by a Mr Lily or Rily (Reilly), the other photographer is unknown.[6] Gareth Linnard and Stephen Sleightholme argue that these images are staged and the tiger is already dead as there is no blurring indicative of motion in them, the tiger is positioned differently in each image yet the surface he rests on shows no sign of change and the sun appears to be low to the horizon.[7] They put the seeming animation of the thylacine down to rigor mortis. My own belief is that the tiger looks alive because he is. It is quite conceivable that local people with cameras arrived before he expired. The need for Linnard and Sleightholme to 'kill' him is understandable given the grotesqueness of the situation, but their arguments, which, although reasoned, are not altogether persuasive, risk mitigating the horror of what unfolded that day. There are two further photographs that also appear to show the tiger in his death throes but these are attributed to O'Halloran and, if the attribution is correct, these must be faked as the photographer arrived at the property the day after the killing.[8]

The two photographs of the tiger on the woodchip pile taken on 13 May indicate that as one of the last of the thylacines lay dying, Batty's neighbours come to rejoice in the spectacle. Batty recalled that: 'People came from all about to look.'[9] From Batty's account, it seems as if word spread rapidly of the expiring tiger, a crowd of people gathering to watch his last breaths. If rural etiquette was followed, perhaps tea and biscuits were served.[10] Offering tea was an important expression of hospitality.[11] The presence of the crowd indicates the rarity of the thylacine. His death constituted an 'event'. It would have formed a talking point in Mawbanna in the days following. The death of a wild dog or a devil, by contrast, would have passed almost unnoted. The killing of the thylacine was sufficiently noteworthy to be reported in the regional press.

The images are grainy and the 'woodchip pile' may also be on a pile of rubble or ash or clinker or sawdust.[12] Whatever surface formed the tiger's 'deathbed', it was clearly human-made rather than a natural feature. In the background, planks are visible. Like O'Halloran's two 'snuff' pictures, which show outhouses and fences, these images also demonstrate how wrong it is to call this a tiger in the 'wild'. This is habitat claimed by humans, occupied. The thylacine's 'natural' territory has already been taken from him, engulfed by rural expansion. He has become part-scavenger. The tiger killed by Batty had entered a hut on Eric Sundquist's property a few nights previously and been discovered 'lapping up some food out of

[a] saucepan'.[13] His nameless discoverer responded by throwing a boot at the 'beast', who sank his fangs into it.[14] This was the 'wilderness' the tiger inhabited, a series of smallholdings and logging and mining camps. Continuing to refer to the thylacine as the last wild-killed of his kind works to obscure the reality that, while not in a zoo, he was already a victim of environmental change and loss of preferred habitat, a creature of an 'unnatural' landscape. The tiger Batty shot was 'out of place', not in the 'wild' but in a backyard. He died not among button grass and eucalypts but beside occupation debris, amid farmyard detritus. His crime was trespass. Wild animals do not 'trespass' in their habitat.

The thylacine was killed because he was on someone's property. Tigers were not conceived of as 'possessing' land, their relation to their surroundings not recognised as rightful. Like the palawa, they were victims of settler colonialism as it manifested in land appropriation. The palawa were subject to racial capitalism, racial oppression bound up with capitalist expansion and resource exploitation, their territories expropriated in the name of 'civilisation'.[15] Colonial capitalism, of which racial capitalism is a major aspect, involved the transformation of flora and fauna into resources.[16] The thylacine was not identified as of value (not seen as a resource that could be exploited) but rather as a threat to sites of agricultural production. Speciesist capitalism (a dimension of capitalism involving the exploitation of animals) in Lutruwita was significantly linked to the pastoral industry.[17] The tiger was seen to threaten this industry and the genocidal gaze was therefore underpinned by economic interests. The thylacine Batty killed was not harrying sheep but was seen as posing a danger to poultry. A newspaper report published the day after the slaying suggests that the carnivore had 'wrought havoc in fowl-pens'.[18] No specifics are given about this plundering so it may be journalistic invention. It nonetheless demonstrates that the thylacine was perceived as a threat to 'commodities', to 'property', a further reminder of how settled the area was.

As well as constructing the tiger as a pillager, the report demonises the animal in other ways, stating that he 'had been the cause of a great deal of trouble in the Mawbanna district' and had 'frightened several children'.[19] Given that thylacines are primarily nocturnal (as the report acknowledges), it is unclear how this fear-provoking occurred. It also goes unmentioned by Batty in a later interview he gave about the event, although he does repeat the hut incident. Like the unsubstantiated suggestion that a Newfoundland wolf had abducted a child, it seems probable that these anecdotal harassments were used to foster antagonism against the animal and discourage post-mortem sympathy. The thylacine is also referred to as a 'great coward', contributing to their construction as a 'wimp species', unfit to survive.[20] The text shows that even in 1930 the tiger was not viewed sympathetically. Although they mention that the species was 'very rare', the journalist displays no sympathy for the thylacine. The killing of a member of the species was justified because of their general 'bad' behaviour and their weakness. As the report demonstrates, the genocidal gaze was still present when tigers were on the verge of extinction.

This gaze is replicated in the two photographs of the dying tiger. There is no compassion present. They act to inflict further violence, forming visual 'kicks' given to the thylacine when he is down. The photographers found a frisson in this occasion of a life ebbing away. It was a cause of some excitement that merited recording. The group of images constitute an archive of uncertain feelings, either hatred or callous curiosity. Whatever the circumstances of their creation, certainly there is no sympathy here, no anger and resentment at Batty's actions. The pictures record the death of a pest and a freak, an eminently 'expendable' animal.[21] As mementoes of this moment – a moment I see as tragic, but the photographers saw differently – the photographs were treasured, kept safe. Care was devoted to their preservation.[22]

Yet was there, perhaps, another photograph taken that day that did not survive? Did a person rush over with their box camera, point it and then pull away, unable to record the thylacine in his death throes or in death? Such an image, blurred earth or blinding sun, once developed, would look an accident and would not be kept, would not be deemed worthy of preservation, seen as an heirloom. Its importance, for important it would be, would subsequently pass unrecognised, unknown to anyone other than the person who knew its backstory. This image – whether it ever saw the light of day – is open to imagining. It would be a photograph that rejected joying in the thylacine's death and archived feelings other than enmity or vicious curiosity. It would be a photograph that registered compassion. Was such a feeling possible in Lutruwita in 1930?

In 1902, a tiger attacked Priscilla Murray in Springfield in north-east Lutruwita. This is one of the rare documented cases of a thylacine harrying a human. The newspaper account of the attack given in *The Examiner* describes Murray, who was on a plot of land owned by her mother, being bitten by 'an old and almost toothless man tiger'.[23] Murray called to a small boy to help her by releasing a kangaroo dog that was nearby, but the child panicked and ran way. She was therefore reported to have pressed on the thylacine's windpipe until he released his hold. The thylacine then 'sneaked off into the scrub, half falling over the small logs in his course'.[24] Two young men who saw the 'hyena' a couple of days before are recorded as describing him in terms that suggested he was 'a very feeble old fellow, and almost in the last stage of starvation'. The news-writer is at pains to emphasise how doddery and decrepit the tiger is. Foregrounding the thylacine's tremulousness works to emasculate him and undermine Murray's achievements in fending off her attacker. There is therefore potentially a sexist subtext to the account, particularly if the text is compared to the report of Clem Penny bludgeoning a female tiger to death with a tree branch and then injuring a male. Both tigers are claimed to have leapt at Penny and the writer casts him as a capable bushman, a heroic fighter. The accompanying photograph shows Penny and an elderly man (probably his grandfather) holding the slaughtered tiger aloft. The thylacine is suspended head downwards, humbled by her human adversaries.

The account of Penny's 'battle royal' seems exaggerated.[25] Murray certainly felt her encounter with a tiger was not reported accurately. When speaking later to Mrs C. F. McLauchlan about the incident, she expressed discontent with how the events were described. Murray noted that she was actually bitten twice (on her right arm and left hand) and she got rid of the tiger, who was one-eyed, not by gripping his windpipe but by treading on his tail and threatening him with a garden hoe. At this, he 'slunk away'.[26] Murray also resented the description of the tiger as toothless as she bled freely from the wounds caused by his teeth and was left permanently scarred. Crucially, despite her injuries, Murray 'tolerantly suggested that [the tiger] was quite old and having difficulty in obtaining winter rations'.[27] Through making light of the thylacine's actions, Murray eschews the common response of violent disdain. Humour here can be understood as an effort to break with the usual demonisation of the thylacine.[28] Troughton, perhaps reflecting Murray's own beliefs, speculates that the tiger 'mistook her moving arm for a small bird or mammal'.[29] Here, then, someone who was physically injured by a tiger refused to embrace the genocidal logic adopted by so many of her contemporaries. She shows instead what can be interpreted as a measure of empathy and, through declining to replicate a language of brutishness, opens the possibility of a more caring outlook.

It is therefore possible to conceive someone thirty years later reacting to the plight of the dying thylacine with concern rather than morbid curiosity or celebration, refusing to reproduce the genocidal gaze that undergirds the extant snapshots.[30] A photograph that portrayed an act of pulling away would, through its mode of indirect reference, index compassion rather than callousness. It would also conform to Rosemarie McGoldrick's call for an aesthetics of absence when representing nonhuman animal subjects.[31] The image would encode the gestures that generated it: compassion for the dying, revulsion at efforts to 'capture' another's final moments. It would also carry an affective dimension. It is improbable that such an image *relating* to the thylacine killed by Batty ever existed, but the feelings that would make such a non-representation of nonhuman animal suffering possible were clearly historically present.[32] Those glimmers of an alternative to the genocidal logic that did for the tiger offer some solace in a history so marked by hatred and apathy: a hatred now often obscured and an apathy that persists by way of that very obscuration.

Photographs such as those of Batty beaming with pride or the two images of the dying animal are regularly reproduced without any sense that they have given the person using them today cause for pause, that the latter has recognised and reflected on their capacity for (re-)enacting violence. The cover image of Samuel Turvey's edited volume *Holocene Extinctions*, for example, is the 1869 photograph of the hunter William George Weaver, a chemist entrepreneur of about fifty years of age, posing facing a dead thylacine. The tiger has been trussed and strung from the ceiling, a morbid mobile in what is clearly a photography studio; a balustrade in the classical style is stencilled on the back wall.[33] This image is a celebration of killing, a souvenir of an act of atrocity. Weaver's rifle rests in the crook of his arm, the tool he used to kill the tiger carefully recorded for posterity, an iconographic choice replicated in the later Batty portraits.[34] The pharmacist was claimed to be 'one of the crack shots of the colony' and the presence of the firearm works to bolster this notion.[35]

The use of the portrait of Weaver for the cover of *Holocene Extinctions* operates to foreground human complicity in species losses and to draw attention to animal genocide as an extinction driver but also risks exposing the thylacine anew to a sadistic gaze, to a visual pleasure in suffering. Weaver's right foot looks set to boot the tiger in the muzzle, to scorn the dead marsupial.[36] Suspending the tiger, as also occurs in the photograph of the tiger Penny killed, has a similar psychological motivation to that of the buggy rug discussed in Chapter 2: the nonhuman animal is brought down, shown to be of a 'lower order'. Writing of trophy photographs taken in the field (rather than the studio), Nigel Rothfels asks what claims such images are intended to make, what sense they are striving to give to an animal's death.[37] Photographs such as Weaver's or Penny's or Batty's appear to glory in a tiger's demise. The thylacine in these images exists primarily to showcase their 'hunting' prowess, historically a manly attribute in Australia, and to reinforce human dominance over nature.[38] Describing these images as 'of' the thylacine is therefore deceptive; these are not portraits of each animal but of their killers. The tigers are props for the personal aggrandisement of Batty, Penny and Weaver. The images are also embodiments of the genocidal gaze, monuments to each tiger's successful extermination. It is for this reason that I do not reproduce them here.

Griselda Pollock rightly emphasises how photographs that document atrocity risk immersing us anew in the voyeurism and sadism that sometimes informed their production.[39] In this sense, images are not passive but act upon us, potentially leading us to adopt undesirable ways of seeing and thinking, this despite our compassion. Sagely, although cognisant of this issue, Pollock does not call for a moratorium on the reproduction of images of atrocity (even when their makers were potentially perpetrators with appalling motives). I too am not an advocate of what might be termed 'image cancel culture'. Removing certain images of atrocity (towards human or nonhuman animals) from circulation risks replicating logics of censorship that have, in the past, been employed by reactionary and repressive regimes to punish the very oppressed and historically disenfranchised that the contemporary calls for image suppression are designed to protect. Such actions also ignore the complexities of reception, the ways images can be read 'against the grain'.[40]

Although I have chosen not to reproduce the two images of the dying thylacine here, they are present through my prose, represented in refracted fashion by words. I necessarily engage in ekphrasis. Giving an image to a reader's eye through words, however, entails a kind of distancing that has positive attributes. Through ekphrasis, the photographs (including their violence) are evoked but not mirrored. Ekphrasis necessarily involves a lack of mastery, a descriptive haziness, that is at odds with the precision of many lens-based images. There is an ethically desirable withholding present that works to minimise the danger of reactivating dormant sadistic and voyeuristic aspects within a given atrocity photograph. The issue of writing, and how it intersects with efforts to attest to animal genocide and its aftermath, thus comes to the fore.

Many authors have explored the difficulties of writing about or alongside nonhuman animals. In the context of nonhuman animal remains, Steve Baker has noted that the task of any writer should be to give the remains their voice alongside the authorial voice, to share the space of the page with the animal other.[41] Baker sees literary style, form, as potentially of comparable importance to content. Attention to style involves giving voice to animal remains while resisting reducing those vestiges to ventriloquist dummies serving the ends of the author. For Baker, considered descriptions suspend judgement, refusing to mobilise remains in the service of an ethical or political argument, resisting raising them to the status of an idea or theory.

I have tried to take a different path here, seeking to safeguard the singularity of those thylacines and wolves I have studied, but with my approach continually openly informed by the horrors of animal genocide, knowingly emotive as well as evocative. Such an approach risks rendering the tiger and wolf ciphers for genocide, depriving them of their multifaceted complexities to service an argument in which they are reduced to symbols of human cruelty.[42] It is for that reason that, alongside discussion of genocide and its aftermath, I have sought to preserve aspects of species and individual singularity, to provide something in excess of the central arguments, a remainder that refuses enclosure within the 'master' narrative that is animal genocide and its aftermath.

The hatred for the thylacine, as Freeman forcefully argues in *Painted Tiger*, was enabled by practices of writing in an extended sense, by words and images that encouraged negative perceptions of the animal. Beyond openly violent modes of representation of this kind, language itself nefariously acts as a nullifying force, reducing the tiger to sign. Elements of Chapter 4 and, particularly, Chapter 5 turn on the role that absence can play in aesthetic terms as a mode of presencing the thylacine (and, by extension, the Newfoundland wolf) that sidesteps the violence accompanying signification or symbolisation. Practices of allusion and evocation such as those employed in the film *The Hunter* cultivate modes of indirect encounter with the tiger that are affective. Through these modes what Bracha Ettinger refers to as wit(h)nessing is encouraged, a bearing witness to the trauma of others that is also a being with those others, encountering them. 'Being with' here involves a kind of inseparable togetherness that is not a unitary becoming; otherness is not evacuated but rather there is a sharing '*with-in-difference*'.[43] The neologisms Ettinger employs foreground the challenge of fostering encounter from within a language and logic that privileges difference and separation.

Anna Johnson describes 'wit(h)nessing' as a means of referring to 'an otherwise occluded mode of (ethical) relation'.[44] It is an ethical relation involving anonymous intimacy in the sense that each 'other' in the encounter is not named and known. Rather, the presence of others is felt and traces of their trauma are affectively transferred. Wit(h)nessing provides a way of theorising Baker's sharing of space within writing as an aesthetic encounter where human and nonhuman animal others can co-affect each other while they are temporarily connected without being fused, the singularities of each preserved within a moment of severality.[45] Wit(h)nessing ensures that nonhuman animals are not reduced to ideas or symbols, voided of their alterity and singularity and rendered familiar and typical. The lives of nonhuman animals should not be humanised, anthropomorphised.[46] Wit(h)nessing enables linking while respecting the other's alterity, refusing to transform them into more familiar terms. Writing in the context of Ettinger's artworks, Butler describes aesthetic encounter as 'without discernment, without full clarity or distinct form'.[47] Encounter in *The Hunter* similarly refuses the cut of differentiation that is form (of which morphology is one manifestation), registering instead as a pressing absence. This absence carries something of the trauma of the animal genocide that was perpetrated against the thylacine and of their ensuing extinction.

My aim in this book has been to define (Chapter 1) and examine (Chapters 2 through 6) the processes and effects of animal genocide. Chapters 4 and 5 also consider aesthetic issues related to non-violent modes of imaging the thylacine and their traumatic history. Animal genocide is not, however, purely a historical phenomenon, it is a current concern. As discussed in Chapter 1, there are also contemporary campaigns of extermination (planned or executed) against fauna. These campaigns have the potential to become genocidal or are so already.[48] This invites reflection on the need to codify animal genocide as a crime under international law so that its perpetration may be prevented or prosecuted.

Animal genocide can act as an extinction driver as occurred in the cases of the thylacine and the Newfoundland wolf. It is enabled by particular ways of seeing – embodied in the genocidal gaze – and thinking. Through studying the nature of this dangerous perceptual framework as it is recorded in archival documents such as the photographs of the tiger shot by Batty, ways of guarding against future genocidal events beyond a legal framework can also be identified. One such way, which Baker gestures towards, is through carefully choosing the words we use to refer to our non-human animal fellows such that description does not become affliction. It is, however, too late for the thylacine and the Newfoundland wolf. In the case of the vanished, our responsibility is instead to know when to cease to speak, when to cede significance, leaving the page open . . .

Notes

1. In 'A Catalogue of the Thylacine Captured on Film', Sleightholme and Campbell give the dates of 13 May (151) and 6 May (164) for the event.
2. Batty alternatively suggests twenty minutes or half an hour, but this duration may be subjective (based on his lived perception of time that day rather than clock time) and either conservative or overstated. Time perception can involve dilation or distension, with the length of an experience perceived to exceed or be less than its duration in clock time. Given the excitement of the situation, time was probably felt to pass quickly so 'half an hour' may well conform to an hour or more in clock time. See Chapter 7 of Witmmann, *Felt Time*.
3. Linnard and Sleightholme, 'The 1930 Mawbanna Thylacine'.
4. Marianne Cooney, 'The Tasmanian Tiger', *Western Tiers*, Thursday 23 November 1989, 14.
5. Sleightholme and Campbell claim that he is already dead in one of the photographs. I do not think this is the case, yet clearly we are at, or close to, the tiger's last breaths.
6. Linnard and Sleightholme propose J. M. Reilly as the photographer in 'The 1930 Mawbanna Thylacine'.
7. Ibid.
8. Ibid. A forensic veterinary pathologist, Ranald Munro, who was consulted by Linnard and Sleightholme about these images, expressed some scepticism that they were staged as it would be difficult to pose a dead tiger with rigor mortis in such a way but

conceded it as a possibility. The four photographs of the thylacine propped against a fence, secured in place by a cord, demonstrate how challenging it was to keep his corpse upright and support Munro's disbelief regarding the contrived nature of the other images. O'Halloran moving the tiger back to the woodchip pile a day after his death to photograph him is also counter-intuitive. All four images featuring the pile were more likely taken on the same day.

9. Wilf Batty cited in Ken Brass, 'The $55,000 Search to Find a Tasmania Tiger', *Australian Women's Weekly*, Wednesday 24 September 1980, 40.
10. For a discussion of tea drinking in rural Australia, see Knight, '"A Poisonous Cup?"' Knight focusses on regular 'tea breaks' as markers of time during the day rather than on impromptu tea drinking caused by the arrival of unexpected visitors.
11. See Griggs, 'Black Poison or Beneficial Beverage?', 33.
12. Woodchips are sometimes created as by-products from chopping timber and can be used as mulch or to cover the ground in kennels. It is therefore quite possible that the smallholding had such a pile.
13. Anonymous, 'A Destructive Animal: Hyena Killed at Mawbanna', *Advocate*, Wednesday 14 May 1930, 6.
14. Ibid.
15. For an analysis of the historical development and nature of racial capitalism that makes some reference to Australia (but not Lutruwita), see Bhandar, *Colonial Lives of Property*.
16. See Ince, *Colonial Capitalism and the Dilemmas of Liberalism*.
17. Speciesist capitalism is discussed in Chapter 1 of Grubb's *Ecoliberation*.
18. Anonymous, 'A Destructive Animal', 6.
19. Ibid.
20. For a discussion of 'wimp species', see Pimm, '"Putting Nature Back Together Again"', 103–4.
21. Baer, 'A Reconsideration of Sexual Violence', 653.
22. The corpse was sold by Batty to James Harrison for £5, who, in turn, is reported to have sold the remains to TMAG where they were taxidermied and then sent on a tour of the mainland, the dead tiger condemned to a 'life' as spectacle. Brass, 'The $55,000 Search to Find a Tasmania Tiger', 40.
23. Anonymous, 'Attacked by a Tiger or Hyena', *Examiner*, Tuesday 8 April 1902, 7. In *Furred Animals of Australia*, Troughton (51) states that the event was also reported in the *North Eastern Advertiser* (a biweekly newspaper published out of Scottsdale). The *North Eastern Advertiser*, however, did not commence publication until 1909, so if the event was discussed in that paper it must have been retrospectively. I have been unable to locate any such discussion, although a 1925 issue of the newspaper does mention a cyclist, Mr Betts, who confronted a tiger near Scottsdale. When they fled, Betts hurled a piece of wood after them. Anonymous, 'Cyclist Meets Tiger', *North Eastern Advertiser*, Friday 11 December 1925, 2. Troughton's phraseology in his summary of the report leads me to think he has the *Examiner* report in mind.
24. Anonymous, 'Attacked by a Tiger or Hyena', 7.
25. The encounter is referred to as a 'battle royal' in the image caption for the picture of Penny and his 'trophy'. See Anonymous, 'Encounter with Tasmanian Tigers', *Weekly Courier*, Thursday 17 January 1924, 26.
26. Troughton, *Furred Animals of Australia*, 52.
27. Ibid.
28. The use of humour in this context is probably overdetermined, also acting as a coping mechanism. For a discussion of the value of humour in mitigating conflict and as a healing mechanism, see Zelizer, 'Laughing Our Way to Peace or War'.

29. Troughton, *Furred Animals of Australia*, 52. Bell interprets Troughton's remarks in this way, attributing them to Murray. See E. A. Bell, *Thylacine*, unpublished manuscript, Tasmanian Archives NS463/1/2, 86–87.
30. For a discussion of a possible example of such a 'pulling away' in the context of photographs of human genocide see Chapter 6 of Chare and Williams, *Matters of Testimony*, and Chapter 3 of Chare and Williams, *The Auschwitz Sonderkommando*.
31. See Chapter 5.
32. This would be a relating to the thylacine that eschews the drive to visual mastery embodied in the extant photographs.
33. See the front cover of Turvey, *Holocene Extinctions*. In their article 'Frank Haes' Thylacine', Sleightholme et al. attribute this photograph to Victor Prout. I have not been able to examine the original but believe it probable Weaver would have used the City Photographic Establishment run by Thomas Nevin. In 1869 it was located at 140 Elizabeth Street. Weaver's pharmacy and chemical plant was also on Elizabeth Street, a short distance away at Wellington Bridge. Weaver imported and manufactured photographic chemicals.
34. This recurring iconography (a hunter showing off their gun alongside their kill) extends to another image in *Holocene Extinctions*, which accompanies the book's dedication to the Yangtze river dolphin or bajii (*Lipotes vexillifer*). It is a black and white photograph of a youth, Charles Hoy, posing with his rifle beside a bajii he has killed, their mouth propped open to reveal their teeth. Hoy shot the dolphin on 18 February 1914 and would take the skull and trophy photograph to the United States. He donated the skull and the cervical vertebrae to the Smithsonian (accession number 61644) on 1 July 1917 and it would subsequently become the holotype for the description (30 March 1918) of the species.
35. Anonymous, 'Local & General', *The Tasmanian*, Saturday 12 February 1887, 23.
36. In reality, Weaver's right foot is below and to the side of the tiger's head.
37. Rothfels, 'Trophies and Taxidermy', 127.
38. The linkage of hunting and manliness may also have extended to First Nations Australians. See Martínez and Lowrie, 'Colonial Constructions of Masculinity', 315.
39. Pollock, *Art in the Time-Space of Memory and Migration*, 110.
40. See, for example, Chapter 4 of Ulrich Baer's *Spectral Evidence*, which engages with photographs that embody a 'Nazi gaze' (139). For a more general overview of how reception can be 'idiosyncratic' or go against the grain of a given text, see Willis, *Reception*.
41. Baker, 'Dead, Dead, Dead, Dead, Dead', 295.
42. See Tyler, *CIFERAE*, 23, 28.
43. Italics in the original. Ettinger, 'Transgressing with-in-to the Feminine', 366.
44. Johnson, 'Nomad-Words', 232.
45. Baker's notion of writing is not informed by psychoanalysis but Ettinger's conception of the aesthetic as a field of ethical encounter seems apposite here.
46. In *CIFERAE*, Tom Tyler rightly cautions against too readily and rapidly employing the term anthropomorphism, noting that using the term involves embracing a number of uninterrogated assumptions about humanness (59). Anthropomorphism is a symptom of anthropocentrism (63). I nonetheless believe there is a danger in writing about the thylacine and the wolf that they are reduced to the familiar – which is to say human cultural frames of reference – and their alterity effaced.
47. Butler, 'Disturbance and Dispersal in the Visual Field', 161.
48. In 2023, for example, Iberian wolves were declared extinct in Andalucía, their disappearance caused, in part, by illegal hunting. As Iberian wolves are a protected species, their rarity well known, the hunters who unlawfully kill them are engaging in genocide.

Bibliography

Adams, Carol J. 'The War on Compassion', in Josephine Donovan and Carol J. Adams (eds), *The Feminist Care Tradition in Animal Ethics* (New York: Columbia University Press, 2007), 21–36.

Adorno, Theodor W. *Negative Dialektik*. Frankfurt am Main: Suhrkamp, 1982 [1966].

———. *Negative Dialectics*, trans. E. B. Ashton. London: Routledge, 1990 [1973].

Aguila-Way, Tania. 'Beyond the Logic of Solidarity as Sameness: The Critique of Animal Instrumentalization in Margaret Atwood's *Surfacing* and Marian Engel's *Bear*', *Isle* 23(1) (2016): 5–29.

Ahmed, Sara. *The Cultural Politics of Emotion*. 2nd edn. Edinburgh: Edinburgh University Press, 2014.

Aldrich, Robert. *Greater France: A History of French Overseas Expansion*. New York: Macmillan, 1996.

Aldworth, Rebecca. 'Welfare Aspects of Commercial Hunting and Climate Change', in Andy Butterworth (ed.), *Marine Mammal Welfare: Human Induced Change in the Marine Environment and Its Impact on Marine Mammal Welfare* (Cham: Springer, 2017), 253–71.

Allen, Don W. 'Notes: The Rights of Nonhuman Animals and World Public Order: A Global Assessment', *New York Law School Law Review* 28 (1983): 377–429.

Allen, Glover M., and Thomas Barbour. 'The Newfoundland Wolf', *Journal of Mammalogy* 18(2) (1937): 229–34.

Allen, Mary. *Animals in American Literature*. Urbana: University of Illinois Press, 1983.

Allen, Steven. 'Tasmania: Australia's Cinematic Landscape of Loss', in Steven Allen and Kirsten Møllegaard (eds), *Narratives of Place in Literature and Film* (New York: Routledge, 2019), 52–67.

Anderson, Rudolph Martin. *Methods of Collecting and Preserving Vertebrate Animals*. Ottawa: F. A. Acland, 1932.

———. 'Summary of the Large Wolves of Canada, with Description of Three New Arctic Races', *Journal of Mammalogy* 24(3) (1943): 386–93.

Andrei, Mary Anne. *Nature's Mirror: How Taxidermists Shaped America's Natural History Museums and Saved Endangered Species*. Chicago: University of Chicago Press, 2020.
Andrews, A. P. 'Thylacine Remains: Florentine Valley', *Southern Caver* 4(1) (1972): 10–11.
Andrews, Peter, and Miranda Armour-Chelu. 'Taphonomic Observations on a Surface Bone Assemblage in a Temperate Environment', *Bulletin de la Société géologique de France* 169(3) (1998): 433–42.
Anonymous. 'Depositions', in *Public Documents of the Senate of the United States: First Session of the Twenty-Third Congress* (Washington, DC: Duff Green, 1834), 1–91.
Anonymous. *Report of the Parliamentary Select Committee on Aboriginal Tribes (British Settlements)*. London: William Ball and Hatchard & Son, 1837.
Anonymous. 'The Wolf Chase', in E. Littell (ed.), *Littell's Living Age: Volume 10* (Boston: Waite, Peirce and Company, 1846), 162–63.
Anonymous. *Tommy Trot and His Loves*. London: Dean & Son, 1861.
Anonymous. 'Decorative Furs', *Decorator & Furnisher* 19(3) (1891): 84–85.
Anonymous. 'Australian Ornithologist's Union: Hobart Congress', *The Emu* 3(3) (1904): 141–63.
Anonymous. 'The Wolf Hunt', *New York Dramatic Mirror* 62 (1909): 17.
Anonymous. 'Hide-and-Seek Tent is Used to Photograph Birds', *Popular Mechanics Magazine* 54(2) (August 1930): 272.
Anonymous. 'George Edwards's Picture of the Dodo', *Nature* 131 (1933): 615.
Anonymous. 'For the Classroom', *English: Journal of the English Association* 11(61) (1956): 36.
Anonymous. '*The Saga of the Buffalo*, by Cy Martin', *Virginia Quarterly Review* 49(4) (1973): clxxiv.
Anthony, David W., and Dorcas R. Brown. 'Looking a Gift Horse in the Mouth: Identification of the Earliest Bitted Equids and the Microscopic Analysis of Wear', in Pam J. Crabtree, Douglas V. Campana and Kathleen Ryan (eds), *Early Animal Domestication and Its Cultural Context* (Philadelphia: MASCA, 1989), 99–116.
Armstrong, Kacie L., and James E. Cutting. 'On the Physical Variables Filmmakers Use to Engage Viewers', *Psychology of Aesthetics, Creativity, and the Arts* (2022). Retrieved 23 February 2025 from https://psycnet.apa.org/fulltext/2022-59629-001.pdf.
Armstrong, Philip. 'The Gaze of Animals', in Nik Taylor and Tania Signal (eds), *Theorizing Animals: Re-Thinking Humanimal Relations* (Leiden: Brill, 2011), 175–99.
Arrighi, Gillian. 'Political Animals: Engagements with Imperial and Gender Discourses in Late-Colonial Australian Circuses', *Theatre Journal* 60(4) (2008): 609–29.
Aschim, Ruth A., and Ryan K. Brook. 'Evaluating Cost-Effective Methods for Rapid and Repeatable National Scale Detection and Mapping of Invasive Species Spread', *Scientific Reports* 9 (2019): 1–13.
Asma, Stephen T. *Stuffed Animals and Pickled Heads: The Culture and Evolution of Natural History Museums*. Oxford: Oxford University Press, 2001.
Atwood, Margaret. *Survival: A Thematic Guide to Canadian Literature*. Toronto: House of Anansi Press, 1972.
———. 'Animal Victims', in Ralph H. Lutts (ed.), *The Wild Animal Story* (Philadelphia: Temple University Press, 1998), 215–24.
Aylward, Christopher P. 'The Beothuk Story: European and First Nations Narratives of the Beothuk of Newfoundland', unpublished PhD thesis. Memorial University of Newfoundland, 2014.
———. *Beothuk: How Story Made a People (Almost) Disappear*. Montréal: McGill-Queen's University Press, 2024.
Backhouse, James. *A Narrative of a Visit to the Australian Colonies*. London: Hamilton, Adams, & Co., 1843.

Baer, Elizabeth R. *The Genocidal Gaze: From German Southwest Africa to the Third Reich*. Detroit: Wayne State University Press, 2017.

———. 'A Reconsideration of Sexual Violence in German Colonial and Nazi Ideology and Its Representation in Holocaust Texts', in Victoria Aarons and Phyllis Lassner (eds), *The Palgrave Handbook of Holocaust Literature and Culture* (Cham: Palgrave Macmillan, 2020), 651–68.

Baer, Ulrich. *Spectral Evidence: The Photography of Trauma*. Cambridge, MA: MIT Press, 2002.

Bailey, Col. *Lure of the Thylacine: True Stories and Legendary Tales of the Tasmanian Tiger*. Melbourne: Echo, 2016.

Bailey, Edgar P. *Introduction of Foxes to Alaskan Islands: History, Effects on Avifauna, and Eradication*. Washington, DC: United States Department of the Interior, Fish and Wildlife Service, 1993.

Bailey, Vernon. *Key to Animals on Which Wolf and Coyote Bounties Are Often Paid: United States Department of Agriculture Bureau of Survey Circular No. 69*. Washington, DC: Government Printing Office, 1909.

Baird, Andy. *Voices of Aboriginal Tasmania: Ningenneh Tunapry Education Guide*. Hobart: Tasmanian Museum and Art Gallery, 2008.

Baker, Andrew, and Chris Dickman. *Secret Lives of Carnivorous Marsupials*. Clayton South: CSIRO, 2018.

Baker, Steve. 'Dead, Dead, Dead, Dead, Dead', in Garry Marvin and Susan McHugh (eds), *Routledge Handbook of Human-Animal Studies* (Abingdon: Routledge, 2014), 290–304.

———. '*Fish Market, Lagos*: Artists Pages and Supporting Statement', in Sarah Bezan and Robert McKay (eds), *Animal Remains* (New York: Routledge, 2021), 264–72.

Ball, Karyn. 'Primal Revenge and Other Anthropomorphic Projections for Literary History', *New Literary History* 39(3) (2008): 533–63.

Ball, Margaret Haig Roosevelt Sewall. 'Grim Commerce: Scalps, Bounties and the Transformation of Trophy-Taking in the Early American Northeast, 1450–1770', unpublished PhD thesis. University of Colorado, 2013.

Baltrušaitis, Jurgis. *Anamorphic Art*, trans. W. J. Strachan. New York: Harry N. Abrams, 1977.

Barad, Karen. 'Posthumanist Performativity: Toward an Understanding of How Matter Comes to Matter', *Signs* 28(3) (2003): 801–31.

Barclay, Paul D. 'A "Curious and Grim Testimony to a Persistent Human Blindness": Wolf Bounties in North America, 1630–1752', *Ethics, Place and Environment* 5(1) (2002): 25–34.

Barker, William. *Narrative of a Journey to Shoa and of an Attempt to Visit Harrar*. Bombay: Education Society's Press, 1868.

Barrington, Richard Manliffe. *The Migration of Birds as Observed by Irish Lighthouses and Lightships*. London: R. H. Porter, 1900.

Barrow, Mark V. 'Teetering on the Brink of Extinction', in Ben A. Minteer, Jane Maienschein and James P. Collins (eds), *The Ark and Beyond: The Evolution of Zoo and Aquarium Conservation* (Chicago: University of Chicago Press, 2018), 51–64.

Barta, Tony. 'Relations of Genocide: Land and Lives in the Colonization of Australia', in Isidor Wallimann and Michael Dobkowski (eds), *Genocide and the Modern Age: Etiology and Case Studies of Mass Death* (Westport, CT: Praeger, 1987), 237–53.

Bartlett, Robert Merrill. *They Stand Invincible: Men Who Are Reshaping Our World*. New York: Thomas Y. Crowell, 1959.

Basso, Luca. *Marx and Singularity: From the Early Writings to the Grundrisse*, trans. Arianna Bove. Leiden: Brill, 2012.

Bath, Alistair J. 'Identification and Documentation of Public Attitudes toward Wolf Reintroduction in Yellowstone National Park', in John D. Varley and Wayne G. Brewster (eds), *Wolves for Yellowstone? A Report to the United States Congress, Volume IV, Research and Analysis* (Yellowstone National Park: National Park Service, 1992), 2-3–2-30.

Batin, Christopher. 'Wolf Hunters', in Jay Cassell (ed.), *The Best Hunting Stories Ever Told* (New York: Skyhorse, 2010), 144–48.

Batty, Philip. 'The *Tywerrenge* as an Artefact of Rule: The (Post)Colonial Life of a Secret/Sacred Aboriginal Object', *History and Anthropology* 25(2) (2014): 296–311.

Baughman, Michael. *Boat: A Memoir of Friendship, Faith, Death and Life Everlasting*. New York: Arcade, 2012.

Bauman, Zygmunt. *Modernity and Ambivalence*. Cambridge: Polity, 1991.

Behan, John. 'The Past', *The Fleur-de-Lys* 5(46) (December 1946): 8–9.

Behrensmeyer, Anna K. 'Taphonomic and Ecologic Information from Bone Weathering', *Paleobiology* 4(2) (1978): 150–62.

Beirne, Piers. 'For a Nonspeciesist Criminology: Animal Abuse as an Object of Study', in Rob White (ed.), *Environmental Crime: A Reader* (Cullompton: Willan, 2009), 175–99.

———. 'Theriocide: Naming Animal Killing', *International Journal for Crime, Justice and Social Democracy* 3(2) (2014): 49–66.

Belich, James. *Making Peoples: A History of New Zealanders from Polynesian Settlement to the End of the Nineteenth Century*. Honolulu: University of Hawai'i Press, 1996.

Belk, R. W., Melanie Wallendorf, John Sherry, Morris Holbrook and Scott Roberts. 'Collectors and Collecting', *Advances in Consumer Research* 15 (1988): 548–53.

Benjamin, Ross, and Eline Lorenzen. 'Thylacine Tales', *Nature Ecology & Evolution* 2 (2018): 7–8.

Berat, Lynn. 'Defending the Right to a Healthy Environment: Toward a Crime of Geocide in International Law', *Boston University International Law Journal* 11(2) (1993): 327–48.

Berger, Dieter A. '"Damn the Mottoe": Scott and the Epigraph', *Anglia* 100 (1982): 373–96.

Berlocher, Stewart H. 'Origins: A Brief History of Research on Speciation', in Daniel J. Howard and Stewart H. Berlocher (eds), *Endless Forms: Species and Speciation* (Oxford: Oxford University Press, 1998), 3–15.

Berryman, Hugh, and Steven Symes. 'Recognizing Gunshot and Blunt Cranial Trauma through Fracture Interpretation', in Kathleen J. Reichs (ed.), *Forensic Osteology: Advances in the Identification of Human Remains*, 2nd edn (Springfield, IL: Charles C. Thomas, 1998), 333–52.

Bewick, Thomas. *A General History of Quadrupeds*, 4th edn. Newcastle: S. Hodgson, R. Beilby and T. Bewick, 1800.

Bhandar, Brenna. *Colonial Lives of Property: Law, Land, and Racial Regimes of Ownership*. Durham, NC: Duke University Press, 2018.

Bienvenue, Valérie. 'From the General to the Particular: Piecing Together the Life and Afterlife of A544, Louis XVI's Quagga', in Valérie Bienvenue and Nicholas Chare (eds), *Animals, Plants and Afterimages: The Art and Science of Representing Extinction* (New York: Berghahn, 2022), 334–54.

Bienvenue, Valérie, and Nicholas Chare. 'Introduction', in Valérie Bienvenue and Nicholas Chare (eds), *Animals, Plants and Afterimages: The Art and Science of Representing Extinction* (New York: Berghahn, 2022), 1–63.

Bishop, Elizabeth. 'The Fish', in *The Complete Poems: 1927–1979* (New York: Farrar, Straus & Giroux, 1984), 42–44.

Black, Sue. *All That Remains: A Life in Death*. London: Doubleday, 2018.
Black, Sue, Ingeborg C. L. Marshall and Andrew C. Kitchener. 'The Skulls of Chief Nonosabasut and His Wife Demasduit: Beothuk of Newfoundland', *International Journal of Osteoarchaeology* 10(6) (2009): 659–752.
Bladen, F. M. (ed.). *Historical Records of New South Wales Volume V: King 1803, 1804, 1805*. Sydney: William Applegate Gullick, 1897.
Blaine, Delabere P. *An Encyclopaedia of Rural Sports*. London: Longmans, Green & Co., 1875.
Blanchot, Maurice. *La part du feu*. Paris: Gallimard, 1949.
Blandy, Doug, and David Cowan. 'Imagine! Yellowstone: Art, Education and the Rehabilitation of Place', *Art Education* 50(6) (1997): 40–46.
Bonyhady, Tim. 'The Art of Pacification', in Tim Bonyhady and Greg Lehman (eds.), *The National Picture: The Art of Tasmania's Black War* (Canberra: National Gallery of Australia, 2018), 73–126.
Bonnycastle, Richard Henry. 'Considerations upon the Political Position and Natural Advantages of Newfoundland [1841]', in *Minutes of Evidence Taken before the Select Committee to Inquire into the State of the Colony of Newfoundland*, Part 6 (London: House of Commons, 1841), 1–23.
———. *Newfoundland in 1842: Volume 2*. London: Henry Colburn, 1842.
Boose, Lynda E. 'Scolding Brides and Bridling Scolds: Taming the Woman's Unruly Member', *Shakespeare Quarterly* 42(2) (1991): 179–213.
Boroditsky, Lera. 'How Language Shapes Thought', *Scientific American* 63 (2011): 62–65.
Borrell, Sally. 'Mercy Killing', *Australian Animal Studies Group News Bulletin* 14 (2011): 53.
Bourdieu, Pierre. *The Logic of Practice*, trans. Richard Nice. Cambridge: Polity, 1990.
Bousé, Derek. *Wildlife Films*. Philadelphia: University of Pennsylvania Press, 2000.
Boyer, P. W. 'Leaders and Helpers: Jane Franklin's Plan for Van Diemen's Land', *Papers and Proceedings: Tasmanian Historical Research Association* 21(2) (1973): 47–65.
Braidotti, Rosi. 'Posthuman, All Too Human: Towards a New Process Ontology', *Theory, Culture and Society* 23(7–8) (2006): 197–208.
Bralliar, Floyd. *Knowing Birds through Stories*. New York: Funk & Wagnalls, 1922.
Brantlinger, Patrick. *Dark Vanishings: Discourse on the Extinction of Primitive Races, 1800–1930*. Ithaca, NY: Cornell University Press, 2003.
Brass, Ken. 'The $55,000 Search to Find a Tasmanian Tiger', *Australian Women's Weekly*, 24 September 1980: 40–41.
Breen, Shayne. 'Extermination, Extinction, Genocide: British Colonialism and Tasmanian Aborigines', in René Lemarchand (ed.), *Forgotten Genocides: Oblivion, Denial and Memory* (Philadelphia: University of Pennsylvania Press, 2011), 71–90.
Breton, William Henry. *Excursions in New South Wales, Western Australia and Van Dieman's Land in 1830, 1831, 1832, and 1833*, 2nd revised and expanded edn. London: Richard Bentley, 1834.
Brinsmead, Hesba Fay. *Echo in the Wilderness*. London: Oxford University Press, 1972.
Brissenden, R. F. 'On the Edge of the Empire: Some Thoughts on Recent Australian Fiction', *Sewanee Review* 87(1) (1979): 142–57.
British Royal Commission. *The British Section at the Vienna Universal Exhibition, 1873: Official Catalogue*. London: J. M. Johnson & Sons, 1873.
Brodie, Alexander Oswald. 'Report of the Governor of Arizona', in *Annual Reports of the Department of the Interior: Miscellaneous Reports Part II* (Washington, DC: Government Printing Office, 1903), 107–366.
Brodie, Nick. *The Vandemonian War*. Richmond: Hardie Grant, 2017.

Brook, Barry W., Stephen R. Sleightholme, Cameron R. Campbell, Ivan Jarić and Jessie C. Buettel. 'Resolving When (and Where) the Thylacine Went Extinct', *Science of the Total Environment* 877 (2023): 162878.

Brower, Matthew. 'Trophy Shots: Early North American Photographs of Nonhuman Animals and the Display of Masculine Prowess', *Society and Animals* 13(1) (2005): 13–31.

Brown, Alys. 'The Reformer', *Christmas Bulletin*, 7 December 1929: 13–14.

Browning, Robert. 'One More Word', in Horace E. Scudder (ed.), *The Complete Poetical Works of Browning* (Cambridge: Riverside Press, 1895), 361–63.

Bryant, Ernest A. 'The Pouched Mammals', in Arthur Mee (ed.), *Popular Science: Volume III* (London: Amalgamated Press, 1912), 2007–15.

Buchanan, Edward A. *Parent/Teacher Handbook: Volume III*. Nashville: Broadman & Holman, 2005.

Buchanan, Gwynneth. *Elements of Animal Morphology*, 4th edn. Melbourne: Lothian, 1944.

Buisson, Olivier, Samia Boukir and Bernard Besserer. 'Motion Compensated Film Restoration', *Machine Visions and Applications* 13 (2003): 201–12.

Burke, Jack E. 'A Brief History of the Taxonomy of Mammals', *MCV Quarterly* 4(2) (1968): 77–80.

Burroughs, John. 'Real and Sham Natural History', *Atlantic Monthly* 91(545) (1903): 298–309.

Burt, Jonathan. *Animals in Film*. London: Reaktion Books, 2002.

Burton, Richard F. *First Footsteps in East Africa or, An Exploration of Harar: Volume II*. London: Tylston and Edwards, 1894.

Burzynski, Michael, Henry Mann and Anne Marceau. *Exploring the Limestone Barrens of Newfoundland and Labrador*. Rocky Harbour: Gros Morne Cooperating Association, 2016.

Buse, Peter. 'The Dog and the Parakeet: Lacan among the Animals', *Angelaki* 22(4) (2017): 133–45.

Busk, George. 'Description of Two Beothuc Skulls', *Journal of the Anthropological Institute of Great Britain and Northern Ireland* 5 (1876): 230–33.

Butler, Judith. *Precarious Life: The Powers of Mourning and Violence*. London: Verso, 2004.

———. 'Disturbance and Dispersal in the Visual Field', in Catherine de Zegher and Griselda Pollock (eds), *Art as Compassion: Bracha L. Ettinger* (Brussels: ASA, 2011), 149–65.

Butterworth, Hezekiah. 'Introduction', in Marshall Saunders, *Beautiful Joe* (Toronto: McClelland and Stuart, 1922), 7–9.

Buxton, Richard. *Myths and Tragedies in their Ancient Greek Contexts*. Oxford: Oxford University Press, 2013.

Byatt, Lucy, Bryndis Snæbjörnsdóttir and Mark Wilson. 'Introduction', in Bryndis Snæbjörnsdóttir and Mark Wilson (eds), *Nanoq: Flat Out and Bluesome: A Cultural Life of Polar Bears* (London: Black Dog, 2006), 13–18.

Cadigan, Sean Thomas. *Newfoundland and Labrador: A History*. Toronto: University of Toronto Press, 2009.

Calarco, Matthew. 'Thinking through Animals: Reflections on the Ethical and Political Stakes of the Question of the Animal in Derrida', *Oxford Literary Review* 29(1–2) (2007): 1–15.

Calhoun, John B. 'Population Density and Human Pathology', *Scientific American* 206(2) (1962): 139–49.

Calhoun, John M. 'The Preservation of Motion-Picture Film', *American Archivist* 30(3) (1967): 517–25.

Calixto, Sergio, Nina Ganzherli, Sergey Gulyaev and Susana Figueroa-Gerstenmaier. 'Gelatin as a Photosensitive Material', *Molecules* 23 (2018). https://doi.org/10.3390/molecules23082064.
Cambefort, Yves. 'Zoological Nomenclature and Speech Act Theory', in Karine Chemla and Jacques Virbel (eds), *Texts, Textual Acts and the History of Science* (Cham: Springer, 2015), 143–81.
Campos, Paula F., Oliver E. Craig, Gordon Turner-Walker, Elizabeth Peacock, Eske Willerslev, M. Thomas and P. Gilbert. 'DNA in Ancient Bone – Where is It Located and How Should We Extract It?', *Annals of Anatomy* 194 (2012): 7–16.
Candlin, Fiona. *Micromuseology: An Analysis of Small Independent Museums*. London: Bloomsbury, 2015.
Caputi, Jane. '*Jaws* as Patriarchal – and Ecocidal – Myth', in Kathy Merlock Jackson and Phillip Simpson (eds), *'This Shark Swallow You Whole': Essays on the Cultural Influence of* Jaws (Jefferson: McFarland, 2023), 9–17.
Carey, Annie. *Autobiographies of a Lump of Coal &c*. London: Cassell, Petter and Galpin, 1870.
Carlson, Colin J., Alexander L. Bond and Kevin R. Burgio. 'Estimating the Extinction Date of the Thylacine with Mixed Certainty Data', *Conservation Biology* 32(2) (2018): 477–83.
Carr, Steven M. 'Evidence for the Persistence of Ancient Beothuk and Maritime Archaic Mitochondrial DNA Genome Lineages among Modern Native American Peoples', *Genome* 63 (2020): 349–55.
Cawte, Mary. 'Craniometry and Eugenics in Australia: R. J. A. Berry and the Quest for Social Efficiency', *Australian Historical Studies* 22(86) (1986): 35–53.
Ceballos, Gerardo, Paul R. Ehrlich and Peter Raven. 'Vertebrates on the Brink as Indicators of Biological Annihilation and the Sixth Mass Extinction', *PNAS* 117(24) (2020): 13596–13602.
Chapman, Andrew. *Woolsheds: A Visual Journey of the Australian Woolshed*. Scoresby: Five Mile Press, 2011.
Chare, Nicholas. *Auschwitz and Afterimages: Abjection, Witnessing and Representation*. London: IB Tauris, 2011.
———. *After Francis Bacon: Synaesthesia and Sex in Paint*. Farnham: Ashgate, 2012.
———. 'Material Witness: Conservation Ethics and the Scrolls of Auschwitz', *symplokē* 24(1–2) (2016): 81–97.
———. 'Vision and Indifference: Writing Perceptions of the Petroglyphs of Murujuga', *Images, Imagini, Images: Journal of Visual and Cultural Studies* 7 (2017): 97–123.
———. 'After the Thylacine: In Pursuit of Cinematic and Literary Improvised Encounters with the Extinct', *Liminalities* 14(1) (2018): 124–68.
———. 'In Her Hands: Affect, Encounter and Gestures of Wit(h)nessing in Shanawdithit's Drawings', *parallax* 26(3) (2020): 286–302.
———. 'Concrete Loss: Attesting to Trauma in Teresa Margolles's *Karla, Hilario Reyes Gallegos*', in Arleen Ionescu and Maria Margaroni (eds), *Arts of Healing: Cultural Narratives of Trauma* (London: Rowman & Littlefield, 2020), 211–32.
———. 'The Beothuk, the Great Auk and the Newfoundland Wolf: Animal and Human Genocide in Canada's Easternmost Province', in Valérie Bienvenue and Nicholas Chare (eds), *Animals, Plants and Afterimages: The Art and Science of Representing Extinction* (New York: Berghahn, 2022), 113–36.
———. 'We Murder to Attest: Conservation Medicine, Environmental Disaster and Bearing Witness to the Effects of the Deepwater Horizon Oil Spill on Eastern Oysters (*Crassostrea virginica*)', in Nicolas Barnett, Nicholas Chare and Dominic Williams

(eds), *The Clinical Witness: Conflict, Catastrophe and Medical Testimony* (New York: Routledge, 2025), pp. 179–91.
Chare, Nicholas, and Mitchell Frank. 'Why History and Art History?', in Nicholas Chare and Mitchell Frank (eds), *History and Art History: Looking Past Disciplines* (New York: Routledge, 2021), 1–22.
Chare, Nicholas, and Dominic Williams. *Matters of Testimony: Interpreting the Scrolls of Auschwitz*. New York: Berghahn, 2016.
———. *The Auschwitz Sonderkommando: Testimonies, Histories, Representations*. Cham: Palgrave Macmillan, 2019.
———. 'Introduction: Testimonies of Resistance', in Nicholas Chare and Dominic Williams (eds), *Testimonies of Resistance: Representations of the Auschwitz-Birkenau Sonderkommando* (New York: Berghahn, 2019), 1–29.
Chare, Nicholas, and Ika Willis. 'Introduction: Trans-: Across/Beyond', *parallax* 22(3) (2016): 267–89.
Chau, Meng Huat, and George M. Jacobs. 'Applied Linguistics, Language Guidelines, and Inclusive Practices: The Case of the Use of *Who* with Nonhuman Animals', *International Journal of Applied Linguistics* 31 (2021): 301–3.
Chaumier, Serge. 'Writing and the Exhibition: The Structures of Museum Revolutions', trans. Nicholas Chare, *RACAR* 45(1) (2020): 23–36.
Cheeke, Stephen. 'Browning, Renaissance Painting, and the Problem of Raphael', *Victorian Poetry* 49(4) (2011): 437–62.
Cheetham, April. 'A Veiling of Identity: Anamorphosis as Double Vision in Contemporary Art Practice', unpublished PhD thesis. Liverpool John Moores University, May 2012.
Cheetham, Mark. 'Analogous Histories? Textual/Visual Constructions of Past and Present', in Nicholas Chare and Mitchell Frank (eds), *History and Art History: Looking Past Disciplines* (New York: Routledge, 2021), 173–86.
Cherchi Usai, Paolo. *Silent Cinema: An Introduction*. London: BFI, 2000.
Chick, Neil Kay. 'Tasmanian Family and Community Reconstitution, with a Case Study of Some Estates and Families of Bothwell, Hamilton and Ouse', unpublished PhD thesis. University of Tasmania, 2006.
Chicone, Sarah J., and Richard A. Kissel. *Dinosaurs and Dioramas: Creating Natural History Exhibitions*. Walnut Creek, CA: Left Coast Press, 2014.
Childress, Andrew Marshall. 'Visualizing the Other: A Phenomenological Analysis of the Objectification of the Body in Biomedical Research Using Human Subjects', unpublished PhD thesis. University of Texas, 2012.
Christie, Michael. 'Computer Databases and Aboriginal Knowledge', *Learning Communities: International Journal of Learning in Social Contexts* 1 (2004): 4–12.
Chrulew, Matthew. 'Managing Love and Death at the Zoo: The Biopolitics of Endangered Species Preservation', *Australian Humanities Review* 50 (2011): 137–57.
Clark, Andrew Hill. *The Invasion of New Zealand by People, Plants and Animals*. Westport, CT: Greenwood Press, 1970 [1949].
Clark, Edward B. 'Real Naturalists on Nature Faking', *Everybody's Magazine* 17 (1907): 423–27.
Clausen, P. H., I. Adeyemi, B. Bauer, M. Breloeer, F. Salchow and C. Staak. 'Host Preferences of Tsetse (Diptera: Glossinidae) Based on Bloodmeal Identifications', *Medical and Veterinary Entomology* 12 (1998): 169–80.
Cline, Rebecca W., and Sarah J. Johnson. 'Mosquitoes, Doorknobs and Sneezing', *Health Communication* 4(4) (1992): 273–89.
Clubb, Ros, and Sophie Vickery. 'Locomotory Stereotypies in Carnivores: Does Pacing Stem from Hunting, Ranging or Frustrated Escape?', in Georgia Mason and Jeff

Rushen (eds), *Stereotypic Animal Behaviour: Fundamentals and Application to Welfare*, 2nd edn (Wallingford: CABI, 2006), 58–85.

Cockerill, R. 'Maydena 21/5/67', *Southern Caver* 68 (2014): 5–6.

Colenso, William. 'Notes, Chiefly Historical, on the Ancient Dog of the New Zealanders', *Transactions and Proceedings of the Royal Society of New Zealand* 10 (1877): 135–55.

Colgan, Don. 'Thylacine Cloning is a Long Shot', *Australasian Science* 23(6) (2002): 15.

Colley, Ann C. *Wild Animal Skins in Victorian Britain: Zoos, Collections, Portraits and Maps*. London: Routledge, 2016 [2014].

Collier, Ashley, and Steven M. Carr. 'On the Persistence and Detectability of Ancient Beothuk Mitochondrial DNA Genomes in Living First Nations Peoples', *Mitochondrial DNA Part A* 30(1) (2019): 68–74.

Combs, Blanton P. 'Great Blue Heron Forages for Fish by Bill-Vibrating', *Western North American Naturalist* 81(2) (2021): 280–84.

Connan, Maëlle, Christopher W. Jones, Michelle M. Risi et al. 'First Evidence of Mouse Predation Killing Adult Great Albatrosses', *Biological Invasions* 26 (2024): 25–31.

Connell, R. W. *Masculinities*, 2nd edn. Cambridge: Polity, 2005.

Cook, W. R. 'Pathophysiology of Bit Control in the Horse', *Journal of Equine Veterinary Science* 19(3) (1999): 196–204.

———. 'Bit-Induced Pain: A Cause of Fear, Flight, Fight and Facial Neuralgia in the Horse', *Pferdeheilkunde* 19 (2003): 1–8.

Cooper, Christine, and Philip Withers. 'Patterns of Body Temperature Variation and Torpor in the Numbat, *Myrmecobius fasciatus* (Marsupialia: Myrmecobiidae)', *Journal of Thermal Biology* 29 (2004): 277–84.

Cooper, Mark G. '"I Saw What You Said": Non-Verbal Communication and the EAP', *Employee Assistance Quarterly* 5(4) (1990): 1–12.

Corbett, Laurance K. 'Social Dynamics of a Captive Dingo Pack: Population Regulation by Dominant Female Infanticide', *Ethology* 78 (1988): 177–98.

Corbin, Benjamin. *Corbin's Advice or The Wolf Hunter's Guide*. Bismarck: Tribune, 1900.

Corkovic, Laura Miroslava. 'La Cultura Indígena en la forografía Mexicana de los 90s', unpublished PhD thesis. Universidad de Salamanca, 2012.

Cottrell, Sariah, Jamie L. Jensen and Steven L. Peck. 'Resuscitation and Resurrection: The Ethics of Cloning Cheetahs, Mammoths, and Neanderthals', *Life Sciences, Society and Policy* 10(3) (2014): 1–17.

Courtney, Bill. 'Cannibal Dogs', *Walkabout* 34(9) (1968): 6.

Coutts, Peter, and Mark Jurisich. 'Canine Passengers in Maori Canoes', *World Archaeology* 5(1) (1973): 72–85.

Cox, G. W., and E. V. R. Ratcliff (eds), *Tasmania Remembered: Early Photographs of H.J. King*. Launceston: Mary Fisher, 1974.

Craenhals, François. *(Les) 4 as et le Loup de Tasmanie*. Tournai: Casterman, 2003.

Crane, Kylie. 'Tracking the Tassie Tiger: Extinction and Ethics in Julia Leigh's *The Hunter*', in Laurenz Volkmann, Nancy Grimm, Ines Detmers and Katrin Thomson (eds), *Local Natures, Global Responsibilities* (Leiden: Brill, 2010), 105–19.

Crane, Ralph, and Danielle Wood (eds). *Island Story: Tasmania in Object and Text*. Melbourne: Text, 2018.

Crofts, Matthew, and Janine Hatter. '"Rats is Bogies I Tell You, and Bogies is Rats": Rats, Repression and the Gothic Mode', in Ruth Heholt and Melissa Edmundson (eds), *Gothic Animals: Uncanny Otherness and the Animal With-Out* (Cham: Palgrave Macmillan, 2020), 127–40.

Croll, Donald A., Kelly M. Newton, Matthew McKown, Nick Holmes, Jeffrey C. Williams et al. 'Passive Recovery of an Island Bird Community after Rodent Eradication', *Biological Invasions* 18 (2016): 703–15.

Crossland, Zoë. 'Of Clues and Signs: The Dead Body and Its Evidential Traces', *American Anthropologist* 111(1) (2009): 69–80.

Crossland, Zoë, and Rosemary A. Joyce. 'Anthropological Perspectives on Disturbing Bodies: An Introduction', in Zoë Crossland and Rosemary A. Joyce, *Disturbing Bodies: Perspectives on Forensic Anthropology* (Santa Fe, NM: School for Advanced Research Press, 2015), 3–27.

Crow, Matthew S., Tara O'Connor Shelley and Paul B. Stretesky. 'Camouflage-Collar Crime: An Examination of Wildlife Crime and Characteristics of Offenders in Florida', *Deviant Behaviour* 34 (2013): 635–52.

Crowley, Terry, and R. M. W. Dixon. 'Tasmanian', in R. M. W. Dixon and Barry J. Blake (eds), *Handbook of Australian Languages: Volume II* (Amsterdam; John Benjamins, 1981), 419–45.

Crowther, Arthur Bingham. 'On Some Points of Interest Connected with the Platypus', *Papers and Proceedings and Report of the Royal Society of Tasmania for 1879* (1880): 96–99.

Crowther, William. 'Appendix', in Vivienne Rae Ellis, *Trucanini* (Canberra: Australian Institute for Aboriginal Studies, 1981), 173.

Cunningham, Sophie. *City of Trees: Essays on Life, Death & the Need for a Forest*. Melbourne: Text, 2019.

Cuttle, Christopher. 'Damage to Museum Objects Due to Light Exposure', *Lighting Research and Technology* 28(1) (1996): 1–9.

Cvetkovich, Ann. *An Archive of Feelings: Trauma, Sexuality and Lesbian Public Cultures*. Durham, NC: Duke University Press, 2003.

Dallas, Kenneth McKenzie. 'Conclusion', in Archibald Lawrence Meston, *The Van Diemen's Land Company: 1825–1842* (Launceston: Launceston City Council, 1958), 58–59.

Daniel, David. *Thalu Sites of the West Pilbara*. Perth: Department of Aboriginal Sites, 1990.

Daniels, Jaret C., Mitch Magdich and Peter Tolson. 'Butterfly Recovery Planning: Determining How to Contribute', in Jaret C. Daniels (ed.), *Butterfly Conservation in North America: Efforts to Help Save Our Charismatic Microfauna* (Dordrecht: Springer, 2015), 1–22.

Darwin, Charles. *Journal of Researches into the Natural History and Geology of the Countries Visited during the Voyage of HMS Beagle Round the World*. New York: D. Appleton & Co., 1896 [1839].

———. *The Variation of Animals and Plants under Domestication: Volume 1*. London: John Murray, 1868.

———. *The Descent of Man: Volume 1*. New York: American Dome Library Company, 1902 [1871].

Davies, Bronwyn. 'Difference and Differentiation', in Bronwyn Davies and Susanne Gannon (eds), *Pedagogical Encounters* (New York: Peter Lang, 2009), 17–30.

Davis, Karen. 'Chicken-Human Relationships: From Procrustean Genocide to Empathic Anthropomorphism', *Spring: The Journal of Archetype and Culture* 83 (2010): 255–80.

Davis, Natalie Zemon. *Slaves on Screen: Film and Historical Vision*. Cambridge, MA: Harvard University Press, 2000.

———. *A Passion for History: Conversations with Denis Crouzet*. Kirksville, MO: Truman State University Press, 2010.

———. '"The People of the Past Come First": Natalie Zemon Davis in Conversation with Nicholas Chare', in Nicholas Chare and Mitchell B. Frank (eds), *History and Art History* (London: Routledge, 2020), 163–72.

Davis, Peter. 'The Wolf Pack', in Ian Convery, Peter Davis, Karen Lloyd, Owen T. Nevin and Erwin van Maanen (eds), *The Wolf: Culture, Nature, Heritage* (Martlesham: Boydell & Brewer, 2023), 119–29.

Davison, Frank Dalby. *Man-Shy*. Sydney: Angus & Roberston, 1944 [1931].

Dawes, Peter R., Magnus Elander and Mats Ericson. 'The Wolf (*Canis lupus*) in Greenland: A Historical Review and Present Status', *Arctic* 39(2) (1986): 119–32.

Dawson, Mary R., Laurent Marivaux, Chuan-kui Li, K. Christopher Beard and Grégoire Métais. '*Leonastes* and the "Lazarus Effect" in Recent Mammals', *Science* 311 (2006): 1456–58.

DeKoven, Marianne. 'Guest Column: Why Animals Now?', *PMLA* 124(2) (2009): 361–69.

De la Mare, William K. 'On the Definition of Threats to the Survival of Species', in Richard Fitter and Maisie Fitter (eds), *The Road to Extinction: Problems of Categorizing the Status of Taxa Threatened with Extinction* (Gland: IUCN, 1987), 113–21.

Deleuze, Gilles. *Cinema 1*, trans. Hugh Tomlinson and Barbara Habberjam. London: Continuum, 2005.

Derrida, Jacques. 'Feu la cendre', *Anima* 5 (1982): 47–99.

———. *The Animal That Therefore I Am*, trans David Wills. New York: Fordham University Press, 2002.

———. *L'animal que donc je suis*. Paris: Galilée, 2006.

Derrida, Jacques, and Élisabeth Roudinesco. *De quoi demain… Dialogue*. Paris: Flammarion, 2001.

———. *For What Tomorrow . . . A Dialogue*, trans. Jeff Fort. Stanford, CA: Stanford University Press, 2004.

Devenish, David C. 'Labelling in Museum Display: A Survey and Practical Guide', *Museum Management and Curatorship* 9 (1990): 63–72.

Dewar, Mickey. 'In Search of the Never-Never', in Ann McGrath (ed.), *In Search of the Never-Never: Mickey Dewar, Champion of History across the Genres* (Acton: Australian National University Press, 2019), 45–266.

Dickman, Chris R., Thomas M. Newsome and Lily M. van Eeden. 'The Dingo Dilemma: A Brief History of Debate', *Australian Zoologist* 41(3) (2021): 298–321.

Digioia, Amanda. 'A Cry in the Dark: The Howls of Wolves in Horror and Heavy Metal Music', *Horror Studies* 7(2) (2016): 293–306.

Dixon, D. S. 'Keyhole Lesions in Gunshot Wounds of the Skull and Direction of Fire', *Journal of Forensic Sciences* 27(3) (1982): 555–66.

Dixon, Joan M. 'Thylacinidae', in D. W. Walton and B. J. Richardson (eds), *Fauna of Australia* (Canberra: Australian Publishing Service, 1989), 1–20.

Dodd, Adam, Karen A. Rader and Liv Emma Thorsen. 'Introduction: Making Animals Visible', in Adam Dodd, Karen A. Rader and Liv Emma Thorsen (eds), *Animals on Display: The Creaturely in Museums, Zoos, and Natural History* (Pennsylvania: University of Pennsylvania Press, 2013), 1–11.

Dominy, Michèle D. *Calling the Station Home: Place and Identity in New Zealand's High Country*. Lanham, MD: Rowman & Littlefield, 2001.

Donà, Giorgia. '"Situated Bystandership" during and after the Rwandan Genocide', *Journal of Genocide Research* 20(1) (2018): 1–19.

Donovan, Josephine. *The Aesthetics of Care: On the Literary Treatment of Animals*. New York: Bloomsbury, 2016.

Dorr, Donal. 'The Scapegoat', *The Furrow* 61(2) (2010): 92–97.

Drake, Samuel Adams. *The Border Wars of New England*. New York: Charles Scribner's Sons, 1897.

Drollette, Daniel. 'On the Tail of a Tiger', *Scientific American* 275(4) (1996): 32–34.

Dubuc, Élise. 'Museum and University Mutations: The Relationship between Museum Practices and Museum Studies in the Era of Interdisciplinarity, Professionalisation,

Globalisation and New Technologies', *Museum Management and Curatorship* 26(5) (2011): 497–508.

Dufresnes, Louis. 'Taxidermie', in *Nouveau dictionnaire d'histoire naturelle: Volume 21: Sol-Teh* (Paris: Imprimerie Crapelet, 1803), 507–65.

Dunlap, Thomas R. 'The Realistic Animal Story: Ernest Thompson Seton, Charles Roberts and Darwinism', in Ralph H. Lutts (ed.), *The Wild Animal Story* (Philadelphia: Temple University Press, 1998), 237–47.

Durkheim, Émile. *Les formes élémentaire de la vie religieuse: Le système totémique en Australie*. Paris: Librairie Félix Alcan, 1912.

———. *The Elementary Forms of Religious Life*, trans. Joseph Ward Swain. London: George Allen & Unwin, 1915.

Eberle, Thomas S. 'The Art of Making Photos: Some Phenomenological Reflections', in Michael Barber and Jochen Dreher (eds), *The Interrelation of Phenomenology, Social Sciences and the Arts* (Cham: Springer, 2014), 311–20.

Eckert, Allan W. *The Great Auk: A Novel*. Boston: Little, Brown, 1963.

Edwards, Austin Burton. 'Differentiation of the Dolerites of Tasmania 1', *The Journal of Geology* 50(5) (1942): 451–80.

Edwards, Elizabeth. 'Book Review: *The Colour of Time: A New History of the World 1850–1960*', *History of Photography* 43(3) (2019): 331–32.

Elbroch, Mark. *Animal Skulls: A Guide to North American Species*. Mechanicsburg, PA: Stackpole, 2006.

Ellis, Vivienne Rae. *Trucanini: Queen or Traitor?* Canberra: Australian Institute of Aboriginal Studies, 1981.

Ellos, William J. 'Ethical Similarities in Human and Animal Social Structures', *Between the Species* 5(4) (1989): 211–18.

Escobedo, Libby Karlinger. 'Holbein's *Mementi Mori*', in Thea Tomaini (ed.), *Dealing with the Dead: Mortality and Community in Medieval and Early Modern Europe* (Leiden: Brill, 2018), 366–92.

Ettinger, Bracha L. 'Matrixial Trans-Subjectivity', *Theory, Culture & Society* 23(2–3) (2006): 218–22.

———. 'Transgressing with-in-to the Feminine', in Bracha L. Ettinger, *Matrixial Subjectivity, Aesthetics, Ethics: Volume 1, 1990–2000*, ed. Griselda Pollock (London: Palgrave Macmillan, 2020), 347–74.

Fang, Janet. 'A World without Mosquitos', *Nature* 466(22) (2010): 432–34.

Farnham, Albert B. *Home Manufacture of Furs and Skins*. Columbus: A. R. Harding, 1916.

———. *Home Taxidermy for Pleasure and Profit*. Columbus: A. R. Harding, 1944.

Farrington, Oliver Cummings. 'The Rise of Natural History Museums', *Science* 42(1076) (1915): 197–208.

Ferencz-Flatz, Christian. 'Humanizing the Animal, Animalizing the Human: Husserl on Pets', *Human Studies* 40 (2017): 217–32.

Ferrett, D. *Dark Sound: Feminine Voices in Sonic Shadow*. New York: Bloomsbury, 2020.

Ferris, Ina. 'The Historical Novel and the Problem of Beginning: The Model of Scott', *Journal of Narrative Technique* 18(1) (1988): 73–82.

Fiennes, Richard. *The Order of Wolves*. London: Hamish Hamilton, 1976.

Figueirido, Borja, and Christine M. Janis. 'The Predatory Behaviour of the Thylacine', *Biology Letters* 7(6) (2011): 937–40.

Fleay, David. 'On the Trail of the Marsupial Wolf', *Victorian Naturalist* 63 (1946): 129–77.

Fletcher, Amy Lynn. 'Bring 'Em Back Alive: Taming the Tasmanian Tiger Cloning Project', *Technology in Society* 30 (2008): 194–201.

———. 'Genuine Fakes: Cloning Extinct Species as Science and Spectacle', *Politics and the Life Sciences* 29(1) (2010): 48–60.

———. *Mendel's Ark: Biotechnology and the Future of Extinction*. Dordrecht: Springer, 2014.

Fletcher, Michael-Shawn, Tegan Hall and Andreas Nicholas Alexandra. 'The Loss of an Indigenous Constructed Landscape Following British Invasion of Australia: An Insight into the Deep Human Imprint on the Australian Landscape', *Ambio* 50(1) (2021): 138–49.

Flint, Christopher. 'Speaking Objects: The Circulation of Stories in Eighteenth-Century Prose Fiction', *PMLA* 113(2) (1998): 212–26.

Flynn, Theodore Thomson. 'Notes on Certain Reproductive Phenomena in Some Tasmanian Marsupials', *Annals and Magazine of Natural History* 10(56) (1922): 225–31.

Foran, Max. *The Subjugation of Canadian Wildlife: Failures of Principle and Policy*. Montreal: McGill-Queen's University Press, 2018.

Forster, Edward Morgan. *Aspects of the Novel*. New York: Harvest, 1954 [1927].

Foster, David R., Glenn Motzkin, Debra Bernardos and James Cardoza. 'Wildlife Dynamics in the Changing New England Landscape', *Journal of Biogeography* 29 (2002): 1337–57.

Foucault, Michel. *Discipline and Punish: The Birth of the Prison*, trans. Alan Sheridan. London: Penguin, 1991 [1975].

Frankel, O. H., and Michael E. Soulé. *Conservation and Evolution*. Cambridge: Cambridge University Press, 1981.

Freccero, Carla. 'A Race of Wolves', *Yale French Studies* 127 (2015): 110–23.

Freeman, Carol. 'Figuring Extinction: Visualizing the Thylacine in Zoological and Natural History Works, 1808–1936', PhD thesis. University of Tasmania, Hobart, 2005.

———. 'The Last Image: Julia Leigh's *The Hunter* as Film', in Jay Johnston and Fiona Probyn-Rapsey (eds), *Animal Death* (Sydney: Sydney University Press, 2013), 189–204.

———. *Paper Tiger: How Pictures Shaped the Thylacine*. Hobart: Forty South, 2014.

Freeman, Carrie Packwood, Marc Bekoff and Sarah M. Bexell. 'Giving Voice to the "Voiceless": Incorporating Nonhuman Animal Perspectives as Journalistic Sources', *Journalism Studies* 12(5) (2011): 590–607.

Frémion, Yves. *Reiser*. Paris: Albin Michel, 1974.

Friend, J. Anthony, and Neil D. Thomas. 'Conservation of the Numbat (*Myrmecobius fasciatus*)', in Menna Jones, Chris Dickman and Mike Archer (eds), *Predators with Pouches: The Biology of Carnivorous Marsupials* (Collingwood: CSIRO, 2003), 452–63.

Fudge, Erica. 'A Left-Handed Blow: Writing the History of Animals', in Nigel Rothfels (ed.), *Representing Animals* (Bloomington: Indiana University Press, 2002), 3–18.

Galbreath, Ross, and Derek Brown. 'The Tale of the Lighthouse Keeper's Cat: Discovery and Extinction of the Stephens Island Wren (*Traversia lyalli*)', *Notornis* 51 (2004): 193–200.

Gammon, Ben. 'Visitors' Use of Computer Exhibits: Findings from Five Gruelling Years of Watching Visitors Getting It Wrong', in Ross Parry (ed.), *Museums in a Digital Age* (Abingdon: Routledge, 2010), 281–90.

Garside, Daniel, Katherine Curran, Capucine Korenberg, Lindsay MacDonald, Kees Teunissen and Stuart Robson. 'How is Museum Lighting Selected? An Insight into Current Practice in UK Museums', *Journal of the Institute of Conservation* 40(1) (2017): 3–14.

Gatschet, Albert. 'The Beothuk Indians', *Proceedings of the American Philosophical Society* 23(123) (1886): 411–32.

Gavitt, John D. 'Disturbance Effect of Free-Running Dogs on Deer Reproduction', unpublished MA thesis. Virginia Polytechnic Institute and State University, August 1973.

Geary, Patrick J. *Furta Sacra: Thefts of Relics in the Central Middle Ages*. Princeton, NJ: Princeton University Press, 1978.

Geiser, Fritz. 'Hibernation and Daily Torpor in Marsupials: A Review', *Australian Journal of Zoology* 42 (1994): 1–16.

Gell, Alfred. *Art and Agency: An Anthropological Theory*. Oxford: Oxford University Press, 1998.

Gibson, Jason. '"You're My Kwertengerl": Transforming Models of Care for Central Australian Aboriginal Museum Collections', *Museum Management and Curatorship* 34(3) (2019): 240–56.

Gibson, Mary Ellis. 'One More Word on Browning's "One Word More"', *Studies in Browning and His Circle* 12 (1984): 76–86.

Gillman, Len Norman, and Shane Donald Wright. 'Restoring Indigenous Names in Taxonomy', *Communications Biology* 3 (2020): 609. https://doi.org/10.1038/s42003-020-01344-y.

Gipson, Philip S., Warren B. Ballard, Ronald M. Nowak and L. David Mech. 'Accuracy and Precision of Estimating Age of Gray Wolves by Tooth Wear', *Journal of Wildlife Management* 64(3) (2000): 752–58.

Girard, René. *Violence and the Sacred*, trans. Patrick Gregory. London: Athlone, 1988.

Girgen, Jen. 'The Historical and Contemporary Persecution of Animals', *Animal Law* 9 (2003): 97–134.

Gittleman, John, and Matthew Gompper. 'Plight of Predators: The Importance of Carnivores for Understanding Patterns of Biodiversity and Extinction Risk', in Pedro Barbosa and Ignacio Castellanos (eds), *Ecology of Predator-Prey Interactions* (Oxford: Oxford University Press, 2005), 370–88.

Glaskin, Katie. 'Extinction, Inscription and the Dreaming: Exploring a Thylacine Connection', *Anthropological Forum* (2021). https://doi.org/10.1080/00664677.2021.1937513.

Glen, Alistair S., and Jeff Short. 'The Control of Dingoes in New South Wales in the Period 1883–1930 and Its Likely Impact on Their Distribution and Abundance', *Australian Zoologist* 31(3) (2000): 432–42.

Gmelch, Sharon Bohn, and Morgan Z. Gmelch. 'Lost Then Loved: The Case of the Tasmanian Tiger', *Natural History* 4 (2017): 36–41.

Goldie, Terry. *Fear and Temptation: The Image of the Indigene in Canadian, Australian, and New Zealand Literatures*. Kingston: McGill-Queen's University Press, 1989.

Goldsmith, Oliver. *A History of the Earth and Animated Nature: Volume 1*. Philadelphia: J. B. Lippincott, 1865.

Goodall, Jane. *Through a Window: My Thirty Years with the Chimpanzees of Gombe*. Boston: Mariner, 2010.

Gordon, Beverly. '"Little Trifles": Understanding Victorian Fancywork', in John Vollmer (ed.), *Textiles as Primary Sources: Proceedings of the First Symposium of the Textile Society of America* (St. Paul, MN: Textile Society of America, 1988), 139–43.

Goss, Adeline, Vidya Viswanathan and Horace DeLisser. 'Not Just a Specimen: A Qualitative Study of Emotion, Morality and Professionalism in One Medical School Gross Anatomy Laboratory', *Anatomical Sciences Education* 12 (2019): 349–59.

Gottier, Richard F. 'The Dominance-Submission Hierarchy in the Social Behavior of the Domestic Chicken', *Journal of Genetic Psychology* 112 (1968): 205–26.

Gould, John. *The Mammals of Australia: Volume I*. London: Taylor & Francis, 1863.

Goyes, David Rodriquez. 'Denying the Harms of Animal Abductions for Biomedical Research', in Ragnhild Aslaug Sollund (ed.), *Green Harms and Crimes: Critical Criminology in a Changing World* (Basingstoke: Palgrave Macmillan, 2015), 170–88.

Graham, Kenneth. 'Wool's Disappearing History', *The Bulletin* 97(4950) (1975): 61.

Grainge, Paul. 'Reclaiming Heritage: Colourization, Culture Wars and the Politics of Nostalgia', *Cultural Studies* 13(4) (1999): 621–38.
Granger, Mary Ethel. *The Merrie Heart: A Collection of Favourite Nursery Rhymes*. London: Cassell, Petter, and Galpin, 1871.
Grau, Günter. *Hidden Holocaust?* New York: Routledge, 2012 [1995].
Greaves, Geoff. *The Circus Comes to Town: Nostalgia of Australian Big Tops*. Sydney: Reed, 1980.
Griggs, Peter. 'Black Poison or Beneficial Beverage? Tea Consumption in Colonial Australia', *Journal of Australian Colonial History* 17 (2015): 23–44.
Grinnell, Joseph. 'The Methods and Uses of a Research Museum', *Popular Science Monthly* 77 (1910): 163–69.
Gripsrud, Birgitta Haga, Ellen Ramvi, Lynn Froggett, Ingvil Hellstrand and Julian Manley. 'Psychosocial and Symbolic Dimensions of the Breast Explored through a Visual Matrix', *NORA: Nordic Journal of Feminist and Gender Research* 26(3) (2018): 210–29.
Grotenhuis, René. *Nation-Building as Necessary Effort in Fragile States*. Amsterdam: Amsterdam University Press, 2016.
Grubb, Jennifer D. *Ecoliberation: Reimagining Resistance and the Green Scare*. Montreal: McGill-Queen's University Press, 2021.
Grube, Kathryn J. 'Detrimental Effects of White Valued Walls in Classrooms', *Educational Settings* 21(2) (2014): 69–82.
Guiler, Eric. *Thylacine: The Tragedy of the Tasmanian Tiger*. Melbourne: Oxford University Press, 1985.
Guiler, Eric, and Philippe Godard. *Tasmanian Tiger: A Lesson to Be Learnt. . .* Perth: Abrolhos, 1998.
Gullbeck, Svein H. 'The Norse Penny Reconsidered: The Goddard Coin – Hoax or Genuine?', *Journal of the North Atlantic* 33 (2017): 1–8.
Gutierrez, A. P., L. Ponti and P. A. Arias. 'Deconstructing the Eradication of New World Screwworm in North America: Retrospective Analysis and Climate Warning Effects', *Medical and Veterinary Entomology* 33 (2019): 282–95.
Haas, Lindsay F. 'Neurological Stamp', *Journal of Neurology, Neurosurgery and Psychiatry* 59(5) (1995): 470.
Halsey, Mark. *Deleuze and Environmental Damage: Violence of the Text*. Aldershot: Ashgate, 2006.
Hancock, Ian. 'Romanies and the Holocaust: A Re-Evaluation and Overview', in Dan Stone (ed.), *The Historiography of the Holocaust* (Basingstoke: Palgrave Macmillan, 2004), 383–96.
Hand, Carol. *Reviving Extinct Species*. New York: Rosen, 2019.
Haque, Omar Sultan, and Adam Waytz. 'Dehumanization in Medicine: Causes, Solutions and Functions', *Perspectives on Psychological Science* 7(2) (2012): 176–86.
Haraway, Donna. *The Companion Species Manifesto: Dogs, People, and Significant Otherness*. Chicago: Prickly Paradigm Press, 2003.
———. *When Species Meet*. Minneapolis: University of Minnesota Press, 2008.
Hardin, Garrett. 'Foreword', in Christopher D. Stone, *Should Trees Have Standing?* (Los Altos: William Kaufmann, 1974), ix–xvii.
Harries, John. 'Of Bleeding Skulls and the Postcolonial Uncanny: Bones and the Presence of Nonosabasut and Demasduit', *Journal of Material Culture* 15(4) (2010): 403–21.
———. 'A Clearing with a View to the Lake, the Bones of a Caribou, and the Sound of Snow Falling on Dead Leaves: Sensing the Presence of the Past in the Wilds of Newfoundland', in Fiona Polack (ed.), *Tracing Ochre: Changing Perspectives on the Beothuk* (Toronto: University of Toronto Press, 2018), 94–114.

Harrington, Fred H., and Cheryl S. Asa. 'Wolf Communication', in David Mech and Luigi Boitani (eds), *Wolves: Behavior, Ecology, and Conservation* (Chicago: University of Chicago Press, 2003), 66–103.

Harris, Chris. 'Junee-Zulu Pot 16/7/72', *Southern Caver* 68 (2014): 50.

Harris, George Prideaux. 'Description of Two New Species of Didelphis from Van Diemen's Land', *Transactions of the Linnean Society of London* 9 (1808): 174–78.

Harris, Samela. 'Hold That Tiger!', *Walkabout* 34(6) (1968): 28–31.

Harris, Steve. 'Florentine Valley 23/7/72', *Southern Caver* 68 (2014): 50–51.

Harty, Kevin J. 'Who's Savage Now?! – The Vikings in North America', in Kevin J. Harty (ed.), *The Vikings on Film: Essays on Depictions of the Nordic Middle Ages* (Jefferson, NC: McFarland, 2011), 106–20.

Harvey, Moses. *Newfoundland in 1900*. New York: South Publishing, 1900.

Harvey-Samuel, Tim, Thomas Ant and Luke Alphey. 'Towards the Genetic Control of Invasive Species', *Biological Invasions* 19 (2017): 1683–1703.

Haswell, William A., and T. Jeffery Parker. *A Text-Book of Zoology: Volume II*. London: Macmillan, 1897.

Hatton, Joseph, and William Harvey. *Newfoundland: The Oldest British Colony*. London: Chapman and Hall, 1883.

Hauser, Kitty. 'Coming Apart at the Seams: Taxidermy and Contemporary Photography', *MAKE* 82 (1998): 8–11.

Hawley, Caitlin R., Chris Beirne, Amelia Maier and John R. Poulsen. 'Conspecific Investigation of a Dead Forest Elephant (*Loxodonta cyclotis*)', *Pachyderm* 59 (2017/18): 97–100.

Haygarth, Nic. 'Theophilus Jones: Tasmania through Anglo-Indian Eyes in the 1880s and 1890s', *Launceston Historical Society Papers and Proceedings* 23 (2011): 51–74.

———. 'The Myth of the Dedicated Thylacine Hunter: Stockman-Hunter Culture and the Decline of the Thylacine (Tasmanian Tiger) in Tasmania during the Late Nineteenth and Early Twentieth Centuries', *Papers and Proceedings of the Tasmanian Historical Research Association* 64(2) (2017): 28–43.

———. 'Mining the Van Diemen's Land Company Holdings 1851–1899: A Case of Bad Luck and Clever Adaptation', *Journal of Australasian Mining History* 16 (2018): 93–110.

———. 'The "Walking Museum": James "Tiger" Harrison, Tasmanian Naturalist and Native Animal Dealer', *Tasmanian Historical Research Association: Papers and Proceedings* 69(1) (2022): 33–54.

Hayward, James S. 'Report of James S. Hayward, Landing Surveyor, St. John's, on a Visit of the Revenue Service, to the Western Portions of Newfoundland, with Various Statistical and Other Information Relative to the Trade and Fisheries of these Localities', in *Journal of the House of Assembly of Newfoundland* (St. John's: James Seaton, 1866), A-624–69.

Hearn, Robert, Ross Balzaretti and Charles Watkins. 'The Wolf in the Landscape: Antonio Cesena and Attitudes to Wolves in Sixteenth-Century Liguria', *Rural History* 26(1) (2015): 1–16.

Hegel, G. W. F. *System of Ethical Life and First Philosophy of Spirit*, trans. H. S. Harris and T. M. Knox. Albany: SUNY Press, 1979.

Heintzman, Kit. 'Material Cultural Studies', in Mieke Roscher, André Krebber and Brett Mizelle (eds), *Handbook of Historical Animal Studies* (Oldenbourg: De Gruyter, 2021), 357–73.

Hendry, Andrew P., Kiyoko M. Gotanda and Erik I. Svensson. 'Human Influences on Evolution, and the Ecological and Societal Consequences', *Philosophical Transactions of the Royal Society B* 372(1712) (2017). http://doi.org/10.1098/rstb.2016.0028.

Herman, David. *Narratology beyond the Human: Storytelling and Animal Life*. Oxford: Oxford University Press, 2018.
Hesketh-Prichard, Hesketh. *Hunting Camps in Wood and Wilderness*. London: William Heinemann, 1910.
Hiatt, Betty. 'Cremation in Aboriginal Australia', *Mankind* 7 (1969): 104–19.
Hickey, Kieran R. 'A Geographical Perspective on the Decline and Extermination of the Irish Wolf *Canis* lupus – An Initial Assessment', *Irish Geography* 33(2) (2000): 185–98.
Hiestand, Nancy Laura. 'A Comparison of Problem-Solving and Spatial Orientation in the Wolf (*Canis lupus*) and the Dog (*Canis familiaris*)', PhD thesis. University of Connecticut, 1989.
Higgins, Polly. *Eradicating Ecocide: Exposing the Corporate and Political Practices Destroying the Planet and Proposing the Laws Needed to Eradicate Ecocide*. London: Shepheard-Walwyn, 2010.
Hill, Joanna F. 'A System Thinking Perspective on the Motivations and Mechanisms that Drive Wildlife Poaching', in Ragnhild Aslaug Sollund (ed.), *Green Harms and Crimes: Critical Criminology in a Changing World* (Basingstoke: Palgrave Macmillan, 2015), 189–219.
Hills, Helen. *The Matter of Miracles: Neapolitan Baroque Architecture and Sanctity*. Manchester: Manchester University Press, 2016.
Hogg, Michael. 'Personality, Individuality, and Social Identity', in Frederick Rhodewalt (ed.), *Personality and Social Behaviour* (New York: Psychology Press, 2010), 177–96.
Holt, Robert D. 'Charismatic Mesofauna: Butterflies as Inspiration and Test for Theory That Integrates Ecology and Evolution', *Israel Journal of Ecology & Evolution* 54 (2008): 1–5.
Holy Bible, King James Version. Cambridge: Cambridge University Press, 1995 [1611].
Hoorn, Jeanette. 'Crocodile Dundee', *Metro Magazine* 185 (2015): 114–23.
Hopkins, John B., III, and Jake M. Ferguson. 'Estimating the Diets of Animals Using Stable Isotopes and a Comprehensive Bayesian Mixing Model', *PLOS ONE* 7(1) (2012). https://doi.org/10.1371/annotation/d222580b-4f36-4403-bb1f-cfd449a5ed74.
Hoquet, Thierry. 'Animal Individuals: A Plea for a Nominalistic Turn in Animal Studies?', *History and Theory* 52(4) (2013): 68–90.
Hornaday, William T. *The Extermination of the American Bison*. Washington, DC: Government Printing Office, 1889.
Hornborg, Anne-Christine. *Mi'kmaq Landscapes: From Animism to Sacred Ecology*. London: Routledge, 2016 [2008].
Hornby, Jim. 'Bear Facts: The History and Folklore of Island Bears, Part 1', *The Island* 22 (1987): 3–9.
———. 'Bear Facts: The History and Folklore of Island Bears, Part 2', *The Island* 23 (1988): 27–31.
Horsley, Reginald. 'Boy Life in Australia', *Boy's Own Paper* 15(739) (1893): 392–95.
Howard, Megan, and Will Hoppitt. 'Ospreys Do Not Teach Offspring How to Kill Prey at the Nest', *Biology Letters* 13 (2017): 20170346.
Howard, Tanya M., Lyndal J. Thompson, Paloma Frumento and Theodore Alter. 'Wild Dog Management in Australia: An Interactional Approach to Case Studies of Community-Led Action', *Human Dimensions of Wildlife* 23(3) (2018): 242–56.
Howarth, Frank. 'The Future of Research in Natural History Museums', in Eric Dorfman (ed.), *The Future of Natural History Museums* (London: Routledge, 2017), 65–81.
Howarth, Robert Guy. 'Short Views', *Southerly* 16(4) (1955): 227–33.
Howley, James Patrick. *Geography of Newfoundland*. London: Edward Stanford, 1876.

———. 'Report of Geological Explorations on the Gambo and Gander Rivers', in *Journal of the House of Assembly of Newfoundland* (St. John's: F. W. Bowden, 1877), 566–92.

———. 'Report of Geological Explorations on the Gambo and Gander Rivers', in Alexander Murray (ed.), *Geological Survey of Newfoundland* (London: Edward Stanford, 1881), 436–62.

———. *The Beothucks or Red Indians: The Aboriginal Inhabitants of Newfoundland.* Cambridge: Cambridge University Press, 1915.

———. *Reminiscences of James P. Howley: Selected Years.* Toronto: Champlain Society, 1997.

Hubbard, Tasha. 'Buffalo Genocide in Nineteenth-Century North America: "Kill. Skin, and Sell"', in Alexander Laban Hinton, Andrew Woolford and Jeff Benvenuto (eds), *Colonial Genocide in Indigenous North America* (Durham: Duke University Press, 2014), 292–305.

Hughes, Anthony. 'Authority, Authenticity and Aura: Walter Benjamin and the Case of Michelangelo', in Anthony Hughes and Erich Ranfft (eds), *Sculpture and Its Reproductions* (London: Reaktion, 1997), 29–45.

Hughes-d'Aeth, Tony. 'Australian Ecology, Deep Ecology and Julia Leigh's *The Hunter*', *JASAL* 1 (2002): 19–31.

Hume, Julian P. 'The History of the Dodo *Raphus cucullatus* and the Penguin of Mauritius', *Historical Biology* 18(2) (2006): 65–89.

Hunt, David. 'The Face of the Wolf is Blessed, or is It? Diverging Perceptions of the Wolf', *Folklore* 119(3) (2008): 319–34.

Huston, Michael. 'Pieces of the Sky', *BioScience* 47(3) (1997): 192–93.

Ince, Onur Ulas. *Colonial Capitalism and the Dilemmas of Liberalism.* Oxford: Oxford University Press, 2018.

Ingold, Tim. 'Humanity and Animality', in Tim Ingold (ed.), *Companion Encyclopedia of Anthropology* (Abingdon: Routledge, 1994), 14–32.

Irvin-Erickson, Douglas. *Raphaël Lemkin and the Concept of Genocide.* Philadelphia: University of Pennsylvania Press, 2017.

Işcan, M. Yaşar. 'Concepts in Teaching Forensic Anthropology', *Medical Anthropology Newsletter* 13(1) (1981): 10–12.

Izatt, Hilary Jan. 'The Precedent of a President: Theodore Roosevelt and Environmental Conservation', *Sigma: Journal of Political and International Studies* 22 (2004): 83–101.

Jackson, Hartley H. T. 'The Care of Museum Specimens of Recent Mammals', *Journal of Mammalogy* 7(2) (1926): 113–18.

James, P. C. 'The Emu Bay and Mount Bischoff Railway', *Papers and Proceedings: Tasmanian Historical Research Society* 10(4) (1963): 90–99.

Janik, V. M., L. S. Sayigh and R. S. Wells. 'Signature Whistle Conveys Identity Information to Bottlenose Dolphins', *PNAS* 103(21) (2006): 8293–97.

Janis, Christine M., and Borja Figueirido. 'Forelimb Anatomy and the Discrimination of the Predatory Behaviour of Carnivorous Mammals', *Journal of Morphology* 275 (2014): 1321–38.

Jansson, Tove. *Moominpappa at Sea*, trans. Kingsley Hart. Harmondsworth: Puffin, 1974 [1966].

———. *Pappan och havet.* Helsinki: Förlaget, 2020 [1965].

Janzen, Daniel H. 'The Future of Tropical Ecology', *Annual Review of Ecology and Systematics* 17 (1986): 305–24.

———. 'Latent Extinction – The Living Dead', in Simon A. Levin (ed.), *Encyclopedia of Biodiversity Volume III*, 2nd edn (San Diego: Academic, 2001), 590–98.

Jeanmard, Jerry. 'Creature Comforts', *Texas Monthly* (November 1989): 110–13.

Jeffries, Michael. *Biodiversity and Conservation*, 2nd edn. London: Routledge, 2006.

Jensen, Howard I. 'Agrichattels: A Glossary and Commentary', *Valuation* 40(1) (1996): 47–52.

Jetson, Tim. 'The High Country', *Papers and Proceedings: Tasmanian Historical Research Society* 34(1) (1987): 15–27.

Johnson, Anna. 'Nomad-Words', in Anthony Bryant and Griselda Pollock (eds), *Digital and Other Virtualities: Renegotiating the Image* (London: IB Tauris, 2010), 217–36.

Johnson, Brian. 'National Species: Ecology, Allegory, and Indigeneity in the Wolf Stories of Roberts, Seton and Mowat', in Janice Fiamengo (ed.), *Other Selves: Animals in the Canadian Literary Imagination* (Ottawa: University of Ottawa Press, 2007), 333–52.

Johnson, Chris. *Australia's Mammal Extinctions: A 50,000 Year History*. Cambridge: Cambridge University Press, 2006.

Johnson, Dianne. *Night Skies of Aboriginal Australia: A Noctuary*. Sydney: University of Sydney, 1998.

Johnson, Samuel. *A Dictionary of the English Language*. London: Longmans, Green & Co., 1876 [1755].

Johnston, Robert M. *Systematic Account of the Geology of Tasmania*. Hobart: J. Walch, 1888.

Joinville, Jean de. *Histoire de Saint Louis*, trans. Natalis de Wailly. Paris: Firmin Didot, 1874 [*c.*1309].

Jolly, Martin. 'Pygmalion in the Archive: The Mythic Desires of Colourisation', *Visual Studies* 40(1) (2025): 41–55.

Jones, Adam. *Genocide: A Comprehensive Introduction*. 3rd Edition. Abingdon & New York: Routledge, 2017.

Jones, Manina. 'Wildlifewriting? Animal Stories and Indigenous Claims in Ernest Seton Thompson's *Wild Animals I Have Known*', *Journal of Canadian Studies* 42(3) (2008): 133–49.

Jones, Menna E., Meri Oakwood, Chris A. Belcher, Keith Morris, Andrew J. Murray, Patricia A. Woolley, Karen B. Firestone, Brent Johnson and Scott Burnett. 'Carnivore Concerns: Problems, Issues and Solutions for Conserving Australasia's Marsupial Carnivores', in Menna Jones, Chris Dickman and Mike Archer (eds), *Predators with Pouches: The Biology of Carnivorous Marsupials* (Collingwood: CSIRO, 2003), 422–34.

Jørgensen, Dolly. 'Erasing the Extinct: The Hunt for Caribbean Monk Seals and Museum Collecting Practices', *História, Ciências, Saùde* 28 (2021): 161–83.

Joy, Jody. 'Reinvigorating Object Biography: Reproducing the Drama of Object Lives', *World Archaeology* 41(4) (2009): 540–56.

Jukes, Joseph Beete. *Excursions In and About Newfoundland During the Years 1839 and 1840: Volume 1*. London: John Murray, 1842.

Kalof, Linda. 'Animal Blood Sport: A Ritual Display of Masculinity and Sexual Virility', *Sociology of Sport Journal* 31 (2014): 438–54.

Kapsalis, Terri. *Public Privates: Performing Gynecology from Both Ends of the Speculum*. Durham, NC: Duke University Press, 1997.

Kearl, Michael C. 'The Proliferation of Skulls in Popular Culture: A Case Study of How the Traditional Symbol of Mortality was Rendered Meaningless', *Mortality* 20(1) (2015): 1–18.

Kennedy, William Robert. *Sport, Travel and Adventure in Newfoundland and the West Indies*. London: William Blackwood, 1885.

Kerezsy, Adam. *Gambusia Control in Spring Wetlands: Final Report, October 2009*. Port Augusta: South Australian Arid Lands Natural Resources Management Board, 2009.

Kershaw, James A. 'The Tasmanian Devil in Victoria', *Victorian Naturalist* 29(5) (1912): 75–76.

Kerwin, Amy M. 'Overcoming the Barriers to the Retirement of Old and New World Monkeys from Research Facilities', *Journal of Applied Animal Welfare Science* 9(4) (2006): 337–47.

Kévorkian, Raymond. *The Armenian Genocide: A Complete History*. London: IB Tauris, 2011.

Kheel, Marti. 'The Killing Game: An Ecofeminist Critique of Hunting', *Journal of the Philosophy of Sport* 23 (1996): 30–44.

Kilham, Lawrence. 'Biology of Young Belted Kingfishers', *American Midland Naturalist* 92(1) (1974): 245–47.

Kilner, Mary Ann. *The Adventures of a Pincushion Designed Chiefly for the Use of Young Ladies: Volumes I & II*. London: John Marshall, 1810 [1783].

Kilroy, Peter. 'Reading the Writing-Machine: Inscription, Loss and the Ethnographic Imagination of A.C. Haddon', unpublished PhD thesis. University of Leeds, 2011.

King, Barbara J. *How Animals Grieve*. Chicago: University of Chicago Press, 2013.

Kite, Marion. 'Collagen Products: Glues, Gelatine, Gut Membrane and Sausage Casings', in Marion Kite and Roy Thomson (eds), *Conservation of Leather and Related Materials* (London: Routledge, 2005), 192–97.

Knight, Jessica. '"A Poisonous Cup"? Afternoon Tea in Australian Society, 1870–1914', unpublished BA dissertation. University of Sydney, 2011.

Knight, John. 'On the Extinction of the Japanese Wolf', *Asian Folklore Studies* 56(1) (1997): 129–59.

Knobel, Johann. 'The Conservation Status of the Wedge-Tailed Eagle in Australian Law and Thoughts on the Value of Early Legal Intervention in the Conservation of a Species', *De Jure* 48(2) (2015): 293–311.

Koçak, Mert. 'Gendering the Holocaust: Gendering Mourning in the Image of Anne Frank', *Kaleidoscope* 7(13) (2016): 542–51.

Koestle-Cate, Jonathan. 'Patrice Moor: Stations of the Cross', *ARTS: The Arts in Religious and Theological Studies* 26(1) (2014): 14–21.

Kohlstedt, Sally Gregory. '"Thoughts in Things": Modernity, History, and North American Museums', *Isis* 96(4) (2005): 586–601.

Koptak, Paul E. *Proverbs: The NIV Application Commentary*. Grand Rapids, MI: Zondervan Academic, 2003.

Kopytoff, Igor. 'The Cultural Biography of Things: Commoditization as Process', in Arjun Appadurai (ed.), *The Social Life of Things: Commodities in Cultural Perspective* (Cambridge: Cambridge University Press, 1984), 64–91.

Kosara, Robert. 'Visualization Criticism – The Missing Link between Information Visualization and Art', in Ebad Banissi (ed.), *Proceedings of the 11th International Conference on Information Visualization (IV 2007)* (Red Hook, NY: Curran Associates, 2009), 631–36.

Kousoulis, Antonis A., Konstantinos S. Mylonas and Konstantinos P. Economopoulos. 'Violent Death and Trauma in Norse Mythology: A Systematic Reading of the *Prose Edda*', *European Surgery* 48 (2016): 304–10.

Kracauer, Siegfried. *Nature of Film: The Redemption of Physical Reality*. London: Dennis Dobson, 1961.

Krajewski, Carey, Amy C. Driskell, Peter R. Baverstock and Michael J. Braun. 'Phylogenetic Relationships of the Thylacine (Mammalia: Thylacinidae) among Dasyuroid Marsupials: Evidence from Cytochrome *b* DNA Sequences', *Proceedings of the Royal Society B: Biological Sciences* 250 (1992): 19–27.

Krebber, André, and Mieke Roscher. 'Introduction: Biographies, Animals and Individuality', in André Krebber and Mieke Roscher (eds), *Animal Biography: Re-Framing Animal Lives* (Cham: Palgrave Macmillan, 2018), 1–15.

Krefft, Gerard. *Notes on the Fauna of Tasmania*. Sydney: F. White, 1868.
———. 'Appendix 6', in W. B. Clarke (ed.), *Australian Museum: Report from Trustees for 1867* (Sydney: Thomas Richards Government Printer, 1868), 9–10.
———. 'Description of a New Species of Thylacine', *Journal of Natural History* 2(10) (1868): 296–97.
———. *The Mammals of Australia*. Sydney: Thomas Richards Government Printer, 1871.
Krien, Anna. *Into the Woods: The Battle for Tasmania's Forests*. Collingwood: Black Inc., 2010.
Landau, Michael. 'The Colorization of Black-and-White Motion Pictures', *Journal of Arts Management and Law* 19(3) (1989): 61–87.
Krog, Chris. 'The Learning Curve: Half-Collared Kingfisher Learns to Fish', *Ornithological Observations* 5 (2014): 114–21.
Kruuk, Hans. 'Surplus Killing by Carnivores', *Journal of Zoology* 166 (1972): 233–44.
Kuch, Melanie, Darren R. Gröcke, Martin C. Nyf, M. T. P. Gilbert, Ban Younghusband, Terry Young, Ingeborg Marshall et al. 'A Preliminary Analysis of the DNA and Diet of the Extinct Beothuk: A Systematic Approach to Ancient Human DNA', *American Journal of Physical Anthropology* 132 (2007): 594–604.
Kuper, Leo. *Genocide: Its Political Use in the Twentieth Century*. New Haven, CT: Yale University Press, 1982.
Lacan, Jacques. *The Four Fundamental Concepts of Psychoanalysis*, trans. Alan Sheridan. London: Penguin, 1979 [1973].
Lambert, Ada M. 'The Structure of an Australian Land Leech', *Proceedings of the Royal Society of Victoria* 10(2) (1898): 211–35.
———. 'Description of Two New Species of Australian Land Leeches, with Notes on their Anatomy', *Proceedings of the Royal Society of Victoria* 11 (1899): 156–63.
Lamoureux, Johanne. 'Le travail de la viande', *Intermédialités* 11 (2008): 13–34.
Landry, Mark William. 'Pokemouche Mi'kmaq and the Colonial Regimes', unpublished MA thesis. Saint Mary's University, Nova Scotia, 2010.
Lang, Berel. 'The Concept of Genocide', in A. Dirk Moses (ed.), *Genocide: Critical Concepts in Historical Studies*, vol. 1 (London: Routledge, 2010), 115–31.
———. *Genocide: The Act as Idea*. Philadelphia: University of Pennsylvania Press, 2017.
Larsson, Mats G. 'The Vinland Sagas and Nova Scotia: A Reappraisal of an Old Theory', *Scandinavian Studies* 64(3) (1992): 305–35.
Laurance, William F., Keith R. McDonald and Richard Speare. 'Epidemic Disease and the Catastrophic Decline of Australian Rainforest Frogs', *Conservation Biology* 10(2) (1996): 406–13.
Lawson, Bryan. 'The Social and Psychological Issues of High-Density City Space', in Edward Ng (ed.), *Designing High-Density Cities: For Social and Environmental Sustainability* (London: Routledge, 2009), 285–92.
Lawson, Tom. *The Last Man: A British Genocide in Tasmania*. London: IB Tauris, 2014.
Lee, Ida. 'Early Days in North Tasmania', in Rosalie Hare, *The Voyage of the Caroline: From England to Van Diemen's Land and Batavia in 1827–28* (London: Longmans, Green & Co., 1927), 81–198.
Leeder, Murray. 'Skeletons Sail an Etheric Ocean: Approaching the Ghost in John Carpenter's *The Fog*', *Journal of Popular Film and Television* 37(2) (2009): 70–79.
Leggio, Angeletta. 'A History of Australia's Kodak Manufacturing Plant', AICCM Symposium, Wellington, New Zealand, 2006. Retrieved 24 February 2025 from https://aiccm.org.au/wp-content/uploads/2013/09/AICCM_BP2006_Leggio_p147-158.pdf.
Leigh, Julia. *The Hunter*. London: Faber & Faber, 2000 [1999].
Leland, Charles G. *The Algonquin Legends of New England*. Boston: Houghton, Mifflin and Company, 1884.

Lemkin, Raphaël. *Axis Rule in Occupied Europe: Laws of Occupation, Analysis of Government, Proposals for Redress*. Washington, DC: Carnegie Endowment for International Peace, 1944.

———. *Lemkin on Genocide*, ed. Steven Leonard Jacobs. Lanham, MD: Lexington, 2012.

———. *Totally Unofficial: The Autobiography of Raphael Lemkin*, ed. Donna-Lee Frieze. New Haven, CT: Yale University Press, 2013.

Lemkin, Raphaël, and Tadeusz Kochanowicz. *Kodeks Karny Republik Sowieckich* [Criminal Code of the Republics of the Soviet Union], Warsaw: Wyd. Sem. Prawa kar., 1926.

Lemoine, Albert. *L'habitude et l'instinct*, 2nd edn. Paris: Germer Baillière, 1880.

Leopold, A. Starker., and T. O. Wolfe. 'Food Habits of Nesting Wedge-Tailed Eagles, *Aquila Audex*, in South-Eastern Australia', *CSIRO Wildlife Research* 15(1) (1970): 1–17.

Le Soeuf, Albert Sherbourne, and Harry James Burrell. *The Wild Animals of Australasia: Embracing the Mammals of New Guinea and the Near Pacific Islands*. London: George Harrap & Co., 1926.

Le Souëf, Dudley. 'Extinct Tasmanian Emu', *The Emu* 3(4) (1904): 229–31.

Lett, William P. 'The Wolf', in George O. Shields (ed.), *The Big Game of North America* (Chicago: Rand, McNally & Company, 1890), 453–74.

Levey, Martin. 'The First American Museum of Natural History', *Isis* 42(1) (1951): 10–12.

Lindner, Evelin G. 'Genocide, Humiliation, and Inferiority: An Interdisciplinary Perspective', in Nicholas A. Robins and Adam Jones (eds), *Genocides by the Oppressed: Subaltern Genocide in Theory and Practice* (Bloomington: Indiana University Press, 2009), 138–58.

Lindsay, Teven W., Michael Davies, Graham Alabaster, Hector Altamirano, Ebrima Jatta et al. 'Recommendations for Building out Mosquito-Transmitted Diseases in Sub-Saharan Africa: The DELIVER Mnemonic', *Philosophical Transactions of the Royal Society B* 376 (2021): 20190814. http://dx.doi.org/10.1098/rstb.2019.0814.

Linnard, Gareth, Mike Williams and Branden Holmes. 'The Parson, the Psychiatrist, the Publican and His Nephews: The Two Final Thylacine Captures in Tasmania', *THRA* 67(3) (2020): 6–22.

Linnard, Gareth, and Stephen L. Sleightholme. 'An Exploration of the Evidence Surrounding the Identity of the Last Captive Thylacine', *Australian Zoologist* 43(2) (2023): 287–338.

———. 'The 1930 Mawbanna Thylacine: The Evidence and Photographic Legacy of the Last Confirmed Wild Kill', *Australian Zoologist* (2024). https://doi.org/10.7882/AZ.2024.036.

———. 'The Rediscovery of the 1931 Stewart Film of the Last Captive Thylacine', *Australian Zoologist* (2025). https://doi.org/10.7882/AZ.2025.006.

Lloyd's Register of British and Foreign Shipping: From 1st July 1838 to the 30th June 1839. Lincoln's Inn Fields: J. L. Cox & Sons, 1838.

Lloyd's Register of British and Foreign Shipping: From 1st July 1839 to the 30th June 1840. Lincoln's Inn Fields: J. L. Cox & Sons, 1839.

Locard, Edmond. *Manuel de technique policière*. Paris: Payot, 1934 [1923].

Long, Rohan. 'Miss Lambert's Thylacine Skull', *Pursuit*, 8 February 2018. Retrieved 24 February 2025 from https://pursuit.unimelb.edu.au/articles/miss-lambert-s-thylacine-skull.

———. 'Five Previously Undescribed Thylacine (*Thylacinus cynocephalus*) Specimens Held in the Museums of the University of Melbourne', *Australian Zoologist* 43(3) (2024): 455–69.

Long, William J. *Secrets of the Woods*. Boston: Ginn and Company, 1901.

———. *School of the Woods: Some Life Studies of Animal Instincts and Animal Training*. Boston: Ginn and Company, 1902.

———. 'The Kingfisher's Kindergarten', *Country Life in America* 2(4) (1902): 125–27.
———. *A Little Brother to the Bear and Other Animal Studies*. Boston: Ginn and Company, 1903.
———. 'The Modern School of Nature Study and Its Critics', *North American Review* 176(558) (1903): 688–98.
———. *Northern Trails: Some Studies of Animal Life in the Far North*. Boston: Ginn and Company, 1905.
Longman, Eleanor D., and Sophy Loch. *Pins and Pincushions*. London: Longman, Green & Co., 1911.
Lord, Clive Errol. 'Existing Tasmanian Marsupials', *Papers and Proceedings of the Royal Society of Tasmania* (1927): 17–24.
Lord, Clive Errol. and Hubert Hedley Scott. *A Synopsis of the Vertebrate Animals of Tasmania*. Hobart: Oldham, Beddome & Meredith, 1924.
Lorimer, Jamie. 'Nonhuman Charisma', *Environment and Planning D: Society and Space* 25 (2007): 911–32.
Lowish, Susan. *Rethinking Australia's Art History: The Challenge of Aboriginal Art*. New York: Routledge, 2018.
Lowther, Eden. 'The Painter and His Pupil', *Metropolitan Magazine* 39 (April 1844): 436–48.
Lucas, Frederic Augustus. *The Expedition to the Funk Island, with Observations upon the History and Anatomy of the Great Auk*. Washington, DC: Government Printing Office, 1890.
Lumholtz, Carl. *Among Cannibals: An Account of Four Years' Travels in Australia and of Camp Life with the Aborigines of Queensland*. London: John Murray, 1889.
Lutts, Ralph H. 'The Wild Animal Story: Animals and Ideas', in Ralph H. Lutts (ed.), *The Wild Animal Story* (Philadelphia: Temple University Press, 1998), 1–21.
———. *The Nature Fakers: Wildlife, Science and Sentiment*. Charlottesville: University Press of Virginia, 2001.
Lyne, A. G. 'Syndactyly in the Manus of the Marsupial', *Papers and Proceedings of the Royal Society of Tasmania* 57 (1953): 31–32.
———. 'The Vibrissae and Pelage Hairs in the Marsupialia', PhD thesis. Department of Zoology, University of Tasmania, 1958.
MacCallum, Monica. 'Georgina Sweet', in John Ritchie (ed.), *Australian Dictionary of Biography*, vol. 12 (Melbourne: Melbourne University Press, 1990), 149–50.
MacDonald, Melissa. 'The Effects of Crowd Size and Keeper Presence on the Stereotypic Pacing of Mexican Gray Wolves (*Canis lupus baileyi*)', unpublished honours thesis. Alfred University, New York, 2016.
MacDonald, Robert H. 'The Revolt Against Instinct: The Animal Stories of Seton and Roberts', in Ralph H. Lutts (ed.), *The Wild Animal Story* (Philadelphia: Temple University Press, 1998), 225–36.
MacDonald, Scott. 'Up Close and Political: Three Short Ruminations on Ideology in the Nature Film', *Film Quarterly* 59(3) (2006): 4–21.
Macgregor, John. *British America: Volume I*. Edinburgh: William Blackwood, 1832.
Mack, John, Wayne Brewster and Steven Fritts. 'A Review of Wolf Depredation on Livestock and Implications for the Yellowstone Area', in John D. Varley and Wayne G. Brewster (eds), *Wolves for Yellowstone? A Report to the United States Congress, Volume IV, Research and Analysis* (Yellowstone National Park: National Park Service, 1992), 5-25–5-43.
Mackinnon, Robert. 'Farming the Rock: The Evolution of Commercial Agriculture around St. John's, Newfoundland, to 1945', *Acadiensis* 20(2) (1991): 32–61.

Mackintosh, Alex. 'Crusoe's Abattoir: Cannibalism and Animal Slaughter in *Robinson Crusoe*', *Critical Quarterly* 53(3) (2011): 24–43.
MacLulich, T. D. 'The Animal Story and the "Nature Faker" Controversy', *Essays on Canadian Writing* 33(1) (1986): 112–24.
MacPhee, Ross D. E. 'Remember the Islands', in Rosamond Purcell (ed.), *Swift as a Shadow: Extinct and Endangered Animals* (Boston: Houghton Mifflin, 1999), 142–49.
Maddams, Tim. *Game: Rover Cottage Handbook 15*. New York: Bloomsbury, 2015.
Mahan, Alfred Thayer. 'The Writing of History', *Atlantic Monthly* 91(545) (1903): 289–98.
Maher, Jennifer, and Tanya Wyatt. 'International Trade in Animals and Animal Parts', in Jennifer Maher, Harriet Pierpoint and Piers Beirne (eds), *The Palgrave International Handbook of Animal Abuse Studies* (London: Palgrave, 2017), 223–48.
Malamud, Randy. *An Introduction to Animals and Visual Culture*. Basingstoke: Palgrave Macmillan, 2012.
Maleuvre, Didier. *Museum Memories: History, Technology, Art*. Stanford, CA: Stanford University Press, 1999.
Mallet, Marie-Louis. 'Avant-propos', in Jacques Derrida, *L'animal que donc je suis* (Paris: Galilée, 2006), 9–14.
Marinkova, Milena. *Michael Ondaatje: Haptic Aesthetics and Micropolitical Writing*. New York: Continuum, 2011.
Marks, Clive. 'Killing Schrödinger's Feral Cat', *Animal Studies Journal* 2(2) (2013): 51–66.
Marshall, Bill. 'Imagined Communities of Prehistory', in Charles Forsdick and Claire Launchbury (eds), *Transnational French Studies* (Liverpool: Liverpool University Press, 2023), 231–46.
Martin, Cy. *The Saga of the Buffalo*. New York: Promontory Press, 1973.
Martin, D. C., and Marie Elaine Danforth. 'An Analysis of Secular Change in the Human Mandible over the Last Century', *American Journal of Human Biology* 21 (2009): 704–6.
Martin, Wendy. *Once upon a Mine: Story of Pre-Confederation Mines on the Island of Newfoundland*. Montreal: Canadian Institute of Mining and Metallurgy, 1983.
Martínez, Julia, and Claire Lowrie. 'Colonial Constructions of Masculinity: Transforming Aboriginal Australian Men into "Houseboys"', *Gender & History* 21(2) (2009): 305–23.
Marvin, Garry. 'Perpetuating Polar Bears: The Cultural Life of Dead Animals', in Bryndis Snæbjörnsdóttir and Mark Wilson (eds), *Nanoq: Flat Out and Bluesome: A Cultural Life of Polar Bears* (London: Black Dog, 2006), 157–65.
———. 'Wolves in Sheep's (and Others') Clothing', in Dorothy Brantz (ed.), *Beastly Natures: Animals, Humans and the Study of History* (Charlottesville: University of Virginia Press, 2010), 59–78.
———. *Wolf*. London: Reaktion, 2012.
Maunder, John E. 'The Newfoundland Wolf', *The Osprey* 13(2) (1982): 36–48.
Mavhunga, Clapperton. 'Big Game Hunters: Bacteriologists, and Tsetse Fly Entomology in Colonial Southeast Africa: The Selous-Austen Debate Revisited, 1905–1940s', *Icon* 12 (2006): 75–117.
Maynard, David. 'Tasmanian Tiger: Precious Little Remains', in Valérie Bienvenue and Nicholas Chare (eds), *Animals, Plants and Afterimages: The Art and Science of Representing Extinction* (New York: Berghahn, 2022), 316–33.
Maynard, David, and Tammy Gordon. *Tasmanian Tiger: Precious Little Remains*. Launceston: Queen Victoria Museum and Art Gallery, 2014.
McCall, Vikki, and Clive Gray. 'Museums and the "New Museology": Theory, Practice and Organisational Change', *Museum Management and Curatorship* 29(1) (2013): 1–17.

McCrea, Robert Barlow. *Lost Amid the Fogs: Sketches of Life in Newfoundland, England's Ancient Colony*. London: Sampson Low, Son & Marston, 1869.
McCure, John Bunyan. *Life in England and Australia: A Memorial*. London: Robert Banks, 1876.
McFarlane, Ian. 'Aboriginal Society in North West Tasmania: Dispossession and Genocide', unpublished PhD thesis. University of Tasmania, 2002.
———. 'Cape Grim', in Robert Manne (ed.), *Whitewash: On Keith Windshuttle's Fabrication of Aboriginal History* (Melbourne: Black Inc. Agenda, 2003), 277–98.
———. 'NJB Plomley's Contribution to North-West Tasmanian Regional History', in Anna Johnston and Mitchell Rolls (eds), *Reading Robinson: Companion Essays to* Friendly Mission (Hobart: Quintus, 2008), 127–42.
McGoldrick, Rosemarie. 'Unscoping Animals', in Ben Stringer (ed.), *Rurality Re-Imagined: Villagers, Farmers, Wanderers and Wild Things* (San Francisco: Applied Research and Design Publishing, 2018), 256–67.
McGregor, Russell. 'The Doomed Race: A Scientific Axiom of the Late Nineteenth Century', *Australian Journal of Politics and History* 39(1) (1993): 14–22.
McHugh, Susan. *Animal Stories: Narrating across Species Lines*. Minneapolis: University of Minnesota Press, 2011.
———. 'Taxidermy's Literary Biographies', in André Krebber and Mieke Roscher (eds), *Animal Biography: Re-Framing Animal Lives* (Cham: Springer, 2018), 141–60.
———. *Love in a Time of Slaughters: Human Animal Stories Against Genocide and Extinction*. Pennsylvania: Pennsylvania State University Press, 2019.
McHuron, Elizabeth A., Rachel R. Holser and Daniel P. Costa. 'What's in a Whisker? Disentangling Ecological and Physiological Isotopic Signals', *Rapid Communications in Mass Spectrometry* 33 (2019): 57–66.
Mcphie, Jamie. 'Nazi Werewolves from Outer Space: Posthuman-Wolf-Multiplicities and their (Mis)appropriations', in Ian Convery, Peter Davis, Karen Lloyd, Owen T. Nevin and Erwin van Maanen (eds), *The Wolf: Culture, Nature, Heritage* (Martlesham: Boydell & Brewer, 2023), 69–81.
Meader, Stephen W. *Trap-Lines Norths: A True Story of the Canadian Woods*. London: George G. Harrap, 1936.
Mech, L. David. *The Wolf: The Ecology and Behavior of an Endangered Species*. Minneapolis: University of Minnesota Press, 1981 [1970].
———. 'The Challenge and Opportunity of Recovering Wolf Populations', *Conservation Biology* 9(2) (1995): 270–78.
———. 'Alpha Status, Dominance, and Division of Labor in Wolf Packs', *Canadian Journal of Zoology* 77 (1999): 1196–1203.
Medlock, Kathryn. 'Exhibiting Extinction: Thylacines in Museum Display', in Valérie Bienvenue and Nicholas Chare (eds), *Animals, Plants and Afterimages: The Art and Science of Representing Extinction* (New York: Berghahn, 2022), 392–406.
Meijaard, E., and V. Nijman. 'Secrecy Considerations for Conserving Lazarus Species', *Biological Conservation* 175 (2014): 21–24.
Meinertzhagen, Richard. 'The Education of Young Ospreys', *Ibis* 96 (1954): 153–55.
Mellor, David. 'Contemporary Racism in Australia: The Experience of Aborigines', *PSPB* 29(4) (2003): 474–86.
Melton, Arthur W. *Problems of Installation in Museums of Art*. Washington, DC: American Association of Museums, 1935.
Mendible, Myra. 'Visualizing Abjection: Gender, Power, and the Culture of Humiliation', *Gender Forum* 11 (2005): 36–54.

Merck, Melinda D. (ed.). *Veterinary Forensics: Animal Cruelty Investigations*. Ames, IA: Blackwell, 2007.

Merck, Melinda D., Doris M. Miller, Robert W. Reisman and Paolo C. Maiorka. 'Blunt Force Trauma', in Melinda D. Merck (ed.), *Veterinary Forensics: Animal Cruelty Investigations*, 2nd edn (Ames, IA: John Wiley & Sons, 2013), 97–122.

Merck, Melinda D., Doris M. Miller and Paolo C. Maiorka. 'Sharp Force Injuries', in Melinda D. Merck (ed.), *Veterinary Forensics: Animal Cruelty Investigations*, 2nd edn (Ames, IA: John Wiley & Sons, 2013), 123–37.

Meredith, Louisa. *My Home in Tasmania during a Residence of Nine Years*. London: John Murray, 1852.

Merle, Sébastien. 'Les Espèces en phase de reconquête: La Cigogne blanche Ciconia ciconia en Centre France; Son retour en Bourgogne', *Revue scientifique Bourgogne Nature* 8 (2008): 231–34.

Meston, Archibald Lawrence. *The Van Diemen's Land Company: 1825–1842*. Launceston: Launceston City Council, 1958.

Mikkonen, Jukka. '"Believe You Me, I Know. I'm Pretty Sharp When it Comes to Things Like That": Little My's Argumentation and Rhetoric', in Kate McLoughlin and Malin Lidström Brock (eds), *Tove Jansson Rediscovered* (Newcastle: Cambridge Scholars, 2007), 97–107.

Milgrom, Melissa. *Still Life: Adventures in Taxidermy*. Boston: Houghton Mifflin Harcourt, 2010.

Miller, D. 'Report of the Newfoundland and Labrador Fisheries: Part 2', *Journal of the House of Assembly of Newfoundland* (1874): 761–71.

Miller, Robert R., James D. Williams and Jack E. Williams. 'Extinctions of North American Fishes during the Past Century', *Fisheries* 14(6) (1989): 22–38.

Miller, Webb, Daniela I. Drautz, Jan E. Janecka, Arthur M. Lesk, Aakrosh Ratan et al. 'The Mitochondrial Genome Sequence of the Tasmanian Tiger (*Thylacinus cynocephalus*)', *Genome Research* 19 (2009): 213–20.

———. 'Supplement: The Mitochondrial Genome Sequence of the Tasmanian Tiger (*Thylacinus cynocephalus*)', *Genome Research* 19 (2009): 1–19.

Milligan, Joseph. 'On the Dialects and Language of the Aboriginal Tribes of Tasmania, and on Their Manners and Customs', *Papers and Proceedings of the Royal Society of Tasmania* 3(2) (1859): 239–82.

———. *Vocabulary of the Dialects of Some of the Aboriginal Tribes of Tasmania*. Hobart: William Thomas Strutt, 1890 [1857].

Milton, Kay. 'Possum Magic, Possum Menace: Wildlife Control and the Demonisation of Cuteness', in Carol Freeman, Elizabeth Leane and Yvette Watt (eds), *Considering Animals: Contemporary Studies in Human-Animal Relations* (Farnham: Ashgate, 2011), 67–78.

Mirzoeff, Nicholas. 'Invisible Again: Rwanda and Representation after Genocide', *African Arts* 38(3) (2005): 36–39, 86–91, 96.

Mitchell, Audra. 'Is IR Going Extinct?', *European Journal of International Relations* 23(1) (2017): 3–25.

———. 'Revitalizing Laws, (Re-)Making Treaties, Dismantling Violence: Indigenous Resurgence Against the 'Sixth Mass Extinction', *Social & Cultural Geography* 21(7) (2020): 909–24.

Mitchell, W. J. T. *Picture Theory*. Chicago: University of Chicago Press, 1994.

———. *What Do Pictures Want? The Lives and Loves of Images*. Chicago: University of Chicago Press, 2005.

Moesnner, Lilo. 'Dog – Man's Best Friend: A Study in Historical Lexicology', in Francisco Fernández, Miguel Fuster and Juan José Calvo (eds), *English Historical Linguistics 1992:*

Papers from the 7th International Conference on English Historical Linguistics (Amsterdam: John Benjamins, 1994), 207–18.
Molloy, Barry P. C., and Dave Grossman. 'Why Can't Johnny Kill? The Psychology and Physiology of Interpersonal Combat', in Barry Molloy (ed.), *The Cutting Edge: Studies in Ancient and Medieval Combat* (Stroud: Tempus, 2007), 188–202.
Montevecchi, William A. 'In Memoriam: Leslie Mills Tuck', *The Auk* 98 (1981): 606–9.
———. 'Influences of Artificial Light on Marine Birds', in Catherine Rich and Travis Longcore (eds), *Ecological Consequences of Artificial Night Lighting* (Washington, DC: Island Press, 2006), 94–113.
Mooney, Nick. 'Risk of Anticoagulant Rodenticides to Tasmanian Raptors', *Tasmanian Bird Report* 38 (2017): 17–25.
———. 'Review of Footprints from the 1938 Jane River Expedition', in Branden Holmes and Gareth Linnard (eds), *Thylacine: The History, Ecology and Loss of the Tasmanian Tiger* (Melbourne: CSIRO, 2023), 139–42.
Morin, Karen M. 'Carceral Spaces: Prisoners and Animals', *Antipode* 48(5) (2016): 1317–36.
Mörner, Torsten, Hanna Erikson, Caroline Bröjer, Kristina Nilsson, Henrik Uhlhorn et al. 'Diseases and Mortality in Free-Ranging Brown Bear (*Ursus arctos*), Gray Wolf (*Canis lupus*) and Wolverine (*Gulo gulo*) in Sweden', *Journal of Wildlife Diseases* 41(2) (2005): 298–303.
Moser, Stephanie. 'The Devil is the Detail: Museum Displays and the Creation of Knowledge', *Museum Anthropology* 33(1) (2010): 22–32.
Moses, A. Dirk. 'Genocide and Settler Society in Australia', in A. Dirk Moses (ed.), *Genocide: Volume III, Colonial and Imperial Genocides* (London: Routledge, 2010), 140–81.
Moses, Michael Valdez. 'The Tragic Novel: Heroism and the Politics of Modernity', unpublished PhD thesis. University of Virginia, 1987.
Mowat, Farley. *Never Cry Wolf*. Toronto: Seal, 1973.
Mudrooroo (Colin Thomas Johnson). *Doctor Wooreddy's Prescription*. Melbourne: Hyland House, 1983.
Muirhead, J., and M. Archer. '*Nimbacinus dicksoni*, a Plesiomorphic Thylacine (Marsupialia: Thylacinidae) from Tertiary Deposits of Queensland and the Northern Territory', *Memoirs of the Queensland Museum* 28(1) (1990): 203–21.
Muirhead, Jeanette. 'A Specialised Thylacinid, *Thylacinus Macknessi* (Marsupialia: Thylacinidae) from Miocene Deposits of Riversleigh, Northwestern Queensland', *Australian Mammalogy* 15 (1992): 67–75.
Muirhead, Jeanette, and Stephen Wroe. 'A New Genus and Species, Badjcinus turnbulli (Thylacinidae: Marsupialia), from the Late Oligocene of Riversleigh, Northern Australia, and an Investigation of Thylacinid Phylogency', *Journal of Vertebrate Paleontology* 18(3) (1998): 612–26.
Mulvaney, Ken. 'Dating the Dreaming: Extinct Fauna in the Petroglyphs of the Pilbara Region, Western Australia', *Archaeology in Oceania* 44 (Supplement) (2009): 40–48.
———. *Murujuga Marni: Rock Art of the Macropod Hunters and Mollusc Harvesters*. Crawley: UWA Publishing, 2015.
Mulvey, Laura. *Death 24x a Second: Stillness and the Moving Image*. London: Reaktion, 2006.
Muñoz Viñas, Salvador. *Contemporary Theory of Conservation*. Oxford: Elsevier, 2005.
Munro, James A. 'A Preliminary Report on the Destruction of Birds at Lighthouses on the Coast of British Columbia', *Canadian Field-Naturalist* 38(8) (1924): 141–45.
———. 'A Preliminary Report on the Destruction of Birds at Lighthouses on the Coast of British Columbia', *Canadian Field-Naturalist* 38(9) (1924): 171–75.
Murphy, John. *The Origins and History of Religions*. Manchester: Manchester University Press, 1949.

Murray, Tim. 'The Childhood of William Lanne: Contact Archaeology and Aboriginality in Tasmania', *Antiquity* 67 (1993): 504–19.

Murray, Wendi Field. 'Feathers, Fasting and the Eagle Complex: A Contemporary Analysis of the Eagle as a Cultural Resource in the Northern Plains', *Plains Anthropologist* 56(218) (2011): 143–53.

Narraway, Guinevere, and Hannah Stark. 'Re-Animating the Thylacine: Narratives of Extinction in Tasmanian Cinema', *Animal Studies Journal* 4(1) (2015): 12–31.

Nelson, Mariah Burton. *The Stronger Women Get, the More Men Love Football: Sexism and the Culture of Sport*. London: Women's Press, 1996.

Newton, Axel H., Frantisek Spoutil, Jan Prochazka, Jay R. Black, Kathryn Medlock et al. 'Letting the "Cat" out of the Bag: Pouch Young Development of the Extinct Tasmanian Tiger Revealed by X-Ray Computed Tomography', *Royal Society Open Science* 5: 171914. http://dx.doi.org/10.1098/rsos.171914.

Newton, Axel H., Vera Weisbecker, Andrew J. Pask and Christy A. Hipsley. 'Ontogenetic Origins of Cranial Convergence between the Extinct Marsupial Thylacine and Placental Gray Wolf', *Communications Biology* 4 (2021): 51. https://doi.org/10.1038/s42003-020-01569-x.

Nickson, Tom. *Toledo Cathedral: Building Histories in Medieval Castile*. Pennsylvania: Penn State University Press, 2015.

Nicol, Christine J. *The Behavioural Biology of Chickens*. Wallingford: CABI, 2015.

Nicol, Stewart, and Michael Maskrey. 'Thermoregulation, Respiration and Sleep in the Tasmanian Devil, *Sarcophilus harrisii* (Marsupialia: Dasyuridae)', *Journal of Comparative Physiology B* 140 (1980): 241–48.

Nochlin, Linda. *Realism*. Harmondsworth: Penguin, 1971.

Norwich, Edward of. *The Master of Game*. Philadelphia: University of Pennsylvania Press, 2005.

Novic, Elisa. *The Concept of Cultural Genocide: An International Law Perspective*. Oxford: Oxford University Press, 2016.

Notton, David G. *Identifying Insect Pests in Museums and Heritage Buildings*, 2nd edn. London: Natural History Museum, 2018.

Nurse, Angus. 'Collecting Wildlife', in Jennifer Maher, Harriet Pierpoint and Piers Beirne (eds), *The Palgrave International Handbook of Animal Abuse Studies* (London: Palgrave, 2017), 203–32.

Nyman, Jopi. *Postcolonial Animal Tale from Kipling to Coetzee*. New Delhi: Atlantic, 2003.

O'Doherty, Brian. *Inside the White Cube: The Ideology of the Gallery Space*. Berkeley: University of California Press, 1999.

Ogilby, William. 'Notice of Certain Australian Quadrupeds, Belonging to the Order Rodentia', *Transactions of the Linnean Society of London* 18(1) (1838): 121–32.

Old, Julie M. 'Immunological Insights into the Life and Times of the Extinct Tasmanian Tiger (*Thylacinus cynocephalus*)', *PLoS ONE* 10(12) (2015): e0144091.

Olsen, Penny. 'Wedge-Tailed Eagle, Australia', in Ruth E. Tingay and Todd E. Katzner (eds), *The Eagle Watchers: Observing and Conserving Raptors around the World* (Ithaca, NY: Comstock, 2010), 61–64.

Onfray, Robert, Ian Ravenwood, Geoff Dean and David De Little. 'Surrey Hills, Northwest Tasmania: The Birthplace of Industrial Scale Eucalypt Plantations in Australia', in Sue Feary and Rob Robinson (eds), *Australia's Ever-Changing Forests VI: Proceedings of the Ninth Australian Forest History Society Conference* (Kingston, ACT: Australian Forest History Society, 2015), 1–16. Retrieved 24 February 2025 from https://www.foresthistory.org.au/2015_conference_papers/08-Onfray.pdf.

Orchard, Christopher, and James Holcombe, 'Photographic Reflections on Landscape Change in Regional Australia', in Nicholas Holm and Sy Taffel (eds), *Ecological Entanglements in the Anthropocene* (Lanham: Lexington Books, 2017), 143–58.

Organ, J. S. 'Quaternary Investigations of the Topsails, North-Central Newfoundland', *Newfoundland and Labrador Department of Natural Resources Geological Survey* 16(1) (2016): 215–26.

Organ, Michael K. *The Illawarra Diary of Lady Jane Franklin, 10th–17th May 1839*. Woonona: Illawarra Historical Publications, 1988.

Osborn, Michael. 'Archetypal Metaphor in Rhetoric: The Light-Dark Family', *Quarterly Journal of Speech* 53(2) (1967): 115–26.

Packard, Jane M. 'Wolf Behavior: Reproductive, Social, and Intelligent', in David Mech and Luigi Boitani (eds), *Wolves: Behavior, Ecology and Conservation* (Chicago: University of Chicago Press, 2003), 35–65.

Paddle, Robert. *The Last Tasmanian Tiger: The History and Extinction of the Thylacine*. Cambridge: Cambridge University Press, 2002.

———. 'The Most Photographed of Thylacines: Mary Roberts' Tyenna Male – Including a Response to Freeman (2005) and a Farewell to Laird (1968)', *Australian Zoologist* 34(4) (2008): 459–70.

———. 'The Thylacine's Last Straw: Epidemic Disease in a Recent Mammalian Extinction', *Australian Zoologist* 36(1) (2012): 75–92.

Paddle, Robert N., and Kathryn M. Medlock. 'The Discovery of the Remains of the Last Tasmanian Tiger (*Thylacinus cynocephalus*)', *Australian Zoologist* 43(1) (2023): 97–108.

Pairault, Amédée. *Nouveau dictionnaire des chasses*. Paris: Librairie Pairault, 1885.

Palmer, Clare. '"Taming the Wild Profusion of Existing Things?": A Study of Foucault, Power, and Human/Animal Relations', *Environmental Ethics* 23(4) (2001): 339–58.

Paquet, Paul C., and Ludwig N. Carbyn. 'Gray Wolf (*Canis lupus* and Allies)', in George A. Feldhamer, Bruce C. Thompson and Joseph A. Chapman (eds), *Wild Mammals of North America: Biology, Management and Conservation* (Baltimore: Johns Hopkins University Press, 2003), 482–510.

Parent-Thivierge, Olivier. 'Relire *Feu la cendre* de Jacques Derrida ou de l'interprétation comme brûlure inextinguible', unpublished MA thesis. Université de Montréal, 2019.

Parker, Emily Anne. 'Singularity in Beauvoir's *The Ethics of Ambiguity*', *Southern Journal of Philosophy* 53(1) (2015): 1–16.

Parker, Rozsika, and Griselda Pollock. *Old Mistresses: Women, Art and Ideology*. London: Pandora, 1981.

Pay, James M., Todd E. Katzner, Clare E. Hawkins, Amelia J. Koch, Jason M. Wiersma et al. 'High Frequency of Lead Exposure in the Population of an Endangered Australian Top Predator, the Tasmanian Wedge-Tailed Eagle (*Aquila audax fleayi*)', *Environmental Toxicology and Chemistry* 40(1) (2021): 219–30.

Peacock, James L. *The Anthropological Lens: Harsh Light, Soft Focus*. Cambridge: Cambridge University Press, 1986.

Pearson, Joseph. 'The Female Urogenital System of the Marsupialia with Special Reference to the Vaginal Complex', *Papers and Proceedings of the Royal Society of Tasmania 1944* (1945): 72–99.

———. 'The Relationships of the Potoridae to the Macropodidae (Marsupialia)', *Papers and Proceedings of the Royal Society of Tasmania 1949* (1950): 211–29.

Pearson, Joseph, and J. M. De Bavay. 'The Urogenital System of the Dasyurinae and the Thylacininae (Marsupialia, Dasyuridae)', *Papers and Proceedings of the Royal Society of Tasmania* 87 (1953): 175–99.

Pearson, Roberta E. *Eloquent Gestures: The Transformation of Performance Style in the Griffith Biograph Films*. Berkeley: University of California Press, 1992.

Pedder-Smith, Rachel Tarn. 'The Glow of Significance: Narrating Stories Using Natural History Specimens', unpublished PhD thesis. London: Royal College of Art, 2011.

Pemberton, David. 'The Tasmanian Devil: A Uniquely Threatened Animal', in Carolyn J. Hogg, Samantha Fox, David Pemberton and Katherine Belov (eds), *Saving the Tasmanian Devil: Recovery through Science-Based Management* (Clayton South: CSIRO, 2019), 11–22.

Penprase, Bryan E. *The Power of Stars: How Celestial Observations Have Shaped Civilization*. New York: Springer, 2011.

peters, charlie. 'Girls, Animals, Fear, and the Iterative Force of the National Pack: Reading the Dear Canada Series', in Mavis Reimer, Nyala Ali, Deanna England and Melanie Dennis Unrau (eds), *Seriality and Young People's Texts: The Compulsion to Repeat* (Basingstoke: Palgrave Macmillan, 2014), 111–28.

Peeters, Benoît. *Derrida*. Paris: Flammarion, 2010.

Philp, Alex. 'Hunted, Now Haunting: The Thylacine as a Gothic Symbol in Julia Leigh's *The Hunter*', in Ruth Heholt and Melissa Edmundson (eds), *Gothic Animals: Uncanny Otherness and the Animal Without* (Cham: Palgrave Macmillan, 2020), 75–90.

Piatti-Farnell, Lorna. *Beef: A Global History*. London: Reaktion, 2013.

Pike, Nicolas. 'The Struggle for Life in the Deep', *Frank Leslie's Popular Monthly* 38(5) (1894): 628–34.

Pimm, Stuart L. 'The Dodo Went Extinct (and Other Ecological Myths)', *Annals of the Missouri Botanical Garden* 89 (2002): 190–98.

———. 'It's a New Century: Do You Know Where Your Orchids Are?' *Selbyana* 26(1/2) (2005): 5–13.

———. '"Putting Nature Back Together Again": Stuart Pimm in Conversation', in Valérie Bienvenue and Nicholas Chare (eds), *Animals, Plants and Afterimages: The Art and Science of Representing Extinction* (New York: Berghahn, 2022), 91–110.

Pimm, Stuart L., Michael P. Moulton and Lenora J. Justice. 'Bird Extinctions in the Central Pacific', in John H. Lawton and Robert M. May (eds), *Extinction Rates* (Oxford: Oxford University Press, 1995), 75–87.

Pinhasi, Ron, Daniel Fernandes, Kendra Sirak, Mario Novak, Sarah Connell et al. 'Optimal Ancient DNA Yields from the Inner Ear Part of the Human Petrous Bone', *PloS ONE* 10(6) (2015): e0129102.

Pitt, William C., Are R. Berentsen, Aaron B. Shiels, Steven F. Volker, John D. Eisemann et al. 'Non-Target Species Mortality and the Measurement of Brodifacoum Rodenticide Residues after a Rat (Rattus Rattus) Eradication on Palmyra Atoll, Tropical Pacific', *Biological Conservation* 185 (2015): 36–46.

Plato. *Cratylus*, in George Burges (ed.), *The Works of Plato: A New and Literal Version: Volume III* (London: George Bell, 1881), 283–96.

Playfair, John. 'Biographical Account of the Late Dr James Hutton', *Transactions of the Royal Society of Edinburgh* 5(3) (1805): 39–99.

Plomley, Norman James Brian. 'Notes on Some of the Tasmanian Aborigines, and on Portraits of Them', *Papers and Proceedings of the Royal Society of Tasmania* 102(2) (1968): 47–54.

———. *Tasmanian Aboriginal Place Names*. Tasmania: Queen Victoria Museum and Art Gallery, 1992.

Polack, Fiona. 'Introduction: De-Islanding the Beothuk', in Fiona Polack (ed.), *Tracing Ochre: Changing Perspectives on the Beothuk* (Toronto: University of Toronto Press, 2018), 3–29.

Poliquin, Rachel. *The Breathless Zoo: Taxidermy and the Cultures of Longing*. Pennsylvania: Penn State University Press, 2012.

Pollock, Griselda. 'Dialogue with Julia Kristeva', *parallax* 4(3) (1998): 5–16.

———. 'The Visual', in Mary Eagleton (ed.), *A Concise Companion to Feminist Theory* (Malden, MA: Blackwell, 2003), 173–94.

———. 'Thinking the Feminine: Aesthetic Practice as Introduction to Bracha Ettinger and the Concepts of Matrix and Metramorphosis', *Theory, Culture & Society* 21(1) (2004): 5–65.

———. 'Introduction: Femininity: Aporia or Sexual Difference?', in Brian Massumi (ed.), *The Matrixial Borderspace* (Minneapolis: University of Minnesota Press, 2006), 1–38.

———. 'Aesthetic Wit(h)nessing in the Era of Trauma', *EurAmerica* 40(4) (2010): 829–86.

———. 'A Matrixial Installation: Artworking in the Freudian Space of Memory and Migration', in Catherine de Zegher and Griselda Pollock (eds), *Bracha Ettinger: Art as Compassion* (Brussels: APA, 2011), 191–241.

———. *After-affects/After-images: Trauma and Aesthetic Transformation in the Virtual Feminist Museum*. Manchester: Manchester University Press, 2013.

———. *Art in the Time-Space of Memory and Migration: Sigmund Freud, Anna Freud, and Bracha L. Ettinger in the Freud Museum: Artwriting after the Event*. Leeds: Wild Pansy Press, 2013.

Portillo, Michael. 'Foreword', in Sue Black and Eilidh Ferguson (eds), *Forensic Anthropology: 2000 to 2010* (Boca Raton, FL: Taylor & Francis, 2011), vii–viii.

Potts, Alex. 'Natural Order and the Call of the Wild: The Politics of Animal Picturing', *Oxford Art Journal* 13(1) (1990): 12–33.

Potts, Thomas H. 'The Kea – Progress of Development', *Nature* 4 (1871): 489.

Preston, Charles R., Hershel Moseley and Carolyn Moseley. 'Green-Backed Heron Baits Fish with Insects', *Wilson Bulletin* 98(4) (1986): 613–14.

Preti, Antonio. 'Animal Suicide: Evolutionary Continuity or Anthropomorphism', *Animal Sentience* 20(10) (2018). http://dx.doi.org/10.51291/2377-7478.1297.

Price, Kenneth M., and Cynthia G. Bernstein. 'Whitman's Sign of Parting: "So Long!" as *L'Envoi*', *Walt Whitman Quarterly Review* 9 (1991): 65–76.

Probyn-Rapsey, Fiona. 'Animals and Extinction', in Adeline Johns-Putra and Kelly Sultzbach (eds), *The Cambridge Companion to Literature and Climate* (Cambridge: Cambridge University Press, 2022), 100–13.

Proops, Leanne, Karen McComb and David Reby. 'Cross-Modal Individual Recognition in Domestic Horses (*Equus Caballus*)', *PNAS* 106(3) (2009): 947–51.

Prahalad, Vishnu, and John Aalders. *Marchwiel Marsh, Marion Bay: Management Report*. Hobart: Geography and Spatial Sciences, University of Tasmania, 2020.

Pugh, Jonathan. 'Driven to Extinction? The Ethics of Eradicating Mosquitoes with Gene-Drive Technologies', *Journal of Medical Ethics* 42(9) (2016): 578–81.

Pyle, Howard. 'The Salem Wolf', *Harpers Monthly Magazine* 120(715) (December 1909): 7–12.

Rabaté, Jean-Michel. *Les guerres de Jacques Derrida*. Montréal: Les Presses de l'Université de Montréal, 2016.

Radic, Thérèse. *Melba: The Voice of Australia*. Melbourne: Macmillan, 1986.

Rajan, Nithya. 'What Do Refugees Want? Reading Refugee Lip-Sewing Protests through a Critical Lens', *International Feminist Journal of Politics* 21(4) (2019): 527–43.

Rausch, Robert A. 'Some Aspects of the Population Ecology of Wolves, Alaska', *American Zoologist* 7 (1967): 253–65.

Ray, Arthur J. *The Canadian Fur Trade in the Industrial Age*. Toronto: University of Toronto Press, 1990.

Ray, Larry. 'Social Theory, Photography and the Visual Aesthetic of Cultural Modernity', *Cultural Sociology* 14(2) (2020): 139–59.

Read, Clio Elizabeth. 'Understanding Attacks by Kea (*Nestor notabilis*), an Endemic Parrot, on Sheep (*Ovis aries*) in the South Island High Country', unpublished PhD thesis. Massey University/Te Kunenga ki Pūrehuroa Manawatū, Aotearoa/New Zealand, 2019.

Reichert, Jeff. 'Headbanger: An Interview with Tony Stone, Director of *Severed Ways: The Norse Discovery of America*', *Reverse Shot*, 13 March 2009. Retrieved 24 February 2025 from http://www.reverseshot.org/interviews/entry/425/tony-stone.

Reichs, Kathleen J. 'Forensic Anthropology in the 1990s', *American Journal of Forensic Medicine and Pathology* 13(2) (1992): 146–53.

Reid, D. J., M. Beveridge and P. Wakefield. 'The Effect of Ability, Colour and Form on Children's Perceptions of Biological Pictures', *Educational Psychology* 6(1) (1986): 9–18.

Reid, Elizabeth Clio. 'Understanding Attacks by Kea (*Nestor notabilis*), an Endemic Parrot, on Sheep (*Ovis aries*) in the South Island High Country', PhD thesis, Massey University/Te Kunenga ki Pūrehuroa, 2019.

Reiser (Jean-Marc Roussillon). *La Vie au grand air*. Paris: Presses de la Bûcherie, 1972.

———. *Vive les Vacances*. Paris: Albin Michel, 1986.

Renshaw, Graham. *More Natural History Essays*. London: Sherratt & Hughes, 1905.

Ressel, Lorenzo, Udo Hetzel and Emanuele Ricci. 'Blunt Force Trauma in Veterinary Forensic Pathology', *Veterinary Pathology* 53(5) (2016): 941–61.

Riber, Anja Brinch. 'Gregarious Nesting – An Anti-Predator Response in Laying Hens', *Applied Animal Behaviour Science* 138(1–2) (2012): 70–78.

Ricco, John Paul. *The Logic of the Lure*. Chicago: University of Chicago Press, 2003.

Richardson, John. *Fauna Boreali-Americana, or, the Zoology of the Northern Parts of British America*, vol. 1. London: John Murray, 1829.

Ridpath, M. G., and M. G. Brooker. 'Age, Movements and the Management of the Wedge-Tailed Eagle, *Aquila audax*, in Arid Western Australia', *Australian Wildlife Research* 13(2) (1986): 245–60.

Riehl, Christina. 'Black-Crowned Night Heron Fishes with Bait', *Waterbirds: The International Journal of Waterbird Biology* 24(2) (2001): 285–86.

Rifkin, Adrian. '"The Answers Are the Question": A Conversation about Art History, Art-writing and Historical Time', in Nicholas Chare and Mitchell B. Frank (eds), *History and Art History: Looking Past Disciplines* (New York: Routledge, 2021), 213–36.

Ritvo, Harriet. 'Animals in Nineteenth-Century Britain: Complicated Attitudes and Competing Categories', in Aubrey Manning and James Serpell (eds), *Animals and Human Society: Changing Perspectives* (London: Routledge, 1994), 106–26.

———. *The Platypus and the Mermaid and Other Figments of the Classifying Imagination*. Cambridge, MA: Harvard University Press, 1997.

———. 'Zoological Nomenclature and the Empire of Victorian Science', in Bernard Lightman (ed.), *Victorian Science in Context* (Chicago: University of Chicago Press, 1997), 334–53.

———. 'Calling the Wild', in Joan B. Landes, Paula Young Lee and Paul Youngquist (eds), *Gorgeous Beasts: Animal Bodies in Historical Perspective* (Pennsylvania: Pennsylvania State University Press, 2012), 105–16.

Robbins, Brent Dean. *The Medicalized Body and Anesthetic Culture*. New York: Palgrave Macmillan, 2018.

Robinson, George Augustus. 'The Flinders Island Journal of George Augustus Robinson, 1835–1839', in N. J. B. Plomley (ed.), *Weep in Silence* (Hobart: Blubber Head Press, 1987), 293–788.

———. *Friendly Mission: The Tasmanian Journals and Papers of George Augustus Robinson, 1829–1834*, ed. Norman James Brian Plomley. Launceston: QVMAG and Quintus, 2008 [1966].

Robinson, Hercules. 'Private Journal Kept on Board HMS Favorite on the Newfoundland Station', *Journal of the Royal Geographic Society of London* 4 (1834): 207–20.

Robisch, Sean Kipling. *Wolves and the Wolf Myth in American Literature*. Reno: University of Nevada Press, 2009.

Roddy, Gareth. 'Mist in "the West": Literatures of Travel and Landscape in the Western British-Irish, c. 1880–1940', in LuAnn McCracken Fletcher (ed.), *Literary Tourism and the British Isles: History, Imagination, and the Politics of Place* (Lanham: Lexington, 2018), 37–66.

Roe, Jill. *Her Brilliant Career: The Life of Stella Miles Franklin*. Cambridge, MA: Belknap Press, 2009.

Roger, Philippe. 'Hallali', *Critique* 34(375–76) (1978): 829–37.

Romanes, George John. *Mental Evolution in Animals*. London: Kegan Paul, Trench & Co., 1883.

Rondinini, Carlo, Luigi Boitani, Ana S. L. Rodrigues, Thomas M. Brooks, Robert L. Pressey et al. 'Reconciling Global Mammal Prioritization Schemes into a Strategy', *Philosophical Transactions of the Royal Society B* 366 (2011): 2722–28.

Roosevelt, Theodore. *The Wilderness Hunter*. New York: G. P. Putnam's Sons, 1893.

———. *The Wilderness Hunter: Volume II*, Philadelphia: Gebbie & Company, 1903.

———. 'Opening Address by the President', in Newton C. Blanchard, John Franklin Fort, James O. Davidson, John C. Cutler and Martin F. Ansel (eds), *Proceedings of a Conference of Governors* (Washington, DC: Government Printing Office, 1909), 3–12.

Rose, Deborah Bird. *Wild Dog Dreaming: Love and Extinction*. Charlottesville: University of Virginia Press, 2011.

Rose, Jacqueline. *Sexuality in the Field of Vision*. London: Verso, 1986.

Rosenberg, Gary, and Michael Khoo. 'Improving Provenance Data in Natural History Collections', *American Malacological Bulletin* 36(2) (2018): 215–31.

Ross, Ann H., and Sarah L. Cunningham. 'Time-Since-Death and Bone Weathering in a Tropical Environment', *Forensic Science International* 204 (2011): 126–33.

Ross, James Clark. *Voyage of Discovery and Research in the Southern and Antarctic Regions, during the Years 1839–43: Volume II*. London: John Murray, 1847.

Rothfels, Nigel. *Savages and Beasts: The Birth of the Modern Zoo*. Baltimore: Johns Hopkins University Press, 2002.

———. 'Immersed with Animals', in Nigel Rothfels (ed.), *Representing Animals* (Bloomington: Indiana University Press, 2002), 199–223.

———. 'Trophies and Taxidermy', in Joan B. Landes, Paula Young Lee and Paul Youngquist (eds), *Gorgeous Beasts: Animal Bodies in Historical Perspective* (Pennsylvania: Pennsylvania State University Press, 2012), 117–36.

Rothschild, Walter. 'On Extinct and Vanishing Birds', in R. Bowdler Sharpe (ed.), *Proceedings of the Fourth International Ornithological Congress, London June 1905* (London: Dulau & Co., 1907), 191–217.

Rovinsky, Douglass S., Alistair R. Evans, Damir G. Martin and Justin W. Adams. 'Did the Thylacine Violate the Costs of Carnivory?', *Proceedings of the Royal Society B* 287 (2020): 20201537.

Rovinsky, Douglass S., Alistair R. Evans and Justin W. Adams. 'Functional Ecological Convergence between the Thylacine and Small-Prey Focussed Canids', *BMC Ecology and Evolution* 21(58) (2021): 1–17.

Rozier, François. 'Loup, Louve', in *Cours complet d'agriculture, théorique, pratique, économique, et de médecin rurale et vétérinaire: Volume VI* (Paris: Rue et Hôtel Serpente, 1785), 298–302.

Russell, Lynette. 'Remembering and Forgetting: Discovering the "Secret History" of Tasmania', *Journal of Genocide Research* 20(3) (2018): 446–50.

Rutchick, Abraham M., Ryan M. McManus, Denise M. Barth, Robert J. Youmans, Andrew T. Ainsworth and H. Johnny Goukassian. 'Technologically Facilitated Remoteness Increases Killing Behavior', *Journal of Experimental Social Psychology* 73 (2017): 147–50.

Ryan, Lyndall. 'The Black Line in Van Diemen's Land: Success or Failure?', *Journal of Australian Studies* 37(1) (2013): 3–18.

———. 'Untangling Aboriginal Resistance and the Settler Punitive Expedition: The Hawkesbury River Frontier in New South Wales, 1794–1810', *Journal of Genocide Research* 15(2) (2013): 219–32.

Ryan, Sean F., Eric Lombaert, Anne Espeset, Roger Vila, Gerard Talavera et al. 'Global Invasion History of the Agricultural Pest Butterfly *Pieris Rapae* Revealed with Genomics and Citizen Science', *PNAS* 116(40) (2019): 20015–24.

S., A. 'Forestry in South Australia', *The Indian Forester: A Monthly Magazine* 18(11) (1892): 436–37.

Sandes, John, and John Stirling. *In Tasman's Land: Gleams and Dreams of the Great North-West*. Melbourne: Emu Bay Railway Company, 1903.

Sands, Philippe. *East West Street: On the Origins of 'Genocide' and 'Crimes Against Humanity'*. New York: Alfred A. Knopf, 2016.

Saunders, Marshall. *Beautiful Joe*. Toronto: McClelland and Stuart, 1922.

Savoi, Camile de Oliveira, and Nicholas Chare. 'Beyond First Impressions: Looking Anew at the Art of Georgina de Albuquerque', *RACAR* 49(1) (2024): 71–93.

Sayer, Karen. 'Rats'. *Rural History Today* 30 (2016): 1–2.

Sayre, Gordon M. *Indian Chief as Tragic Hero: Native Resistance and the Literatures of America, from Moctezuma to Tecumseh*. Chapel Hill: University of North Carolina Press, 2005.

———. 'Rare Books and Rare Birds: The Species as Work of Art', in Valérie Bienvenue and Nicholas Chare (eds), *Animals, Plants and Afterimages: The Art and Science of Representing Extinction* (New York: Berghahn, 2022), 191–210.

Scales, G. H., R. N. Burton and R. A. Moss. 'Lamb Mortality, Birthweight, and Nutrition in Late Pregnancy', *New Zealand Journal of Agricultural Research* 29(1) (1986): 75–82.

Schilthuizen, Menno. *The Loom of Life: Unravelling Ecosystems*. Berlin: Springer, 2008.

Schlunke, Katrina. 'Thylacine Buggy Rug', in Cameron Muir, Kirsten Wehner and Jenny Newell (eds), *Living With the Anthropocene: Love, Loss and Hope in the Face of Environmental Crisis* (Sydney: NewSouth Publishing, 2020), 130–33.

———. 'The Practices of Care: Extinction and De-colonization in the Natural History Museum', *Continuum: Journal of Media and Cultural Studies* 39(1) (2025): 184–95.

Schoppmann, Claudia. 'National Socialist Policies towards Female Homosexuality', trans. Elizabeth Harvey, in Lynn Abrams and Elizabeth Harvey (eds), *Gender Relations in German History: Power, Agency and Experience from the Sixteenth to the Twentieth Century* (Durham, NC: Duke University Press, 1996), 177–87.

Schreiber, Reinhard, and Herbert Gareis. *Gelatine Handbook: Theory and Industrial Practice*. Weinheim: Wiley-VCH, 2007.

Schwenger, Peter. 'Words and the Murder of the Thing', *Critical Inquiry* 28 (2001): 99–113.

Sclocchi, Maria Carla, Elisa Damiano, Donatella Matè, Piero Colaizzi and Flavia Pinzari. 'Fungal Biosorption of Silver Particles on 20th-Century Photographic Documents', *International Biodeterioration & Biodegradation* 84 (2013): 367–71.

Scott, Peter. 'Transhumance in Tasmania', *New Zealand Geographer* 11(2) (1955): 155–72.

Scott, Walter. 'Legend of Montrose', in *The Bride of Lammermoor and A Legend of Montrose* (Boston: Aldine, n.d.).
Scrope, William. *The Art of Deer-Stalking*. London: John Murray, 1839 [1838].
Segal, Nancy L. 'News Reports: First Report of Identical Twin Puppies', *Twin Research and Human Genetics* 20(1) (2017): 93–94.
Segerdahl, Pär. 'Can Natural Behaviour Be Cultivated? The Farm as Local Human/Animal Culture', *Journal of Agricultural and Environmental Ethics* 20 (2007): 167–93.
Sellheim, Nikolas. 'Black and Viking Metal: How Two Extreme Music Genres Depict, Construct and Transfigure the (Sub-)Arctic', *Polar Record* 52(266) (2016): 509–17.
Seton, Ernest Thompson. *Wild Animals I Have Known*. Toronto: G.N. Morang, 1899 [1898].
———. 'Lobo: The King of Currumpaw', in *Wild Animals I Have Known* (Toronto: G. N. Morang, 1899 [1898]), 17–54.
———. 'Introduction', in *The Shagganappi* (Toronto: Ryerson Press, 1913), 7–9.
———. *Lives of Game Animals: Volume 1: Cats, Wolves, and Foxes*. New York: Doubleday, Doran & Company, 1929 [1925].
———. *Lives of Game Animals: Volume 3: Hoofed Animals*. New York: Literary Guild of America, 1937 [1927].
Sharland, Michael Stanley Reid. 'In Search of the Thylacine: Society's Interest in the Preservation of a Unique Marsupial', *Royal Zoological Society of New South Wales* 59 (1938–39): 20–38.
———. 'Tracking the Thylacine: Quest Reveals Footprints in Southern Tasmania', *Wild Life* (June 1939): 7–9.
———. 'Tasmania's Rare "Tiger"', *Bulletin, New York Zoological Society* 44(3) (1941): 84–88.
Shearman, John. *Raphael in the Early Modern Sources: 1483–1602*. New Haven, CT: Yale University Press, 2003.
Shepherd, R. R., C. B. Winkler and R. Jones. 'Management of the Central Plateau of Tasmania as a Conservation Area', Attachment 2 in 'The Central Plateau of Tasmania: A Resource Survey and Management Plan', unpublished MA dissertation (University of Tasmania, 1974), 1–39.
Short, Jeff. 'The Extinction of Rat-Kangaroos (Marsupialia; Potoridae) in New South Wales, Australia', *Biological Conservation* 86 (1998): 365–77.
Silva, Alexandre Ulisses, and Luis Corte-Real. 'Removal of Blotches and Scratches from Film and Video Sequences Using a Digital Restoration Chain', in A. Enis Çetin, Lale Akarun, Aysin Ertuzun, Metin N. Gurcan and Yasemin Yardimci (eds), *Conference Proceedings of the IEEE-EURASIP Workshop on Nonlinear Signal and Image Processing (NSIP'99)* (Istanbul: Boğaziçi University, 1999), 826–29.
Silverman, Kaja. *The Subject of Semiotics*. Oxford: Oxford University Press, 1983.
Sleightholme, Stephen R. 'Confirmation of the Gender of the Last Captive Thylacine', *Australian Zoologist* 35(4) (2011): 953–56.
Sleightholme, Stephen R., and Nicholas P. Ayliffe. *The International Thylacine Specimen Database*, 5th Revision. Westbury, UK: 2013. [Digital resource].
Sleightholme, Stephen R., and Cameron R. Campbell. 'The Earliest Motion Picture Footage of the Last Captive Thylacine?', *Australian Zoologist* 37(3) (2015): 282–87.
———. 'A Retrospective Assessment of 20th Century Thylacine Populations', *Australian Zoologist* 38(1) (2016): 102–29.
———. 'The International Thylacine Specimen Database (6th Revision – Project Summary & Final Report)', *Australian Zoologist* 39(3) (2018): 480–512.
———. 'Stripe Pattern Variation in the Coat of the Thylacine', *Australian Zoologist* 40(2) (2019): 290–307.

———. 'Reverend George H. Judd's Thylacines', *Australian Zoologist* 41(1) (2020): 74–79.
———. 'A Catalogue of the Thylacine Captured on Film', *Australian Zoologist* 41(2) (2021): 143–78.
———. 'A Catalogue of Motion Picture Films of the Thylacine (*Thylacinus cynocephalus*)', *Australian Zoologist* 41(4) (2021): 778–93.
———. 'The Thylacine: A Wolf in Name Only', in Ian Convery, Peter Davis, Karen Lloyd, Owen T. Nevin and Erwin van Maanen (eds), *The Wolf: Culture, Nature, Heritage* (Martlesham: Boydell & Brewer, 2023), 231–36.
Sleightholme, Stephen R., Cameron R. Campbell and Andrew C. Kitchener. 'Frank Haes' Thylacine', *Australian Zoologist* 38(2) (2016): 203–11.
Sleightholme, Stephen R., Tammy J. Gordon and Cameron R. Campbell. 'The Kaine Capture: Questioning the History of the Last Thylacine in Captivity', *Australian Zoologist* 41(1) (2020): 1–11.
Sleightholme, Stephen, Jan Robovský and Vladimír Vohralík. 'Description of Four Newly Discovered Thylacine Pouch Young and a Comparison with Boardman (1945)', *Zoologist* 36(2) (2012): 232–38.
Smalley, Andrea L. *Wild by Nature: North American Mammals Confront Colonization*. Baltimore: Johns Hopkins University Press, 2017.
Smith, Nicholas. 'The Return of the Living Dead: Unsettlement and the Tasmanian Tiger', *Journal of Australian Studies* 36(3) (2012): 269–89.
Smith, O. C., Elayne J. Pope and Steven A. Symes. 'Look Until You See: Identification of Trauma in Skeletal Material', in Dawnie Wolf Steadman (ed.), *Hard Evidence: Case Studies in Forensic Anthropology* (Upper Saddle River, NJ: Prentice Hall, 2003), 138–54.
Smith, Steven. *The Tasmanian Tiger – 1980: A Report on an Investigation of the Current Status of the Thylacine (*Thylacinus *cynocephalus).* Hobart: National Parks and Wildlife Service, 1981.
Smith-Blackmore, Martha, and Jonathan D. Bethard. 'A Multidisciplinary Investigation of Chronic Animal Abuse: Collaboration between Veterinary Forensics and Forensic Anthropology', *Journal of Forensic Sciences* 66 (2021): 389–92.
Sobey, Douglas S. 'An Analysis of the Historical Records for the Native Mammalian Fauna of Prince Edward Island', *Canadian Field Naturalist* 121(4) (2007): 384–96.
Sofaer, Joanna R. *The Body as Material Culture: A Theoretical Osteoarchaeology*. Cambridge: Cambridge University Press, 2006.
———. 'Touching the Body: The Living and the Dead in Osteoarchaeology and the Performance Art of Marina Abramović', *Norwegian Archaeological Review* 45(2) (2012): 135–50.
Sollund, Ragnhild Aslaug. *The Crimes of Wildlife Trafficking: Issues of Justice, Legality and Morality*. London: Routledge, 2019.
Sontag, Susan. *On Photography*. London: Penguin, 1979.
Sorlin, Pierre. *The Film in History: Restaging the Past*. Oxford: Basil Blackwell, 1980.
Soulé, Michael E. 'What is Conservation Biology?', *BioScience* 35(11) (1985): 727–34.
Southey, Thomas. *A Treatise on Sheep: Addressed to the Flock-Masters of Australia, Tasmania, and Southern Africa*. London: Smith and Elder, 1840.
Spaulding, R. L., Paul R. Krausman and Warren B. Ballard. 'Summer Diet of Gray Wolves, *Canis lupus* in Northwestern Alaska', *Canadian Field-Naturalist* 112(2) (1998): 262–66.
Springer, John W. 'Annual Address of the President, John W. Springer', in Charles F. Martin (ed.), *Proceedings of the Sixth Annual Convention and Annual Report of the National Live Stock Association* (Denver: The Association, 1903), 129–42.
Springer, Keith. 'Eradication of Invasive Species on Macquarie Island to Restore the Natural Ecosystem', in Stephen Garnett, Peter Latch, David Lindenmayer and John Woi-

narski (eds), *Recovering Australian Threatened Species: A Book of Hope* (Clayton: CSIRO, 2018), 13–22.

Sramek, Joseph. '"Face Him Like a Briton": Tiger Hunting, Imperialism, and British Masculinity in Colonial India, 1800–1875', *Victorian Studies* 48(4) (2006): 659–80.

Stark, Hannah. 'The Cultural Politics of Mourning in the Era of Mass Extinction: Thylacine Specimen P762', *Australian Humanities Review* 63 (2018): 65–79.

State of Oregon. *General and Special Laws and Joint Resolutions and Memorials Enacted and Adopted by the Twentieth Regular Session of the Legislative Assembly, 1899*. Salem: W. H. Leeds, 1899.

Staub, Ervin. *The Roots of Evil: The Origins of Genocide and Other Group Violence*. Cambridge: Cambridge University Press, 1989.

———. 'The Psychology of Bystanders, Perpetrators and Heroic Helpers', *International Journal of Intercultural Relations* 17(3) (1993): 315–41.

Stephanou, Aspasia. 'Playing Wolves and Red Riding Hoods in Black Metal', in Nicola Masciandaro (ed.), *Hideous Gnosis: Black Metal Theory Symposium 1* (Charleston: Createspace, 2010), 157–70.

Stevenson, Caryn, and Eric J. Woehler. 'Population Decreases in Little Penguins *Eudyptula minor* in Southeastern Tasmania, Australia, over the Past 45 Years', *Marine Ornithology* 35 (2007): 71–76.

Stoker, Bram. 'The Lair of the White Worm', in Kate Hebblethwaite (ed.), *Dracula's Guest and Other Weird Stories* (London: Penguin, 2006 [1914]), 151–369.

Stokes, H. J. W. 'The Settlement and Development of the Van Diemen's Land Company's Grants in North-Western Van Diemen's Land, 1824–1860', unpublished honours dissertation. University of Tasmania, 1964.

Stolzenburg, William. *Rat Island: Predators in Paradise and the World's Greatest Wildlife Rescue*. New York: Bloomsbury, 2011.

Stone, Christopher D. *Should Trees Have Standing? Toward Legal Rights for Natural Objects*. Los Altos: William Kaufmann, 1974.

Stone, Dan. 'Genocide as Transgression', *European Journal of Social Theory* 7(1) (2004): 45–65.

Storey, William K. 'Big Cats and Imperialism: Lion and Tiger Hunting in Kenya and Northern India, 1898–1930', *Journal of World History* 2(2) (1991): 135–73.

Stuart, Amanda Graham. 'The Dingo in the Colonial Imagination', unpublished PhD thesis. Canberra: Australian National University, 2013.

Swainson, William. 'Australasia: Natural Geography', in Hugh Murray (ed.), *An Encyclopedia of Geography, 1478–1485* (London: Longman, Rees, Orme, Brown, Green and Longman, 1834), 1476–506.

Swinney, Geoffrey N. 'What Do We Know about What We Know? The Museum "Register" as Museum Object', in Sandra Dudley, Amy Jane Barnes, Jennifer Binnie, Julia Petrov and Jennifer Walklate (eds), *The Thing about Museums: Objects and Experience, Representation and Contestation* (London: Routledge, 2012), 31–46.

Szanto, Thomas. 'In Hate We Trust: The Collectivization and Habitualization of Hatred', *Phenomenology and the Cognitive Sciences* 19(3) (2020): 453–80.

Tassell, Margaret. *Rural Launceston Heritage Study*. Launceston: QVMAG, 2000.

Taylor, William Timothy Treal, et al. 'Early Dispersal of Domestic Horses into the Great Plains and Northern Rockies', *Science* 379 (2023): 1316–23.

Temminck, Coenraad Jacob. *Monographies de mammalogie: Volume 1*. Paris: G. Dufour & Éditions D'Ocagne, 1827.

Terrie, Philip G. 'Urban Man Confronts the Wilderness: The Nineteenth-Century Sportsman in the Adirondacks', *Journal of Sport History* 5(3) (1978): 7–20.

Thomas, Richard H., Walter Schaffner, Allan C. Wilson and Svante Pääbo. 'DNA Phylogeny of the Extinct Marsupial Wolf', *Nature* 340 (1989): 465–67.
Thresher, Ann C. 'When Extinction is Warranted: Invasive Species, Suppression-Drives and the Worst-Case Scenario', *Ethics, Policy & Environment* 25(2) (2022): 132–52.
Thureau, Gustav. 'Report on the Corinna Goldfield', in *Mount Cleveland and Corinna Goldfields* (Hobart: Parliament of Tasmania, 1884), unpaginated.
Timm, Robert M., Suzanne B. McLaren and Hugh H. Genoways. 'Innovations That Changed Mammalogy: Museum Study Skins', *Journal of Mammalogy* 102(2) (2021): 367–71.
Tindol, Robert. 'The Best Friend of the Murderers: Guard Dogs and the Holocaust', in Ryan Hediger (ed.), *Animals and War: Studies of Europe and North America* (Leiden: Brill, 2013), 105–21.
Tocque, Philip. *Wandering Thoughts, or Solitary Hours*. London: Thomas Richardson, 1846.
Torres, Walter J., and Raymond M. Bergner. 'Humiliation: Its Nature and Consequences', *Journal of the American Academy of Psychiatry and Law* 38(2) (2010): 195–204.
Townsend, Charles Haskins. 'An Account of Recent Captures of the California Sea Elephant, and the Statistics Relating to the Present Abundance of the Species', *Proceedings of the United States National Museum* 8(6) (1885): 90–93.
———. *Pelagic Sealing with Notes on the Fur Seals of Guadalupe, the Galapagos, and Lobos Islands*. Washington, DC: Government Printing Office, 1899.
———. 'The Northern Elephant Seal', *Zoologica* 1(8) (1912): 159–73.
———. *The USS 'Albatross' in Lower Californian Seas: Cruise of 1911*. New York: American Museum of Natural History, 1926.
Trafford, Simon. 'Viking Metal', in Stephen Meyer and Kirsten Yri (eds), *The Oxford Handbook of Music and Medievalism* (Oxford: Oxford University Press, 2020), 564–85.
Troughton, Ellis. *Furred Animals of Australia*. Sydney: Angus & Roberston, 1962 [1941].
Trustees of the Tasmanian Museum and Art Gallery. *Tasmanian Museum and Art Gallery Annual Report 2002–2003*. Hobart: Tasmanian Museum and Art Gallery, 2003.
———. *Tasmanian Museum and Art Gallery Annual Report 2003–2004*. Hobart: Tasmanian Museum and Art Gallery, 2004.
———. *Tasmanian Museum and Art Gallery Annual Report 2005–2006*. Hobart: Tasmanian Museum and Art Gallery, 2006.
Tuck, Leslie. 'The Newfoundland Wolf', *The Newfoundlander* (September 1947): 4–6.
———. *The Murres, their Distribution, Populations and Biology: A Study of the Genus Uria*. Ottawa: Canadian Wildlife Service, 1960.
———. 'The Newfoundland Wolf', *Newfoundland Quarterly* 11(1) (1979): 22–26.
Turner, Annette. *The Children of the Mist, The Conqueror, and Other Poems*. London: G. Lutz, 1827.
Turner, Hannah. *Cataloguing Culture: Legacies of Colonialism in Museum Documentation*. Vancouver: UBC Press, 2021.
Turvey, Samuel T. (ed.). *Holocene Extinctions*. Oxford: Oxford University Press, 2009.
———. 'Holocene Mammal Extinctions', in Samuel T. Turvey (ed.), *Holocene Extinctions* (Oxford: Oxford University Press, 2009), 41–61.
Turvey, Samuel T., and Joanne H. Cooper. 'The Past is Another Country: Is Evidence for Prehistoric, Historical and Present-Day Extinction Really Comparable?', in Samuel T. Turvey (ed.), *Holocene Extinctions* (Oxford: Oxford University Press, 2009), 193–212.
Tyler, Tom. *CIFERAE: A Bestiary in Five Fingers*. Minneapolis: University of Minnesota Press, 2012.
U.S. Department of Agriculture. *Cattle and Calves Predator Death Loss in the United States, 2005*. Fort Collins, CO, 2008.

———. *Death Loss in U.S. Cattle and Calves Due to Predator and Nonpredator Causes, 2015*. Fort Collins, CO, 2017.

Van Eeden, Lily M., Bradley P. Smith, Matthew S. Crowther, Chris R. Dickman and Thomas M. Newsome. '"The Dingo Menace": An Historic Survey on Graziers' Management of an Australian Carnivore', *Pacific Conservation Biology* 25(3) (2018): 245–56.

Van Essen, Susanna. 'Denying Extinction? The Thylacine in Children's Literature', *Reading Time* 40(4) (1996): 12–15.

Van Hecke, Peter. 'Metaphor in the Hebrew Bible: An Introduction', in Peter Van Hecke (ed.), *Metaphor in the Hebrew Bible* (Leuven: Leuven University Press, 2005), 1–18.

Vass, Stan. 'The Black Bear on Prince Edward Island: A Natural History', *The Island* 22 (1987): 8–9.

Vaughan, Christopher. 'Change in Dense Forest Habitat for Endangered Wildlife Species in Costa Rica from 1940 to 1977', *Cuadernos de Investigación UNED* 3(1) (2011): 99–161.

Vaughan, Herbert Millingchamp. *An Australasian Wander-Year*. London: Martin Secker, 1914.

———. 'The Tasmanian Tiger', *Country Life* (13 September 1946): 495–96.

Vedel, Helge, and Johan Lange. *Trees and Bushes in Wood and Hedgerow*, trans. C. H. R. Hillman. London: Methuen, 1960 [1958].

Verhoeff, Nanna. *The West in Early Cinema: After the Beginning*. Amsterdam: Amsterdam University Press, 2006.

Vivian, Helen. *Tasmania's Chinese Heritage: An Historical Record of Chinese Sites in North East Tasmania*. Unpublished report for the Australian Heritage Commission and the Queen Victoria Museum and Art Gallery, University of Tasmania, 1985.

Von Helden, Imke. 'Barbarians and Literature: Viking Metal and Its Links to Old Norse Mythology', in Niall Scott and Imke Von Helden (eds), *The Metal Void: First Gatherings* (Oxford: Inter-Disciplinary Press, 2010), 257–64.

Wajcman, Gérard. *Les Experts: La police des morts*. Paris: Presses Universitaires de France, 2012.

Wajsberg, Jeffrey. 'Notes on the Failures of Salvage Linguistics', PhD thesis. York University, Canada, 2018.

Walch, J. & Sons. *Walch's Tasmanian Almanac*. J. Walch & Sons: Hobart, 1879.

———. *Walch's Tasmanian Almanac*. J. Walch & Sons: Hobart, 1886.

Waldkoenig, Kirstin Marie. 'Can Nonhumans be Victims of Genocide?', MA thesis. University of Montana, USA, 2019.

Walker, Brett. 'Meiji Modernization, Scientific Agriculture and the Destruction of Japan's Hokkaido Wolf', *Environmental History* 9 (2004): 248–74.

Walker, James Backhouse. 'The Discovery and Occupation of Port Dalrymple', *Papers and Proceedings of the Royal Society of Tasmania* (1890): 155–75.

Wallace, Birgitta Linderoth. 'An Archaeologist's Interpretation of the *Vinland Sagas*', in William W. Fitzhugh and Elisabeth I. Ward (eds), *Vikings: The North Atlantic Saga* (Washington, DC: Smithsonian Books, 2000), 225–31.

Waller, James. *Becoming Evil: How Ordinary People Commit Genocide and Mass Killing*. Oxford: Oxford University Press, 2007.

Walsh, Ashley. '"A Great Heathen Fist from the North": Vikings, Norse Mythology, and Medievalism in Nordic Extreme Metal Music', MA dissertation. University of Oslo, 2013.

Ward, Philip S. 'Ants', *Current Biology* 16(5) (2006): R152–R155.

Warner, Charles Dudley. 'The Adirondacks Verified: A-Hunting of the Deer', *Atlantic Monthly* 41(246) (1878): 522–29.

———. *A-Hunting of the Deer and Other Essays*. New York: Houghton, Mifflin and Company, 1888.
Waters, Kenneth H. 'Solon's "Price-Equalisation"', *Journal of Hellenic Studies* 80 (1960): 181–90.
Watkins, Liz. 'The Politics of Nostalgia: Colorization, Spectatorship and the Archive', *Comparative Cinema* 9(17) (2021): 123–45.
Watkins, Liz, and Dominic Williams. 'The *Faces of Auschwitz*: Digital Colourisation, Ethics and the Archive', *Visual Studies* 40(1) (2025): 85–100.
Wayt Gibbs, W. 'On the Termination of Species', *Scientific American* 285(5) (2001): 40–49.
Weaver, Hilary N. 'The Colonial Context of Violence: Reflections on Violence in the Lives of Native American Women', *Journal of Interpersonal Violence* 24(9) (2009): 1552–63.
Weinman, David. 'The Human Trypanosomiases', in David Weinman and Miodrag Ristic (eds), *Infectious Blood Diseases of Man and Animals: Volume II: The Pathogens, the Infections and the Consequences* (New York: Academic Press, 1968), 98–142.
Weitz, Bernard. 'Feeding Habits of Bloodsucking Anthropods', *Experimental Parasitology* 9(1) (1960): 63–82.
Wendt, Herbert. *Ils n'étaient pas dans l'arche* [They were not on the ark], trans. André Pougetoux. Paris: Denoël, 1959.
West, John. *The History of Tasmania: Volume 1*. Launceston: Henry Dowling, 1852.
Westell, W. Percival. *The Book of the Animal Kingdom*. London: Dent, 1910.
Wey, Tina, Daniel T. Blumenstein, Weiwei Shen and Ferenc Jordán. 'Social Network Analysis of Animal Behaviour: A Promising Tool for the Study of Sociality', *Association for the Study of Animal Behaviour* 75(2) (2008): 333–44.
Whitbourne, Richard. *A Discourse and Discovery of New-found-land*. London: Felix Kyngston, 1622.
White, Gilbert. *The Natural History of Selborne*. London: Penguin, 1977.
White, Hayden. *Metahistory: The Historical Imagination in 19th-Century Europe*. Baltimore: Johns Hopkins University Press, 1973.
White, Lauren C., Kieren J. Mitchell and Jeremy J. Austin. 'Ancient Mitochondrial Genomes Reveal the Demographic History and Phylogeography of the Extinct, Enigmatic Thylacine (*Thylacinus Cyncocephalus*)', *Journal of Biogeography* 45 (2018): 1–13.
Whitehead, Hal. *Sperm Whales: Social Evolution in the Ocean*. Chicago: University of Chicago Press, 2003.
Whitehouse, Simon. 'Bounty Systems in Vermin Control', *Journal of Agriculture* 17(3) (1976): 85–89.
Whitley, G. P. 'George Masters, Naturalist', *Proceedings of the Royal Zoological Society of New South Wales* 16(2) (1971): 25–32.
Wicks, Santhi, Kasia Masur, Patricia Please, Saan Eckar and Benjamin Buetre. *An Integrated Assessment of the Impact of Wild Dogs in Australia*. ABARES Research Report 14.4. Canberra, 2014.
Widmar, Nicole J. Olynk, Courtney Bir, Evan Long and Audrey Ruple. 'Public Perceptions of Threats from Mosquitoes in the U.S. Using Online Media Analytics', *Pathogens and Global Health* 115 (1) (2021): 40–52.
Wilde, William H., Joy Hooton and Barry Andrews. 'Wilson, Erle', in *The Oxford Companion to Australian Literature* (online version) (Oxford: Oxford University Press, 2005).
Willey, P., and Paulette Leach. 'The Skull on the Lawn: Trophies, Taphonomy, and Forensic Anthropology', in Dawnie Wolf Steadman (ed.), *Hard Evidence: Case Studies in Forensic Anthropology* (Upper Saddle River, NJ: Prentice Hall, 2003), 176–88.
Williams, Paul. 'Witnessing Genocide: Vigilance and Remembrance at Tuol Seng and Choeung Ek', *Holocaust and Genocide Studies* 18(2) (2004): 234–54.

Willis, Ika. *Reception*. Abingdon: Routledge, 2018.
Wilson, Edward O. 'Ants', *Bulletin of the American Academy of Arts and Sciences* 45(3) (1991): 13–23.
———. *Half-Earth: Our Planet's Fight for Life*. New York: Liveright, 2016.
Wilson, Elisabeth. '"Wandering Stars": The Impact of British Evangelists in Australia 1870s–1900', unpublished PhD thesis. University of Tasmania, 2011.
Wilson, Erle. 'Ten Thousand Yen', *The Bulletin* (17 April 1929): 51, 53.
———. 'Marava', *The Bulletin* (22 March 1939): 7.
———. 'The Shoes of the Negus-Negusti', in *The Green Frog and Other Stories* (London: W. & R. Chambers, 1939), 98–109.
———. *The Grey Seals and Other Stories*. London: W. & R. Chambers, 1940.
———. 'Combined Operations', *Australian Journal* (1 March 1945): 236–38.
———. 'Haeremai!', *Australian Journal* (1 September 1945): 673–76.
———. 'The Squealing Devil of Kitori', *Australian Journal* (1 April 1946): 248, 252.
———. 'Jungle Record', *Australian Journal* (1 May 1946): 360–63.
———. 'The Pay-Back', *Australian Journal* (1 June 1946): 488, 490–95.
———. *Churinga Tales: Stories of Alchuringa – The Dream-Time of the Australian Aborigines*. London: George G. Harrap & Co., 1950.
———. 'The Beginning of Languages', in *Churinga Tales: Stories of Alchuringa – The Dream-Time of the Australian Aborigines* (London: George G. Harrap & Co., 1950), 13–26.
———. *Coorinna: A Novel of the Tasmanian Uplands*. New York: Random House, 1954 [1953].
———. *Minado: A Tale of the Quebec Wilderness*. New York: Appleton Century Crofts, 1956.
Wilson, Ross J. *Natural History: Heritage, Place and Politics*. Abingdon: Routledge, 2017.
Winters, Ria, and Julian P. Hume. 'The Dodo, the Deer and a 1647 Voyage to Japan', *Historical Biology* 27(2) (2014): 258–64.
Wittmann, Marc. *Felt Time: The Psychology of How We Perceive Time*, trans. Erik Butler. Cambridge, Mass: MIT Press, 2016.
Wolfe, Cary. 'From *Dead Meat* to Glow in the Dark Bunnies: Seeing "the Animal Question" in Contemporary Art', *parallax* 12(1) (2006): 95–109.
Wolfe, Patrick. 'Settler Colonialism and the Elimination of the Native', *Journal of Genocide Research* 8(4) (2006): 387–409.
———. 'Settler Colonialism and the Elimination of the Native', in A. Dirk Moses (ed.), *Genocide: Critical Concepts in Historical Studies*, vol. 1 (London: Routledge, 2010), 132–58.
Wonders, Karen. 'Habit Dioramas and the Issue of Nativeness', *Landscape Research* 28(1) (2003): 89–100.
Wood, Danielle. 'Fabrication on a Thylacine Pincushion', in Ralph Crane and Danielle Wood (eds), *Island Story: Tasmania in Object and Text* (Melbourne: Text, 2018), 183–88.
Woodruff, David S. 'Declines of Biomes and Biotas and the Future of Evolution', *PNAS* 98 (10) (2001): 5471–76.
Wooley, John. *Shot in Oklahoma: A Century of Sooner State Cinema*. Norman: University of Oklahoma Press, 2011.
Woolley, Patricia. 'Reproductive Biology of Antechinus stuartii Macleay (Marsupialia: Dasyuridae)', PhD thesis. Canberra: Australian National University, 1966.
Wright, Judith. 'Bullocky', in Judith Wright (ed.), *The Moving Image: Poems* (Melbourne: Meanjin, 1953 [1946]), 25.
Wroe, Stephen. 'Australian Marsupial Carnivores: Recent Advances in Palaeontology', in Menna Jones, Chris Dickman and Mike Archer (eds), *Predators with Pouches: The Biology of Carnivorous Marsupials* (Collingwood: CSIRO, 2003), 102–23.

Wylie, Caitlin Donahue. 'Overcoming the Underdetermination of Specimens', *Biology and Philosophy* 34(24) (2019). https://doi.org/10.1007/s10539-019-9674-2.

Wylie, James Aitken. *History of the Scottish Nation, Volume I: Pre-Historic, Druidic, Roman and Early Christian Scotland*. London: Hamilton, Adams & Co., 1886.

Yates, Adam M. 'New Craniodental Remains of *Thylacinus Potens* (Dasyuromorphia Thylacinidae), a Carnivorous Marsupial from the Late Miocene Alcoota Local Fauna of Central Australia', *PeerJ* 2 (2014): e547.

Yellowstone National Park, US Fish and Wildlife Service, University of Wyoming, University of Idaho, Interagency Grizzly Bear Study Team and University of Minnesota Cooperative Park Studies Unit. *Wolves for Yellowstone? A Report to the United States Congress, Volume I, Executive Summary*. Yellowstone National Park: National Park Service, 1990.

Young, Henry J. 'A Note on Scalp Bounties in Pennsylvania', *Pennsylvania History: A Journal of Mid-Atlantic Studies* 24(3) (1957): 207–18.

Young, Stanley P., and Edward A. Goldman. *The Wolves of North America: Part 1*. New York: Dover, 1964 [1944].

———. *The Wolves of North America: Part 2*. New York: Dover, 1964 [1944].

Zavagno, Daniele. 'The Perception of Light from a Phenomenological Perspective', *Psychology of Consciousness: Theory, Research, and Practice* 8(4) (2021): 346–66.

Žeger, Ivana, Sonja Grgic, Josip Vuković and Gordon Ŝiŝul. 'Greyscale Image Colorization Methods: Overview and Evaluation', *IEEE Access* 9 (2021): 113326–113346.

Zelizer, Craig. 'Laughing Our Way to Peace or War: Humour and Peacebuilding', *Journal of Conflictology* 1(2) (2010): 1–8.

Zickefoose, Julie, and William E. Davis. 'Great Blue Heron (*Ardea herodias*) Uses Bread as Bait for Fish', *Colonial Waterbirds* 21(1) (1998): 87–88.

Zimmermann, Michael F. 'Corinth et la chair de la peinture', in Serge Lemoine and Marie-Amélie zu Salm-Salm (eds), *Lovis Corinth 1858–1925: Entre Impressionnisme et Expressionnisme* (Paris: Musée d'Orsay, 2008), 320–31.

Index

C

Q

R

T

www.ingramcontent.com/pod-product-compliance
Ingram Content Group UK Ltd.
Pitfield, Milton Keynes, MK11 3LW, UK
UKHW021837150926
8656IPUK00004B/32

9 781836 951254